Current Topics in Behavioral Neurosciences

Volume 43

Current Topics in Behavioral Neurosciences provides critical and comprehensive discussions of the most significant areas of behavioral neuroscience research, written by leading international authorities. Each volume in the series represents the most informative and contemporary account of its subject available, making it an unrivalled reference source. Each volume will be made available in both print and electronic form.

With the development of new methodologies for brain imaging, genetic and genomic analyses, molecular engineering of mutant animals, novel routes for drug delivery, and sophisticated cross-species behavioral assessments, it is now possible to study behavior relevant to psychiatric and neurological diseases and disorders on the physiological level. The Behavioral Neurosciences series focuses on translational medicine and cutting-edge technologies. Preclinical and clinical trials for the development of new diagnostics and therapeutics as well as prevention efforts are covered whenever possible. Special attention is also drawn on epigenetical aspects, especially in psychiatric disorders.

CTBN series is indexed in PubMed and Scopus.

More information about this series at http://www.springer.com/series/7854

Lique M. Coolen • David R. Grattan
Editors

Neuroendocrine Regulation of Behavior

Springer

Editors
Lique M. Coolen
Department of Biological Sciences
Brain Health Research Institute, Kent State University
Kent, USA

David R. Grattan
Department of Anatomy
Centre for Neuroendocrinology, University of Otago
Dunedin, New Zealand

ISSN 1866-3370 ISSN 1866-3389 (electronic)
Current Topics in Behavioral Neurosciences
ISBN 978-3-030-38719-8 ISBN 978-3-030-38720-4 (eBook)
https://doi.org/10.1007/978-3-030-38720-4

This Springer imprint is published by the registered company Springer Nature Switzerland AG.
The registered company address is: Gewerbestrasse 11, 6330 Cham, Switzerland

Preface

Behavioral neuroendocrinology has a proud history of advancing our understanding of how brain circuits integrate both environmental and interoceptive information and drive complex behaviors that are necessary for survival. The latest technologies that are revolutionizing neuroscience provide an opportunity for a renaissance in behavioral neuroendocrinology. These technologies include the widespread application of genetic manipulation and gene knockouts, which have resulted in a variety of behavioral phenotypes. Genetically encodable markers improved our capacity to map functional brain circuits controlling behavior. Optogenetics and chemogenetics have offered the opportunity to manipulate neuronal activity in conscious animals in a highly targeted way, offering opportunities to study the behavioral outcomes of these manipulations. And methods such as fiber photometry and implantable mini-microscopes can be used to monitor neuronal activity in freely behaving animals, allowing high-fidelity measurements of neuronal activity to be taken while an animal is performing specific tasks. Neuroscientists are recognizing that the ultimate expression of brain function is, in fact, behavior. There has never been a more important time for careful, experienced researchers who understand the normal behavior of an animal, and how it changes with age, sex differences and changes in behavior across a reproductive cycle, modification of behavior in response to social interactions, or in response to physiological and environmental changes.

In this volume, we have drawn together a number of established and emerging experts across a broad range of topics within the field of behavioral neuroendocrinology and asked them to review some of the key paradigms in their area and to highlight future directions incorporating novel approaches. The first three chapters describe the latest research on hormonal and neural regulation of female sexual behavior (Snoeren; Female Reproductive Behavior), sexual differentiation in the human brain (Bakker; The Sexual Differentiation of the Human Brain: Role of Sex Hormones Versus Sex Chromosomes), and sexual differentiation in neural development (Turano, Osborne, and Schwarz; Sexual Differentiation and Sex Differences in Neural Development). A series of three chapters then highlight the recent findings in the areas of paternal behavior (Horrell, Hickmott, and Saltzman; Neural Regulation of Paternal Behavior in Mammals: Sensory, Neuroendocrine, and Experiential

Influences on the Paternal Brain), maternal behavior (Smiley, Ladyman, Gustafson, Grattan, and Brown; Neuroendocrinology and Adaptive Physiology of Maternal Care), and social affiliation (Lee and Beery; Neural Circuits Underlying Rodent Sociality: A Comparative Approach). The next four chapters describe the integration of hypothalamic and neuroendocrine signals for the regulation of food intake (Klockars, Levine, and Olszewski; Hypothalamic Integration of the Endocrine Signaling Related to Food Intake), anxiety and fear (Hassell, Nguyen, Gates, and Lowry; The Impact of Stressor Exposure and Glucocorticoids on Anxiety and Fear), circadian rhythms (Karatsoreos; Circadian Regulation of the Brain and Behavior: A Neuroendocrine Perspective), and sleep (Smith and Mong; Neuroendocrine Control of Sleep). Finally, neuroendocrine contributions to neural plasticity are highlighted in the last two chapters describing hormonal regulation of hippocampal neurogenesis (Gheorghe, Qiu, and Galea; Hormonal Regulation of Hippocampal Neurogenesis: Implications for Depression and Exercise) and the neuroimmune impacts of early life stress (Brenhouse, Danese, and Grassi-Oliveira; Neuroimmune Impacts of Early-Life Stress on Development and Psychopathology).

Together, these review chapters provide a comprehensive background highlighting the current state of knowledge, and the important considerations in the study of various aspects of behavior, to serve as a resource for the current and future studies that are revisiting these behaviors and furthering understanding with state-of-the-art modern techniques. As a field, we have a fantastic opportunity to not only contribute to but also lead in high-quality in vivo neuroscience research in conscious, freely behaving animals.

Kent, USA — Lique M. Coolen
Dunedin, New Zealand — David R. Grattan

Contents

Female Reproductive Behavior 1
Eelke M. S. Snoeren

The Sexual Differentiation of the Human Brain: Role of Sex Hormones Versus Sex Chromosomes 45
Julie Bakker

Sexual Differentiation and Sex Differences in Neural Development 69
Alexandra Turano, Brittany F. Osborne, and Jaclyn M. Schwarz

Neural Regulation of Paternal Behavior in Mammals: Sensory, Neuroendocrine, and Experiential Influences on the Paternal Brain 111
Nathan D. Horrell, Peter W. Hickmott, and Wendy Saltzman

Neuroendocrinology and Adaptive Physiology of Maternal Care 161
Kristina O. Smiley, Sharon R. Ladyman, Papillon Gustafson, David R. Grattan, and Rosemary S. E. Brown

Neural Circuits Underlying Rodent Sociality: A Comparative Approach 211
Nicole S. Lee and Annaliese K. Beery

Hypothalamic Integration of the Endocrine Signaling Related to Food Intake 239
Anica Klockars, Allen S. Levine, and Pawel K. Olszewski

The Impact of Stressor Exposure and Glucocorticoids on Anxiety and Fear 271
J. E. Hassell Jr., K. T. Nguyen, C. A. Gates, and C. A. Lowry

Circadian Regulation of the Brain and Behavior: A Neuroendocrine Perspective 323
Ilia N. Karatsoreos

Neuroendocrine Control of Sleep 353
Philip C. Smith and Jessica A. Mong

Hormonal Regulation of Hippocampal Neurogenesis: Implications for Depression and Exercise 379
Ana Gheorghe, Wansu Qiu, and Liisa A. M. Galea

Neuroimmune Impacts of Early-Life Stress on Development and Psychopathology 423
Heather C. Brenhouse, Andrea Danese, and Rodrigo Grassi-Oliveira

Female Reproductive Behavior

Eelke M. S. Snoeren

Contents

1 Introduction 3
2 The Estrous Cycle 4
  2.1 Follicle Growth: Premenstrual Phase 5
  2.2 Early Follicle Phase 5
  2.3 Late Follicle Phase 7
  2.4 Luteal Phase 9
3 Mating Behavior 10
4 Brain Regions Involved in Female Sexual Behavior 13
5 Chemical Messengers Involved in Female Sexual Behavior 17
  5.1 Estrogen and Progesterone 17
  5.2 Serotonin 20
  5.3 Dopamine 23
  5.4 Noradrenaline 27
  5.5 Oxytocin 29
  5.6 GABA 30
6 General Discussion 31
References 32

Abstract Reproductive behavior is the behavior related to the production of offspring and includes all aspects from the establishment of mating systems, courtship, sexual behavior, and parturition to the care of young. In this chapter, I outline the hormonal regulation of the estrous cycle, followed by a description of the neural regulation of female sexual behavior. Ovarian hormones play an important role in the induction of ovulation and behavioral estrus, in which they interact closely with several neurotransmitters and neuropeptides to induce sexual behavior. This chapter discusses the latest research on the role of estrogen, progesterone, serotonin, dopamine, noradrenaline, oxytocin, and GABA in female mating behavior. In addition, the most relevant brain areas, such as the preoptic area and the ventromedial nucleus of the hypothalamus, in which these regulations take place, are discussed.

E. M. S. Snoeren (✉)
Department of Psychology, UiT the Arctic University of Norway, Tromsø, Norway
e-mail: eelke.snoeren@uit.no

Curr Topics Behav Neurosci (2019) 43: 1–44
DOI 10.1007/7854_2018_68
Published Online: 19 February 2019

Keywords Dopamine · Estrogen · Estrous cycle · Females · Mating · Neural regulation · Noradrenaline · Oxytocin · Progesterone · Reproductive behavior · Serotonin · Sexual behavior

Abbreviations

5-HT	Serotonin
ARC	Arcuate nucleus of the hypothalamus
ARC-VM	Arcuate-ventromedial area of the hypothalamus
BNST	Bed nucleus of the stria terminalis
BNSTpm	Posteromedial part of BNST
CTF	Central tegmental field
DA	Dopamine
DRN	Dorsal raphé nucleus
E	Estrogen
ER	Estrogen receptor
Fos-IR	Fos-immunoreactivity
FSH	Follicle-stimulating hormone
GABA	Gamma-aminobutyric acid
GnRH	Gonadotropin-releasing hormone
GnSAF	Gonadotrophin surge-attenuating factor
Icv	Intracerebroventricular
LH	Luteal hormone
LHA	Lateral hypothalamic area
MeA	Medial amygdala
MeApd	Posterodorsal part of MeA
MPN	Medial preoptic nucleus
MRN	Medial raphé nucleus
NAc	Nucleus accumbens
NPY	Neuropeptide Y
OT	Oxytocin
P	Progesterone
PAG	Periaqueductal gray
PD	Posterodorsal preoptic nucleus
PMV	Ventral premammillary nucleus
POA	Preoptic area
PR	Progesterone receptor
PVN	Paraventricular hypothalamic nucleus
SERT	Serotonin transporter
SPFp	Subparafiscular nucleus
SSRI	Selective serotonin reuptake inhibitor
VMN	Ventromedial hypothalamus

VMNcv	Caudoventral part of VMN
VMNvl	Ventrolateral part of VMN
VTA	Ventral tegmental area

1 Introduction

Reproductive behavior can be described as behavior related to the production of offspring. The combination of behavior leading to the union of male and female gametes and behavior facilitating or ensuring the survival and development of the young is required for an optimal reproductive success. Therefore, reproductive behavior includes basically all aspects from the establishment of mating systems, courtship, sexual behavior, and parturition to the care of young (McGraw-Hill concise encyclopedia of bioscience 2005). The timing and patterning of reproductive activity varies among species and is coordinated by an integration of both overt behavioral and internal physiological events leading to a complex set of behavioral adaptations. Although care of the young is an important part of reproductive behavior, here I will solely focus on the aspects of female mating behavior and the neural mechanisms that are involved in the regulation of these behaviors. In search for explanations to the neurobiological and endocrine bases for behavior, the field has been enormously enriched by the studies carried out on sexual behavior in rodent species. This chapter, therefore, mentions mostly examples of rat sexual behavior, sometimes accompanied by information of some other species.

Female mating behavior is often considered more complicated than male mating behavior, because most animals only start sexual encounters with mates when the females are in a so-called estrus. Females can get highly motivated in this period and will, for instance, cross a highly charged electrified floor or poke obsessively a lever to gain physical access to a male. In contrast, when the female is in the opposite phase of her estrous cycle, in *anestrus*, no effort will be undertaken to get access to a potential mate. Women, and some other primates, are probably the only mammals in which their sexual drive is not directly associated to their estrous cycle and mate with their partners throughout the whole cycle (Wallen and Zehr 2004).

The estrous phase has several consequences. As already mentioned, females in estrus will seek males, remain in close proximity, and initiate copulation. Estrus, thus, affects the attraction female have toward males. However, at the same time, it also influences their own attractivity: males prefer to seek proximity to or mount females in estrus over their conspecifics in anestrus. In the whole process, it is important that the females are able to detect the potential mate and that this elicits particular behavior leading to successful copulation. Reproductive hormones play an important role in the regulation of this process. They are elevated during estrus and cause an enhancement of the acuity, sensitivity, and efficiency of the sensory system, thereby improving the detection and response to potential mates when in estrus. In addition, the neuroendocrine changes affect the central nervous system: the higher hormone levels enhance the female's motivation, attention, perception, and behavior. Thus, an environment is created during estrus in which the conditions are most optimal for successful reproduction.

In this chapter, I will first touch upon the concept of the estrous cycle, which leads to the induction of the estrous phase. It is important to highlight this event first, because the major changes in hormones that are occurring as part of the estrous cycle do not only induce fertility but also have parallel behavioral consequences. Secondly, I will evaluate the principles of mating behavior and the neural regulation of the female reproductive behavior in more detail.

2 The Estrous Cycle

The basis of the estrous cycle is already present in females at the time of birth, since they are born with a finite number of germ cells that will later become oocytes. When puberty starts, the estrous cycle can be divided into the follicle phase and the luteal phase. The follicle phase occurs prior to ovulation, when the follicles are developing, while the luteal phase is the part of the cycle after ovulation during which the corpus luteum is active. Only when no fertilization of the oocyte takes place, the corpus luteum degenerates, and the estrous cycle starts over. Under normal conditions, this process continues until all germ cells are released and the females will enter the menopause.

The length of the estrous cycle and the intervals between these cycles vary from species to species. For example, in women this cycle takes about 25–35 days, cows have a cycle of 21 days, and rats and mice complete the phases within 4–5 days. Interestingly, some species, like mice, rats, and hamsters, have the capacity to skip the luteal phase when no mating has been taken place. This can be very beneficial because it ensures a rapid return to estrus and thus a new faster opportunity to be impregnated and reproduce.

In all cases, the estrous cycle is under regulation of the well-known hypothalamic-pituitary-gonadal axis. The principle of this axis is that the hypothalamus releases gonadotropin-releasing hormone (GnRH), which induces the pituitary to release follicle-stimulating hormone (FSH) and luteal hormone (LH). The release of FSH and LH, in turn, stimulates the ovaries to release steroid hormones, like estrogen (E) and progesterone (P), which subsequently leads to negative or positive feedback mechanisms acting back on the hypothalamus or pituitary. Although E and P are probably the most important regulators of the estrous cycle, other peptides like inhibin, kisspeptin, and gonadotrophin surge-attenuating factor (GnSAF) also play a significant role. It is the interplay between endocrine hormones and neuropeptides, causing the release or decline of other substances, which regulates the phases of the estrous cycle and thus the female reproductive behavior. We will discuss the occurring events and the hormonal regulations per phase.

2.1 Follicle Growth: Premenstrual Phase

As mentioned before, females have a finite number of *immature oocytes* at birth that are packed in a layer of granulosa cells and form the so-called *primordial follicles*. Only the surviving follicles will finally enter the estrous cycle as *tertiary follicles*. At the primordial stage, follicles are biologically inactive which can last for about 50 years (in women). After an initial recruitment, some primordial follicles are "awaken" during puberty and enter the primary follicle stage. In this stage, the granulosa cells change shape, and the oocyte genome is activated and starts transcribing genes. An important step in this phase is that these primary follicles start expressing FSH receptors (as shown in Fig. 1), although they remain gonadotropin-independent until later. As soon as the follicle develops a theca layer in addition to the granulosa cells, the follicle is called a *secondary follicle*. These theca cells will now express LH receptors. The presence of LH induces these cells to produce androgens, which will be transformed into estrogens by the granulosa cells: the follicle is now called a *tertiary follicle*.

2.2 Early Follicle Phase

The follicle phase is the start of the estrous cycle. Only a fraction of the follicles have survived and developed into tertiary follicles, containing granulosa and theca cells surrounding the mature oocyte. The follicles now also contain a fluid-filled cavity adjacent to the oocyte, also called *antrum*: the follicle is now called *antrum follicle* or *Graafian follicle*. Besides the support function for the oocyte, the granulosa and theca cell layers play an important role in the establishment of a high estrogen microenvironment needed for the development of the follicle. The follicles now express more FSH receptors and become responsive to FSH which in turn stimulates follicle growth. At the same time, increased LH levels stimulate the theca cells to produce androgens, which are transformed into E by the granulosa cells. Together, this creates the high estrogen microenvironment that is necessary for follicle development. The high levels of E now trigger a negative feedback loop downregulating the release of FSH from the anterior pituitary. The decrease in FSH levels causes the death of follicles that express less FSH receptors leading to the selection of one *dominant follicle*.

It should be mentioned, though, that this is a very simplified description of the process. In reality, it is much more complicated (reviewed in Messinis et al. 2014). The role of E in this negative feedback system is well known (Messinis and Templeton 1987; Messinis et al. 1994, 1998), but more factors are involved. When both E and P are administered to healthy postmenopausal women (who normally have elevated LH and FSH levels), LH levels decrease in response to these steroids, while FSH decreases but remains higher than usual during the normal follicular phase. This suggests that the gonadotrophins LH and FSH are differentially controlled by the

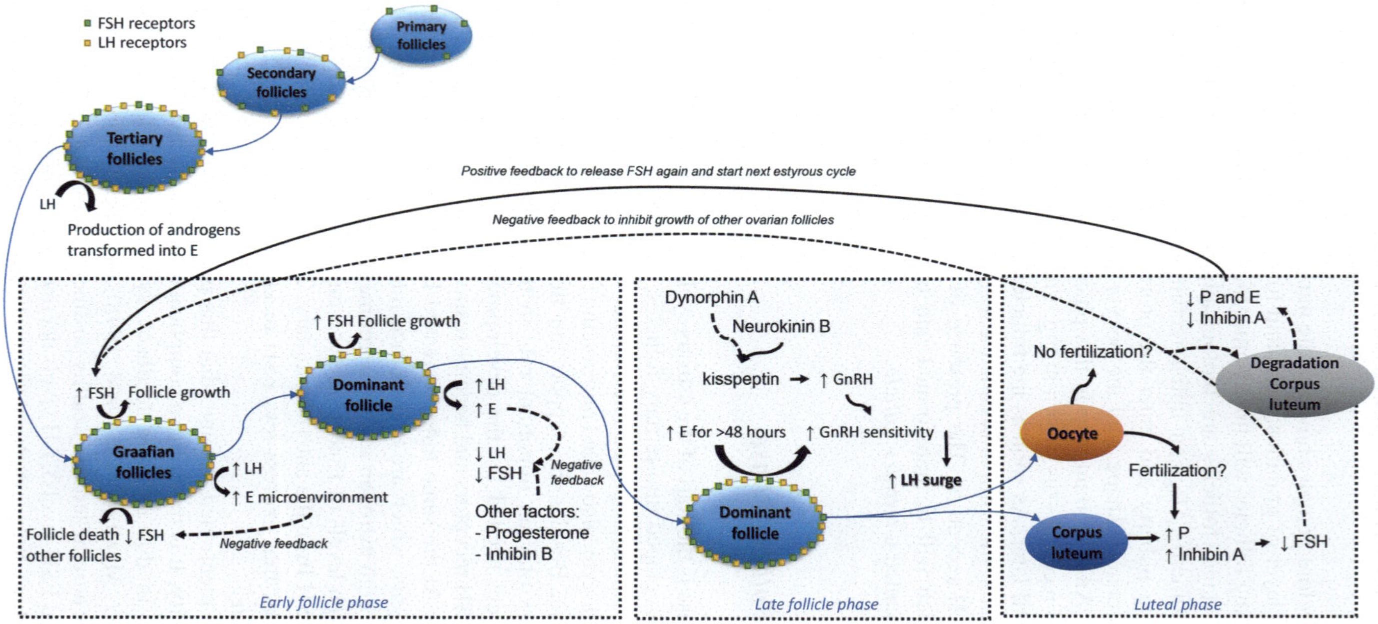

Fig. 1 (Simplified) schematic drawing of the estrous cycle

ovaries that release E and P and that FSH is also regulated by other substances. In addition, LH must also have other regulators, since the decline in LH is not sustained by estrogens administration alone (Dafopoulos et al. 2004). A candidate for this control function is progesterone. Progesterone levels increase during the luteal phase, but also in the follicular phase some, although a low levels of, progesterone is present. The administration of anti-progesterone drugs also increases LH levels during the follicular phase (Kazem et al. 1996), suggesting that P is indeed controlling the LH levels in this phase as well.

Another factor that is involved in this whole regulation is the nonsteroidal substance inhibin. Inhibin B is mainly released by the growing follicles during the early follicular phase of the cycle, while inhibin A is secreted especially by the corpus luteum in the luteal phase. The precise role of these substances is not yet known, but inhibin B seems to be involved in the negative feedback effect of the ovaries on the FSH secretion, because increased levels of inhibin B are related with decreased levels in FSH (de Koning et al. 2008).

2.3 Late Follicle Phase

Now the follicles produce high levels of E, the late follicle phase starts, in which the role of E changes from a negative feedback regulator into a positive feedback controller. The earlier slight increase in E resulted in a decrease in FSH and LH secretion, but now the E levels are increased dramatically for a longer time (when it exceeds a certain threshold and last for more than 48 h); the pituitary instead starts to secrete a large amount of LH (Lasley et al. 1975).

The *LH surge* is most likely caused by a combination of GnRH release and sensitization of the pituitary. The high levels of E trigger the pituitary to express more GnRH receptors, thereby sensitizing the pituitary to hypothalamic GnRH (Laws et al. 1990). At the same time, the high levels of E (and P) stimulate the release of GnRH in the hypothalamus. This effect, though, is regulated via an indirect action because GnRH neurons do not express estrogen and progesterone receptors (Huang and Harlan 1993). An important mediator in this process is the neuropeptide kisspeptin, co-released with neurokinin B and dynorphin A (Clarke et al. 2015). The kisspeptin neurons situated closely to the GnRH neurons in the hypothalamus do express estrogen receptors (Franceschini et al. 2006), and, therefore, it is believed that kisspeptin is the direct trigger of GnRH secretion. Kisspeptin most likely binds to the KISS1 receptors expressed on the cell body of most GnRH neurons to induce this effect (Clarke et al. 2015; Herbison et al. 2010). This hypothesis is strengthened by the fact that selective deletion of KISS1 receptors on GnRH neurons does indeed induce a hypogonadal phenotype (Novaira et al. 2014). In addition, kisspeptin levels rise during the preovulatory phase (Latif and Rafique 2015; Smith et al. 2006), and kisspeptin neurons get activated around ovulation (Clarkson et al. 2008), suggesting a direct role of kisspeptin in the LH surge.

Inhibition of kisspeptin action abolished the proestrous LH surge and inhibits the estrous cycle in rats (Adachi et al. 2007; Kinoshita et al. 2005).

Interestingly kisspeptin is not only involved in the regulation of the LH surge. A very recent study provided evidence that kisspeptin neurons in the hypothalamus are an essential part of a motivational neural circuit as well, that is triggered by male olfactory cues and leads to female lordosis behavior. Deletion of KISS1 receptors did not disrupt lordosis itself, but it did cause changes in modulating mate preferences (Hellier et al. 2018). This suggests that kisspeptin does not only regulate the estrous cycle by inducing the LH surge but is also involved in the next step of the behavioral consequences of receptivity.

As mentioned above, kisspeptin is actually co-released with neurokinin B and dynorphin A in the so-called KNDy neurons (Clarke et al. 2015). An increase in neurokinin B is thought to initiate a positive feedback loop resulting in activation of neurokinin-3 receptors on the KNDy neurons and thereby releasing kisspeptin (Goodman et al. 2014). Dynorphin A, on the other hand, is thought to regulate the inhibitory control on the GnRH pulse by acting directly on the KNDy neurons. Thus, neurokinin B and dynorphin A control the synchronized activity of KNDy neurons and kisspeptin release, which in turn can activate GnRH neurons to regulate pulsatile GnRH secretion (Goodman et al. 2014).

Interestingly, the LH surge in response to the GnRH release and sensitivity does occur solely during the late follicular phase, and not during the early phase, suggesting that additional factors must be involved in the regulation of this specific effect. Interestingly, administration of GnRH during the early phase, in normal cycling women, does not induce an increase in LH (Messinis et al. 1994, 1998). This suggests that the factors that are involved must have a suppressing effect on the pituitary.

One potential candidate for this antagonizing effect could be the nonsteroidal substance gonadotropin surge-attenuating factor (GnSAF), which is mainly produced by the granulosa cells in the ovaries in response to FSH (and not inhibin, E, P, or any other steroid hormone) (reviewed in Fowler et al. 2003). The production of GnSAF in the follicles is clearly related to the follicle size, in which the smaller follicles contain the highest levels of GnSAF (Fowler et al. 2001). As soon as the number of small follicles declines, as happens toward the late follicle phase, the bioactive levels of GnSAF decline. Whereas GnSAF normally negatively affects the pituitary responsiveness to GnRH and keeps LH levels low, it is, thus, the low of absent negative feedback by GnSAF that could contribute to increase LH secretion during the late follicle phase (De Koning et al. 2001), providing E the possibility to overcome the inhibitory GnSAF effects. E levels increase already during the middle follicle phase, but only in the late follicle phase, after which FSH levels have declined and a dominant follicle has been selected, GnSAF production stops. GnSAF, thus, seems to regulate the timing of the LH surge.

Another important players in the regulation of the LH surge is P, in which P probably also functions via sensitizing the pituitary to GnRH (Kazem et al. 1996).

The concentrations of P remain low during the largest part of the follicle phase, until the onset of the LH surge. The increase in P functions as an amplifier of the positive feedback mechanism of E on the pituitary. P seems to advance the onset of the LH surge and augments the amplitude of the LH surge (Messinis and Templeton 1990; March et al. 1979) but only in the presence of certain concentrations of E. One study suggested that it is actually the neuroprogesterone synthesized in the hypothalamus that plays a role as obligatory mediator for the onset of E-induced LH surge (Micevych et al. 2003).

At last, GABA and glutamate seem also to be involved in the regulation of the GnRH/LH surge. Subpopulations of estrogen receptor-expressing cells in the hypothalamus are GABAergic or glutamatergic (Cheong et al. 2015). It is hypothesized that these E-related GABAergic and glutamatergic transmissions are important for both the negative and positive feedback mechanism on the GnRH release. GABA levels are increased in the hypothalamus during the moment of negative feedback (Herbison and Dyer 1991) but then fall just before the GnRH/LH surge (Jarry et al. 1992; Robinson et al. 1991), suggesting that estrogen modulates GABAergic signaling to suppress GnRH neuron activity and thereby regulate the negative feedback (Petersen et al. 2003; Herbison 1998). For positive feedback, on the other hand, an increase in glutamate levels close to the GnRH neuron cell bodies occurs at the time of the GnRH/LH surge (Ping et al. 1994; Jarry et al. 1995) and glutamate receptor antagonists abolish the surge (Ping et al. 1997), suggesting the glutamate is involved in the positive feedback loop to release GnRH.

In conclusion, the amplitude of the LH surge is a result of a balance between the positive actions of E and P and the negative action of GnSAF. In addition, the role of kisspeptin in the regulation of this LH surge should not be forgotten ultimately resulting in the rupture of the ovarian follicle, causing the oocyte to be released from the ovary via the oviduct toward the uterus for fertilization: *ovulation*.

2.4 Luteal Phase

To induce the start of the luteal phase, the LH surge should now be terminated. This process is probably controlled by ovarian hormones as well, because a decrease in LH is seen before but not after ovariectomy (Dafopoulos et al. 2006). This time it is P that is mostly involved in the regulation via a negative feedback system. When the oocyte is expelled from the follicle, the remaining follicle cells become luteinized and form the corpus luteum. In this luteal phase, the corpus luteum secretes progesterone and inhibin A which act on the pituitary to suppress the release of FSH and thereby inhibit the growth of other ovarian follicles. Only when no fertilization of the oocyte takes place, the corpus luteum degenerates, and the levels of inhibin, estrogen, and progesterone decline. Now FSH can be released again which results in the start of the next estrous cycle (reviewed in Hawkins and Matzuk 2008).

Box 1 Vaginal Estrus
The vaginal lumen also undergoes cellular changes during the estrous cycle. When observing vaginal smears under the microscope, four different stages can be differentiated: vaginal estrus, diestrus I, diestrus II, and proestrus. The first vaginal estrous stage is characterized by cornified epithelial cells, which look like cornflakes under the microscope. In rats, this stage last for about 36 h, after which the cornified cells reduce in number and are replaced by leukocytes (white blood cells) and a few nucleated epithelial cells. This stage, called vaginal diestrus, persists for approximately 48 h and is divided in diestrus I (the first day) and diestrus II (the second day). After diestrus, the vaginal lumen enters the vaginal proestrous stage. This stage (which takes about 12 h) can be recognized by the presence of many nucleated epithelial cells, in addition to a dramatic reduction in the number of leukocytes.

These stages of the vaginal cycle are driven by ovarian hormones and therefore coincide with the phases of the ovarian estrous cycle. The vaginal estrus is correlated with the presence of recently ruptured follicles following ovulation and the formation of the corpora luteum. In addition, tertiary follicles begin to develop from secondary follicles at this time. As soon as the tertiary follicles become larger and the granulosa cells become more numerous, the diestrous stage started. The tertiary follicles are now called Graafian follicles, and the corpus luteum is fully formed from the postovulatory follicle. At the time the Graafian follicles are about to ovulate, the vaginal proestrous stage has started.

It is important to mention that the term estrus is rather confusing here, because the behavioral estrus (the moment in which female are ready to mate) coincides with the night of the vaginal proestrous stage, and not with the vaginal estrous stage. Therefore, we always refer to the vaginal estrus and behavioral estrus to clarify this difference.

3 Mating Behavior

As soon as females enter behavioral estrus, they are ready to participate in mating behavior. In order to understand the neurobiological regulation of this mating behavior, it is important to recognize the elements and phases of the sexual interactions. Interestingly, there is an enormous similarity in the patterns of mating behavior among species. The typical behaviors could be different between species, but the goals of the behaviors are comparable. Mating behavior in general can be divided into three different phases: the introductory (precopulatory), copulatory, and executive phase (in males ejaculations, in females orgasms). The completion of all phases is needed to have a chance for reproductive success. In human males, sexual activity normally ends at ejaculation: multiple ejaculations are quite infrequent. Some women, on the other hand, are able to reach multiple orgasms. In other species, like rats and mice, ejaculation is not the end of the sexual encounter, resulting in that most rats display multiple ejaculations in rapid succession before ceasing to copulate (Ågmo 1997).

The interplay between males and females starts with behaviors like approaching and sniffing each other's anogenital regions to obtain pheromonal cues of sexual readiness to mate: *the introductory phase*. Without an intrinsic state of sexual motivation, no approach behavior will occur. Sexual motivation could thus be seen as another important component of the introductory phase. In order to have reproductive success, individuals must be more attractive within a population or preferring more attractive partners to produce offspring. Intrasexual competition for access to a mate is believed to be common among many mammals (Darwin 1859), with the exception for, e.g., wild rats (Barnett 1975; Mcclintock 1984). Though, even if intrasexual competition were unusual, also rats still have to make a choice of partner with whom to initiate copulatory activity whenever there is more than one potential partner available. It was already in 1976 that Beach proposed that *attractivity* was an important component of female sexual behavior (Beach 1976). He defined attractivity as a female's value as a sexual stimulus and included behavioral as well as nonbehavioral components such as olfactory cues that stimulated the male to engage in sexual behavior with the female.

The introductory phase is followed by the *copulatory phase* in which female rats in behavioral estrus display a variety of paracopulatory behaviors (also called solicitation or proceptive behavior), e.g., hopping and darting (reviewed in Heijkoop et al. 2018). Hopping can be recognized by a typical jumping behavior of the female with four legs off the ground, while darting is a runaway behavior that suddenly stops with the female presenting her body to the male rat for mounting. Darting and hopping occur frequently and in a random order and seem to have a similar implication. Therefore, these behaviors are scored together as "paracopulatory behavior," which are species-typical and signal the readiness to mate, thereby functioning as an index of feminine sexual motivation (Beach 1976). However, in order to distinguish this motivation from the intrinsic motivation of the introductory phase, it could be hypothesized that the paracopulatory behaviors might represent the motivational level of keeping participating in the sexual intercourse rather than of the female's intrinsic sexual motivation to start sexual interaction.

In close relation to the paracopulatory behaviors of females, male rats show repeated intromissions and mounts during the copulatory phase. Intromissions are characterized as mounts including pelvic thrusting. It was always believed that copulation occurred upon initiation of the female rats (Mcclintock and Adler 1978). However, a recent study by Bergheim et al. (2015), performed in a seminatural environment, showed that the copulatory acts were a consequence of a subtle interaction between the male and female. This indicates that the behavior of both rats is equally important in the initiation of copulation and thus not controlled by solely the female.

In response to the male copulatory behaviors, the females display lordoses in which a posture of a hollow back and deflected tail is presented to give the male access to her vagina. The reflexive behavior lordosis reflects the female's receptivity and is very much depending on the hormonal state of the female. The presence of estrogen alone is sufficient to induce receptivity, but progesterone facilitates the estrogen-induced lordosis response (Edwards et al. 1968). Another important female sexual behavior shown in response to the male copulatory behaviors is "pacing" or solicitation. This is the ability of the female rat to control the timing of the receipt of sexual stimulation, as a pattern of approach and withdrawal from the male. The display of this behavior is directly dependent upon the intensity of the coital stimulation (mounts, intromissions, and ejaculations) received immediately prior to the solicitation behaviors. The rate of approaches toward the male is negatively correlated with the intensity of the stimulus from the male (Erskine 1985).

After a series of mounts and intromissions, ejaculation (the executive phase) is reached. In rats, usually 10–20 intromissions are needed during a short period (ca. 2–10 min) to reach an ejaculation. After an ejaculation, male rats need a break for about 5 min (the post ejaculatory interval) before preceding with the next ejaculatory cycle. The copulatory and executive phase can be repeated until the rats obtain sexual satisfaction.

The different phases of the sexual cycle are probably regulated via different mechanisms, because sexual motivation and copulatory behavior can vary independently upon a variety of treatments (Kondo and Sachs 2002; Ågmo 2002). However, the phases are not independent. When β-endorphin is infused in the medial preoptic area of the hypothalamus before introduction to a mate, only introductory behavior in terms of investigating the anogenital region of the female is displayed. But when β-endorphin is infused after the first mounts, and thus after the start of the copulatory phase, copulation until ejaculation succeeded normally (Stavy and Herbert 1989). This suggests the existence of different mechanisms in the sexual behavioral cycle that are closely interconnected with a specific transition point (threshold) preceding the mounts and intromissions. Similar findings do not exist in females yet, but we hypothesize that comparable mechanisms also exist in females.

Unfortunately, most studies on female sexual behavior have solely focused on lordosis behavior; sexual motivation and paracopulatory behaviors were often forgotten. The use of lordosis as a simple laboratory measure of sexual behavior has been useful in addressing many questions regarding sexual responsiveness. However, the approach and intrinsic motivation and paracopulatory behaviors displayed by the female during mating may be equally important for the understanding of the mechanisms behind behavior (Heijkoop et al. 2018). In order to draw conclusions on the neural basis of female sexual behavior, it is important to evaluate the full spectrum of behaviors shown by females. Still, most results described in this chapter are from previous articles in which only lordosis behavior was investigated. This means that this chapter is incomplete in describing the full picture of female sexual behavior. Fortunately, the additional behaviors gain interest among scientists; more and more investigators evaluate paracopulatory behaviors alongside to lordosis behavior. Hopefully, future studies will lead to additional insights into the neural regulation of all facets of female sexual behavior.

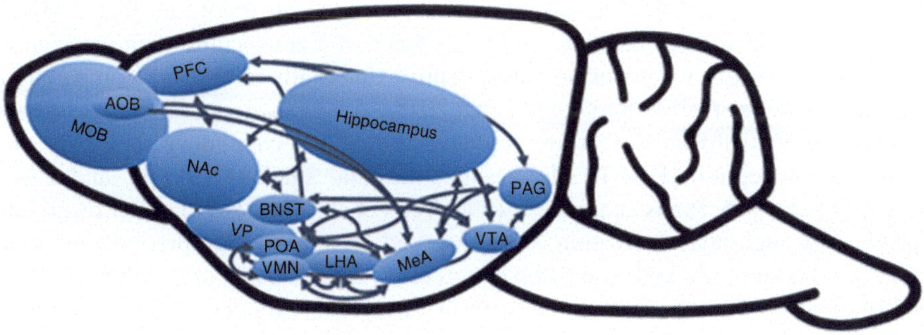

Fig. 2 A schematic drawing of the most important brain areas involved in female reproductive behavior and their connectivity. *NAc* nucleus accumbens, *BNST* bed nucleus of stria terminalis, *POA* preoptic areas, *LHA* lateral hypothalamus, *VMN* ventromedial nucleus of the hypothalamus, *MeA* medial amygdala, *VTA* ventral tegmental area, *PAG* periaqueductal gray, *MOB/AOB* olfactory bulb, *PFC* prefrontal cortex, *VP* ventral pallidum

4 Brain Regions Involved in Female Sexual Behavior

Sexual behavior is regulated by several brain areas. Studies with lesions, electrical stimulation, tract-tracing, and Fos-immunoreactivity (Fos-IR), performed in different species, give a nice overview of these functional regions (previously reviewed in Snoeren et al. 2014a). Fos-IR activation studies have shown that several brain areas are activated after mating behavior in female rats, e.g., the preoptic area (POA), the bed nucleus of the stria terminalis (BNST), the medial amygdala (MeA), the central tegmental field (CTF), the ventromedial hypothalamic nucleus (VMN), and the periaqueductal gray (PAG) (Fig. 2; Erskine 1993; Pfaus et al. 1993; Polston and Erskine 1995; Rowe and Erskine 1993; Tetel et al. 1993). Additional studies showed in more detail that different subregions of the brain areas become activated during the different phases of sexual behavior. Chemosensory investigation (introductory phase), for instance, induced Fos-IR in the female posteromedial part of the BNST (BNSTpm) (Coolen et al. 1996) and posterodorsal part of MeA (MeApd) (Coolen et al. 1996; Tetel et al. 1993), whereas other brain subregions become active during the copulatory phase. The VMN is known as the major site for the regulation of lordosis behavior. Lesions of this area dramatically reduce lordosis (Pfaff and Sakuma 1979a), while electrical stimulation facilitates the expression of lordosis (Pfaff and Sakuma 1979b). Fos-IR techniques have shown that lordosis behavior in response to mounts activates especially the ventrolateral part of the VMN (VMNvl) (Coolen et al. 1996), while intromissions cause stronger induction of Fos-IR in the same brain area without a higher expression of lordosis behavior (Pfaus et al. 1993; Rowe and Erskine 1993). This suggests that the intensity of Fos-IR induction in the VMNvl is not exclusively a reflection of vaginocervical sensory stimulation nor a translation of motor activity related to lordosis behavior but might reflect some aspects of the internal motivational state of the female concerning the display of lordosis behavior. In mice, it was shown with single-unit recording that VMNvl neurons are indeed more activated in the presence of a conspecific with preference for the male stimulus when hormonally receptive (internally motivated) compared to the non-receptive state (Nomoto and Lima 2015).

Other brain (sub) areas that are activated during the copulatory phase are the POA, lateral septum, BNST, MeA, subparafiscular nucleus (SPFp), and PAG (Coolen et al. 1996; Pfaus et al. 1993, 1996; Tetel et al. 1993, 1994). After receiving ejaculations, neuronal activation was observed in the medial preoptic nucleus (MPN), BNSTpm, MeApd, and SPFp (Pfaus et al. 1993; Coolen et al. 1996; Polston and Erskine 1995; Rowe and Erskine 1993). After receiving ejaculations, neuronal activation was observed in the medial preoptic nucleus (MPN), BNSTpm, MeApd and SPFp (Pfaus et al. 1993; Coolen et al. 1996; Polston and Erskine 1995; Rowe and Erskine 1993), in addition to the VMNvl, the caudoventral part of VMN (VMNcv) and ventral premammillary nucleus (PMV) (Coolen et al. 1996). Interestingly, most of the areas activated after mating (MPN, BNSTpm, posterodorsal preoptic nucleus (PD), VMNcv, and PMV) also showed Fos-IR after treatment with estrogen and progesterone (Coolen et al. 1996), suggesting a close link between hormonal treatment and sexual behavior.

These Fos-IR studies show which brain areas become activated under certain circumstances, but it does not explain how essential these regions are in the control of female sexual behavior. For this kind of information, lesion and electrical stimulation studies are more useful. Unfortunately, only a few brain regions have been extensively studied in females. Several studies demonstrate that lesions of the VMN result in dramatic decreases in lordosis and paracopulatory behaviors (Mathews and Edwards 1977; Rajendren et al. 1991), while electrical stimulation results in a facilitation of lordosis (Pfaff and Sakuma 1979b). VMN lesions also abolished sexual incentive motivation, expressed as approach to a potential mate, in the female rat (Emery and Moss 1984). This suggests that the role of the VMN in the regulation of sexual motivation and behavior in female rats is mainly stimulatory.

In contrast to the VMN, the POA might play an inhibitory role in the control of lordosis behavior, because POA lesions cause an increase in lordosis responses in females (Powers and Valenstein 1972a; Hoshina et al. 1994). Electrical stimulation of this brain area, on the other hand, reduces lordosis behavior (Moss et al. 1974). It should be mentioned, though, that some studies did not find these facilitating effects on receptive behavior after POA lesions (Guarraci et al. 2004; Yang and Clements 2000). The regulatory role of the POA on lordosis might be context specific, in which the lesions only affect lordosis behavior when the females are in a non-paced mating situation (Whitney 1986). In contrast to the stimulatory effects on lordosis, paracopulatory behavior seems to get abolished by POA lesions (Hoshina et al. 1994; Guarraci et al. 2004), just as paced mating behavior seen as prolonging contact-return latencies and percentage of exits declines (Yang and Clements 2000; Guarraci et al. 2004). Together, this indicates that the POA plays a dual role in the copulatory phase of sexual behavior, an inhibiting role in the reflexive lordosis response and a stimulating role in paracopulatory behaviors. During the introductory phase, the POA plays a stimulatory role as well, since

POA lesions abolished the female's preference for sexually receptive males (Guarraci and Clark 2006).

Single-cell recordings showed that different subsets of neurons in the POA are involved in the regulation of the different behaviors (Kato and Sakuma 2000). Nevertheless, neuronal links for both paracopulatory and receptive components have been found to originate in the POA, indicating an important role of this hypothalamic nucleus in the regulation of sexual motivation and behavior.

The VMN and POA receive intensive neural inputs from the MeA and the BNST (Canteras et al. 1995; Gu et al. 2003; Shimogawa et al. 2015), which in turn receive projections from both the main and accessory olfactory systems (Baum and Cherry 2015). It is thought that the olfactory stimulation will reach the VMN and POA via the MeA (Canteras et al. 1995; Gu et al. 2003), mainly through the BNST. Olfactory stimuli are crucial for the activation of approach behaviors (Bergvall et al. 1991; Thor and Flannelly 1977), and without approach copulation will never occur. It would, therefore, be logic that these brain areas play an important role in female sexual behavior as well. In mice, neural activation in the MeA is found in response to pheromonal stimuli (Samuelsen and Meredith 2009), with a different response to same sex and opposite sex chemosensory stimuli (Bergan et al. 2014). In line with this finding, lesion studies have shown that the MeA reduces approach behavior of sexually receptive females toward male rats (Kondo and Sakuma 2005), suggesting that the MeA plays a stimulatory role on the introductory phase of female sexual behavior. Interestingly, in terms of effects on paracopulatory or receptive behaviors, MeA lesions studies show conflicting results. Some have shown no changes in darts and lordosis behavior upon MeA lesions (Guarraci and Clark 2006; Kondo and Sakuma 2005), while others found increases in lordosis intensity and paracopulatory behaviors (Polston and Erskine 2001; Masco and Carrer 1980; Rajendren and Moss 1993), thereby suggesting an inhibitory role of the MeA during copulation. To cause even more confusion, when the state-of-the-art technique optogenetics was used in mice to silence the specific projections from the olfactory system to the MeA, sexually experiences female mice showed reduced levels of lordosis responses (Mccarthy et al. 2017a). This would suggest that chemosensory inputs from the olfactory areas are required to show full receptivity in female mice. When all neurons in the MeA were inhibited with a chemogenetic approach, a decrease in lordosis responses was found in CNO-treated females versus controls, suggesting that the MeA plays a stimulatory on receptive behavior. Interestingly, when the females were tested multiple times during inhibition of the MeA, the lordosis responses returned to a normal level after six tests, suggesting that it was more the "learning curve" in improvement of receptivity after obtaining experience that was attenuated rather than the ability to perform a lordosis (Mccarthy et al. 2017b). They hypothesized that the combination of hormone priming and the experience of the sensory input of a mounting male is required for a progressive improvement in lordosis responsiveness. In that perspective, it could be hypothesized that a minimal number of MeA neurons should be inhibited in order to completely block the lordosis throughout sexual experience (Mccarthy et al. 2017b). This idea is supported by a previous lesion study in which bilaterial deletion of the MeA indeed attenuated the expression of lordosis in mice (Dibenedictis et al. 2012). These differences in findings could be caused by the use of different species, either mice or rats, which show complete different patterns of sexual behavior. However, more research is needed to understand the function of MeA on female reproductive behavior.

For the BNST, it was found that this brain area is not involved in the regulation of sexual behavior, whereas lesions did not change lordosis, paracopulatory, and pacing behavior in females (Guarraci and Clark 2006; Gray et al. 1978). This was in line with a study performed in hamsters that showed that BNST lesions do not affect male-odor preference and lordosis behavior in females (Martinez and Petrulis 2011). However, other studies did suggest a stimulatory role for the BNST in approach behavior (Jenkins and Becker 2001; Yang and Clements 2000), but this was then probably caused by collateral damage to other brain regions. Although surprising, the BNST does not seem to play an essential role in the execution of female sexual behavior. The expression of c-Fos in this area could instead reflect the integration of sensory information and not the execution of sexual behavior.

A few studies have been evaluating the role of some other brain regions in female sexual behavior, e.g., the anterior hypothalamus, nucleus accumbens (NAc), ventral tegmental area (VTA), and periventricular gray. Lesions in the NAc seem to increase the number of rejections from the female rat while leaving the lordosis behavior intact (Rivas and Mir 1990). Still, conclusions could not really be drawn from these studies, whereas the author claimed that increase in rejections might just as well be a consequence of an increase in general hyper-reactivity reducing the female's tolerance to sexual interactions from the male. The results of the studies investigating the other brain regions are too limited or controversial to outline them any further, but an overview can be found in Paredes and Ågmo (2004).

Unfortunately, this is still an important gap in knowledge when exploring the neural regulatory circuitry of female reproductive behavior. The studies mentioned above have explored the role of different brain areas in regulating the different phases of female sexual behavior, but they have not investigated the connection between these brain areas. As mentioned above, and shown in Fig. 2, all brain areas are closely connected to one and another, meaning that they are should work together to get from the introductory phase to a successful executive phase. None of the studies mentioned above was able to explore the role of specific projections between brain areas on female sexual behavior. In the last decade, many new state-of-the-art techniques have entered the field of behavioral neuroscience which could be very helpful in answering these questions, e.g., optogenetics, chemogenetics, and calcium imaging. Lesion studies might have revealed the structures and projections needed for a behavior, but it has proven to be extremely difficult to manipulate the precise connections that are involved: lesions tend to affect both the direct and indirect pathways. Fiber photometry, as example, works by injecting adeno-associated viral (AAV) vectors expressing calcium indicators in neurons locally in brain area A, these indicators will be expressed across all axons of neurons originating in this area, including the downstream projections to other areas (Girven and Sparta 2017). Activation of neurons corresponds with increasing levels of intracellular calcium. This calcium then binds to the calcium indicators in the axons, resulting in the emission of a fluorescent signal that can be detected by the implanted fiber (Girven and Sparta 2017). By implanting an optic fiber into the projection area B, fiber photometry can measure the neural activity of the specific projections from the area A to area B within a millisecond timescale. Using the same principle with AAV vectors, but now designed to express clozapine-N-oxide (CNO)-sensitive receptors in neurons originating in the brain area, chemogenetics can turn on or off (depending on the type of AAV vector) the specific brain projections by inserting CNO intracranially in the area B. In summary, fiber photometry can study the neural activity linked to certain behaviors, whereas chemogenetics can be used as proof of concept to show that disinhibition/activation of the brain projections have direct consequences for certain behaviors (Roth 2016). As one can imagine, this kind of novel techniques have been important for the understanding of the neural regulation of behavior. These methods could contribute to a large extend to the understanding of the complexity of female sexual behavior and give insight into the connectivity between brain areas. Unfortunately, the use of these state-of-the-art techniques has not yet entered the field of reproductive behavior and especially not in female sexual behavior. Except for the few example that were given above, all our knowledge still comes from lesion and pharmacological studies only, but hopefully future studies will also implement the new techniques to unravel the role of specific projections in female reproductive behavior.

5 Chemical Messengers Involved in Female Sexual Behavior

The brain regions involved in female sexual behavior use several different types of chemical messengers to regulate mating behavior. We have already discussed how ovarian hormones such as E and P regulate the estrous cycle and thereby induce behavioral estrus, but these hormones also play an important role in mating behavior itself. In addition, several types of neurotransmitters and neuropeptides are involved in the regulation mechanisms behind mating behavior, e.g., serotonin, dopamine, noradrenaline, oxytocin, and gamma-aminobutyric acid (GABA). A selection of the most important chemical messengers will be discussed in more detail.

5.1 Estrogen and Progesterone

Female sexual behavior is highly dependent on the ovarian hormones E and P (Pfaff and Schwartz-Giblin 1988). Ovariectomy causes a robust disruption of lordosis and paracopulatory behaviors in females, but this decline can be partly restored by treatment with E, or completely following sequential treatment with E and then P (Powers and Valenstein 1972b; Fadem et al. 1979; Jones et al. 2013). Several studies have shown that there is a dose-dependent effect of E on female sexual behavior, in which the lordosis and paracopulatory behavior increase with a rising dose of E (Snoeren et al. 2011a; Powers and Valenstein 1972b; Davidson et al. 1968; Meyerson 1964; Jones et al. 2013). P has a similar dose-dependent effect when added to E in terms of paracopulatory behavior and time spent with a male (Powers and Valenstein 1972b; Snoeren et al. 2011a; Meyerson 1964). The lordosis quotient seems to be less dose-dependent of P but requires a minimal hormonal level to be induced.

The mechanisms behind the hormonal regulation are not yet completely understood, but the behavioral effects of E and P rely, at least in part, on their regulatory actions on neurotransmitter synthesis, release, and/or receptors (reviewed in Kow et al. 1994). When E levels increase, the expression of additional P receptors (PR) is seen in the POA and the VMN (Maclusky and Mcewen 1978). Changes is co-localization of E receptors (ER) and PR are also found during the estrous cycle with high co-localization at proestrus (Sa and Fonseca 2017). Classically, it was though that E and P affect transcription mechanism in the nucleus, but with the recent discovery of membrane-associated steroid receptors, it now is assumed that E and P have also neurotransmitter-like actions activating intracellular events and influence transcription (reviewed in Micevych et al. 2015). It is a complicated mechanism in which many peptides play a role (Micevych et al. 2015), but principally within minutes of treatment with E, an active inhibition of lordosis is initiated in the arcuate nucleus (ARC). In this brain region, E induces the release if neuropeptide Y (NPY) that act on neurons expressing NPY-Y1 and $GABA_B$ receptors (Mills et al. 2004; Sinchak et al. 2013). The activated neurons project to the POA where they release another peptide called ß-endorphin that, in turn, activates and induces internalization of μ-opioid receptors. These μ-opioid neurons subsequently innervate VMN neurons regulating the descending output of the hypothalamus that controls lordosis behavior (reviewed in Micevych and Dewing 2011). For approximately 20 h, this activation of μ-opioid receptors inhibits the display of lordosis behavior, until P is added to the system and reverses the E-induced inhibition, allowing for lordosis (Sinchak and Micevych 2001). How this P-regulated process works is still unclear, but P-knockout mice are not able to show lordosis (Lydon et al. 1995), and PR antagonists and PR antisense oligonucleotide block the induction of lordosis by P (Pfaff and Schwartz-Giblin 1988). The mechanism that P uses for the stimulating effect in lordosis is probably a different mechanism than the E-circuit.

Besides the direct relationship between E and P, there is also a link between ovarian hormones and neurotransmitters and peptides. Both E and P, for example, change the levels of GABA and glutamate in the VMN. Whereas E causes a rise in GABA and glutamate levels in the VMN, P results in a decline in neurotransmitter levels (Luine et al. 1997). This indicates that ovarian hormones regulate sexual behavior via a control on side-specific neurotransmitter levels. A similar relationship exists between ovarian hormones and other neurotransmitters. ER and PR agonists, for instance, also modify noradrenaline levels (Lubbers et al. 2010; Nagle and Rosner 1980; Janowsky and Davis 1970), and an increase in E levels within the VMN causes a rise in oxytocin binding in this region (Schumacher et al. 1989). Levels of serotonin levels, in addition, vary during the estrous cycle (Gundlah et al. 1998). In addition, P treatment results in a serotonin turnover in the VMN of rats treated with E directly in the VMN (Gereau et al. 1993). Important in this situation is that P does not have an effect when E is missing, suggesting that there is a direct link between both hormones and serotonin and that the facilitating effects of P on female sexual behavior could be modulated via a serotonin turnover. This suggests the existence of a mechanism by which two ovarian hormones act synergically in a specific brain region to promote the right conditions to induce a behavioral response.

The E-induced lordosis behavior is depending on ER activation. Estrogens act via two different estrogen receptors, the estrogen receptor α (ERα) and the estrogen receptor β (ERβ). Studies in both rats and mice have shown that the ERα is important for the activation of sexual behaviors (Spiteri et al. 2010a, 2012; Ogawa et al. 1998, 1999; Mazzucco et al. 2008). The ERβ, on the other hand, is not necessary to induce receptivity (Mazzucco et al. 2008) and might be more involved in other behaviors in which E plays a role, such as fear and anxiety (Spiteri et al. 2010b, 2012), social recognition (Spiteri and Ågmo 2009; Spiteri et al. 2010b), and aggression (Albert et al. 1992; Spiteri et al. 2010b).

5.1.1 Estrogen in Specific Brain Areas

One of the main sites of action of E is the VMN. The ERα plays an important role in the stimulatory control of the VMN on female sexual behavior, since the VMN shows a high expression of this receptor (Simerly et al. 1990) and has many ERα positive neurons (Yamada et al. 2009). Besides, E treatment promotes the development of axodendritic synapses in the VMN (Frankfurt and Mcewen 1991). Infusion of E directly in the VMN facilitates lordosis behavior in female rats (Barfield and Chen 1977), which is probably regulated via ERα. This was shown by several studies using site-specific silencing of ERα (via the infusion of adeno-associated viral vector directed against the ERα gene) within the VMN. Reduced levels of ERα in the VMN cause a decline in sexual receptivity and paracopulatory behaviors in rats and mice (Spiteri et al. 2010a; Musatov et al. 2006; Snoeren et al. 2015). In addition, local infusions of antiestrogens in the VMN decrease lordosis in rats (Meisel et al. 1987).

Another important area for E-mediated effects is the POA. As mentioned before, POA lesions have been shown to abolish paracopulatory behavior while promoting lordosis (Hoshina et al. 1994). The role of the ERα in the POA, however, is rather confusing. Site-specific silencing of ERα in the POA resulted in increased levels of lordosis responses, while paracopulatory behaviors remained unaffected (Spiteri et al. 2012). Sexual motivation for a sexually attractive male rat was also reduced in these females with reduced ERα levels in the POA (Spiteri et al. 2012).

This suggests that ERα could play a role in the inhibitory function of the POA in lordosis, but not in the regulation of paracopulatory behaviors.

However, my study in which we used a seminatural environment to study the effects of ERα silencing in the POA and VMN on female sexual behavior revealed more surprising results. Females with fewer ERα in the POA and VMN showed lower levels of both lordosis responses and paracopulatory behaviors compared to the control females (Snoeren et al. 2015). However, the lordosis quotients were left unaffected, meaning that the reduction in lordosis responses was caused by a decrease in received mounts and intromissions, and not by the incapability to perform lordosis (Snoeren et al. 2015). In conclusion, ERα in both the POA and VMN are *not* essential to induce lordosis responses. The decline in lordosis responses and paracopulatory behaviors could instead reflect a reduction in sexual motivation.

A reduction in ERα levels in the MeA or BNST did not affect the number of paracopulatory behaviors or lordosis responses compared to controls (Spiteri et al. 2010a; Snoeren et al. 2015). As mentioned before, the MeA might play an inhibitory role on female copulatory behavior, but ERα expression seems to be not essential in this regulation.

5.2 Serotonin

The sexual behavior system is under constant inhibitory control to assure that copulation occurs only under the proper circumstances. Serotonin is involved in this inhibition and in the disinhibition to induce the sexual behaviors. Therefore, serotonin release is regulated via a negative feedback mechanism (Aghajanian 1978; Gothert and Weinheimer 1979) that is controlled by different serotonergic receptors. The serotonergic neurotransmitter system consists of the endogenous ligand, 5-hydroxytryptamine (5-HT, serotonin), and 14 functionally distinct 5-HT receptor subtypes. The receptors can be divided into seven families, namely, 5-$HT_{1–7}$, with all different and limited distributions in the nervous system (Hoyer et al. 1994). Except for the 5-HT_3 receptor subtype, which is a ligand-gated ion channel, 5-HT receptors are 7-transmembrane receptors and act via G-proteins. The last mechanism involved in the maintenance of serotonin levels is the serotonin transporter (SERT) which is responsible for the active transport of serotonin into neurons (Murphy et al. 2004).

5-HT neural activity is most likely tonically elevated during the execution of behavior, thereby facilitating the whole behavior instead of subparts like specific muscle groups or motor programs (Jacobs and Fornal 1997; Muller and Jacobs 2010). Therefore, 5-HT probably acts as modulator or facilitator in sexual behavior, rather than playing a role as central mediator. The increased firing of 5-HT neurons during sexual behaviors is under control of several feedback systems located somatodendritically (5-HT_{1A} receptors) or presynaptically (5-HT_{1B} receptors) and also postsynaptic (5-HT_{1A}, $5\text{-HT}_{2A/2C}$, and 5-HT_4 receptors (Sharp 2010)). The role of the 5-HT system in sexual behavior is mostly studied in male rats (reviewed in Snoeren et al. 2014b), although this neurotransmitter is regularly studied in female sexual behavior as well (reviewed in Snoeren et al. 2014a).

5-HT seems to play a dual role in the control of female sexual behavior with 5-HT_{1A} receptors acting to inhibit and 5-HT_2 receptors to facilitate female sexual behavior upon activation. It, therefore, depends on which receptor subtype becomes activated on what kind of effects is induced. Many studies have shown an inhibiting effect of 5-HT_{1A} receptor agonists on paracopulatory behavior and lordosis in female rats (Mendelson and Gorzalka 1986; Kishitake and Yamanouchi 2003; Ahlenius et al. 1986, 1989; Fernandez-Guasti et al. 1987; Snoeren et al. 2011b, c). The inhibiting effects were found in both ovariectomized females primed with E alone or with E in combination with P, although some studies could not find the effects in females primed with only E (Snoeren et al. 2011c; Ahlenius et al. 1986). The inhibiting effects of 5-HT_{1A} receptor agonists are antagonized by specific 5-HT_{1A} receptor antagonists (Snoeren et al. 2010; Johansson et al. 1991). Likewise, 5-HT_{1B} receptor agonists also induce inhibiting effects on lordosis behavior (Uphouse et al. 2009, 2010). In conclusion, 5-HT_{1A} and 5-HT_{1B} receptors are involved in the inhibition of female sexual behavior during the copulatory phase.

5-HT_2 and 5-HT_3 receptors, on the other hand, seem to play a stimulatory role on sexual functioning. Agonists of these receptors facilitate lordosis behavior in female rats (Mendelson and Gorzalka 1985; Wolf et al. 1998a), while antagonists for 5-HT_2 and 5-HT_3 receptors inhibit lordosis and paracopulatory behavior (Gonzalez et al. 1997; Maswood et al. 1997; Miryala et al. 2013). Agonists for the $5\text{-HT}_{2A/2C}$ receptor subtypes stimulate lordosis and paracopulatory behavior (Nedergaard et al. 2004; Rossler et al. 2006; Wolf et al. 1999), while antagonists inhibit sexual behavior (Sinclair-Worley and Uphouse 2004; Uphouse et al. 2003; Kaspersen and Ågmo 2012).

Drugs that increase serotonin levels, like selective serotonin reuptake inhibitors (SSRIs), could disrupt the balance between activation of 5-HT receptors that inhibit and those that facilitate sexual behavior. The SSRI fluoxetine, for instance, is known to cause sexual dysfunctions in women (Clayton 2002) and female rats (Ventura-Aquino and Fernandez-Guasti 2013; Guptarak et al. 2010). This inhibition in sexual behavior could be caused by activation of the 5-HT_{1A} receptor (Guptarak et al. 2010), or its blocking effect on 5-HT_2 receptors (Palvimaki et al. 1996). Interestingly, studies with chronic treatment of another SSRI (paroxetine) did not affect lordosis and paracopulatory behavior in estrous females (Snoeren et al. 2011c; Kaspersen and Ågmo 2012), although it did reduce the sexual incentive motivation (Kaspersen and Ågmo 2012). In females primed with only E, however, 7 days of paroxetine treatment induced a decrease in sexual behavior (Snoeren et al. 2011c), but this effect disappeared after chronic treatment with the antidepressant. The same lack of effects on copulatory behavior was found in SERT knockout female rats

(Snoeren et al. 2010). In all studies it was shown that the 5-HT_{1A} receptor was desensitized, which might explain the lack of sexual dysfunctions (Snoeren et al. 2010, 2011c). The 5-HT_{1B} and 5-HT_{2C} receptors, on the other hand, do not play a role in the paroxetine effects (Kaspersen and Ågmo 2012).

5.2.1 Serotonin in Specific Brain Areas

The inhibiting role of 5-HT_{1A} receptors in female sexual behavior is without doubt, but the specific location of the responsible 5-HT_{1A} receptors is unknown. The dorsal raphé nucleus (DRN) is one of the brain areas with an intense distribution of 5-HT_{1A} receptors. Studies have shown that lesions in the DRN increase lordosis (Kakeyama et al. 1997; Kakeyama and Yamanouchi 1996; Arendash and Gorski 1983), suggesting an inhibiting role for this brain area. Nevertheless, local injections of 5-HT_{1A} receptor agonists in the DRN do not affect lordosis or solicitation behavior (Uphouse et al. 1992a), suggesting that the 5-HT_{1A} receptors do not play a role in the inhibiting effects of the DRN. Infusion of the 5-HT_{1A} receptor agonist in the medial raphé nucleus (MRN), on the other hand, decreases lordosis responses.

The MRN projects to the VMN, which is another brain area with high levels of 5HT_{1A} receptors. Local 5-HT_{1A} receptor agonist injections in the VMN suppress lordosis and paracopulatory behavior (Uphouse et al. 1992b, 1993, 1996a, 2000; Wolf et al. 1998b; Gonzalez et al. 1997; Trevino et al. 1999). These inhibitory effects can be attenuated by several 5-HT_{1A} receptor antagonists (Uphouse et al. 1996a). Together, this suggests a role for 5-HT_{1A} receptors in the VMN in the regulation of lordosis and paracopulatory behavior in females.

However, the VMN contains also other 5-HT receptor subtypes which could play a role on the regulation of female sexual behavior. Local infusions of 5-$HT_{2A/2C}$ receptor antagonists, for instance, inhibit lordosis behavior (Uphouse et al. 1996b; Wolf et al. 1998a), while a 5-HT_2 receptor agonist increases lordosis responses in suboptimally hormone-primed females (Maswood et al. 1997; Wolf et al. 1998a). In the VMN, it is the 5-HT_{2C} that have been implicated in lordosis modulation (Wolf et al. 1999), while 5-HT_{2A} receptors may be involved more in the POA (Gonzalez et al. 1997). It was also shown that bilateral infusions of a 5-HT_3 receptor antagonist in the VMN disrupt the lordosis responses, an effect that could be attenuated by the administration of a receptor agonist (Maswood et al. 1997). The expected stimulatory role of the VMN is thus probably regulated via the 5-$HT_{2A/2C}$ receptors.

Another possible brain area that is involved in female sexual behavior is the POA. As mentioned before in Sect. 4, the POA is assumed to be inhibitory in the control of lordosis behavior while playing a stimulatory role in the regulation of paracopulatory behavior. That this effect might be regulated via 5-HT_{1A} receptors in the POA was suggested by a reduction in lordosis behavior in response to local 5-HT_{1A} receptor agonist infusions (Uphouse and Caldarola-Pastuszka 1993). Interestingly, the local infusions did not change the amount of paracopulatory behavior, suggesting that these different sexual behaviors might be regulated via the interplay of different neurotransmitters and/or brain areas.

In conclusion, 5-HT seems to play a dual role in the control of female sexual behavior with 5-HT_1 receptors playing an inhibitory and 5-$HT_{2/3}$ receptors a stimulatory role in female sexual behavior. The hypothalamus is an essential player in the serotonergic control of sexual functioning. The VMN and POA are already studied, but hopefully more brain areas and precise brain projections will be investigated in the future. In addition, very few studies are available upon the role of 5-HT in the introductory phase. I hope that this gap in research will be acknowledged and explored in future studies as well.

5.3 Dopamine

Dopamine (DA) has received extensive attention on the role it plays in male and female sexual behavior. Regrettably, most studies on dopamine and female sexual behavior have been focusing solely on lordosis behavior and neglected the paracopulatory behaviors. Despite the single focus, though, the role of dopamine in lordosis remains rather controversial (also reviewed in Melis and Argiolas 1995). DA receptor agonists and antagonists have been reported to have both inhibitory and facilitatory effects on female receptivity. On one hand, DA receptor agonists have been shown to suppress lordosis behavior in hormonally primed females, while antagonists stimulate receptivity (Everitt and Fuxe 1977; Everitt et al. 1975; Michanek and Meyerson 1977b, 1982; Fernandez-Guasti et al. 1987). In addition, agents that destruct dopaminergic neurons cause stimulation of lordosis (Ahlenius et al. 1972; Everitt et al. 1975). To the contrary, however, some other studies showed beneficial effects of dopamine on receptive behavior (Foreman and Moss 1979; Hamburger-Bar and Rigter 1975) or no effect on lordosis (Ellingsen and Ågmo 2004). Infusion of dopaminergic agonists in the third ventricles in the brain also resulted in increased lordosis behavior (Ma et al. 2010). The few studies that did investigate the effects of nonselective DA receptor agonists on paracopulatory behaviors reported inhibitory effects for dopamine in hormonally fully primed females (Ellingsen and Ågmo 2004; Snoeren et al. 2011b).

Two main details should be considered to explain the differences in the dopaminergic effects: hormones and dosage. As mentioned before, ovarian hormones play an important role in female sexual behavior. In general, it is easier to facilitate female sexual behavior from a low baseline of receptivity and to inhibit fully primed females than the other way around. In addition, besides the direct role of ovarian hormones on sexual functioning, they also play an indirect role by changing the balance and sensitivity of receptor expressions. Ovariectomy changes the dopamine content and turnover, as well as the density in DA receptors and their affinity for agonists and antagonists in different brain areas (Gunnet et al. 1986; Levesque and Di Paolo 1988; Hruska 1986; Hruska and Nowak 1988). This suggests that differences in hormonal priming in the experiments could influence the outcomes of dopaminergic agent priming. When analyzing the studies mentioned above, they suggest that low doses of DA receptor agonists facilitate lordosis in low-primed females, while high doses inhibit receptivity in fully primed females.

The actions of dopamine are mediated by five different receptor subtypes, which are members of the large G protein-coupled receptor superfamily. The dopamine receptor subtypes are divided into two major subclasses, the D1-like and D2-like receptors, which typically couple to Gs- and Gj-mediated transduction systems. In the brain, the various receptor subtypes display specific anatomical distributions, with D1-like receptors being mainly postsynaptic and D2-like receptors being both pre- and postsynaptic (reviewed in Jaber et al. 1996). The different receptors could also explain the different role of dopamine on female sexual behavior.

Grierson et al. suggested that low doses of dopaminergic agents act via presynaptic receptors and therefore inhibit dopamine release and stimulate female sexual behavior, while high doses inhibit lordosis via postsynaptic receptors (Grierson et al. 1988). When agents were administered that cause a small increase in DA levels, lordosis behavior was stimulated, while the induction of large amounts of dopamine resulted in inhibitory effects (Stoof and Kebabian 1984). Similar differences in effects with low and high doses of agents were found with selective D_2 receptor agonists. It was suggested that these effects acted solely on presynaptic receptors (Titus et al. 1983). Interestingly, no effects were seen with D_1 receptor agonists and antagonists (O'connor and Brown 1982). In addition, the stimulating effects of dopaminergic agents could only be inhibited by D_2 receptor antagonists and not by antagonists for D_1 receptors (Grierson et al. 1988). This suggests that dopamine D_2 receptors are more (or only) involved in lordosis behaviors than dopamine D_1 receptors.

5.3.1 Dopamine in Specific Brain Areas

When dopamine or DA receptor agonists are administered directly in the hypothalamic area, it exerts mostly stimulatory effects on female sexual behavior, while DA receptor antagonists inhibit lordosis (Foreman and Moss 1979). These effects were found in the POA and ARC, but not the lateral hypothalamus (LHA), and suggest that these brain regions play a role in the dopaminergic regulation of female sexual behavior. Another important brain region with dopaminergic control on female sexual behavior is the VMN. Dopaminergic agents that were locally infused in the VMN had also a stimulatory effect on female sexual behavior (Mani et al. 1994).

The results of these studies, however, are rather confusing. Infusion of low doses of the nonselective dopamine receptor agonist apomorphine directly in the POA causes an increase in paracopulatory behavior (Graham and Pfaus 2010) in E-primed females. High doses of the same drug, however, seem to have no effect on this behavior. Other components of female sexual behavior such as lordosis, solicitation, and pacing behavior are also not affected by both the low and high doses of the agonist (Graham and Pfaus 2010). Another study, however, did show an increase in lordosis behavior in females primed with low doses of hormones (Foreman and Moss 1979). In fully primed females, on the other hand, the lordosis responses were left unaffected by the DA receptor agonist (Foreman and Moss 1979). Administration of nonselective DA receptor antagonists, on the other hand, resulted in disrupted lordosis (Foreman and Moss 1979), solicitation, pacing, and paracopulatory behaviors in fully primed females (Graham and Pfaus 2012).

Whether these effects are regulated via the dopamine D_1 or D_2 receptors remains unclear. On one hand, it was suggested that the dopaminergic effects in the POA are also regulated via the dopamine D_2 receptors, because administration of a selective D_2 receptor agonist induced the same stimulatory effects in low-primed females (Graham and Pfaus 2010). However, when the data was analyzed in more detail, this effect was only found on solicitation behaviors and not in the number of darts and hops, lordosis, or pacing behavior (Graham and Pfaus 2010). Locally infused D_1 receptor agonists, on the other hand, resulted in normal lordosis behavior (Apostolakis et al. 1996; Graham and Pfaus 2010) but a decrease in paracopulatory behaviors (only in low doses of the dopaminergic agent) (Graham and Pfaus 2010). While the D_2 receptor agonist solely affected solicitation behavior, the D_1 receptor agonist affected only darts and hops. This indicates that both receptors might play a role in the regulation of female sexual behavior, with each receptor being involved in different aspects of behavior. Graham and Pfaus suggested that the ratio of DA receptor subtypes within the POA is critical for the display of sexual behavior (Graham and Pfaus 2010).

Surprisingly, when DA receptor *antagonists* were injected within the POA, unexpected results were found. Whereas one would expect that antagonists would have no or opposing effects to agonists, high doses of D_2 receptor antagonist increased solicitation behavior in fully primed females (Graham and Pfaus 2012), just as agonists have shown to do in low-primed females. The D_2 receptor antagonist did also increase pacing behavior, but had no effect on lordosis and paracopulatory behavior (Graham and Pfaus 2012). Local administration of a D_1 receptor antagonist, on the other hand, caused a slight decrease in solicitation and pacing behavior while leaving other components of female sexual behavior unaffected (Graham and Pfaus 2012).

Could we, therefore, conclude that infusions of DA receptor antagonists in the POA have similar effects as DA receptor agonist? The elements of female sexual behavior were unequally affected by the dopaminergic agents: sometimes the pacing behavior was affected, while another time paracopulatory behavior was influenced. Since no clear pattern was found, it can be suggested that the different elements were not regulated by a certain DA receptor.

A possible explanation for the differences in results is that the stimulatory effects of the antagonists were found in E + P-primed females, whereas the effects of the agonists were found in E-primed females. The authors, therefore, suggested that female primed with E + P shows a tilt toward D_1 receptor stimulation, thereby increasing sexual behavior. In E-primed females, on the other hand, shifting activity toward the D_2 receptors has the same effect (Graham and Pfaus 2012). These effects are in line with previously described studies of Grierson et al. (1988), who showed that D_2 receptor activation facilitates lordosis in E-primed females but inhibits sexual behavior in E + P-primed females (Grierson et al. 1988). The switch in D_1/D_2 balance in the POA after different hormonal priming was later confirmed by using immunohistochemistry, Western blots, and autoradiography: E and P affected the dopaminergic receptors in opposite ways in the POA in which E causes a lower and P a higher D_1:D_2 ratio, respectively (Graham et al. 2015). This suggests that E act via a D_2 receptor mediated system, while E + P work via the D_1 receptors.

Extracellular DA levels in the POA fluctuate in response to circulating hormones. P injections increase extracellular DA levels and stimulate sexual behavior in females primed with a low dose of E, while it does not affect de levels and behavior of females primed with higher doses of E (Matuszewich et al. 2000). This suggests that DA in the POA may indeed be important for the facilitation of sexual behavior by P. As mentioned above, DA seem to interact with a P-dependent mechanism in the VMN to promote lordosis via the D_1 receptor (Mani et al. 1994). Therefore, a possible mechanism could exist in the POA in which E upregulates P receptors in the POA and other brain areas, which in turn increases D_1 receptor activation (like previously shown in the ventral tegmental area (VTA)) (Petralia and Frye 2006). In this perspective, D_1 receptor activation becomes only relevant when P is added to the circulation and causes facilitation for paracopulatory behaviors in female rats. D_2 receptor activation, on the other hand, would have an inhibitory role under the control of P. DA could, thus, have two different effects on female sexual behavior by acting on different receptors on certain neuron populations with an altered balance in D_1 and D_2 receptors (Graham and Pfaus 2012).

More research is needed to determine whether these hypotheses on the role of dopamine in female sexual behavior are correct. According to a very elegant review written by Paredes and Ågmo (2004), the role of dopamine might be less essential than always assumed. They nicely describe how the effects of the dopaminergic agents on female sexual behavior might be linked more to the dopaminergic effect on motoric aspects than sexual behavior (Paredes and Ågmo 2004). Dopamine is very important in the control of movement, and many effects of dopaminergic agents can be explained by the effects on locomotor activity: e.g., amphetamine, a dopaminergic agonist, is shown to produce inhibitory effects on lordosis behavior in female rats (Michanek and Meyerson 1977a). However, in the same study, it was shown that the same dose that affected the lordosis also increased stereotyped activity, which could be blocked by a DA antagonist. In addition, Parades and Ågmo described a study in which a DA receptor antagonist could induce prolonged lordosis responses, which could be interpreted as an increase in lordosis intensity. However, the drug produced at the same time a dramatic decrease in motor execution (Paredes and Ågmo 2004). They recognized that there might be a link between the effects on lordosis and motor activity, an idea that is also supported by a study in males in which it was shown that whenever motor execution was impaired, sexual behavior was disrupted as well (Ågmo et al. 1987). In summary, caution is needed when drawing conclusions on the role of dopamine in female sexual behavior.

5.4 *Noradrenaline*

The role of noradrenaline (NA) in female sexual behavior is not yet clear. Most studies performed in this field are studies that administered adrenoceptor agonists and antagonists locally in different brain areas. It is, therefore, difficult to determine what general effect NA has on female sexual functioning.

The NA system consists of different receptor types, including α_1, α_2, β adrenoceptors, and noradrenaline transporters. Adrenoceptors are located in the brain, spinal cord, and periphery (Frankhuyzen and Mulder 1982; Nasseri and Minneman 1987) and are localized both post- and presynaptically, as inhibitory receptors on non-adrenergic neurons (heteroceptors) and on the terminals and dendrites of the noradrenergic neurons themselves (autoreceptors) (Frankhuyzen and Mulder 1982; Nasseri and Minneman 1987). Again, the different adrenoceptors seem to play diverse roles in female reproductive behavior (reviewed in Snoeren 2015).

Unfortunately, no studies are available in which the effect of systemically administered NA was investigated on female sexual behavior. Only a few studies are known in which agents acting on specific adrenoceptors have been investigated. Systematic administration of an α2-adrenoceptor agonist, for instance, does not have an effect on lordosis in female rats (Davis and Kohl 1977). Similar lack of results was found with the administration of nonselective and selective α2-adrenoceptor antagonists: no effects were found on sexual incentive motivation, paracopulatory behavior, and lordosis (Snoeren 2015; Gonzalez et al. 1996; Davis and Kohl 1977; Ventura-Aquino and Fernandez-Guasti 2013). This suggests that, systemically, α2-adrenoceptors are not involved in the regulation of female sexual behavior.

Thus, if NA is involved in the regulation of sexual behavior in females, it must involve other adrenoceptors, like the α_1- or β-adrenoceptors. Studies in which α1-adrenoceptor agonists were administered in the cerebral ventricles show that α_1-adrenoceptors might play a stimulatory role on lordosis (Kow et al. 1992). α_1-Adrenoceptor antagonists, on the other hand, attenuate lordosis and paracopulatory behavior in females when administered in the ventricles (Gonzalez-Flores et al. 2007). The role of the β-adrenoceptors is still rather unclear. While one study found that the β-adrenoceptor agonist isoproterenol facilitated lordosis (Kow et al. 1992), another study did not find any effects (Gonzalez-Flores et al. 2007).

In summary, these few studies suggest that NA has in general a stimulatory effect on lordosis behavior, an effect that is probably regulated via α_1- and/or β-adrenoceptors, and definitely not via α_2-adrenoceptors. However, all these studies were performed in ovariectomized females primed with only E. Therefore, we can only conclude that α_1- and/or β-adrenoceptors are involved in sexual behavior of hormonally low-primed females. In addition, it is unknown how these adrenoceptors are involved in the introductory phase of female sexual behavior.

5.4.1 Noradrenaline in Specific Brain Areas

Interestingly, it seems that the adrenoceptors play more defined roles in specific brain areas. Several studies have been performed in which adrenergic agents were administered locally in the POA, VMN, arcuate-ventromedial area of the hypothalamus (ARC-VM), lateral hypothalamic area (LHA), and median eminence.

The role of NA and adrenoceptors in the POA on female sexual behavior is rather unclear: both, a stimulatory and an inhibitory function on sexual behavior, have been suggested. On one hand, it was suggested that NA had a stimulatory effects on lordosis, an effect that was regulated via β-adrenoceptors (Foreman and Moss 1978), while on the other hand, inhibitory effects on lordosis behavior were found when NA was injected in the POA (Caldwell and Clemens 1986). The inhibition was probably regulated via the α_2-adrenoceptors, instead of the α1- and β-adrenoceptors, since administration of α_1- and β-adrenoceptor antagonists did not affect lordosis behavior (Etgen 1990) or attenuated the inhibitory effects on NA in the POA (Caldwell and Clemens 1986). Only α_2-adrenoceptor antagonists attenuated the effect of NA in the POA (Caldwell and Clemens 1986).

These differences in results are pretty peculiar but might be caused (again) by the hormonal state of the females. The stimulatory effects were found in ovariectomized females treated with only E, while the inhibitory effects were seen in females primed with both E and P. Because natural cycling females have both E and P, it is most reasonable to conclude that NA in the POA plays an inhibitory role on lordosis behavior. This inhibitory effect is most likely regulated via α2-adrenoceptors and not α_1- or β-adrenoceptors in the POA. It should be mentioned, though, that α_2-adrenoceptor antagonists by itself do not affect lordosis when locally injected into the POA (Gonzalez et al. 1996; Etgen 1990). This indicates that under normal basal circumstances, α_2-adrenoceptors in the POA do not play a crucial role in sexual behavior, but with elevated levels of NA, α_2-adrenoceptors become more important.

The stimulating effect of NA could then be regulated, for instance, via the ventromedial nucleus of the hypothalamus. Local NA injections in this area stimulate lordosis behavior in E-primed females (Fernandez-Guasti et al. 1985a). The role of α_2-adrenoceptors in this region is rather unclear, but α_1- and β-adrenoceptor seem to be involved in the stimulatory effects in the VMN. Systemic co-administration of both an α_1-adrenoceptor antagonist and nonselective β-adrenoceptor antagonist prevent the effects of locally injected NA in the VMN (Fernandez-Guasti et al. 1985a). α_1-Adrenoceptor antagonists by itself decrease lordosis quotients in most studies (Etgen 1990; Fernandez-Guasti et al. 1985b; Kow et al. 1992).

Other brain areas that are involved in the noradrenergic system regulating female sexual behavior are the ARC-VM and the median eminence. However, it remains unclear via which receptors NA regulates sexual behavior in the ARC-VM. β-Adrenoceptors might be involved in the stimulatory effects, while α_1-adrenoceptors might inhibit lordosis in this brain area (Foreman and Moss 1978). In the median eminence, β-adrenoceptors, and not the α_1-adrenoceptors, are involved in the stimulatory effects of NA (Scimonelli et al. 2000). The adrenoceptors in the LHA, on the other hand, are clearly not involved in the regulation of female sexual behavior (Foreman and Moss 1978).

As mentioned before, hormones play an important role in the role of NA on sexual behavior. To date, it appears that inhibitory effects can only be found in rats primed with both E and P, while stimulatory effects are mainly found in females primed with only E (as reviewed in Snoeren 2015). As discussed previously under Sect. 5.1, it is obvious that the hormonal status of the females is important for their sexual functioning. Interestingly, there seems to be a close relation between hormones and the adrenergic system. For example, P has a direct stimulating effect on NA levels (Nagle and Rosner 1980; Janowsky and Davis 1970), but also ER agonists modify NA levels in the rat brain (Lubbers et al. 2010). Interestingly, E modifies activity of both β- and α_1-adrenoceptors in the hypothalamus and POA, attenuating β-adrenoceptors while augmenting α_1-adrenoceptor responses (Etgen et al. 1992; Petitti et al. 1992; Ungar et al. 1993). It is tempting to speculate that attenuation of NA action at hypothalamic β-adrenoceptor along with the potentiation of NA action at the α_1-adrenoceptors is functionally related to E priming of lordosis behavior. More research is needed to discover the exact relationship between ovarian hormones and NA.

5.5 *Oxytocin*

Oxytocin (OT) is a neuropeptide with a remarkable variety of physiological functions, especially in pair bonding, reproductive behavior, and conditions during and after childbirth. OT is produced in magno- and parvocellular neurons of the paraventricular hypothalamic nucleus (PVN), in the supraoptic hypothalamic nucleus, as well as in the BNST and POA (reviewed in Veening et al. 2015). OT plays most likely a stimulatory role in sexual desire, part of the introductory phase, and the expectancy of future reward (Bancroft 2005; Pfaus 2009). Though the presence of OT is not essential for the coordination and performance of sexual behavior in female rodents, normal patterns of sexual behavior were seen in OT knockout mice (Nishimori et al. 1996).

However, when OT is injected in the cerebral ventricles, it induces an increase in paracopulatory behaviors and lordosis in females (Pedersen and Boccia 2006). Intracerebroventricular (icv) infusions of OT receptor antagonists, on the other hand, have an inhibitory effect on female sexual behavior (Pedersen and Boccia 2002). Mice and rats could be a little bit different in these aspects. The studies in mice are not conclusive. Some claim that OT may be unnecessary for the induction of lordosis in mice (Lee et al. 2010); others showed impaired lordosis behavior in oxytocin gene knockout mice (Zimmermann-Peruzatto et al. 2017). In rats, though, OT clearly facilitates lordosis (Arletti and Bertolini 1985; Caldwell et al. 1986; Schumacher et al. 1989). Interestingly, these stimulatory effects seem to depend on the light/dark schedule, whereas rats respond stronger on OT during the dark period (Schumacher et al. 1991).

In addition, the timing of OT receptor antagonist infusion is relevant. Both lordosis and paracopulatory behavior are suppressed by OT receptor antagonist icv infusions before P administration (Pedersen and Boccia 2002). However, when infused after P administration, female sexual behavior was not affected at first (4–6 h after P) but was reduced 8–12 h after P (Pedersen and Boccia 2002). This suggests that OT receptors are involved in mediating the onset of female sexual behavior during the first hour following P treatment. In addition, OT contributes in the maintenance of sexual behavior many more hours.

The VMN and POA have been shown to be involved in the oxytocinergic control of female sexual behavior. Infusion of OT into the VMN and POA results in increased lordosis in females treated with E and P (Schulze and Gorzalka 1991; Schumacher et al. 1989). In the VMN, an interesting interplay takes place between OT, E, and P. An increase in E levels within the VMN causes a fourfold rise in OT binding in this region. Additionally, within 4 h of co-treatment of P, OT receptors are distributed over a zone surrounding the ventrolateral part of the VMN (Schumacher et al. 1989), on the location of the neuroactive substance. This suggests a mechanism by which two ovarian hormones act synergically in a specific brain region to promote the right conditions to induce a behavioral response.

5.6 GABA

GABA is known as an inhibitory neurotransmitter in the brain. Increased GABA activity in the brain results in less lordosis responses (Mcginnis et al. 1980), suggesting an inhibitory role for GABA in female sexual behavior. Interestingly, results from infusions of GABAergic agents near the VMN or mid-central gray are associated with facilitating effects on lordosis, which probably is regulated via $GABA_A$ receptors (Donoso and Zarate 1981; Mccarthy et al. 1991; Kow and Pfaff 1988). When $GABA_A$ receptor agonists were infused in the VMN, they also facilitated sexual behavior (Mccarthy et al. 1990), whereas local infusion of antagonists had inhibitory effects (Luine et al. 1999). The hypothesis is that GABA might facilitate behavior via inhibiting the action of another inhibitory transmitter, a disinhibitory mechanism. Serotonin might be a good candidate for this regulation (Luine et al. 1997; Ogawa et al. 1991).

To the contrary, infusion of GABAergic antagonists in the POA seems to enhance lordosis (Mccarthy et al. 1990), suggesting an inhibitory role of GABA in the POA. Thus, GABA seems to have an opposite effect in the POA and the VMN. GABA levels in the VMN are significantly higher during proestrus than diestrus, while in the POA the levels are lowest at proestrus (Frankfurt et al. 1984). This suggests that the GABAergic facilitatory effect of lordosis is regulated mostly in the VMN via disinhibitory mechanisms, whereas GABA in the POA might be involved in the termination of the behavior.

6 General Discussion

In summary, it is clear the female reproductive behavior is regulated via a close interplay between different hormones, neurotransmitters, and neuropeptides, of which estrogen, progesterone, serotonin, dopamine, and oxytocin have received most attention in research. In addition, several brain areas are important in this regulation. The VMN and POA are mostly studied, but more brain areas must be involved in female sexual behavior. It is therefore very crucial that more research will be done of the neural regulation of female sexual behavior in order to unravel the complete mechanism by which paracopulatory behaviors and lordosis are regulated. In addition, it would be helpful if future research would also focus on the communication and interplay between brain areas in regulating the different phases of sexual behavior.

Females are often considered difficult to study, because ovarian hormones play an important role and make the results rather inconclusive. In this chapter we have often seen different effects of agents on sexual behavior tested in females with different levels of receptivity. It is not clear how an agent in the same dose can facilitate female sexual behavior in E-primed rats and inhibit this behavior in E- and P-primed females. A logic explanation would be an interaction effect of the agent with P, but the question remains whether this effect is only pharmacological and if they have any physiological relevance. In any case, it should be clear that the baseline level of female sexual behavior is a crucial factor that needs to be considered when evaluating the effects of any agent upon this behavior. It is, therefore, very disappointing that many studies in this field have only explored one or the other condition. When the effects of agents would have been investigated in low- and high-primed females within the same study, the hypothesis could have evolved into a real proven theory. When more research will be done on the interactions between ovarian hormones and neurotransmitters, we might be able to unravel this complexity and come to clear conclusions of how female sexual behavior is regulated. At least we are able to control for the hormonal effects in females by using ovariectomy and manually change the hormonal level. Females could, therefore, also been seen as an excellent study object, in comparison to males where testosterone levels might also influence results but are neglected.

The context in which females are tested seems to be very relevant as well. In this chapter, we have described several studies in which they have shown that different results were found in pace versus non-paced mating conditions or in a standard testing situation versus a seminatural environment. These differences in effects are clearly seen in the POA. As mentioned before, Whitney (1986) showed that lesioned females allowed fewer copulatory contacts, exhibited less paracopulatory behaviors, and spent less time with the males than controls in a paced mating setup, while it increased lordosis under non-paced mating conditions (Whitney 1986). Similar results were found after deletion of ERα in the POA: in a regular test setup, females show normal sexual behavior, whereas they show declined sexual activity in a seminatural environment in which they can pace their interactions (Snoeren et al. 2015). This leads to the conclusion that when given the opportunity, complete or partial POA lesioned females will not interact with a male. Using a different and more unnatural test setup can lead to different conclusions and should, therefore, be avoided as much as possible. For sure, more attention should be given to the type of test setups that are used to produce certain results.

However, the scientists' worst enemy is probably the complexity of the neural regulation of female sexual behavior. The behavior is clearly regulated via a close interplay between different neurotransmitter systems and brain regions, and the current research methods were not sufficient to unravel this complex system. Until recently, brain areas were seen as structures that could be divided into a few subregions that probably could play different role in regulating behaviors. However, nowadays, new research can prove that even within such subregions, different subtypes of neurons can be differently involved in the same mechanisms, or that neurons co-release different neurotransmitters. Besides, there is a tight line between different behaviors: e.g., Lee et al. (2014) showed how neurons in the VMN of male mice can easily switch from stimulating mounting and attacking another mouse (Lee et al. 2014). Interestingly, they have used the state-of-the-art technique optogenetics for this study.

So far, only lesion and pharmacological approaches have been used to study female sexual behavior. As discussed before, these methods are limited in their interpretations. Lesion studies will not only affect a certain brain area but will also affect the direct and indirect pathways. Pharmacology, on the other hand, is a "dirty" method in that drugs can act on multiple receptors in multiple cell types and brain areas at the same time. They are, therefore, not sufficient for the understanding of the complexity of the neural regulation. Therefore, it is highly recommended that researchers in the field of female reproductive behavior would modernize their research and start to use the novel technique like fiber photometry and optogenetics to study the neural circuitries involved in regulating female sexual behavior. This would open up for opportunities to study the precise interplay of brain areas and neurotransmitters in regulating sexual behaviors. In addition, the temporal resolution of these techniques would also allow us to study the neural regulation of the different phases in more detail and to study what is needed to the switch from one phase to the other.

References

Adachi S, Yamada S, Takatsu Y, Matsui H, Kinoshita M, Takase K, Sugiura H, Ohtaki T, Matsumoto H, Uenoyama Y, Tsukamura H, Inoue K, Maeda K (2007) Involvement of anteroventral periventricular metastin/kisspeptin neurons in estrogen positive feedback action on luteinizing hormone release in female rats. J Reprod Dev 53:367–378

Aghajanian GK (1978) Feedback regulation of central monoaminergic neurons: evidence from single cell recording studies. Essays Neurochem Neuropharmacol 3:1–32

Ågmo A (1997) Male rat sexual behavior. Brain Res Brain Res Protoc 1:203–209

Ågmo A (2002) Copulation-contingent aversive conditioning and sexual incentive motivation in male rats: evidence for a two-stage process of sexual behavior. Physiol Behav 77:425–435
Ågmo A, Paredes R, Fernandez H (1987) Differential effects of GABA transaminase inhibitors on sexual behavior, locomotor activity, and motor execution in the male rat. Pharmacol Biochem Behav 28:47–52
Ahlenius S, Engel J, Eriksson H, Sodersten P (1972) Effects of tetrabenazine on lordosis behaviour and on brain monoamines in the female rat. J Neural Transm 33:155–162
Ahlenius S, Fernandez-Guasti A, Hjorth S, Larsson K (1986) Suppression of lordosis behavior by the putative 5-HT receptor agonist 8-OH-DPAT in the rat. Eur J Pharmacol 124:361–363
Ahlenius S, Larsson K, Fernandez-Guasti A (1989) Evidence for the involvement of central 5-HT1A receptors in the mediation of lordosis behavior in the female rat. Psychopharmacology 98:440–444
Albert DJ, Jonik RH, Walsh ML (1992) Hormone-dependent aggression in male and female rats: experiential, hormonal, and neural foundations. Neurosci Biobehav Rev 16:177–192
Apostolakis EM, Garai J, Fox C, Smith CL, Watson SJ, Clark JH, O'malley BW (1996) Dopaminergic regulation of progesterone receptors: brain D5 dopamine receptors mediate induction of lordosis by D1-like agonists in rats. J Neurosci 16:4823–4834
Arendash GW, Gorski RA (1983) Suppression of lordotic responsiveness in the female rat during mesencephalic electrical stimulation. Pharmacol Biochem Behav 19:351–357
Arletti R, Bertolini A (1985) Oxytocin stimulates lordosis behavior in female rats. Neuropeptides 6:247–253
Bancroft J (2005) The endocrinology of sexual arousal. J Endocrinol 186:411–427
Barfield RJ, Chen JJ (1977) Activation of estrous behavior in ovariectomized rats by intracerebral implants of estradiol benzoate. Endocrinology 101:1716–1725
Barnett SA (1975) The rat: a study in behavior. University of Chicago Press, Chicago
Baum MJ, Cherry JA (2015) Processing by the main olfactory system of chemosignals that facilitate mammalian reproduction. Horm Behav 68:53–64
Beach FA (1976) Sexual attractivity, proceptivity, and receptivity in female mammals. Horm Behav 7:105–138
Bergan JF, Ben-Shaul Y, Dulac C (2014) Sex-specific processing of social cues in the medial amygdala. elife 3:e02743
Bergheim D, Chu X, Ågmo A (2015) The function and meaning of female rat paracopulatory (proceptive) behaviors. Behav Processes 118:34–41
Bergvall AH, Vega Matuszczyk J, Dahlof LG, Hansen S (1991) Peripheral anosmia attenuates female-enhanced aggression in male rats. Physiol Behav 50:33–40
Caldwell JD, Clemens LG (1986) Norepinephrine infusions into the medial preoptic area inhibit lordosis behavior. Pharmacol Biochem Behav 24:1015–1023
Caldwell JD, Prange AJ Jr, Pedersen CA (1986) Oxytocin facilitates the sexual receptivity of estrogen-treated female rats. Neuropeptides 7:175–189
Canteras NS, Simerly RB, Swanson LW (1995) Organization of projections from the medial nucleus of the amygdala: a PHAL study in the rat. J Comp Neurol 360:213–245
Cheong RY, Czieselsky K, Porteous R, Herbison AE (2015) Expression of ESR1 in glutamatergic and GABAergic neurons is essential for normal puberty onset, estrogen feedback, and fertility in female mice. J Neurosci 35:14533–14543
Clarke H, Dhillo WS, Jayasena CN (2015) Comprehensive review on kisspeptin and its role in reproductive disorders. Endocrinol Metab 30:124–141
Clarkson J, D'Anglemont DE, Tassigny X, Moreno AS, Colledge WH, Herbison AE (2008) Kisspeptin-GPR54 signaling is essential for preovulatory gonadotropin-releasing hormone neuron activation and the luteinizing hormone surge. J Neurosci 28:8691–8697
Clayton AH (2002) Female sexual dysfunction related to depression and antidepressant medications. Curr Womens Health Rep 2:182–187
Coolen LM, Peters HJ, Veening JG (1996) Fos immunoreactivity in the rat brain following consummatory elements of sexual behavior: a sex comparison. Brain Res 738:67–82

Dafopoulos K, Kotsovassilis CG, Milingos S, Kallitsaris A, Galazios G, Zintzaras E, Sotiros P, Messinis IE (2004) Changes in pituitary sensitivity to GnRH in estrogen-treated post-menopausal women: evidence that gonadotrophin surge attenuating factor plays a physiological role. Hum Reprod 19:1985–1992

Dafopoulos K, Mademtzis I, Vanakara P, Kallitsaris A, Stamatiou G, Kotsovassilis C, Messinis IE (2006) Evidence that termination of the estradiol-induced luteinizing hormone surge in women is regulated by ovarian factors. J Clin Endocrinol Metab 91:641–645

Darwin C (1859) On the origin of species by means of natural selection. J. Murray, London

Davidson JM, Smith ER, Rodgers CH, Bloch GJ (1968) Relative thresholds of behavioral and somatic responses to estrogen. Physiol Behav 3:227

Davis GA, Kohl R (1977) The influence of alpha-receptors on lordosis in the female rat. Pharmacol Biochem Behav 6:47–53

De Koning J, Lambalk CB, Helmerhorst FM, Helder MN (2001) Is GnRH self-priming an obligatory feature of the reproductive cycle? Hum Reprod 16:209–214

de Koning CH, Mcdonnell J, Themmen AP, De Jong FH, Homburg R, Lambalk CB (2008) The endocrine and follicular growth dynamics throughout the menstrual cycle in women with consistently or variably elevated early follicular phase FSH compared with controls. Hum Reprod 23:1416–1423

Dibenedictis BT, Ingraham KL, Baum MJ, Cherry JA (2012) Disruption of urinary odor preference and lordosis behavior in female mice given lesions of the medial amygdala. Physiol Behav 105:554–559

Donoso AO, Zarate MB (1981) Enhancement of lordotic behavior by injection GABA into the ventromedial nucleus of hypothalamus. IRCS Med Sci Biochem 9:1039

Edwards DA, Whalen RE, Nadler RD (1968) Induction of estrus – estrogen-progesterone interactions. Physiol Behav 3:29–33

Ellingsen E, Ågmo A (2004) Sexual-incentive motivation and paced sexual behavior in female rats after treatment with drugs modifying dopaminergic neurotransmission. Pharmacol Biochem Behav 77:431–445

Emery DE, Moss RL (1984) Lesions confined to the ventromedial hypothalamus decrease the frequency of coital contacts in female rats. Horm Behav 18:313–329

Erskine MS (1985) Effects of paced coital stimulation on estrus duration in intact cycling rats and ovariectomized and ovariectomized-adrenalectomized hormone-primed rats. Behav Neurosci 99:151–161

Erskine MS (1993) Mating-induced increases in FOS protein in preoptic area and medial amygdala of cycling female rats. Brain Res Bull 32:447–451

Etgen AM (1990) Intrahypothalamic implants of noradrenergic antagonists disrupt lordosis behavior in female rats. Physiol Behav 48:31–36

Etgen AM, Ungar S, Petitti N (1992) Estradiol and progesterone modulation of norepinephrine neurotransmission: implications for the regulation of female reproductive behavior. J Neuroendocrinol 4:255–271

Everitt BJ, Fuxe K (1977) Dopamine and sexual behaviour in female rats. Effects of dopamine receptor agonists and sulpiride. Neurosci Lett 4:209–213

Everitt BJ, Fuxe K, Hokfelt FT, Jonsson G (1975) Role of monoamines in the control by hormones of sexual receptivity in the female rat. J Comp Physiol Psychol 89:556–572

Fadem BH, Barfield RJ, Whalen RE (1979) Dose-response and time-response relationships between progesterone and the display of patterns of receptive and proceptive behavior in the female rat. Horm Behav 13:40–48

Fernandez-Guasti A, Larsson K, Beyer C (1985a) Potentiative action of alpha- and beta-adrenergic receptor stimulation in inducing lordosis behavior. Pharmacol Biochem Behav 22:613–617

Fernandez-Guasti A, Larsson K, Beyer C (1985b) Prevention of progesterone-induced lordosis behavior by alpha or beta adrenergic antagonists in ovariectomized estrogen-primed rats. Pharmacol Biochem Behav 22:279–282

Fernandez-Guasti A, Ahlenius S, Hjorth S, Larsson K (1987) Separation of dopaminergic and serotonergic inhibitory mechanisms in the mediation of estrogen-induced lordosis behaviour in the rat. Pharmacol Biochem Behav 27:93–98

Foreman MM, Moss RL (1978) Role of hypothalamic alpha and beta adrenergic receptors in the control of lordotic behavior in the ovariectomized-estrogen primed rat. Pharmacol Biochem Behav 9:235–241

Foreman MM, Moss RL (1979) Role of hypothalamic dopaminergic receptors in the control of lordosis behavior in the female rat. Physiol Behav 22:283–289

Fowler PA, Sorsa T, Harris WJ, Knight PG, Mason HD (2001) Relationship between follicle size and gonadotrophin surge attenuating factor (GnSAF) bioactivity during spontaneous cycles in women. Hum Reprod 16:1353–1358

Fowler PA, Sorsa-Leslie T, Harris W, Mason HD (2003) Ovarian gonadotrophin surge-attenuating factor (GnSAF): where are we after 20 years of research? Reproduction 126:689–699

Franceschini I, Lomet D, Cateau M, Delsol G, Tillet Y, Caraty A (2006) Kisspeptin immunoreactive cells of the ovine preoptic area and arcuate nucleus co-express estrogen receptor alpha. Neurosci Lett 401:225–230

Frankfurt M, Mcewen BS (1991) Estrogen increases axodendritic synapses in the VMN of rats after ovariectomy. Neuroreport 2:380–382

Frankfurt M, Fuchs E, Wuttke W (1984) Sex differences in gamma-aminobutyric acid and glutamate concentrations in discrete rat brain nuclei. Neurosci Lett 50:245–250

Frankhuyzen AL, Mulder AH (1982) Pharmacological characterization of presynaptic alpha-adrenoceptors modulating [3H]noradrenaline and [3H]5-hydroxytryptamine release from slices of the hippocampus of the rat. Eur J Pharmacol 81:97–106

Gereau RWT, Kedzie KA, Renner KJ (1993) Effect of progesterone on serotonin turnover in rats primed with estrogen implants into the ventromedial hypothalamus. Brain Res Bull 32:293–300

Girven KS, Sparta DR (2017) Probing deep brain circuitry: new advances in in vivo calcium measurement strategies. ACS Chem Neurosci 8:243–251

Gonzalez MI, Patmore L, Wilson CA (1996) Effect of delequamine (RS15385) on female sexual behaviour in the rat. Eur J Pharmacol 312:1–6

Gonzalez MI, Greengrass P, Russell M, Wilson CA (1997) Comparison of serotonin receptor numbers and activity in specific hypothalamic areas of sexually active and inactive female rats. Neuroendocrinology 66:384–392

Gonzalez-Flores O, Beyer C, Lima-Hernandez FJ, Gomora-Arrati P, Gomez-Camarillo MA, Hoffman K, Etgen AM (2007) Facilitation of estrous behavior by vaginal cervical stimulation in female rats involves alpha1-adrenergic receptor activation of the nitric oxide pathway. Behav Brain Res 176:237–243

Goodman RL, Coolen LM, Lehman MN (2014) A role for neurokinin B in pulsatile GnRH secretion in the ewe. Neuroendocrinology 99:18–32

Gothert M, Weinheimer G (1979) Extracellular 5-hydroxytryptamine inhibits 5-hydroxytryptamine release from rat brain cortex slices. Naunyn Schmiedeberg's Arch Pharmacol 310:93–96

Graham MD, Pfaus JG (2010) Differential regulation of female sexual behaviour by dopamine agonists in the medial preoptic area. Pharmacol Biochem Behav 97:284–292

Graham MD, Pfaus JG (2012) Differential effects of dopamine antagonists infused to the medial preoptic area on the sexual behavior of female rats primed with estrogen and progesterone. Pharmacol Biochem Behav 102:532–539

Graham MD, Gardner Gregory J, Hussain D, Brake WG, Pfaus JG (2015) Ovarian steroids alter dopamine receptor populations in the medial preoptic area of female rats: implications for sexual motivation, desire, and behaviour. Eur J Neurosci 42:3138–3148

Gray GD, Soderstein P, Tallentire D, Davidson JM (1978) Effects of lesions in various structures of the suprachiasmatic-preoptic region on LH regulation and sexual behavior in female rats. Neuroendocrinology 25:174–191

Grierson JP, James MD, Pearson JR, Wilson CA (1988) The effect of selective D1 and D2 dopaminergic agents on sexual receptivity in the female rat. Neuropharmacology 27:181–189

Gu G, Cornea A, Simerly RB (2003) Sexual differentiation of projections from the principal nucleus of the bed nuclei of the stria terminalis. J Comp Neurol 460:542–562
Guarraci FA, Clark AS (2006) Ibotenic acid lesions of the medial preoptic area disrupt the expression of partner preference in sexually receptive female rats. Brain Res 1076:163–170
Guarraci FA, Megroz AB, Clark AS (2004) Paced mating behavior in the female rat following lesions of three regions responsive to vaginocervical stimulation. Brain Res 999:40–52
Gundlah C, Simon LD, Auerbach SB (1998) Differences in hypothalamic serotonin between estrous phases and gender: an in vivo microdialysis study. Brain Res 785:91–96
Gunnet JW, Lookingland KJ, Moore KE (1986) Comparison of the effects of castration and steroid replacement on incertohypothalamic dopaminergic neurons in male and female rats. Neuroendocrinology 44:269–275
Guptarak J, Sarkar J, Hiegel C, Uphouse L (2010) Role of 5-HT(1A) receptors in fluoxetine-induced lordosis inhibition. Horm Behav 58:290–296
Hamburger-Bar R, Rigter H (1975) Apomorphine: facilitation of sexual behaviour in female rats. Eur J Pharmacol 32:357–360
Hawkins SM, Matzuk MM (2008) The menstrual cycle: basic biology. Ann N Y Acad Sci 1135:10–18
Heijkoop R, Huijgens PT, Snoeren EMS (2018) Assessment of sexual behavior in rats: the potentials and pitfalls. Behav Brain Res 352:70–80
Hellier V, Brock O, Candlish M, Desroziers E, Aoki M, Mayer C, Piet R, Herbison A, Colledge WH, Prevot V, Boehm U, Bakker J (2018) Female sexual behavior in mice is controlled by kisspeptin neurons. Nat Commun 9:400
Herbison AE (1998) Multimodal influence of estrogen upon gonadotropin-releasing hormone neurons. Endocr Rev 19:302–330
Herbison AE, Dyer RG (1991) Effect on luteinizing hormone secretion of GABA receptor modulation in the medial preoptic area at the time of proestrous luteinizing hormone surge. Neuroendocrinology 53:317–320
Herbison AE, de Tassigny X, Doran J, Colledge WH (2010) Distribution and postnatal development of Gpr54 gene expression in mouse brain and gonadotropin-releasing hormone neurons. Endocrinology 151:312–321
Hoshina Y, Takeo T, Nakano K, Sato T, Sakuma Y (1994) Axon-sparing lesion of the preoptic area enhances receptivity and diminishes proceptivity among components of female rat sexual behavior. Behav Brain Res 61:197–204
Hoyer D, Clarke DE, Fozard JR, Hartig PR, Martin GR, Mylecharane EJ, Saxena PR, Humphrey PP (1994) International Union of Pharmacology classification of receptors for 5-hydroxytryptamine (Serotonin). Pharmacol Rev 46:157–203
Hruska RE (1986) Elevation of striatal dopamine receptors by estrogen: dose and time studies. J Neurochem 47:1908–1915
Hruska RE, Nowak MW (1988) Estrogen treatment increases the density of D1 dopamine receptors in the rat striatum. Brain Res 442:349–350
Huang X, Harlan RE (1993) Absence of androgen receptors in LHRH immunoreactive neurons. Brain Res 624:309–311
Jaber M, Robinson SW, Missale C, Caron MG (1996) Dopamine receptors and brain function. Neuropharmacology 35:1503–1519
Jacobs BL, Fornal CA (1997) Serotonin and motor activity. Curr Opin Neurobiol 7:820–825
Janowsky DS, Davis JM (1970) Progesterone-estrogen effects on uptake and release of norepinephrine by synaptosomes. Life Sci 9:525–531
Jarry H, Hirsch B, Leonhardt S, Wuttke W (1992) Amino acid neurotransmitter release in the preoptic area of rats during the positive feedback actions of estradiol on LH release. Neuroendocrinology 56:133–140
Jarry H, Leonhardt S, Schwarze T, Wuttke W (1995) Preoptic rather than mediobasal hypothalamic amino acid neurotransmitter release regulates GnRH secretion during the estrogen-induced LH surge in the ovariectomized rat. Neuroendocrinology 62:479–486

Jenkins WJ, Becker JB (2001) Role of the striatum and nucleus accumbens in paced copulatory behavior in the female rat. Behav Brain Res 121:119–128
Johansson CE, Meyerson BJ, Hacksell U (1991) The novel 5-HT1A receptor antagonist (S)-UH-301 antagonizes 8-OH-DPAT-induced effects on male as well as female rat copulatory behaviour. Eur J Pharmacol 202:81–87
Jones SL, Farrell S, Gregory JG, Pfaus JG (2013) Sensitization of sexual behavior in ovariectomized rats by chronic estradiol treatment. Horm Behav 64:8–18
Kakeyama M, Yamanouchi K (1996) Inhibitory effect of baclofen on lordosis in female and male rats with dorsal raphe nucleus lesion or septal cut. Neuroendocrinology 63:290–296
Kakeyama M, Negishi M, Yamanouchi K (1997) Facilitatory effect of ventral cut of dorsal raphe nucleus on lordosis in female rats. Endocr J 44:589–593
Kaspersen H, Ågmo A (2012) Paroxetine-induced reduction of sexual incentive motivation in female rats is not modified by 5-HT1B or 5-HT2C antagonists. Psychopharmacology 220:269–280
Kato A, Sakuma Y (2000) Neuronal activity in female rat preoptic area associated with sexually motivated behavior. Brain Res 862:90–102
Kazem R, Messinis LE, Fowler P, Groome NP, Knight PG, Templeton AA (1996) Effect of mifepristone (RU486) on the pituitary response to gonadotrophin releasing hormone in women. Hum Reprod 11:2585–2590
Kinoshita M, Tsukamura H, Adachi S, Matsui H, Uenoyama Y, Iwata K, Yamada S, Inoue K, Ohtaki T, Matsumoto H, Maeda K (2005) Involvement of central metastin in the regulation of preovulatory luteinizing hormone surge and estrous cyclicity in female rats. Endocrinology 146:4431–4436
Kishitake M, Yamanouchi K (2003) Effects of highly or relatively selective 5-HT1A receptor agonists on lordosis in female rats. Zool Sci 20:1133–1138
Kondo Y, Sachs BD (2002) Disparate effects of small medial amygdala lesions on noncontact erection, copulation, and partner preference. Physiol Behav 76:443–447
Kondo Y, Sakuma Y (2005) The medial amygdala controls the coital access of female rats: a possible involvement of emotional responsiveness. Jpn J Physiol 55:345–353
Kow LM, Pfaff DW (1988) Transmitter and peptide actions on hypothalamic neurons in vitro: implications for lordosis. Brain Res Bull 20:857–861
Kow LM, Weesner GD, Pfaff DW (1992) Alpha 1-adrenergic agonists act on the ventromedial hypothalamus to cause neuronal excitation and lordosis facilitation: electrophysiological and behavioral evidence. Brain Res 588:237–245
Kow LM, Mobbs CV, Pfaff DW (1994) Roles of second-messenger systems and neuronal activity in the regulation of lordosis by neurotransmitters, neuropeptides, and estrogen: a review. Neurosci Biobehav Rev 18:251–268
Lasley BL, Wang CF, Yen SS (1975) The effects of estrogen and progesterone on the functional capacity of the gonadotrophs. J Clin Endocrinol Metab 41:820–826
Latif R, Rafique N (2015) Serum kisspeptin levels across different phases of the menstrual cycle and their correlation with serum oestradiol. Neth J Med 73:175–178
Laws SC, Webster JC, Miller WL (1990) Estradiol alters the effectiveness of gonadotropin-releasing hormone (GnRH) in ovine pituitary cultures: GnRH receptors versus responsiveness to GnRH. Endocrinology 127:381–386
Lee HJ, Pagani J, Young WS (2010) Using transgenic mouse models to study oxytocin's role in the facilitation of species propagation. Brain Res 1364:216–224
Lee H, Kim DW, Remedios R, Anthony TE, Chang A, Madisen L, Zeng H, Anderson DJ (2014) Scalable control of mounting and attack by Esr1+ neurons in the ventromedial hypothalamus. Nature 509:627–632
Levesque D, Di Paolo T (1988) Rapid conversion of high into low striatal D2-dopamine receptor agonist binding states after an acute physiological dose of 17 beta-estradiol. Neurosci Lett 88:113–118

Lubbers LS, Zafian PT, Gautreaux C, Gordon M, Alves SE, Correa L, Lorrain DS, Hickey GJ, Luine V (2010) Estrogen receptor (ER) subtype agonists alter monoamine levels in the female rat brain. J Steroid Biochem Mol Biol 122:310–317
Luine VN, Grattan DR, Selmanoff M (1997) Gonadal hormones alter hypothalamic GABA and glutamate levels. Brain Res 747:165–168
Luine VN, Wu V, Hoffman CS, Renner KJ (1999) GABAergic regulation of lordosis: influence of gonadal hormones on turnover of GABA and interaction of GABA with 5-HT. Neuroendocrinology 69:438–445
Lydon JP, Demayo FJ, Funk CR, Mani SK, Hughes AR, Montgomery CA Jr, Shyamala G, Conneely OM, O'malley BW (1995) Mice lacking progesterone receptor exhibit pleiotropic reproductive abnormalities. Genes Dev 9:2266–2278
Ma ST, Maier EY, Ahrens AM, Schallert T, Duvauchelle CL (2010) Repeated intravenous cocaine experience: development and escalation of pre-drug anticipatory 50-kHz ultrasonic vocalizations in rats. Behav Brain Res 212:109–114
Maclusky NJ, Mcewen BS (1978) Oestrogen modulates progestin receptor concentrations in some rat brain regions but not in others. Nature 274:276–278
Mani SK, Allen JM, Clark JH, Blaustein JD, O'malley BW (1994) Convergent pathways for steroid hormone- and neurotransmitter-induced rat sexual behavior. Science 265:1246–1249
March CM, Goebelsmann U, Nakamura RM, Mishell DR Jr (1979) Roles of estradiol and progesterone in eiliciting the midcycle luteinizing hormone and follicle-stimulating hormone surges. J Clin Endocrinol Metab 49:507–513
Martinez LA, Petrulis A (2011) The bed nucleus of the stria terminalis is critical for sexual solicitation, but not for opposite-sex odor preference, in female Syrian hamsters. Horm Behav 60:651–659
Masco DH, Carrer HF (1980) Sexual receptivity in female rats after lesion or stimulation in different amygdaloid nuclei. Physiol Behav 24:1073–1080
Maswood N, Caldarola-Pastuszka M, Uphouse L (1997) 5-HT3 receptors in the ventromedial nucleus of the hypothalamus and female sexual behavior. Brain Res 769:13–20
Mathews D, Edwards DA (1977) The ventromedial nucleus of the hypothalamus and the hormonal arousal of sexual behaviors in the female rat. Horm Behav 8:40–51
Matuszewich L, Lorrain DS, Hull EM (2000) Dopamine release in the medial preoptic area of female rats in response to hormonal manipulation and sexual activity. Behav Neurosci 114:772–782
Mazzucco CA, Walker HA, Pawluski JL, Lieblich SE, Galea LA (2008) ERalpha, but not ERbeta, mediates the expression of sexual behavior in the female rat. Behav Brain Res 191:111–117
Mccarthy MM, Malik KF, Feder HH (1990) Increased GABAergic transmission in medial hypothalamus facilitates lordosis but has the opposite effect in preoptic area. Brain Res 507:40–44
Mccarthy MM, Pfaff DW, Schwartz-Giblin S (1991) Midbrain central gray GABA A receptor activation enhances, and blockade reduces, sexual behavior in the female rat. Exp Brain Res 86:108–116
Mccarthy EA, Kunkhyen T, Korzan WJ, Naik A, Maqsudlu A, Cherry JA, Baum MJ (2017a) A comparison of the effects of male pheromone priming and optogenetic inhibition of accessory olfactory bulb forebrain inputs on the sexual behavior of estrous female mice. Horm Behav 89:104–112
Mccarthy EA, Maqsudlu A, Bass M, Georghiou S, Cherry JA, Baum MJ (2017b) DREADD-induced silencing of the medial amygdala reduces the preference for male pheromones and the expression of lordosis in estrous female mice. Eur J Neurosci 46:2035–2046
Mcclintock MK (1984) Group mating in the domestic rat as a context for sexual selection: consequences for the analysis of sexual behavior and neuroendocrine responses. In: Advances in the study of behavior. Academic Press, New York
Mcclintock MK, Adler NT (1978) The role of the female during copulation in wild and domestic Norway rats (*Rattus Norvegicus*). Behaviour 67:67–96

Mcginnis MY, Gordon JH, Gorski RA (1980) Influence of gamma-aminobutyric acid on lordosis behavior and dopamine activity in estrogen primed spayed female rats. Brain Res 184:179–197

McGraw-Hill concise encyclopedia of bioscience (2005) McGraw-Hill, New York

Meisel RL, Dohanich GP, Mcewen BS, Pfaff DW (1987) Antagonism of sexual behavior in female rats by ventromedial hypothalamic implants of antiestrogen. Neuroendocrinology 45:201–207

Melis MR, Argiolas A (1995) Dopamine and sexual behavior. Neurosci Biobehav Rev 19:19–38

Mendelson SD, Gorzalka BB (1985) A facilitatory role for serotonin in the sexual behavior of the female rat. Pharmacol Biochem Behav 22:1025–1033

Mendelson SD, Gorzalka BB (1986) 5-HT1A receptors: differential involvement in female and male sexual behavior in the rat. Physiol Behav 37:345–351

Messinis IE, Templeton AA (1987) Endocrine and follicle characteristics of cycles with and without endogenous luteinizing hormone surges during superovulation induction with pulsatile follicle-stimulating hormone. Hum Reprod 2:11–16

Messinis IE, Templeton AA (1990) Effects of supraphysiological concentrations of progesterone on the characteristics of the oestradiol-induced gonadotrophin surge in women. J Reprod Fertil 88:513–519

Messinis IE, Lolis D, Zikopoulos K, Tsahalina E, Seferiadis K, Templeton AA (1994) Effect of an increase in FSH on the production of gonadotrophin-surge-attenuating factor in women. J Reprod Fertil 101:689–695

Messinis IE, Milingos S, Zikopoulos K, Hasiotis G, Seferiadis K, Lolis D (1998) Luteinizing hormone response to gonadotrophin-releasing hormone in normal women undergoing ovulation induction with urinary or recombinant follicle stimulating hormone. Hum Reprod 13:2415–2420

Messinis IE, Messini CI, Dafopoulos K (2014) Novel aspects of the endocrinology of the menstrual cycle. Reprod Biomed Online 28:714–722

Meyerson BJ (1964) Effect of neuropharmacological agents on hormone-activated estrus behaviour in ovariectomised rats. Arch Int Pharmacodyn Ther 150:4–33

Micevych PE, Dewing P (2011) Membrane-initiated estradiol signaling regulating sexual receptivity. Front Endocrinol 2:26

Micevych P, Sinchak K, Mills RH, Tao L, Lapolt P, Lu JK (2003) The luteinizing hormone surge is preceded by an estrogen-induced increase of hypothalamic progesterone in ovariectomized and adrenalectomized rats. Neuroendocrinology 78:29–35

Micevych PE, Wong AM, Mittelman-Smith MA (2015) Estradiol membrane-initiated signaling and female reproduction. Compr Physiol 5:1211–1222

Michanek A, Meyerson BJ (1977a) A comparative study of different amphetamines on copulatory behavior and stereotype activity in the female rat. Psychopharmacology 53:175–183

Michanek A, Meyerson BJ (1977b) The effects of different amphetamines on copulatory behaviour and stereotype activity in the female rat, after treatment with monoamine depletors and synthesis inhibitors. Arch Int Pharmacodyn Ther 229:301–312

Michanek A, Meyerson BJ (1982) Influence of estrogen and progesterone on behavioral effects of apomorphine and amphetamine. Pharmacol Biochem Behav 16:875–879

Mills RH, Sohn RK, Micevych PE (2004) Estrogen-induced mu-opioid receptor internalization in the medial preoptic nucleus is mediated via neuropeptide Y-Y1 receptor activation in the arcuate nucleus of female rats. J Neurosci 24:947–955

Miryala CS, Hiegel C, Uphouse L (2013) Comparison of female Fischer and Sprague-Dawley rats in the response to ketanserin. Pharmacol Biochem Behav 114–115:32–37

Moss RL, Paloutzian RF, Law OT (1974) Electrical stimulation of forebrain structures and its effect on copulatory as well as stimulus-bound behavior in ovariectomized hormone-primed rats. Physiol Behav 12:997–1004

Muller CP, Jacobs BL (2010) Activity of brain serotonergic neurons in relation to physiology and behavior. In: Handbook of behavioral neurobiology of serotonin. Academic Press, Amsterdam

Murphy DL, Lerner A, Rudnick G, Lesch KP (2004) Serotonin transporter: gene, genetic disorders, and pharmacogenetics. Mol Interv 4:109–123

Musatov S, Chen W, Pfaff DW, Kaplitt MG, Ogawa S (2006) RNAi-mediated silencing of estrogen receptor {alpha} in the ventromedial nucleus of hypothalamus abolishes female sexual behaviors. Proc Natl Acad Sci U S A 103:10456–10460
Nagle CA, Rosner JM (1980) Rat brain norepinephrine release during progesterone-induced LH secretion. Neuroendocrinology 30:33–37
Nasseri A, Minneman KP (1987) Relationship between alpha 2-adrenergic receptor binding sites and the functional receptors inhibiting norepinephrine release in rat cerebral cortex. Mol Pharmacol 32:655–662
Nedergaard P, Sanchez C, Mellerup E (2004) Different roles of 5-HT2A and 5-HT2C receptors in regulation of female rat paced mating behaviour. Behav Brain Res 149:151–157
Nishimori K, Young LJ, Guo Q, Wang Z, Insel TR, Matzuk MM (1996) Oxytocin is required for nursing but is not essential for parturition or reproductive behavior. Proc Natl Acad Sci U S A 93:11699–11704
Nomoto K, Lima SQ (2015) Enhanced male-evoked responses in the ventromedial hypothalamus of sexually receptive female mice. Curr Biol 25:589–594
Novaira HJ, Sonko ML, Hoffman G, Koo Y, Ko C, Wolfe A, Radovick S (2014) Disrupted kisspeptin signaling in GnRH neurons leads to hypogonadotrophic hypogonadism. Mol Endocrinol 28:225–238
O'connor SE, Brown RA (1982) The pharmacology of sulpiride--a dopamine receptor antagonist. Gen Pharmacol 13:185–193
Ogawa S, Kow LM, Pfaff DW (1991) Effects of GABA and related agents on the electrical activity of hypothalamic ventromedial nucleus neurons in vitro. Exp Brain Res 85:85–92
Ogawa S, Eng V, Taylor J, Lubahn DB, Korach KS, Pfaff DW (1998) Roles of estrogen receptor-alpha gene expression in reproduction-related behaviors in female mice. Endocrinology 139:5070–5081
Ogawa S, Chan J, Chester AE, Gustafsson JA, Korach KS, Pfaff DW (1999) Survival of reproductive behaviors in estrogen receptor beta gene-deficient (betaERKO) male and female mice. Proc Natl Acad Sci U S A 96:12887–12892
Palvimaki EP, Roth BL, Majasuo H, Laakso A, Kuoppamaki M, Syvalahti E, Hietala J (1996) Interactions of selective serotonin reuptake inhibitors with the serotonin 5-HT2c receptor. Psychopharmacology 126:234–240
Paredes RG, Ågmo A (2004) Has dopamine a physiological role in the control of sexual behavior? A critical review of the evidence. Prog Neurobiol 73:179–226
Pedersen CA, Boccia ML (2002) Oxytocin maintains as well as initiates female sexual behavior: effects of a highly selective oxytocin antagonist. Horm Behav 41:170–177
Pedersen CA, Boccia ML (2006) Vasopressin interactions with oxytocin in the control of female sexual behavior. Neuroscience 139:843–851
Petersen SL, Ottem EN, Carpenter CD (2003) Direct and indirect regulation of gonadotropin-releasing hormone neurons by estradiol. Biol Reprod 69:1771–1778
Petitti N, Karkanias GB, Etgen AM (1992) Estradiol selectively regulates alpha 1B-noradrenergic receptors in the hypothalamus and preoptic area. J Neurosci 12:3869–3876
Petralia SM, Frye CA (2006) In the ventral tegmental area, G-proteins mediate progesterone's actions at dopamine type 1 receptors for lordosis of rats and hamsters. Psychopharmacology 186:133–142
Pfaff DW, Sakuma Y (1979a) Deficit in the lordosis reflex of female rats caused by lesions in the ventromedial nucleus of the hypothalamus. J Physiol 288:203–210
Pfaff DW, Sakuma Y (1979b) Facilitation of the lordosis reflex of female rats from the ventromedial nucleus of the hypothalamus. J Physiol 288:189–202
Pfaff DW, Schwartz-Giblin S (1988) The physiology of reproduction. Raven, New York
Pfaus JG (2009) Pathways of sexual desire. J Sex Med 6:1506–1533
Pfaus JG, Kleopoulos SP, Mobbs CV, Gibbs RB, Pfaff DW (1993) Sexual stimulation activates c-fos within estrogen-concentrating regions of the female rat forebrain. Brain Res 624:253–267

Pfaus JG, Marcangione C, Smith WJ, Manitt C, Abillamaa H (1996) Differential induction of Fos in the female rat brain following different amounts of vaginocervical stimulation: modulation by steroid hormones. Brain Res 741:314–330
Ping L, Mahesh VB, Wiedmeier VT, Brann DW (1994) Release of glutamate and aspartate from the preoptic area during the progesterone-induced LH surge: in vivo microdialysis studies. Neuroendocrinology 59:318–324
Ping L, Mahesh VB, Bhat GK, Brann DW (1997) Regulation of gonadotropin-releasing hormone and luteinizing hormone secretion by AMPA receptors. Evidence for a physiological role of AMPA receptors in the steroid-induced luteinizing hormone surge. Neuroendocrinology 66:246–253
Polston EK, Erskine MS (1995) Patterns of induction of the immediate-early genes c-fos and egr-1 in the female rat brain following differential amounts of mating stimulation. Neuroendocrinology 62:370–384
Polston EK, Erskine MS (2001) Excitotoxic lesions of the medial amygdala differentially disrupt prolactin secretory responses in cycling and mated female rats. J Neuroendocrinol 13:13–21
Powers B, Valenstein ES (1972a) Sexual receptivity: facilitation by medial preoptic lesions in female rats. Science 175:1003–1005
Powers JB, Valenstein ES (1972b) Individual differences in sexual responsiveness to estrogen and progesterone in ovariectomized rats. Physiol Behav 8:673–676
Rajendren G, Moss RL (1993) The role of the medial nucleus of amygdala in the mating-induced enhancement of lordosis in female rats: the interaction with luteinizing hormone-releasing hormone neuronal system. Brain Res 617:81–86
Rajendren G, Dudley CA, Moss RL (1991) Role of the ventromedial nucleus of hypothalamus in the male-induced enhancement of lordosis in female rats. Physiol Behav 50:705–710
Rivas FJ, Mir D (1990) Effects of nucleus accumbens lesion on female rat sexual receptivity and proceptivity in a partner preference paradigm. Behav Brain Res 41:239–249
Robinson JE, Kendrick KM, Lambart CE (1991) Changes in the release of gamma-aminobutyric acid and catecholamines in the preoptic/septal area prior to and during the preovulatory surge of luteinizing hormone in the ewe. J Neuroendocrinol 3:393–399
Rossler AS, Bernabe J, Denys P, Alexandre L, Giuliano F (2006) Effect of the 5-HT receptor agonist DOI on female rat sexual behavior. J Sex Med 3:432–441
Roth BL (2016) DREADDs for neuroscientists. Neuron 89:683–694
Rowe DW, Erskine MS (1993) c-Fos proto-oncogene activity induced by mating in the preoptic area, hypothalamus and amygdala in the female rat: role of afferent input via the pelvic nerve. Brain Res 621:25–34
Sa SI, Fonseca BM (2017) Dynamics of progesterone and estrogen receptor alpha in the ventromedial hypothalamus. J Endocrinol 233:197–207
Samuelsen CL, Meredith M (2009) Categorization of biologically relevant chemical signals in the medial amygdala. Brain Res 1263:33–42
Schulze HG, Gorzalka BB (1991) Oxytocin effects on lordosis frequency and lordosis duration following infusion into the medial pre-optic area and ventromedial hypothalamus of female rats. Neuropeptides 18:99–106
Schumacher M, Coirini H, Frankfurt M, Mcewen BS (1989) Localized actions of progesterone in hypothalamus involve oxytocin. Proc Natl Acad Sci U S A 86:6798–6801
Schumacher M, Coirini H, Pfaff DW, Mcewen BS (1991) Light-dark differences in behavioral sensitivity to oxytocin. Behav Neurosci 105:487–492
Scimonelli T, Medina F, Wilson C, Celis ME (2000) Interaction of alpha-melanotropin (alpha-MSH) and noradrenaline in the median eminence in the control of female sexual behavior. Peptides 21:219–223
Sharp T (2010) Serotonergic feedback control. In: Muller CP, Jacobs BL (eds) Handbook of the behavioral neurobiology of serotonin. Elsevier, Amsterdam

Shimogawa Y, Sakuma Y, Yamanouchi K (2015) Efferent and afferent connections of the ventromedial hypothalamic nucleus determined by neural tracer analysis: implications for lordosis regulation in female rats. Neurosci Res 91:19–33

Simerly RB, Chang C, Muramatsu M, Swanson LW (1990) Distribution of androgen and estrogen receptor mRNA-containing cells in the rat brain: an in situ hybridization study. J Comp Neurol 294:76–95

Sinchak K, Micevych PE (2001) Progesterone blockade of estrogen activation of mu-opioid receptors regulates reproductive behavior. J Neurosci 21:5723–5729

Sinchak K, Dewing P, Ponce L, Gomez L, Christensen A, Berger M, Micevych P (2013) Modulation of the arcuate nucleus-medial preoptic nucleus lordosis regulating circuit: a role for GABA B receptors. Horm Behav 64:136–143

Sinclair-Worley L, Uphouse L (2004) Effect of estrogen on the lordosis-inhibiting action of ketanserin and SB 206553. Behav Brain Res 152:129–135

Smith JT, Popa SM, Clifton DK, Hoffman GE, Steiner RA (2006) Kiss1 neurons in the forebrain as central processors for generating the preovulatory luteinizing hormone surge. J Neurosci 26:6687–6694

Snoeren EM (2015) The role of adrenoceptors in the central nervous system in male and female rat sexual behavior. Eur J Pharmacol 753:229–245

Snoeren E, Chan J, Bovens A, Cuppen E, Waldinger M, Olivier B, Oosting R (2010) Serotonin transporter null mutation and sexual behavior in female rats: 5-HT1A receptor desensitization. J Sex Med 7:2424–2434

Snoeren EM, Bovens A, Refsgaard LK, Westphal KG, Waldinger MD, Olivier B, Oosting RS (2011a) Combination of testosterone and vardenafil increases female sexual functioning in sub-primed rats. J Sex Med 8:989–1001

Snoeren EM, Chan JS, De Jong TR, Waldinger MD, Olivier B, Oosting RS (2011b) A new female rat animal model for hypoactive sexual desire disorder; behavioral and pharmacological evidence. J Sex Med 8:44–56

Snoeren EM, Refsgaard LK, Waldinger MD, Olivier B, Oosting RS (2011c) Chronic paroxetine treatment does not affect sexual behavior in hormonally sub-primed female rats despite 5-HT(A) receptor desensitization. J Sex Med 8:976–988

Snoeren EM, Veening JG, Olivier B, Oosting RS (2014a) Serotonin 1A receptors and sexual behavior in female rats: a review. Pharmacol Biochem Behav 121:43–52

Snoeren EM, Veening JG, Olivier B, Oosting RS (2014b) Serotonin 1A receptors and sexual behavior in male rats: a review. Pharmacol Biochem Behav 121:102–114

Snoeren EM, Antonio-Cabrera E, Spiteri T, Musatov S, Ogawa S, Pfaff DW, Agmo A (2015) Role of oestrogen alpha receptors in sociosexual behaviour in female rats housed in a seminatural environment. J Neuroendocrinol 27:803–818

Spiteri T, Ågmo A (2009) Ovarian hormones modulate social recognition in female rats. Physiol Behav 98:247–250

Spiteri T, Musatov S, Ogawa S, Ribeiro A, Pfaff DW, Ågmo A (2010a) Estrogen-induced sexual incentive motivation, proceptivity and receptivity depend on a functional estrogen receptor alpha in the ventromedial nucleus of the hypothalamus but not in the amygdala. Neuroendocrinology 91:142–154

Spiteri T, Musatov S, Ogawa S, Ribeiro A, Pfaff DW, Ågmo A (2010b) The role of the estrogen receptor alpha in the medial amygdala and ventromedial nucleus of the hypothalamus in social recognition, anxiety and aggression. Behav Brain Res 210:211–220

Spiteri T, Ogawa S, Musatov S, Pfaff DW, Ågmo A (2012) The role of the estrogen receptor alpha in the medial preoptic area in sexual incentive motivation, proceptivity and receptivity, anxiety, and wheel running in female rats. Behav Brain Res 230:11–20

Stavy M, Herbert J (1989) Differential effects of beta-endorphin infused into the hypothalamic preoptic area at various phases of the male rat's sexual behaviour. Neuroscience 30:433–442

Stoof JC, Kebabian JW (1984) Two dopamine receptors: biochemistry, physiology and pharmacology. Life Sci 35:2281–2296

Tetel MJ, Getzinger MJ, Blaustein JD (1993) Fos expression in the rat brain following vaginal-cervical stimulation by mating and manual probing. J Neuroendocrinol 5:397–404
Tetel MJ, Getzinger MJ, Blaustein JD (1994) Estradiol and progesterone influence the response of ventromedial hypothalamic neurons to tactile stimuli associated with female reproduction. Brain Res 646:267–272
Thor DH, Flannelly KJ (1977) Peripheral anosmia and social investigatory behavior of the male rat. Behav Biol 20:128–134
Titus RD, Kornfeld EC, Jones ND, Clemens JA, Smalstig EB, Fuller RW, Hahn RA, Hynes MD, Mason NR, Wong DT, Foreman MM (1983) Resolution and absolute configuration of an ergoline-related dopamine agonist, trans-4,4a,5,6,7,8,8a,9-Octahydro-5-propyl-1H(or 2H)-pyrazolo[3,4-g]quinoline. J Med Chem 26:1112–1116
Trevino A, Wolf A, Jackson A, Price T, Uphouse L (1999) Reduced efficacy of 8-OH-DPAT's inhibition of lordosis behavior by prior estrogen treatment. Horm Behav 35:215–223
Ungar S, Makman MH, Morris SA, Etgen AM (1993) Estrogen uncouples beta-adrenergic receptor from the stimulatory guanine nucleotide-binding protein in female rat hypothalamus. Endocrinology 133:2818–2826
Uphouse L, Caldarola-Pastuszka M (1993) Female sexual behavior following intracerebral infusion of the 5-HT1A agonist, 8-OH-DPAT, into the medial preoptic area. Brain Res 601:203–208
Uphouse L, Caldarola-Pastuszka M, Droge M (1992a) 8-OH-DPAT in the midbrain central gray inhibits lordosis behavior. Pharmacol Biochem Behav 43:833–838
Uphouse L, Caldarola-Pastuszka M, Montanez S (1992b) Intracerebral actions of the 5-HT1A agonists, 8-OH-DPAT and buspirone and of the 5-HT1A partial agonist/antagonist, NAN-190, on female sexual behavior. Neuropharmacology 31:969–981
Uphouse L, Caldarola-Pastuszka M, Moore N (1993) Inhibitory effects of the 5-HT1A agonists, 5-hydroxy- and 5-methoxy-(3-di-n-propylamino)chroman, on female lordosis behavior. Neuropharmacology 32:641–651
Uphouse L, Andrade M, Caldarola-Pastuszka M, Jackson A (1996a) 5-HT1A receptor antagonists and lordosis behavior. Neuropharmacology 35:489–495
Uphouse L, Colon L, Cox A, Caldarola-Pastuszka M, Wolf A (1996b) Effects of mianserin and ketanserin on lordosis behavior after systemic treatment or infusion into the ventromedial nucleus of the hypothalamus. Brain Res 718:46–52
Uphouse L, Maswood S, Jackson A (2000) Factors elevating cAMP attenuate the effects of 8-OH-DPAT on lordosis behavior. Pharmacol Biochem Behav 66:383–388
Uphouse L, White S, Harrison L, Hiegel C, Majumdar D, Guptarak J, Truitt WA (2003) Restraint accentuates the effects of 5-HT2 receptor antagonists and a 5-HT1A receptor agonist on lordosis behavior. Pharmacol Biochem Behav 76:63–73
Uphouse L, Hiegel C, Guptarak J, Maswood N (2009) Progesterone reduces the effect of the serotonin 1B/1D receptor antagonist, GR 127935, on lordosis behavior. Horm Behav 55:169–174
Uphouse L, Guptarak J, Hiegel C (2010) Progesterone reduces the inhibitory effect of a serotonin 1B receptor agonist on lordosis behavior. Pharmacol Biochem Behav 97:317–324
Veening JG, de Jong TR, Waldinger MD, Korte SM, Olivier B (2015) The role of oxytocin in male and female reproductive behavior. Eur J Pharmacol 753:209–228
Ventura Aquino E, Fernandez-Guasti A (2013) The antidepressants fluoxetine and bupropion differentially affect proceptive behavior in the naturally cycling female rat. J Sex Med 10:2679–2687
Wallen K, Zehr JL (2004) Hormones and history: the evolution and development of primate female sexuality. J Sex Res 41:101–112
Whitney JF (1986) Effect of medial preoptic lesions on sexual behavior of female rats is determined by test situation. Behav Neurosci 100:230–235
Wolf A, Caldarola-Pastuszka M, Uphouse L (1998a) Facilitation of female rat lordosis behavior by hypothalamic infusion of 5-HT(2A/2C) receptor agonists. Brain Res 779:84–95

Wolf A, Jackson A, Price T, Trevino A, Caldarola-Pastuszka M, Uphouse L (1998b) Attenuation of the lordosis-inhibiting effects of 8-OH-DPAT by TFMPP and quipazine. Brain Res 804:206–211

Wolf A, Caldarola-Pastuszka M, Delashaw M, Uphouse L (1999) 5-HT2C receptor involvement in female rat lordosis behavior. Brain Res 825:146–151

Yamada S, Noguchi D, Ito H, Yamanouchi K (2009) Sex and regional differences in decrease of estrogen receptor alpha-immunoreactive cells by estrogen in rat hypothalamus and midbrain. Neurosci Lett 463:135–139

Yang LY, Clements LG (2000) MPOA lesions affect female pacing of copulation in rats. Behav Neurosci 114:1191–1202

Zimmermann-Peruzatto JM, Lazzari VM, Agnes G, Becker RO, de Moura AC, Guedes RP, Lucion AB, Almeida S, Giovenardi M (2017) The impact of oxytocin gene knockout on sexual behavior and gene expression related to neuroendocrine systems in the brain of female mice. Cell Mol Neurobiol 37:803–815

The Sexual Differentiation of the Human Brain: Role of Sex Hormones Versus Sex Chromosomes

Julie Bakker

Contents

1 Introduction 46
  1.1 Sex Determination and Sexual Differentiation 46
  1.2 Sexual Differentiation of the Brain: Animal Studies 47
  1.3 Sexual Differentiation of the Human Brain 48
  1.4 Indirect Measures of Prenatal Hormone Levels 49
  1.5 Postmortem Studies 49
  1.6 Clinical Studies 50
  1.7 The Sexual Differentiation of the Human Brain: Neuroimaging Studies in CAIS ... 52
2 Concluding Remarks 61
References 61

Abstract Men and women differ, not only in their anatomy but also in their behavior. Research using animal models has convincingly shown that sex differences in the brain and behavior are induced by sex hormones during a specific, hormone-sensitive period during early development. Thus, male-typical psychosexual characteristics seem to develop under the influence of testosterone, mostly acting during early development. By contrast, female-typical psychosexual characteristics may actually be organized under the influence of estradiol during a specific prepubertal period. The sexual differentiation of the human brain also seems to proceed predominantly under the influence of sex hormones. Recent studies using magnetic resonance imaging have shown that several sexually differentiated aspects of brain structure and function are female-typical in women with complete androgen insensitivity syndrome (CAIS), who have a 46 XY karyotype but a female phenotype due to complete androgen resistance, suggesting that these sex differences most likely reflect androgen action, although feminizing effects of estrogens or female-typical socialization cannot be ruled out. By contrast, some male-typical neural characteristics were also observed in women with CAIS suggesting direct effects of sex chromosome genes in the sexual differentiation of the human brain.

J. Bakker (✉)
Laboratory of Neuroendocrinology, GIGA Neurosciences, Liège University, Liège, Belgium
e-mail: jbakker@uliege.be

Curr Topics Behav Neurosci (2019) 43: 45–68
DOI 10.1007/7854_2018_70
Published Online: 1 January 2019

In conclusion, the sexual differentiation of the human brain is most likely a multifactorial process including both sex hormone and sex chromosome effects, acting in parallel or in combination.

Keywords Androgens · Brain function · Brain structure · Complete androgen insensitivity syndrome · Estrogens · Magnetic resonance imaging · Sex differences · Sexual development

1 Introduction

Although men and women might be more similar than different, notable sex differences exist in cognitive abilities (Maccoby and Jacklin 1974), brain morphology (Cosgrove et al. 2007), emotion processing (Schirmer et al. 2004), and vulnerability to psychiatric disorders (Bao and Swaab 2011). The origin of these sex differences in the brain and behavior remains the subject of heated debate. In particular, whether sex differences observed in adulthood result from specific biological processes occurring during early development ("nature") or from the (social) environment ("nurture") has been contested. There is increasing evidence, however, that the organization of the brain is affected by circulating sex hormones and the expression of specific genes during early development and thus that sex differences in brain structure and organization are already present as early as birth. The biological bases of sex differences in brain and behavior are becoming better known, but many questions remain about how sex hormones induce such different patterns of neural and behavioral differentiation in men and women. This chapter will provide a short overview of our current knowledge of how the sexual differentiation of the human brain proceeds under the influence of organizing actions of sex hormones. For obvious ethical reasons, many of our ideas about the sexual differentiation of the human brain are derived from (clinical) research on patients with disorders of sex development (DSD).

1.1 Sex Determination and Sexual Differentiation

In mammals, including humans, the developing organism has the potential to become either male or female. The first step in the process of sex determination is the establishment of the genetic sex at conception. Each cell nucleus contains a species-specific number of paired autosomes and two sex chromosomes. Humans have 22 pairs of autosomes and one pair of sex chromosomes (the X and Y). Autosomes are numbered roughly in relation to their sizes, i.e., chromosome 1 has approximately 2,800 genes, whereas chromosome 22 has approximately 750 genes. The sex chromosomes will determine the sex of the fetus: females have two X chromosomes, whereas males have one X and Y chromosome. The latter contains the *Sry* gene, which will induce the formation of testes from the undifferentiated gonads in males (Koopman et al. 1990). The Leydig cells in the testes will produce testosterone which will promote the development of the Wolffian ducts into the internal male genital structures such as the epididymis, the vas deferens, and the seminal vesicles, whereas anti-Müllerian hormone secreted by the Sertoli cells in the testes causes regression of the female-typical Müllerian ducts. The penis and scrotum develop under the influence of dihydrotestosterone which is formed from testosterone by the enzyme, 5α-reductase. In typical female differentiation, i.e., in the absence of the Y chromosome and consequently the *Sry* gene, the undifferentiated gonads will develop into ovaries. The Müllerian ducts develop without any apparent hormonal input into the uterus, fallopian tubes, and the distal portion of the vagina, whereas the Wolffian ducts regress and disappear in the absence of androgenic stimulation.

1.2 Sexual Differentiation of the Brain: Animal Studies

Animal studies have convincingly shown that the sexual differentiation of the brain generally follows the same pattern as that of the genitals, i.e., exposure to testosterone induces male-typical psychosexual and neural characteristics, whereas in the absence of testosterone, female-typical psychosexual and neural characteristics develop, i.e., by default. This is mainly based on studies in rodents in which the testes were removed directly after birth or conversely, when testosterone was administered to newborn females. It was found that neonatally castrated male rats showed very few male-typical sexual behaviors even following administration of testosterone in adulthood (Grady et al. 1965), whereas testosterone-treated females showed increased levels of male-typical sexual behaviors, such as mounting behavior (reviewed in Baum 1979). In addition, when treated with "female-typical" sex hormones, i.e., estradiol and progesterone, neonatally castrated male rats showed female-typical sexual behaviors, like the expression of the typical female rodent mating posture, lordosis (Feder and Whalen 1964). This led to the conclusion that the female-typical differentiation of the brain proceeds in the absence of any hormonal secretion, i.e., by default. This was further supported by the finding that the ovaries are functionally quiescent during early development: no significant amounts of estradiol could be detected before postnatal day 7 (Lamprecht et al. 1976), whereas the testes start to produce testosterone about 1 week before birth in rodents. However, more recent studies have shown that estradiol may be required for the development of the female brain. Female aromatase knockout (ArKO) mice which carry a targeted mutation in the aromatase gene and as a result cannot synthesize estrogens from androgens, show reduced levels of female sexual behavior in adulthood, even following ovariectomy and subsequent treatment with estradiol and progesterone (Bakker et al. 2002). Interestingly, administration of estradiol over a specific prepubertal period (postnatal days P15–P25) almost completely restored female sexual behavior in female ArKO mice (Brock et al. 2011). This result clearly challenges the classical theory of a default organization of the female brain. It also challenges the idea that sex differences are established before or directly after birth and that sex hormones beyond the perinatal period only have so-called "activational" effects on the brain. Hence, it might be well possible that the brain remains plastic and sensitive to any organizational actions of sex hormones for a much longer period as initially thought.

Although very little doubt remains on the pivotal role of gonadal hormones in establishing sex differences in the brain and behavior, some evidence has been emerging that genes on the sex chromosomes might also contribute to the sexual differentiation of the brain (Arnold et al. 2004). By using a core-cross transgenic mouse model in which the *Sry* gene was deleted from the Y chromosome and inserted into an autosome to create XX and XY male and female phenotypes, it was possible to distinguish between the contribution of gonadal hormones and sex chromosome genes to the development of sex differences in the brain and behavior. It was shown, for instance, that the male-typical profile of vasopressin innervation of the lateral septum depends on the presence of a Y chromosome (de Vries et al. 2002). XY males and XY female mice (i.e., females with a deletion of the *Sry* gene) were more masculine than XX mice in the density of vasopressin-immunoreactive fibers in the lateral septum. Based on these findings, it has been proposed that the mechanism of sexual differentiation is multifactorial and includes both sex hormone and sex chromosome effects, acting in parallel or in combination (Arnold 2017).

1.3 Sexual Differentiation of the Human Brain

An important question raised by animal studies is that if sex hormones play such a pivotal role in masculinizing or feminizing the brains of nonhuman species, do they have similar actions in our own species? In other words, do men and women behave differently because men have been exposed to higher concentrations of testosterone during development or conversely, women to higher levels of estrogens? In humans the sexual differentiation of brain is thought to occur between 8 and 24 weeks of gestation when testosterone levels are higher in male than female fetuses (Reyes et al. 1974; Nagamani et al. 1979). Furthermore, the first months after birth are also marked by a testosterone surge in boys, peaking at around 3 months after birth, and increased estradiol levels in girls, which decrease more gradually during the second year of life (Winter et al. 1976; Kuiri-Hänninen et al. 2014). This period is often referred to as "mini-puberty" (Kuiri-Hänninen et al. 2014), is most likely caused by an increased gonadotropin secretion since there is no longer any negative feedback by estrogens after birth, and might be an additional critical period for organizational actions of sex hormones on the brain and behavior.

1.4 Indirect Measures of Prenatal Hormone Levels

For obvious ethical reasons, indirect markers of prenatal hormone levels have been used to study potential organizational effects of sex hormones in the human brain. The 2D:4D ratio, i.e., the relative length of the second to the fourth digit, which is larger in women than men (Hönekopp and Watson 2010), is the most extensively used marker (Morris et al. 2004), because it is so easy to obtain, a simple photocopy of the hand suffices. However, its validity as an indicator of prenatal androgen exposure has been criticized (Berenbaum et al. 2009; van Hemmen et al. 2017a). Firstly, two independent studies (Lutchmaya et al. 2004; Ventura et al. 2013) have actually attempted to measure fetal androgen levels through amniocentesis and to correlate them with digit ratios, but their results were inconclusive. Secondly, the results obtained in women with complete androgen insensitivity syndrome (CAIS) strongly suggest that other non-androgenic factors are also involved (van Hemmen et al. 2017a). Otoacoustic emissions (OAE) are sounds produced by the cochlea, which can be measured in the inner ear canal (Kemp 1978, 2008; Davis 1983), and are another marker used for assessing prenatal hormone exposure retrospectively. Spontaneous OAEs are more frequent and stronger in women than in men. Likewise, OAEs evoked by click stimuli have larger amplitudes in women than in men (McFadden and Pasanen 1998; Shinur and Hampson 2011). These two measures have in particularly been used to assess whether sexual orientation might reflect changes in prenatal androgen exposure (for comprehensive review, see Breedlove 2017). It was found, however, that homosexual men showed very similar 2D:4D ratios and OAEs as heterosexual men. By contrast, lesbian women showed "masculinized" 2D:4D ratios and OAEs suggesting that they might have been exposed to increased levels of androgens during early development. Although these markers might be useful to determine early androgen exposure, they have many limitations since they can only be used retrospectively and might also be affected by circulating hormone levels in adulthood. This has particularly been suggested for OAEs (Shinur and Hampson 2011).

1.5 Postmortem Studies

Postmortem analyses of the brain have been an important method to determine whether sex differences in human behavior, but also differences related to sexual orientation and gender identity, could be explained by structural differences in the brain. Many of these studies have been inspired by observations made in animal studies. For instance, the discovery of a sexually dimorphic nucleus (SDN) in the rat preoptic area led to a close examination of the human preoptic area where a similar sexually dimorphic nucleus was observed (SDN or interstitial nucleus of the anterior hypothalamus (INAH-1 and 2)) which is larger in men than in women. However, no variations in the size of this nucleus have been observed in relation to sexual orientation (Swaab 2007). In addition, three independent groups have shown that a different nucleus in the anterior hypothalamus, INAH-3, is significantly larger in heterosexual men than in heterosexual women (Allen et al. 1989; Byne et al. 2001; Levay 1991), and in a highly publicized study (LeVay 1991), it was reported that the volume of INAH-3 was greater in heterosexual than in homosexual men, although this latter finding still awaits full replication (Byne et al. 2001). A rather unexpected finding (Swaab and Hofman 1990) was that the suprachiasmatic nucleus (SCN), the clock of the brain, was twice as large in homosexual men compared to heterosexual men, in particularly because no sex differences were observed. This finding suggests that homosexual men do not have a "female-typical" hypothalamus as has been proposed (Dörner 1988).

By contrast, female-typical volumes of the central nucleus of the bed nucleus of the stria terminalis (Zhou et al. 1995; Kruijver et al. 2000) and the INAH-3 (Garcia-Falgueras and Swaab 2008) have been observed in male-to-female (MtF) transsexuals. Likewise, a female-typical expression of two neuropeptides important in regulating GnRH secretion, i.e., neurokinin B (Taziaux et al. 2012) and kisspeptin (Taziaux et al. 2016), has been found in the infundibular nucleus of the hypothalamus of MtF transsexuals. These results may suggest that transsexual people, which are now referred to as people diagnosed with gender dysphoria (defined as a strong incongruence between their gender assigned at birth and the gender that they identify with; DSM-5), have undergone an atypical sexual differentiation of the brain. However, some caution is warranted in interpreting the results because some of these effects might be due to adult hormone treatment (for instance, estrogen treatment in MtF transsexuals) or other not-yet-identified causes.

1.6 Clinical Studies

Perhaps more valuable insights into the mechanisms underlying the sexual differentiation of the human brain might be obtained by studying "disorders of sex development" also known as "disorders of sex differentiation" (DSD). The term DSD refers to congenital conditions in which the development of chromosomal, gonadal, or anatomical sex is atypical (as cited in Hughes et al. 2006). Although many DSDs are not well studied in this context because they are extremely rare, several DSDs provide a unique opportunity to assess the different players involved in sexual differentiation. Examples include, but are not limited to, congenital adrenal hyperplasia (CAH), sex chromosome aneuploidies, and complete androgen insensitivity syndrome (CAIS).

Congenital adrenal hyperplasia (CAH) is the most common DSD (1:10,000) which is characterized in about 95% of the cases by a mutation in the gene encoding the enzyme 21-hydroxylase which is important for the conversion of progesterone to deoxycorticosterone. As a result, progesterone can only be converted to 17-α-hydroxyprogesterone (which in turn cannot be converted to 11-deoxycortisol). The lack of cortisol will lead to an increased release of adrenocorticotropic hormone

(ACTH) by the anterior pituitary since cortisol normally has a negative feedback action on ACTH as well as on the secretion of its stimulator corticotropin-releasing hormone (CRH) from the paraventricular nucleus of the hypothalamus. This increased stimulation by ACTH induces overgrowth (hyperplasia) and hyperactivity of the steroid-producing cells of the adrenal cortex ultimately leading to an increased production of androgens during fetal life. At birth, 46, XX females often have ambiguous genitalia depending on the degree of exposure to androgens during fetal development. Starting early in childhood, girls with CAH also typically show masculinized patterns of sex-typed behavior such as toy and activity preferences, i.e., spending more time playing with boy's toys such as cars than girls' toys such as dolls, showing more rough and tumble play, and preferring boys over girls as playmates (Collaer and Hines 1995; Pasterski et al. 2005). In adulthood, CAH is often associated with polycystic ovarian syndrome (oligomenorrhea, polycystic ovaries, hirsutism) and increased incidence of bisexual orientation and gender dysphoria (Meyer-Bahlburg et al. 2008). In contrast, 46, XY boys with CAH do not show any clear symptoms other than those related to decreased cortisol production. These observations in girls/women with CAH suggest that prenatal androgens indeed might organize male-typical psychosexual and neurobehavioral characteristics.

Individuals with sex chromosome aneuploidies (SCAs) have an atypical number of sex chromosomes and often show deficits in certain cognitive domains (Printzlau et al. 2017), some of which showing sex differences in the general population. Therefore, SCAs have been suggested as a method to study possible effects of sex chromosome genes on the development of the brain. Klinefelter syndrome (KS; incidence, 1:500–1,000), characterized by two or more X chromosomes and a Y chromosome, thus male phenotype, and Turner syndrome (TS; incidence, 1:2,500), characterized by one X chromosome and a lack of (all or part) of the second X chromosome, thus female phenotype, have been mostly studied in this context. However most SCAs also result in atypical sex hormone levels, such as hypogonadism in men with KS prior to being supplemented with testosterone (Davis et al. 2015) and decreased estradiol production in women with TS due to premature ovarian failure (Modi et al. 2003). Consequently, it has been challenging to establish whether any findings obtained in these two SCAs represent direct genetic effects related to sex chromosome gene dosage, sex hormone effects, or both.

Androgen insensitivity syndrome (AIS) might be one of the most interesting DSDs to study the respective roles of sex hormones versus sex chromosome genes in the sexual differentiation of the human brain. AIS has an estimated incidence of 1:40,800 to 1:99,000 (Boehmer et al. 2001) and is characterized by a mild (MAIS), partial (PAIS), or complete (CAIS) defect in androgen action caused by mutation(s) in the X-chromosome-linked androgen receptor gene (Hughes et al. 2012), resulting in decreased or completely abolished AR function. The degree of androgen resistance determines the phenotypical presentation, ranging from a male phenotype with fertility problems in MAIS to mild or severe hypomasculinization and ambiguous genitalia in PAIS and a female phenotype in CAIS. In a fetus with CAIS, the gonads develop into testes under the influence of the SRY gene and start producing androgens and AMH. Due to the inability of androgens to activate the AR, the external genitalia develop in the female direction, whereas AMH causes regression of the Müllerian duct, resulting in a blind-ending vagina and absent uterus. CAIS is detected in infancy in case of an inguinal hernia or in adolescence in case of primary amenorrhea (Hughes et al. 2012). The assigned gender at birth and gender upbringing is typically female. With regard to secondary sex characteristics, pubic and axillary hair is either sparse or absent (Tadokoro-Cuccaro and Hughes 2014) since they depend on androgens. There is spontaneous breast development since testosterone is converted into estradiol by the enzyme aromatase, and CAIS might be more sensitive to estrogens (Zachmann et al. 1986). Because of an increased risk of gonadal tumor development in DSDs, including CAIS, the general medical advice is to surgically remove the gonads (Cools et al. 2006; Lee et al. 2016). Following gonadectomy, estrogen replacement therapy is initiated to induce puberty in case of prepubertal gonadectomy and to optimize bone health later on (Bertelloni et al. 2011). Studies on the psychosexual development of CAIS have generally shown an androphilic sexual orientation (i.e., sexual attraction to men), a female gender identity, and female-typical gender role behavior (Masica et al. 1971; Wisniewski et al. 2000; Hines et al. 2003) which is in line with a hypothesized role of androgens in the sexual differentiation of these psychosexual characteristics. Recent work (T'Sjoen et al. 2011; Brunner et al. 2016), however, showed that some individuals with CAIS reported other-than-female gender roles or neither-female-nor-male genders. In addition, not all CAIS reported an exclusively androphilic sexual orientation. Gender development might thus not always be typically female in CAIS suggesting a potential contribution of the sex chromosome genes as well. However, for the rest of this chapter when discussing the results of the different neuroimaging studies, we will refer to women with CAIS because they all identified themselves as women in these studies and expressed a very strong desire to be called women.

1.7 The Sexual Differentiation of the Human Brain: Neuroimaging Studies in CAIS

1.7.1 Brain Function

The introduction of neuroimaging techniques such as magnetic resonance imaging (MRI) has made it possible to study sex differences in brain structure and function in the general population but also in clinical populations. Functional MRI (fMRI) studies, which measure the activity of the brain while performing a task, have, for instance, shown that neural activity while viewing sexual images, emotional stimuli, or during the performance of spatial tasks such as a three-dimensional mental rotation task (MRT) differs between men and women (reviewed in Sacher et al. 2013). Furthermore, MRI studies of brain structure have shown sex differences in the volume of and structural connectivity between brain regions (Gong et al. 2011; Ruigrok et al. 2014). Women with CAIS provide a unique opportunity to study the origin of these sex differences and in particular to analyze the respective roles of sex chromosome genes versus androgens in the sexual differentiation of the human brain. At present, three independent groups (Hamann et al. 2014; van Hemmen et al. 2016, 2017b; Savic et al. 2017) have examined brain structure and function in CAIS using neuroimaging techniques. The fMRI study by Hamann et al. (2014) focused in particularly on brain responses to sexually arousing stimuli in light of the robust sex differences observed in this domain (e.g., Gizewski et al. 2009; Hamann et al. 2004). Men showed greater activation in the amygdala compared to control women and women with CAIS with the latter two groups not being any different. These results suggest that the male-typical activation most likely reflects androgen actions and thus no direct effects of genes of the Y chromosome. In addition, these data argue against an important role for brain aromatization in the masculinization of the brain as has been reported in some animal species (reviewed in Balthazart and Ball 2012) since women with CAIS have presumably higher brain estrogen levels derived from aromatization of testosterone but showed no signs of brain masculinization with regard to their responses to sexually arousing stimuli. However, all women with CAIS in this study (Hamann et al. 2014) identified themselves as women (gender identity and gender role) and were androphilic. They were also all raised as girls and received a female-typical socialization. Finally, it should be noted that group sizes were rather small ($n = 13$/group) and there was no information available on whether the women with CAIS were gonadectomized and/or received any hormone replacement therapy (HRT) nor a confirmation of their diagnosis other than self-report.

As mentioned previously, men and women differ in cognitive abilities with the mental rotation task, a visuospatial task, often showing the greatest differences, with men generally outperforming women (Linn and Petersen 1985). Accordingly, fMRI studies have demonstrated sex differences in neural activation while performing this task, with generally higher levels of activation in parietal regions in men (Jordan et al. 2002; Weiss et al. 2003; Butler et al. 2006; Gizewski et al. 2006; Schöning et al. 2007; Hoppe et al. 2012). By contrast, several studies showed a greater activation in frontal and temporal brain regions in women (Thomsen et al. 2000; Jordan et al. 2002; Weiss et al. 2003; Seurinck et al. 2004; Kucian et al. 2005; Butler et al. 2006; Gizewski et al. 2006; Schöning et al. 2007; Hoppe et al. 2012). Based on these findings from both behavioral and neuroimaging studies, it has been proposed that men and women use different strategies for solving this task: women are thought to use a serial and analytical approach, whereas men are thought to rely on a more automatic "gestalt" strategy (Thomsen et al. 2000; Jordan et al. 2002; Butler et al. 2006). It has been hypothesized that these sex differences most likely reflect early androgen actions since it has been shown that women diagnosed with CAH outperformed their unaffected sisters on spatial tasks in childhood, adolescence, and adulthood (Puts et al. 2008; Berenbaum et al. 2012). In addition, men who have had low androgen levels throughout life due to idiopathic hypogonadotropic hypogonadism showed impaired spatial abilities compared with control men (Hier and Crowley 1982). Androgen replacement therapy in these men did not provide any consistent results (Hier and Crowley 1982; Zitzmann et al. 2001), which could also have been due to small sample sizes. The Wechsler Intelligence Scale for Adults (WAIS) and Wechsler Intelligence Scale for Children (WISC) have been used to study spatial abilities in women with CAIS. It was found that CAIS overall performed worse than control men and women (Imperato-McGinley et al. 1991; Masica et al. 1969). However, they were classified as being "feminine" based on their superior performance on verbal compared to spatial ability subtests. These results suggest that androgens and not sex chromosome genes play a predominant role in the masculinization of these behavioral and cognitive domains. To confirm this hypothesis, we compared brain activation during the performance of a 3D mental rotation task (MRT) in women with CAIS to control men and women (van Hemmen et al. 2016). We were able to recruit a total of 21 CAIS women, who were all gonadectomized and taking HRT (estrogens or combined estrogens/progestins) at the time of our study, and 30 control women and 30 control men. Groups were matched for age and educational level. The diagnosis of CAIS was based on both clinical characteristics and mutation analysis of the androgen receptor gene using genomic DNA. All participants received clear instructions for the MRT at the day of testing and performed a practice trial before the actual MRI session was started. The stimuli used were colored 3D objects (Shephard and Metzler 1971) with varying degrees of rotation ranging from 45° to 315°. The stimuli were presented in an alternating block design with five rotation or control trials in each block. For the control trails, the participants just had to answer the question whether an arrow was pointing to the left or to the right. Response latency (reaction time) and accuracy were recorded. Significant differences were observed with control men responding faster than women with CAIS, but there were no group differences in accuracy scores. At the neural level, an overall similar activation pattern during mental rotation after subtraction of activation during the control condition was found: all groups showed bilateral activations in the parietal lobe, predominantly in inferior and superior regions, extending into the occipital lobe. Significant activations were also observed in frontal areas, to a large extent in the precentral and superior and middle frontal gyrus. The overall pattern observed is consistent with results from a meta-analysis on neuroimaging studies during mental rotation (Zachs 2008). Between group analyses using region of interest (ROIs) revealed sex differences with control men showing significantly more activation than control women in the left inferior parietal lobe and a trend for the right inferior lobe as well. Women with CAIS resembled control women in neural activation patterns and thus differed significantly from control men. We found no significant correlations between circulating hormone levels (testosterone, estradiol) and brain activation patterns. These results thus suggest that reported sex differences in brain functioning while performing a spatial ability task is not directly driven by genetic sex but might be attributable to gonadal hormone exposure, most likely androgens. The absence of male-typical results in CAIS further supports the notion that in humans, androgens, and not estrogens, are the masculinizing hormones. However, we cannot distinguish between organizational and activational effects of these androgens as the insensitivity to androgens is already present prenatally and remains continuous throughout life. Furthermore, women with CAIS have supposedly higher serum estrogen levels than men due to aromatization of testosterone when the gonads are still in situ (Hughes and Deeb 2006) and due to estrogen replacement therapy after gonadectomy. Thus the female-typical pattern in neural activation observed in CAIS can also be attributed to estradiol, which would be in line with studies that have proposed a role for estrogens in mental rotation-related neural activation and performance (e.g., Maki et al. 2002; Schöning et al. 2007), although it should be noted that others have not found these effects (e.g., Peters et al. 1995; Halari et al. 2005).

Finally, it cannot be ruled out that socialization has had an effect on brain activity as well. Exposure to typically masculine toys and activities is thought to have enhancing effects on performance on spatial tasks (e.g., Connor and Serbin 1977). Since recalled childhood toy and activity preferences were sex typical in our groups, i.e., there was a greater preference of masculine toys and activities in control men, and for feminine toys and activities in control women and women with CAIS. Gender stereotypes about male superiority on spatial tasks have also been proposed to affect the behavioral sex difference observed in the MRT (Hausmann et al. 2009). However, to minimize potential gender stereotype effects on the sex differences in neural activation, the participants were not informed about the sex difference in MRT performance before participation.

1.7.2 Brain Structure

Numerous neuroimaging studies have focused on macro- and mesoanatomical sex differences, such as in overall or regional gray (GM) and white matter (WM) volumes derived from structural MRI scans (for meta-analysis, see Ruigrok et al. 2014). Overall, it has been shown that men have larger bilateral GM volumes in limbic regions, including the amygdala, hippocampus, parahippocampal and cingulate gyrus, the temporal pole, precuneus, putamen, and cerebellum, whereas women have larger GM volumes in the bilateral thalamus and precuneus, right planum temporale/parietal operculum, insula, Heschl's and anterior cingulate gyrus, parts of the frontal cortex, and left parahippocampal gyrus and lateral occipital cortex. These sex differences are however not related to sex differences in total brain size (the male brain is on average 11% larger than the female brain). Evidence of early sex hormone effects on regional GM volumes seems to be rather inconsistent. In one study of boys (Lombardo et al. 2012), testosterone levels in amniotic fluid were associated with GM volume at age 8–11 in some, but not all brain regions showing sex differences. By contrast, no masculinizing effects of fetal testosterone on brain structure have been observed in girls diagnosed with CAH, although a decrease in amygdala volume was observed in both boys and girls with CAH which is probably caused by their glucocorticoid deficiency (Merke et al. 2003). Some effects of circulating gonadal hormones on adult GM volumes have been reported (e.g., Lessov-Schlaggar et al. 2005; Witte et al. 2010; Lentini et al. 2012), but results varied among studies probably due to methodological differences.

In a recent study (Savic et al. 2017), cortical thickness and subcortical GM volumes were compared between a group of 16 women diagnosed with CAIS and control groups of men and women ($n = 32$/sex). It was found that both women with CAIS and control women displayed thicker parietal and occipital cortices and a thinner left temporal cortex than control men. Interestingly, women with CAIS also displayed a "male" pattern, i.e., a significantly thinner cortex in the precentral gyrus and to some extent, in the postcentral gyrus compared to female control women, and thus similar to control men. Furthermore, caudate volumes were significantly smaller, i.e., "male-like," in women with CAIS compared to control women, but hippocampus volumes were female-like and thus significantly larger than in control men. Thus women with CAIS showed a mixed male and female pattern suggesting direct effects of sex chromosome genes in addition to sex hormone effects.

The MRI technique of diffusion tensor imaging (DTI), which measures the diffusion of water molecules, was recently developed to study sex differences in white matter microstructure. The most used quantitative measure that can be derived from DTI is fractional anisotropy (FA) (Pierpaoli and Basser 1996). The FA value provides information about the degree of diffusion anisotropy. A low FA value reflects isotropic diffusion, i.e., equal diffusion in all directions, as, for instance, in cerebrospinal fluid. A high degree of anisotropy is found in WM fiber bundles, in which water diffusion is restricted in the direction perpendicular to the axon. The majority of DTI studies have found higher FA values in major WM regions in men compared to women (e.g., Chou et al. 2011; Inano et al. 2011; Schoonheim et al. 2014; Takao et al. 2014). Lower FA values in WM tracts have been observed in women with Turner syndrome (Holzapfel et al. 2006) and men with Klinefelter syndrome (DeLisi et al. 2005), but these two DSDs are characterized by sex chromosome aneuploidies and subsequent sex hormone deficiencies, which makes it impossible to determine the relative contribution of sex chromosomes versus sex hormones in the sexual differentiation of WM characteristics. Therefore, to address this particular question, we acquired DTI scans of women with CAIS and compared them with groups of control men and women (van Hemmen et al. 2017b). The final sample consisted of 20 women with CAIS, 30 control men, and 30 control women.

Using analyses based on tract-based spatial statistics, widespread sex differences were observed in FA with control men showing higher FA values than control women in a single cluster covering a large part of the skeleton, including major WM tracts; subcortical regions, such as the bilateral thalamus and basal ganglia; and the brain stem. By contrast, control women did not show any regions with higher FA values than control men. Similar differences were found between control men and women with CAIS with control men having higher FA values in a large part of the skeleton, whereas the reverse contrast did not reveal any significant differences. Furthermore, no differences in FA values were observed between control women and women with CAIS. These findings thus suggest a more important role for sex hormones, most likely masculinizing androgen and/or feminizing estrogen effects, than for genetic effects related to sex chromosome genes in the sexual differentiation of WM microstructure.

A limitation of these studies (Hamann et al. 2014; van Hemmen et al. 2016, 2017b; Savic et al. 2017) conducted in women with CAIS that no inferences can

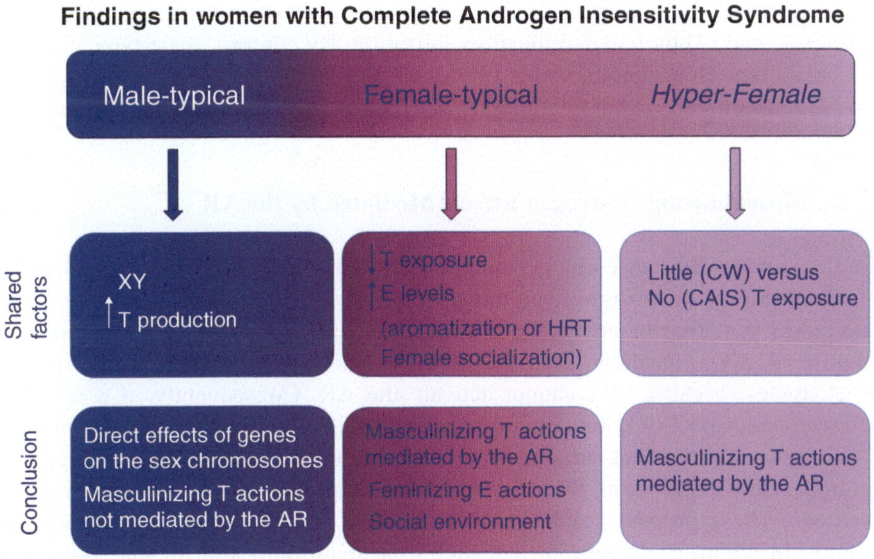

Fig. 1 Implications of male- and female-typical findings in women with CAIS. The upper boxes summarize the factors that women with CAIS share with the group of reference. In the lower boxes, the conclusions that can be drawn on these shared factors. *T* testosterone, *AR* androgen receptor, *E* estrogen, *HRT* hormone replacement therapy, *CW* control women, *CAIS* women with complete androgen insensitivity syndrome

be made about the exact timing of the proposed sex hormone effects. Since women with CAIS are insensitive to androgens throughout life, any finding obtained in adulthood can reflect organizational and/or activational sex hormone effects during the perinatal phase, adolescence, and/or adulthood. Longitudinal studies in women with CAIS, starting before adolescence, could provide important information regarding the timing of sex hormone effects on the sexual differentiation of the human brain.

To summarize, several sexually differentiated aspects of brain structure and function are female-typical in women with CAIS, although there is also evidence for male-typical neural characteristics. Female-typical findings in women with CAIS with respect to brain function were observed in the left inferior parietal lobe while performing a mental rotation task (van Hemmen et al. 2016) and in the amygdala when visualizing sexual images (Hamann et al. 2014). Regarding brain structure, a female-typical pattern was observed in regional GM volume of the hippocampus, as well as the parietal and occipital cortices (Savic et al. 2017), and in WM microstructure throughout extensive WM regions. By contrast, a male-typical caudate nucleus volume and pre- and postcentral gyrus cortex were observed in women with CAIS (Savic et al. 2017). These neuroimaging findings can be explained by several mechanisms (summarized in Fig. 1). Overall, neural development in the female direction in women with CAIS can be explained in three ways: (1) absence of masculinizing androgen effects following activation of the AR, (2) by feminizing estrogen effects derived from aromatization of androgens, and (3) by female-typical socialization. By contrast, male-typical neural and behavioral characteristics might be explained by (1) sex chromosome effects and (2) masculinizing androgen effects not mediated by the AR.

1.7.3 Masculinizing Androgen Effects Mediated by the AR

Nonfunctional ARs, caused by genetic mutations in the AR gene, result in a lack of effective androgen exposure. Even though the production of testosterone in women with CAIS is within or above the male range when their gonads are still in situ (Melo et al. 2003; Doehnert et al. 2015), these androgens have no direct effect on target tissues, because they cannot activate the AR. Consequently, if a sexually differentiated aspect of brain structure or function is female-typical in women with CAIS, this might reflect an important role for masculinizing androgen effects through AR activation in the sex-typical development of these structures and functions. These proposed androgen effects could be organizational, activational, or both. In addition, the role of the AR in the masculinization of the adolescent brain has been studied by looking specifically at a functional polymorphism of the AR gene: a low number of CAG repeats has been associated with stronger androgen signaling and vice versa (Hsiao et al. 1999; Irvine et al. 2000). It was found to modulate relative GM and WM volumes (Paus et al. 2010), cortical thickness development (Raznahan et al. 2010), and WM growth (Perrin et al. 2008). It should be noted that long CAG repeats of the AR have also been associated with male-to-female transsexuality of which it has been hypothesized to reflect reduced androgen action during development (Hare et al. 2009).

Finally, since control women produce low amounts of androgens and have a functional AR, the level of androgen exposure differs between control women and women with CAIS. Therefore, in theory, the absence of any masculinizing androgen effects through the AR might result in "ultra-feminine" characteristics in women with CAIS (Fig. 1).

1.7.4 Masculinizing Androgen Effects Not Mediated by the AR

In general, it is difficult to determine whether potential androgen effects are the result of direct AR activation or indirect activation of the ER by androgen-derived estrogens upon aromatization, as is the predominant prenatal masculinizing pathway in rodent species (Bakker et al. 2006). It is, however, assumed that in humans masculinizing androgen effects are mediated by the AR and not the ER based on male-typical psychosexual development in men with aromatase deficiency or estrogen insensitivity due to a mutation in the estradiol receptor (Baum 2006), as well as on studies conducted in nonhuman primates (Wallen 2005). Previously reported predominant female-typical psychosexual characteristics in women with CAIS (Masica et al. 1971; Wisniewski et al. 2000; Hines et al. 2003; but see

T'Sjoen et al. 2011; Brunner et al. 2016) also argue against a critical role for estrogens in brain masculinization in humans. Interestingly, it should be noted that there is some evidence for an activational role for estradiol on the brain in adult men. Indeed, a large clinical study (Finkelstein et al. 2013) showed that treatment with an aromatase inhibitor led to a significant decline in sexual desire in adult men.

1.7.5 Feminizing Estrogen Effects

Women with CAIS are thought to have higher estrogen levels than men because androgens produced by the gonads can be aromatized to estrogens and following gonadectomy, women with CAIS generally take estrogens from puberty onwards to induce puberty (in case of prepubertal gonadectomy) and to maintain their health. Therefore, female-typical neural characteristics might reflect feminizing effects of estrogens.

Recent studies in mice have shown that female-typical neural and behavioral characteristics develop under the influence of estradiol during a specific prepubertal period (Brock et al. 2010, 2011). These results thus challenged the classical view of a default organization of the female brain. To date, in humans, there is only evidence for a role of testosterone (and not estradiol) in the development of the human brain as mentioned above (e.g., Baum 2006). However, there is some indirect evidence that estradiol might play a role in the development of the female brain. Several studies (Downey et al. 1989; Rolstad et al. 2007; Shaeffer et al. 2008) have shown that TS women reported that aspects of heterosexual function (e.g., ever engaging in genital petting or sexual intercourse, ever having had a boyfriend) were significantly lower compared to control women. Furthermore, Ross et al. (1998) reported some beneficial effects of treatment with lose doses of estrogen at prepubertal ages on cognitive function. Twenty-four TS girls exhibited a significant improvement in their motor function and nonverbal processing speed after estradiol treatment, when compared to their TS peers who received a placebo treatment. Clearly more research is needed to determine whether estradiol feminizes the brain in humans.

1.7.6 Socialization Effects

A major challenge when studying potential biological factors underlying the sexual differentiation of the human brain and behavior is that from the moment someone is born, his/her social environment is gender-biased. Since women with CAIS are raised as girls, they share a female-typical socialization with control women. Thus, female-typical findings in women with CAIS might reflect effects from the environment related to the gender of rearing. It has been thought that experience can alter the brain throughout life (e.g., Maguire et al. 2006), and gender-typical experiences might have an effect on sex differences found in brain structure and function. For example, exposure to male-typical toys and activities, such as playing action video games, might result in better performance on spatial tasks (Connor and Serbin 1977; Feng et al. 2007). However, there is also a very strong suggestion of early androgen effects in toy preferences: studies in girls with CAH have shown that they were more interested in playing with male-typical instead of female-typical toys (Hines et al. 2016).

1.7.7 Sex Chromosome Effects

The presence of some male-typical neural characteristics in women with CAIS might be explained by the fact that they have a Y chromosome. Although hormone-independent effects of genes located on the sex chromosomes have long been overlooked as being relevant in brain sexual differentiation, recent animal studies have provided evidence for a role of sex chromosome genes in addition to sex hormone effects (e.g., Arnold and Chen 2009). Sexual differentiation as result of sex chromosome effects might reflect (1) direct effects of genes on the Y chromosome or (2) effects related to having one versus two X chromosomes. Studies in rodents have shown neural sex differences related to differences in *Sry* expression in the brain (Dewing et al. 2006). In individuals with two X chromosomes, one of the two X chromosomes is silenced to prevent higher expression of X-linked genes in XX versus XY cells (Chang et al. 2006). This silencing is referred to as X-inactivation and serves as a mechanism to reduce sex differences. However, the gene responsible for X-inactivation, *Xist*, is only expressed in XX cells and has now been suggested as a potential sex-differentiating gene (Arnold 2017). Approximately 10–15% of the X-linked genes escape X-inactivation (Carrel et al. 1999; Carrel and Willard 2005), of which some are located on the pseudoautosomal region (PAR), i.e., have Y-linked homologues, while others are outside the PAR (Disteche 2012). Higher expression of X-linked escapee genes located outside the PAR in XX versus XY cells may result in neural or behavioral sex differences. Furthermore, maternal versus paternal imprinting of the X chromosome might also influence sexual differentiation, as it might cause differences in expression of X-linked genes between men and women (Babak et al. 2015).

These proposed mechanisms of direct sex chromosome gene effects have not been adequately studied in humans yet, and even though animal studies have already provided some valuable information, many questions remain. A recent study using the FCG mouse model has shown that brain structure was related to sex hormone actions in 16, and sex hormone-independent effects in 11 brain regions (Corre et al. 2016) suggesting that the contribution of sex chromosome genes might still be underestimated (Arnold 2017). Human studies on cognitive abilities and brain structure and function in SCAs such as TS and KS have revealed a potential contribution of sex chromosome complement on, for example, verbal and spatial abilities and GM volume (reviewed in Printzlau et al. 2017), but these results remain difficult to interpret as they might also reflect sex hormone effects, since sex hormone levels are also affected in most SCAs.

2 Concluding Remarks

Sex differences exist in many aspects of human behavior, cognition, and brain structure and function. It is of great importance to identify these factors causing sex differences, not only to increase our understanding of the development of the healthy brain but also to provide valuable information on the ontogeny of several neuropsychiatric disorders with an important sex difference in their prevalence. Neuroimaging studies in women with CAIS suggest that sex differences in the human brain results from a combination of sex hormone-, sex chromosome-, and socialization-related effects and that the relative contribution of each factor might vary throughout the brain. Nevertheless, androgens acting through the AR seem to play a major role in inducing male-typical neural and psychosexual characteristics in humans. By contrast, whether female-typical neural and psychosexual characteristics develop under the influence of estrogens remains to be elucidated.

References

Allen LS, Hines M, Shryne JE, Gorski RA (1989) Two sexually dimorphic cell groups in the human brain. J Neurosci 9:497–506

Arnold AP (2017) A general theory of sexual differentiation. J Neurosci Res 95:291–300

Arnold AP, Chen X (2009) What does the "four core genotypes" mouse model tell us about sex differences in the brain and other tissues? Front Neuroendocrinol 30:1–9

Arnold AP, Xu J, Grisham W, Chen X, Kim YH, Itoh Y (2004) Minireview: sex chromosomes and brain sexual differentiation. Endocrinology 145:1057–1062

Babak T, DeVeale B, Tsang EK, Zhou Y, Li X, Smith KS, Kukurba KR, Zhang R, Li JB, van der Kooy D, Montgomery SB, Fraser HB (2015) Genetic conflict reflected in tissue-specific maps of genomic imprinting in human and mouse. Nat Genet 47:544–549

Bakker J, Honda S, Harada N, Balthazart J (2002) The aromatase knockout mouse provides new evidence that estradiol is required during development for the expression of sociosexual behaviors in adulthood. J Neurosci 22:9104–9112

Bakker J, De Mees C, Douhard Q, Balthazart J, Gabant P, Szpirer J, Szpirer C (2006) Alpha-fetoprotein protects the developing female mouse brain from masculinization and defeminization by estrogens. Nat Neurosci 9:220–226

Balthazart J, Ball G (2012) Brain aromatase, estrogens, and behavior. Oxford University Press, Oxford

Bao A-M, Swaab DF (2011) Sexual differentiation of the human brain: relation to gender identity, sexual orientation and neuropsychiatric disorders. Front Neuroendocrinol 32:214–226

Baum MJ (1979) Differentiation of coital behavior in mammals: a comparative analysis. Neurosci Biobehav Rev 3:265–284

Baum MJ (2006) Mammalian animal models of psychosexual differentiation: when is translation to the human situation possible? Horm Behav 50:579–588

Berenbaum SA, Bryk KK, Nowak N, Quigley CA, Moffat S (2009) Fingers as a marker of prenatal androgen exposure. Endocrinology 150:5119–5124

Berenbaum SA, Bryk KLK, Beltz AM (2012) Early androgen effects on spatial and mechanical abilities: evidence from congenital adrenal hyperplasia. Behav Neurosci 126:86–96

Bertelloni S, Dati E, Baroncelli GI, Hiort O (2011) Hormonal management of complete androgen insensitivity syndrome from adolescence onward. Horm Res Paediatr 76:428–433

Boehmer AL, Brüggenwirth H, van Assendelft C, Otten BJ, Verleun-Mooijman MC, Niermeijer MF, Brunner HG, Rouwé CW, Waelkens JJ, Oostdijk W, Kleijer WJ, van der Kwast TH, de Vroede MA, Drop SL (2001) Genotype versus phenotype in families with androgen insensitivity syndrome. J Clin Endocrinol Metab 86:4151–4160
Breedlove SM (2017) Prenatal influences on human sexual orientation: expectations versus data. Arch Sex Behav 46:1583–1592
Brock O, Douhard Q, Baum MJ, Bakker J (2010) Reduced prepubertal expression of progesterone receptor in the hypothalamus of female aromatase knockout mice. Endocrinology 151:1814–1821
Brock O, Baum MJ, Bakker J (2011) The development of female sexual behavior requires prepubertal estradiol. J Neurosci 31:5574–5578
Brunner F, Fliegner M, Krupp K, Rall K, Brucker S, Richter-Appelt H (2016) Gender role, gender identity and sexual orientation in CAIS ("XY-women") compared with subfertile and infertile 46,XX women. J Sex Res 53:109–124
Butler T, Imperato-McGinley J, Pan H, Voyer D, Cordero J, Zhu Y-S, Stern E, Silbersweig D (2006) Sex differences in mental rotation: top-down versus bottom-up processing. NeuroImage 32:445–456
Byne W, Tobet S, Mattiace LA, Lasco MS, Kemether E, Edgar MA, Morgello S, Buchsbaum MS, Jones LB (2001) The interstitial nuclei of the human anterior hypothalamus: an investigation of variation with sex, sexual orientation, and HIV status. Horm Behav 40:86–92
Carrel L, Willard HF (2005) X-inactivation profile reveals extensive variability in X-linked gene expression in females. Nature 434:400–404
Carrel L, Cottle AA, Goglin KC, Willard HF (1999) A first-generation X-inactivation profile of the human X chromosome. Proc Natl Acad Sci U S A 96:14440–14444
Chang SC, Tucker T, Thorogood NP, Brown CJ (2006) Mechanisms of X-chromosome inactivation. Front Biosci 11:852–866
Chou K-H, Cheng Y, Chen I-Y, Lin C-P, Chu W-C (2011) Sex-linked white matter microstructure of the social and analytic brain. NeuroImage 54:725–733
Collaer ML, Hines M (1995) Human behavioral sex differences: a role for gonadal hormones during early development? Psychol Bull 118:55–107
Connor JM, Serbin LA (1977) Behaviorally based masculine- and feminine-activity-preference scales for preschoolers: correlates with other classroom behaviors and cognitive tests. Child Dev 48:1411–1416
Cools M, Drop SLS, Wolffenbuttel KP, Oosterhuis JW, Looijenga LHJ (2006) Germ cell tumors in the intersex gonad: old paths, new directions, moving frontiers. Endocr Rev 27:468–484
Corre C, Friedel M, Vousden DA, Metcalf A, Spring S, Qiu LR, Lerc JP, Palmert MR (2016) Separate effects of sex hormones and sex chromosomes on brain structure and function revealed by high-resolution magnetic resonance imaging and spatial navigation of the Four Core Genotype mouse model. Brain Struct Funct 221:997–1016
Cosgrove KP, Mazure CM, Staley JK (2007) Evolving knowledge of sex differences in brain structure, function, and chemistry. Biol Psychiatry 62:847–855
Davis H (1983) An active process in cochlear mechanics. Hear Res 9:79–90
Davis SM, Rogol AD, Ross JL (2015) Testis development and fertility potential in boys with klinefelter syndrome. Endocrinol Metab Clin N Am 44:843–865
de Vries GJ, Rissman EF, Simerly EB, Yang LY, Scordalakes EM, Auger CJ, Swain A, Lovell-Badge R, Burgoyne PS, Arnold AP (2002) A model system for sex chromosome effects on sexually dimorphic neural and behavioral traits. J Neurosci 22:9005–9014
DeLisi LE, Maurizio AM, Svetina C, Ardekani B, Szulc K, Nierenberg J, Leonard J, Harvey PD (2005) Klinefelter's syndrome (XXY) as a genetic model for psychotic disorders. Am J Med Genet B Neuropsychiatr Genet 135B:15–23
Dewing P, Chiang CWK, Sinchak K, Sim H, Fernagut P-O, Kelly S, Chesselet M-F, Micevych PE, Albrecht KH, Harley VR, Vilain E (2006) Direct regulation of adult brain function by the male specific factor SRY. Curr Biol 16:415–420
Disteche CM (2012) Dosage compensation of the sex chromosomes. Annu Rev Genet 46:537–560

Doehnert U, Bertelloni S, Werner R, Dati E, Hiort O (2015) Characteristic features of reproductive hormone profiles in late adolescent and adult females with complete androgen insensitivity syndrome. Sex Dev 9:69–74

Dörner G (1988) Neuroendocrine response to estrogen and brain differentiation in heterosexuals, homosexuals, and transsexuals. Arch Sex Behav 17:57–75

Downey J, Ehrhardt AA, Gruen R, Bell JJ, Morishima A (1989) Psychopathology and social functioning in women with Turner syndrome. J Nerv Ment Dis 177:191–201

Feder HH, Whalen RE (1964) Feminine behavior in neonatally castrated and estrogen-treated male rats. Science 147:306–307

Feng J, Spence I, Pratt J (2007) Playing an action video game reduces gender differences in spatial cognition. Psychol Sci 18:850–855

Finkelstein JS, Lee H, Burnett-Bowie SA, Pallais JC, Yu EW, Borges LF, Jones BF, Barry CV, Wulczyn KE, Thomas BJ, Leder BZ (2013) Gonadal steroids and body composition, strength, and sexual function in men. N Engl J Med 369:1011–1022

Garcia-Falgueras A, Swaab DF (2008) A sex difference in the hypothalamic uncinate nucleus: relationship to gender identity. Brain 131:3132–3146

Gizewski ER, Krause E, Wanke I, Forsting M, Senf W (2006) Gender-specific cerebral activation during cognitive tasks using functional MRI: comparison of women in mid-luteal phase and men. Neuroradiology 48:14–20

Gizewski ER, Krause E, Schlamann M, Happich F, Ladd ME, Forsting M, Senf W (2009) Specific cerebral activation due to visual erotic stimuli in male-to female transsexuals compared with male and female controls: an fMRI study. J Sex Med 6:440–448

Gong G, He Y, Evans AC (2011) Brain connectivity: gender makes a difference. Neuroscientist 17:575–591

Grady KL, Phoenix CH, Young WC (1965) Role of the developing rat testis in differentiation of the neural tissues mediating mating behavior. J Comp Physiol Psychol 59:176–182

Halari R, Hines M, Kumari V, Mehrotra R, Wheeler M, Ng V, Sharma T (2005) Sex differences and individual differences in cognitive performance and their relationship to endogenous gonadal hormones and gonadotropins. Behav Neurosci 119:104–117

Hamann S, Herman RA, Nolan CL, Wallen K (2004) Men and women differ in amygdala response to visual sexual stimuli. Nat Neurosci 7:411–416

Hamann S, Stevens J, Vick JH, Bryk K, Quigley CA, Berenbaum SA, Wallen K (2014) Brain responses to sexual images in 46XY, women with complete androgen insensitivity syndrome are female-typical. Horm Behav 66:724–730

Hare L, Bernard P, Sanchez FJ, Baird PN, Vilain E, Kennedy T, Harley VR (2009) Androgen receptor repeat length polymorphism associated with male-to-female transsexualism. Biol Psychiatry 65:93–96

Hausmann M, Schoofs D, Rosenthal HES, Jordan K (2009) Interactive effects of sex hormones and gender stereotypes on cognitive sex differences – a psychobiosocial approach. Psychoneuroendocrinology 34:389–401

Hier DB, Crowley WF (1982) Spatial ability in androgen-deficient men. N Engl J Med 306:1202–1205

Hines M, Ahmed SF, Hughes IA (2003) Psychological outcomes and gender-related development in complete androgen insensitivity syndrome. Arch Sex Behav 32:93–101

Hines M, Pasterski V, Spencer D, Neufeld S, Patalay P, Hindmarsh PC, Hughes IA, Acerini CL (2016) Prenatal androgen exposure alters girls' responses to information regarding gender-appropriate behaviour. Philos Trans R Soc B Biol Sci 371:20150125

Holzapfel M, Barnea-Goraly N, Eckert MA, Kesler SR, Reiss AL (2006) Selective alterations of white matter associated with visuospatial and sensorimotor dysfunction in Turner syndrome. J Neurosci 26:7007–7013

Hönekopp J, Watson S (2010) Meta-analysis of digit ratio 2D:4D shows greater sex difference in the right hand. Am J Hum Biol 22:619–630

Hoppe C, Fliessbach K, Stausberg S, Stojanovic J, Trautner P, Elger CE, Weber B (2012) A key role for experimental task performance: effects of math talent, gender and performance on the neural correlates of mental rotation. Brain Cogn 78:14–27
Hsiao PW, Lin DL, Nakao R, Chang C (1999) The linkage of Kennedy's neuron disease to ARA24, the first identified androgen receptor polyglutamine region-associated coactivator. J Biol Chem 274:20229–20234
Hughes IA, Deeb A (2006) Androgen resistance. Best Pract Res Clin Endocrinol Metab 20:577–598
Hughes IA, Houk C, Ahmed SF, Lee PA (2006) Consensus statement on management of intersex disorders. J Pediatr Urol 2:148–162
Hughes IA, Davies JD, Bunch TI, Pasterski V, Mastroyannopoulou K, MacDougall J (2012) Androgen insensitivity syndrome. Lancet 380:1419–1428
Imperato-McGinley J, Pichardo M, Gautier T, Voyer D, Bryden MP (1991) Cognitive abilities in androgen-insensitive subjects: comparison with control males and females from the same kindred. Clin Endocrinol 34:341–347
Inano S, Takao H, Hayashi N, Abe O, Ohtomo K (2011) Effects of age and gender on white matter integrity. AJNR Am J Neuroradiol 32:2103–2109
Irvine RA, Ma H, Yu MC, Ross RK, Stallcup MR, Coetzee GA (2000) Inhibition of p160-mediated coactivation with increasing androgen receptor polyglutamine length. Hum Mol Genet 9:267–274
Jordan K, Wüstenberg T, Heinze HJ, Peters M, Jäncke L (2002) Women and men exhibit different cortical activation patterns during mental rotation tasks. Neuropsychologia 40:2397–2408
Kemp DT (1978) Stimulated acoustic emissions from within the human auditory system. J Acoust Soc Am 64:1386–1391
Kemp DT (2008) Otoacoustic emissions: concepts and origins. In: Manley GA, Fay RR, Popper AN (eds) Active processes and otoacoustic emissions in hearing. Springer handbook of auditory research. Springer, New York, pp 1–38
Koopman P, Munsterberg A, Capel B, Vivian N, Lovell-Badge R (1990) Expression of a candidate sex-determining gene during mouse testis differentiation. Nature 348:450–452
Kruijver FP, Zhou JN, Pool CW, Hofman MA, Gooren LJ, Swaab DF (2000) Male-to-female transsexuals have female neuron numbers in a limbic nucleus. J Clin Endocrinol Metab 85:2034–2041
Kucian K, Loenneker T, Dietrich T, Martin E, von Aster M (2005) Gender differences in brain activation patterns during mental rotation and number related cognitive tasks. Psychol Sci 47:112–131
Kuiri-Hänninen T, Sankilampi U, Dunkel L (2014) Activation of the hypothalamic-pituitary-gonadal axis in infancy: minipuberty. Horm Res Paediatr 82:73–80
Lamprecht SA, Kohen F, Ausher J, Zor U, Lindner HR (1976) Hormonal stimulation of oestradiol-17 beta release from the rat ovary during early postnatal development. J Endocrinol 68:343–344
Lee PA, Nordenström A, Houk CP, Ahmed SF, Auchus R, Baratz A, Baratz Dalke K, Liao L-M, Lin-Su K, Looijenga LHJ, Mazur T, Meyer-Bahlburg HFL, Mouriquand P, Quigley CA, Sandberg DE, Vilain E, Witchel S (2016) Global disorders of sex development update since 2006: perceptions, approach and care. Horm Res Paediatr 85:158–180
Lentini E, Kasahara M, Arver S, Savic I (2012) Sex differences in the human brain and the impact of sex chromosomes and sex hormones. Cereb Cortex 23:2322–2336
Lessov-Schlaggar CN, Reed T, Swan GE, Krasnow RE, DeCarli C, Marcus R, Holloway L, Wolf PA, Carmelli D (2005) Association of sex steroid hormones with brain morphology and cognition in healthy elderly men. Neurology 65:1591–1596
LeVay S (1991) A difference in hypothalamic structure between heterosexual and homosexual men. Science 253:1034–1037
Linn MC, Petersen AC (1985) Emergence and characterization of sex differences in spatial ability: a meta-analysis. Child Dev 56:1479–1498

Lombardo MV, Ashwin E, Auyeung B, Chakrabarti B, Taylor K, Hackett G, Bullmore ET, Baron-Cohen S (2012) Fetal testosterone influences sexually dimorphic gray matter in the human brain. J Neurosci 32:674–680

Lutchmaya S, Baron-Cohen S, Raggatt P, Knickmeyer R, Manning JT (2004) 2nd to 4th digit ratios, fetal testosterone and estradiol. Early Hum Dev 77:23–28

Maccoby EE, Jacklin CN (1974) The psychology of sex differences. Stanford University Press, Palo Alto

Maguire EA, Woollett K, Spiers HJ (2006) London taxi drivers and bus drivers: a structural MRI and neuropsychological analysis. Hippocampus 16:1091–1101

Maki PM, Rich JB, Shayna Rosenbaum R (2002) Implicit memory varies across the menstrual cycle: estrogen effects in young women. Neuropsychologia 40:518–529

Masica DN, Money J, Ehrhardt AA, Lewis VG (1969) IQ, fetal sex hormones and cognitive patterns: studies in the testicular feminizing syndrome of androgen insensitivity. Johns Hopkins Med J 124:34–43

Masica DN, Money J, Ehrhardt AA (1971) Fetal feminization and female gender identity in the testicular feminizing syndrome of androgen insensitivity. Arch Sex Behav 1:131–142

McFadden D, Pasanen EG (1998) Comparison of the auditory systems of heterosexuals and homosexuals: click-evoked otoacoustic emissions. Proc Natl Acad Sci U S A 95:2709–2713

Melo KFS, Mendonca BB, Billerbeck AEC, Costa EMF, Inacio M, Sliva FAQ, Leal AMO, Latronico AC, Arnhold IJP (2003) Clinical, hormonal, behavioral, and genetic characteristics of androgen insensitivity syndrome in a Brazilian cohort: five novel mutations in the androgen receptor gene. J Clin Endocrinol Metab 88:3241–3250

Merke DP, Fields JD, Keil MF, Vaituzis AC, Chrousos GP, Giedd JN (2003) Children with classic congenital adrenal hyperplasia have decreased amygdala volume: potential prenatal and postnatal hormonal effects. J Clin Endocrinol Metab 88:1760–1765

Meyer-Bahlburg HFL, Dolezal C, Baker SW, New MI (2008) Sexual orientation in women with classical or non-classical congenital adrenal hyperplasia as a function of degree of prenatal androgen excess. Arch Sex Behav 37:85–99

Modi DN, Sane S, Bhartiya D (2003) Accelerated germ cell apoptosis in sex chromosome aneuploid fetal human gonads. Mol Hum Reprod 9:219–225

Morris JA, Jordan CL, Breedlove SM (2004) Sexual differentiation of the vertebrate nervous system. Nat Neurosci 7:1034–1039

Nagamani M, McDonough PG, Ellegood JO, Mahesh VB (1979) Maternal and amniotic fluid steroids throughout human pregnancy. Am J Obstet Gynecol 134:674–680

Pasterski VL, Geffner ME, Brain C, Hindmarsh P, Brook C, Hines M (2005) Prenatal hormones and postnatal socialization by parents as determinants of male-typical toy play in girls with congenital adrenal hyperplasia. Child Dev 76:264–278

Paus T, Nawaz-Khan I, Leonard G, Perron M, Pike GB, Pitiot A, Richer L, Susman E, Weillette S, Pausova Z (2010) Sexual dimorphism in the adolescent brain: role of testosterone and androgen receptor in global and local volumes of grey and white matter. Horm Behav 57:63–75

Perrin JS, Hervé P-Y, Leonard G, Perron M, Pike GB, Pitiot A, Richer L, Veillette S, Pausova Z, Paus T (2008) Growth of white matter in the adolescent brain: role of testosterone and androgen receptor. J Neuro-Oncol 28:9519–9524

Peters M, Laeng B, Latham K, Jackson M, Zaiyouna R, Richardson C (1995) A redrawn Vandenberg and Kuse mental rotation test: different versions and factors that affect performance. Brain Cogn 28:39–58

Pierpaoli C, Basser PJ (1996) Toward a quantitative assessment of diffusion anisotropy. Magn Reson Med 36:893–906

Printzlau F, Wolstencroft J, Skuse DH (2017) Cognitive, behavioral and neural consequences of sex chromosome aneuploidy. J Neurosci Res 95:311–319

Puts DA, McDaniel MA, Jordan CL, Breedlove SM (2008) Spatial ability and prenatal androgens: meta-analyses of congenital adrenal hyperplasia and digit ratio (2D:4D) studies. Arch Sex Behav 37:100–111

Raznahan A, Lee Y, Stidd R, Long R, Greenstein D, Clasen L, Addington A, Gogtay N, Rapoport JL, Giedd JN (2010) Longitudinally mapping the influence of sex and androgen signaling on the dynamics of human cortical maturation in adolescence. Proc Natl Acad Sci U S A 107:16988–16993

Reyes FI, Boroditsky RS, Winter JS, Faiman C (1974) Studies on human sexual development. II. Fetal and maternal serum gonadotropin and sex steroid concentrations. J Clin Endocrinol Metab 38:612–617

Rolstad SG, Moller A, Bryman I, Boman UW (2007) Sexual functioning and partner relationships in women with Turner syndrome: some empirical data and theoretical considerations regarding sexual desire. J Sex Marital Ther 33:231–247

Ross JL, Roeltgen D, Feuillan P, Kushner H, Cutler GB (1998) Effects of estrogen on nonverbal processing speed and motor function in girls with Turner's Syndrome. J Clin Endocrinol Metab 83:3198–3204

Ruigrok ANV, Salimi-Khorshidi G, Lai M-C, Baron-Cohen S, Lombardo MV, Tait RJ, Suckling J (2014) A meta-analysis of sex differences in human brain structure. Neurosci Biobehav Rev 39:34–50

Sacher J, Neumann J, Okon-Singer H, Gotowiec S, Villringer A (2013) Sexual dimorphism in the human brain: evidence from neuroimaging. Magn Reson Imaging 31:366–375

Savic I, Frisen L, Manzouri A, Nordenstrom A, Lindén Hirschberg A (2017) Role of testosterone and Y chromosome genes for the masculinization of the human brain. Hum Brain Mapp 38:1801–1814

Schirmer A, Zysset S, Kotz SA, Yves von Cramon D (2004) Gender differences in the activation of inferior frontal cortex during emotional speech perception. NeuroImage 21:1114–1123

Schöning S, Engelien A, Kugel H, Schäfer S, Schiffbauer H, Zwitserlood P, Pletziger E, Beizai P, Kersting A, Ohrann P (2007) Functional anatomy of visuo-spatial working memory during mental rotation is influenced by sex, menstrual cycle, and sex steroid hormones. Neuropsychologia 45:3203–3214

Schoonheim MM, Rueda Lopes FC, Pouwels PJW, Polman CH, Barkhof F, Geurts JJG (2014) Sex-specific extent and severity of white matter damage in multiple sclerosis: implications for cognitive decline. Hum Brain Mapp 35:2348–2358

Seurinck R, Vingerhoets G, de Lange FP, Achten E (2004) Does egocentric mental rotation elicit sex differences? NeuroImage 23:1440–1449

Shaeffer AT, Lange E, Bondy CA (2008) Sexual function in women with Turner syndrome. J Womens Health 17:27–33

Shephard RN, Metzler J (1971) Mental rotation of three-dimensional objects. Science 171:701–703

Shinur AWK, Hampson E (2011) Sex and ear differences in spontaneous and click-evoked otoacoustic emissions in young adults. Brain Cogn 77:40–47

Swaab DF (2007) Sexual differentiation of the brain and behavior. Best Pract Res Clin Endocrinol Metab 21:431–444

Swaab DF, Hofman MA (1990) An enlarged suprachiasmatic nucleus in homosexual men. Brain Res 537:141–148

T'Sjoen G, de Cuypere G, Monstrey S, Hoebeke P, Freedman FK, Appari M, Holterhus P-M, van Borsel J, Cools M (2011) Male gender identity in complete androgen insensitivity syndrome. Arch Sex Behav 40:635–638

Tadokoro-Cuccaro R, Hughes IA (2014) Androgen insensitivity syndrome. Curr Opin Endocrinol Diabetes Obes 21:499–503

Takao H, Hayashi N, Ohtomo K (2014) Sex dimorphism in the white matter: fractional anisotropy and brain size. J Magn Reson Imaging 39:917–923

Taziaux M, Swaab DF, Bakker J (2012) Sex differences in the neurokinin B system in the human infundibular nucleus. J Clin Endocrinol Metab 97:E2210–E2220

Taziaux M, Staphorsius AS, Ghatei MA, Bloom SR, Swaab DF, Bakker J (2016) Kisspeptin expression in the human infundibular nucleus in relation to sex, gender identity, and sexual orientation. J Clin Endocrinol Metab 101:2380–2389

Thomsen T, Hugdahl K, Ersland L, Barndon R, Lundervold A, Smievoll AI, Roscher BE, Sundberg H (2000) Functional magnetic resonance imaging (fMRI) study of sex differences in a mental rotation task. Med Sci Monit 6:1186–1196

van Hemmen J, Veltman DJ, Hoekzema E, Cohen-Kettenis PT, Dessens AB (2016) Neural activation during mental rotation in complete androgen insensitivity syndrome: the influence of sex hormones and sex chromosomes. Cereb Cortex 26:1036–1045

van Hemmen J, Cohen-Kettenis PT, Steensma TD, Veltman DJ, Bakker J (2017a) Do sex differences in CEOAEs and 2D:4D ratios reflect androgen exposure? A study in women with complete androgen insensitivity syndrome. Biol Sex Differ 8:11

van Hemmen J, Saris IMJ, Cohen-Kettenis PT, Veltman DJ, Pouwels PJW, Bakker J (2017b) Sex differences in white matter microstructure in the human brain predominantly reflect differences in sex hormone exposure. Cereb Cortex 27:2994–3001

Ventura T, Gomes MC, Pita A, Neto MT, Taylor A (2013) Digit ratio (2D,4D) in newborns: influences of prenatal testosterone and maternal environment. Early Hum Dev 89:107–112

Wallen K (2005) Hormonal influences on sexually differentiated behavior in nonhuman primates. Front Neuroendocrinol 26:7–26

Weiss E, Siedentopf CM, Hofer A, Deisenhammer EA, Hoptman MJ, Kremser C, Golaszewski S, Felber S, Fleischhacker WW, Delazer M (2003) Sex differences in brain activation pattern during a visuospatial cognitive task: a functional magnetic resonance imaging study in healthy volunteers. Neurosci Lett 344:169–172

Winter JS, Hughes IA, Reyes FI, Faiman C (1976) Pituitary-gonadal relations in infancy: 2. Patterns of serum gonadal steroid concentrations in man from birth to two years of age. J Clin Endocrinol Metab 42:679–686

Wisniewski AB, Migeon CJ, Meyer-Bahlburg HF, Gearhart JP, Berkovitz GD, Brown TR, Money J (2000) Complete androgen insensitivity syndrome: long-term medical, surgical, and psychosexual outcome. J Clin Endocrinol Metab 85:2664–2669

Witte AV, Savli M, Holik A, Kasper S, Lanzenberger R (2010) Regional sex differences in grey matter volume are associated with sex hormones in the young adult human brain. NeuroImage 49:1205–1212

Zachmann M, Prader A, Sobel EH, Crigler JF Jr, Ritzen EM, Atares M, Ferrandez A (1986) Pubertal growth in patients with androgen insensitivity: indirect evidence for the importance of estrogens in pubertal growth of girls. J Pediatr 108:694–697

Zachs JM (2008) Neuroimaging studies of mental rotation: a meta-analysis and review. J Cogn Neurosci 20:1–19

Zhou JN, Hofman MA, Gooren LJ, Swaab DF (1995) A sex difference in the human brain and its relation to transsexuality. Nature 378:68–70

Zitzmann M, Weckesser M, Schober O, Nieschlag E (2001) Changes in cerebral glucose metabolism and visuospatial capability in hypogonadal males under testosterone substitution. Exp Clin Endocrinol Diabetes 109:302–304

Sexual Differentiation and Sex Differences in Neural Development

Alexandra Turano, Brittany F. Osborne, and Jaclyn M. Schwarz

Contents

1 Introduction 70
2 Sex Determination and Sexual Differentiation 71
 2.1 Biological Sex is Determined at the Time of Conception 71
 2.2 Sex Hormones Induce Sexual Differentiation of the Body 73
 2.3 Sex Hormones Induce Sexual Differentiation of the Brain 74
 2.4 Genetic Sex Differences Impact the Developing Brain 78
3 Sex Differences in the Size of Brain Nuclei 80
 3.1 The Spinal Nucleus of the Bulbocavernosus 80
 3.2 The Anteroventral Periventricular Nucleus of the Hypothalamus 81
 3.3 The Bed Nucleus of the Stria Terminalis 83
4 Sexual Differentiation of the Brain Requires Communication Across and Involvement of Multiple Cell Types in the Brain 84
 4.1 An Immune Molecule and a Neurotransmitter Coordinate Masculinization of the Preoptic Area 85
 4.2 Sexual Differentiation Across the Synapse in the Ventromedial Nucleus of the Hypothalamus 86
 4.3 Development of Neuropeptides Involved in Sex-Specific Behaviors 87
 4.4 Microglia Induce Sex-Specific Neuronal Development 90
 4.5 Sex Differences in Cell Genesis and Fate During the Development of Cognitive Circuits 91
5 A Whole Body Perspective on Sex Differences in Neural Development 93
 5.1 The Gut Microbiome 94
 5.2 The Liver 94
 5.3 The Placenta 95
6 Epigenetics and the Sexual Differentiation of the Brain 96
7 Conclusions 101
References 102

Abstract Sex determination occurs at the moment of conception, as a result of XX or XY chromosome pairing. From that point, the body undergoes the process of

A. Turano, B. F. Osborne, and J. M. Schwarz (✉)
Department of Psychological and Brain Sciences, University of Delaware, Newark, DE, USA
e-mail: jschwarz@psych.udel.edu

Curr Topics Behav Neurosci (2019) 43: 69–110
DOI 10.1007/7854_2018_56
Published Online: 3 July 2018

sexual differentiation, inducing the development of physical characteristics that are easily distinguishable between the sexes and are often reflected in one's physical appearance and gender identity. Although less apparent, the brain *also* undergoes sexual differentiation. Sex differences in the brain are organized during a critical period of neural development and have an instrumental role in determining the physiology and behavior of an individual throughout the lifespan. Understanding the extent of sex differences in neurodevelopment also influences our understanding of the potential risk for a number of neurodevelopmental, neurological, and mental health disorders that exhibit strong sex biases. Advances made in our understanding of sexually dimorphic brain nuclei, sex differences in neural cell communication, and sex differences in the communication between the brain and peripheral organs are all research fields that have provided valuable information related to the physiological and behavioral outcomes of sex differences in brain development. More recently, investigations into the impact of epigenetic mechanisms on sexual differentiation of the brain have indicated that changes in gene expression, via epigenetic modifications, also contribute to sexual differentiation of the developing brain. Still, there are a number of important questions and ideas that have arisen from our current understanding of sex differences in neurodevelopmental processes that necessitate more time and attention in this field.

Keywords Hormones · Neural development · Sexual determination · Sexual differentiation

1 Introduction

The brain develops very early in prenatal life via a number of important processes that produce the cells, shape the structures, and refine the connections within the central nervous system. The goal of these neurodevelopmental processes is to create a brain that controls our physiology and behavior for the rest of our lives; but also, these processes produce the identifying characteristics and personality traits that we associate with each and every one of us. One of the most notable and arguably more important defining characteristics of ourselves is our sex. The sex of many mammals, including humans, is determined at the moment of conception, via the simple process of sex determination that almost belies its importance; however, that single moment of sex determination subsequently shapes the physiology, brain, behavior, and health outcomes of the individual for a lifetime.

The physical characteristics that define our sex are often quite obvious and are usually enhanced in our appearance by our gender identity, which is reflected in our personality, our mannerisms, and our personal style. What may be less obvious, but equally fascinating, are the sex differences that are established in our brains during the process of neural development. The primary goal of these sex differences in the brain is to control our physiology and certain behaviors in a sex-specific manner, and in humans these sex differences also define our gender identity and influence our sexuality. Just as important, sex differences in the brain can influence the risk for certain neurodevelopmental disorders, neurological disorders, and mental health disorders; thus, it is important that we understand how sex differences in the brain are established during early brain development and maintained throughout the lifespan in order to better understand how these differences influence our mental health and well-being.

The goal of this chapter is to summarize our understanding of how sex differences in the brain are established during neural development in the context of how these sex differences regulate sex-specific behaviors later in life. We will identify historical and recent findings that highlight the various mechanisms by which sex differences in the brain are established during development. Another goal throughout this chapter will be to identify important future questions or areas of research in the field of sexual differentiation of the brain and sex differences in neural development.

2 Sex Determination and Sexual Differentiation

2.1 Biological Sex is Determined at the Time of Conception

Sex determination occurs at the moment of conception, when a female's egg is fertilized by a male's sperm. The egg and sperm each contribute one half of the genetic material to the resulting zygote. Each human zygote contains 23 chromosomes, including 22 pairs of autosomes and 1 pair of sex chromosomes. The egg contributes one X chromosome to the zygote, whereas the sperm may contribute either one X *or* one Y chromosome. If the sperm contributes an X chromosome, the resulting zygote will become a female. She will possess two of the *same* allosomal (sex) chromosomes, one X chromosome from the egg and one X chromosome from the sperm. Conversely, if the sperm contributes a Y chromosome, the zygote will become a male (Fig. 1). He will have two *different* allosomes, an X chromosome from the egg and a Y chromosome from the sperm, thereby differentiating him genetically and phenotypically from a developing female (Painter 1923). As cells multiply and the zygote matures into an embryo, then fetus, and eventually a newborn baby, the presence of either XX or XY chromosomes is maintained within each and every cell, and therefore each cell sustains an autonomous chromosomal sex throughout the lifespan that is established at the moment of conception.

Sexual differentiation is the process by which the developing embryo *becomes* male or female after sex determination; and this important process occurs quite early in embryonic development. Sexual differentiation is the process by which primary sex characteristics develop, including the development of the gonads and sex organs, but it is also the process by which many sex differences in the brain are established (Fig. 1). While it is a distinct process, sexual differentiation is dependent upon the process of sex determination. In particular, the sex-determining region of the Y chromosome, known as the *Sry* gene, is critical for the process of sexual differentiation of males. The *Sry* gene encodes for the protein known as testis-determining factor (TDF). TDF initiates the differentiation of steroidogenic precursor cells and primordial germ cells within the developing gonads into Sertoli cells, resulting in the

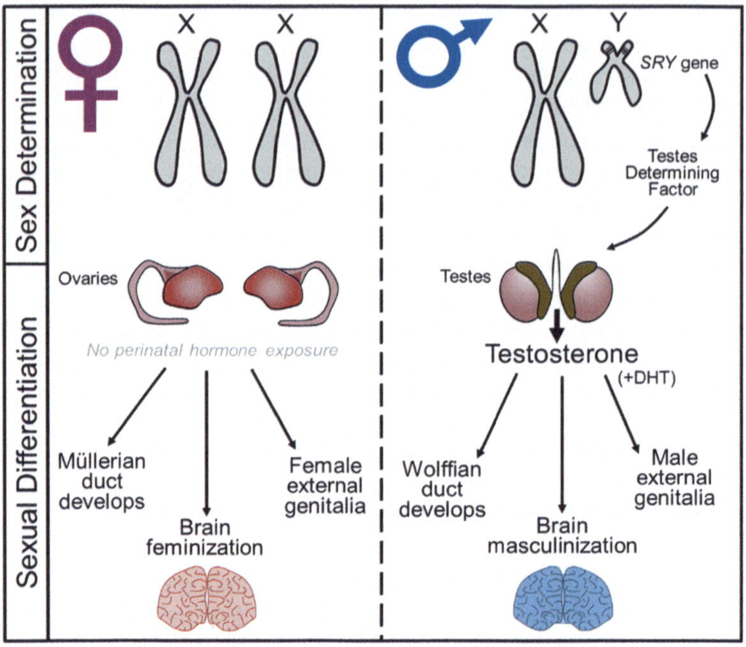

Fig. 1 Sex determination and sexual differentiation of the brain. Sex determination occurs at the moment of conception, when the embryo receives either two X chromosomes (female embryo) or an X and a Y chromosome (male embryo) from the mother and the father. If the developing embryo has a Y chromosome, this chromosome contains the sex-determining region of the Y chromosome (*SRY* gene) that is expressed early in development within specific tissues, in particular the undifferentiated gonads. The SRY gene encodes for a protein known as the testes-determining factor, which differentiates the gonads into a testis early in development. In the absence of the Y chromosome, in females, the gonads differentiate into ovaries. Later, the testes begin to secrete testosterone and dihydrotestosterone (DHT), which differentiate the body and the brain into a male phenotype. In particular, the external genitalia are masculinized; the Wolffian duct develops into the ducts, epididymis, vas deferens, and seminal vesicles; and, importantly for this chapter, the brain is masculinized. In the absence of any testosterone secretion, the external genitalia develops as female, the Müllerian duct develops as the uterus and fallopian tubes, and the brain is feminized

formation of testes (Fig. 1). Developing females (XX) do not have Y chromosomes and therefore, do not have the SRY gene. Consequently, devoid of TDF expression, the precursor cells of the developing gonadal organs differentiate into granulosa cells, resulting in the formation of ovaries (Sekido 2014; Fig. 1).

Prior to sexual differentiation and *Sry* gene expression, the developing gonads are "bipotential," meaning they have the ability to develop into either testes *or* ovaries. Importantly, expression of the *Sry* gene in developing males *must* occur within a sensitive window during embryogenesis in order for the bipotential gonads to differentiate into testes as opposed to ovaries (Taketo et al. 2005). The long-held belief was that without the presence of the *Sry* gene, the precursor cells of the developing gonads *passively* become ovaries (Haseltine and Ohno 1981;

Jost 1947). Contrary to this long-held belief, however, sex reversal studies investigating the result of XX, loss-of-function and XY, gain-of-function mutations on specific sex-linked genes including *Wnt4*, *β-catenin*, *R-spondin1*, and *Foxl2* have found that differentiation of the bipotential gonads into ovaries also requires the *active* repression of testis formation (Chassot et al. 2008; Maatouk and Capel 2008; Pailhoux et al. 2001; Sekido 2014; Vainio et al. 1999). These findings in particular highlight the complex process of gonadal differentiation in males and females. Importantly, the brain is also differentiated as either male or female via *active* and *distinct* processes; however, it has proven to be more difficult to discern the mechanisms by which the developing brain actively differentiates as female, which we will discuss in further detail in Sect. 6.

2.2 Sex Hormones Induce Sexual Differentiation of the Body

Once sexually differentiated, the male testes begin to secrete important molecules for the continued sexual differentiation of the body and the brain; thus, the process of gonadal differentiation is not concurrent with the process of brain sexual differentiation, but rather one process sequentially follows the other (Fig. 1). Because the process of sexual differentiation of the body and brain occurs at various points after sex determination and the formation of the gonads, disruption in one or more of these processes can result in disorders of sexual development wherein the "sex" of the differentiated body or brain does not "match the sex" of the already differentiated gonads. This may result in ambiguous genitalia or a dysmorphic gender identity, both of which comprise an important, but distinct area of research from what we will cover here.

In males, the testes produce an early surge of androgens, including testosterone and dihydrotestosterone (DHT), that leads to the development of sex-related organs, such as the epididymis, vas deferens, and male genitalia (Fig. 1). The testes also secrete anti-Müllerian hormone, which is responsible for actively repressing the Müllerian duct. The Müllerian duct is a primordial structure necessary for the development of the female sex organs; thus its suppression is necessary for the complete and correct development of male as opposed to female sex organs (Josso et al. 1993). As the Müllerian duct regresses in developing males, the Wolffian duct grows into the male accessory organs (Fig. 1). In contrast to developing males, developing females possess ovaries. Like the testes, the ovaries also secrete sex hormones; however, sex hormone secretion from the ovaries occurs at a much later time – at the onset of puberty (Sekido 2014). Thus, in the complete absence of *perinatal* androgen exposure, the Wolffian duct regresses, and the Müllerian duct develops, opposite to the ongoing processes that occur in developing males. Specifically, in the absence of early-life androgens and anti-Müllerian hormone, female sex organs, such as the oviduct, uterus, and female genitalia, begin to develop (Jost 1947; Fig. 1). As the appearance of these outwardly distinct male or female sex organs begins to develop, we are able to determine the sex of a developing fetus in the womb during an ultrasound – often a monumental event when anticipating the arrival of a soon-to-be bumbling baby.

2.3 Sex Hormones Induce Sexual Differentiation of the Brain

Sexual differentiation does not end at the level of the gonads and sex organs. This process of *becoming* male or female also involves the sexual differentiation of the brain (Fig. 1). Like the gonads, the brain is initially a bipotential organ, meaning that it has the potential to be either masculinized or feminized. The presence (or absence) of androgens during perinatal development "organizes" certain structures and functions of the brain in a sexually dimorphic manner. The consequences of this organization become most apparent following "activation" of these sexually dimorphic brain regions during puberty, which is induced by the production of the sex hormones testosterone (males) and estrogen and progesterone (females). As a result, this theory of sexual differentiation of the brain became known as the organizational-activational hypothesis (Phoenix et al. 1959; Fig. 2) that sex differences in the brain are organized during early brain development and then activated by circulating

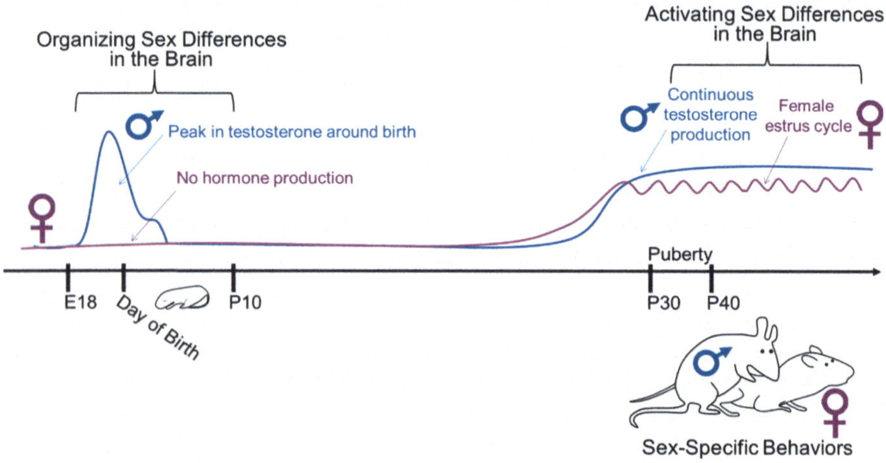

Fig. 2 The organizational-activational hypothesis of sex differences in the rodent brain. Testosterone levels peak around birth in the developing male rodent and naturally decrease within a few days after birth. During that time, the testosterone permanently masculinizes the brain, organizing the size and structure of certain brain regions necessary to perform male sex behavior. In the absence of testosterone, the brain is feminized, permanently organizing the size and structures of the brain regions necessary to perform female sex behavior. At puberty, the testes and the ovaries begin to secrete sex steroid hormones. Males gradually increase production of testosterone, the levels of which are maintained at a relatively consistent point. Females show varying levels of estradiol and progesterone throughout the estrous cycle necessary for ovulation. The production of sex steroid hormones at puberty activates brain regions that were organized during development, thereby producing sex-specific sexual behaviors necessary for reproduction

hormones after puberty. Demonstrating this theory, Phoenix et al. (1959) first ovariectomized female neonatal guinea pigs and then treated them with testosterone during the "organizational" period of brain development. Following what would have been the onset of puberty, female guinea pigs were treated with a second injection of testosterone, and they subsequently exhibited male-typical, as opposed to female-typical, sex behavior (Phoenix et al. 1959). Testosterone given to the ovariectomized neonatal female guinea pig was sufficient to masculinize the brain of the genetically female (XX) guinea pig, causing her to display male-typical sex behaviors following a second testosterone injection later in life. Notably, this process of testosterone-induced sexual differentiation of the brain is conserved across many species including other rodents, birds, and primates, including humans.

The mechanisms by which certain brain regions are organized in a sexually dimorphic manner, resulting in sex-specific behavior, involve specific processes that must occur at specific stages of neural development. In accordance with the organizational-activational hypothesis and the experiments by Phoenix and colleagues in 1959, studies by Gorski and colleagues in 1978 indicated that ovariectomized female rats also exhibited male-typical sex behavior following a pubertal testosterone injection, *if* they had been exposed to testosterone early in life. Interestingly, they proposed that this behavior was due to *increases* in the number of cells within a brain region known as the sexually dimorphic nucleus of the preoptic area (SDN-POA; Gorski et al. 1978). The SDN-POA is larger in males than in females, indicative of its role in male-typical sex behavior. Concurrently, castrated males (i.e., devoid of perinatal testosterone exposure) exhibited no sex behavior following an injection with testosterone in puberty and showed *decreases* in the number of cells within the SDN-POA (Murakami and Arai 1989). Therefore, androgens are necessary to prevent ongoing programmed cell death, or apoptosis, in the SDN-POA during early postnatal brain development, and in doing so, androgens organize the size and structure of the SDN-POA in a sexually dimorphic manner (Dodson and Gorski 1993). The presence or absence of early-life androgens impacts the size of the SDN-POA, as well as a number of other sexually dimorphic brain regions (Fig. 3); and in doing so, it permanently affects the behavioral output of the animal in a sex-specific manner later in life.

Notably, testosterone secreted by the testes is converted to estradiol in the developing male *rodent* brain, by the p450 enzyme, aromatase. While testosterone's action on the androgen receptor contributes to masculinization of the brain, *estradiol's* action on estrogen receptors (in particular, estrogen receptor α) is also important for the masculinization and defeminization of the developing male *rodent* brain (Arnold and Gorski 1984). During this critical period of sexual differentiation of the brain, female rodents may be exposed to exogenous estrogens from the mother, but, female rodents are protected from any possible estradiol-induced changes in the developing brain by the high affinity alpha-fetoprotein (AFP) binding protein, which binds any possible circulating estrogens during the early developmental period, thus protecting females against potential masculinization and defeminization of the brain by maternal estrogens (Bakker et al. 2006).

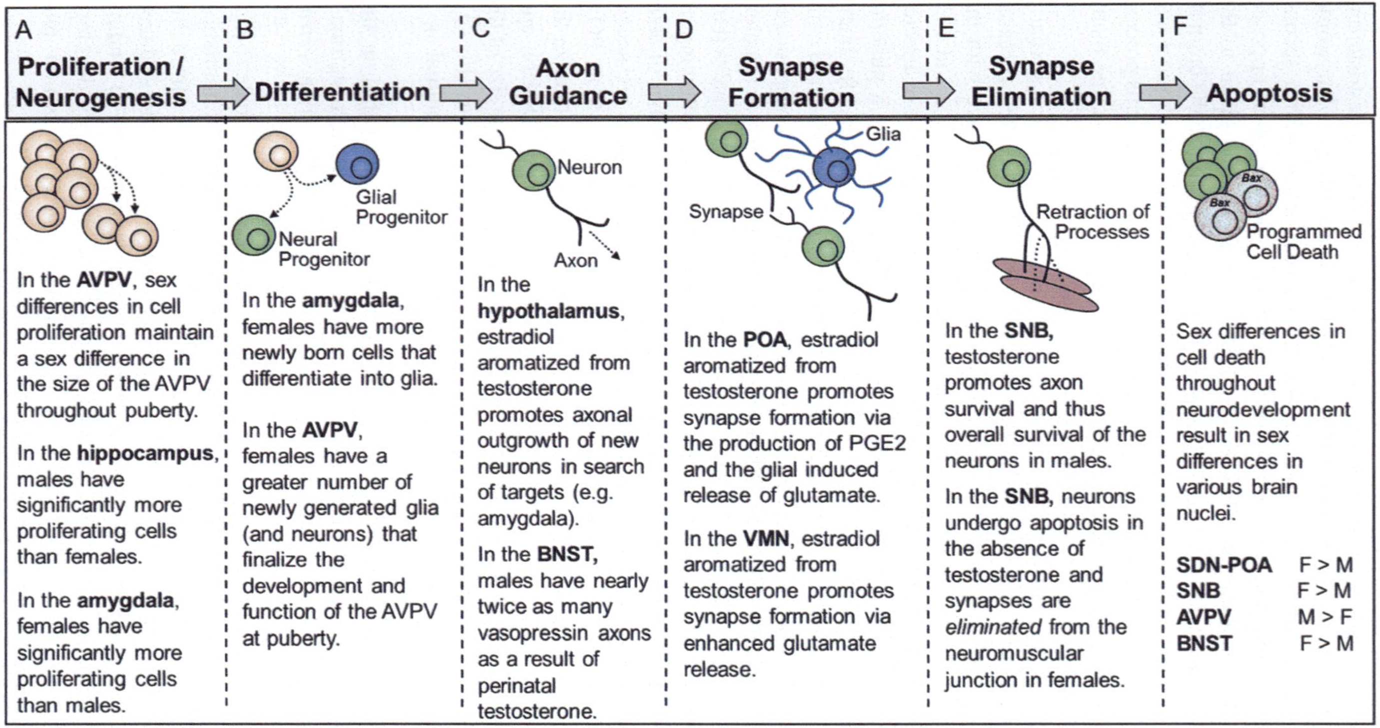

Fig. 3 The basic processes of neural development that occur in a sex-specific manner. (**a**) Proliferation of cells or neurogenesis (the generation of new neurons) occurs in a sexually dimorphic manner in brain regions including the anteroventral periventricular nucleus (AVPV), the hippocampus, and the amygdala. (**b**) Differentiation of neural progenitor cells into either neurons or glia can occur in a sex-dependent manner in brain regions such as the amygdala and the anteroventral periventricular nucleus (AVPV). (**c**) Testosterone or estradiol converted from testosterone promotes the outgrowth of axons seeking new synaptic connections in the amygdala and the AVPV. (**d**) In the preoptic area (POA) and the ventromedial nucleus of the hypothalamus (VMN), estradiol converted from

testosterone promotes dendritic spine synapses during postnatal brain development. (**e**) In the spinal nucleus of the bulbocavernosus (SNB) muscle, testosterone promotes axon survival of neurons projecting to the muscle. In the absence of testosterone, in females, the synapses are naturally eliminated during postnatal development. (**f**) Naturally occurring cell death, apoptosis occurs in a sexually dimorphic manner to produce sex differences in the size of certain brain nuclei. In brain regions where the size is larger in males than in females (M > F), testosterone promotes the survival of the neurons and prevents apoptosis from occurring. In brain regions where the size is larger in females than in males (F > M), testosterone promotes naturally occurring cell death during neural development

The seminal study by Phoenix and colleagues in 1959, and subsequent studies by Gorski and colleagues, provided the first evidence of gonadal hormone-induced sexual differentiation of the brain and behavior. This work led to an intense investigation into the cellular mechanisms by which hormones can produce sex-specific phenotypes of the brain and behavior during early brain development. One of the best-known examples of sex differences in brain structure producing sex differences in behavior is observed in birds that produce song. Male birds that produce song as a way to attract female mates have significantly larger nuclei compared to females within two main neural pathways important for the learning and the production of song (Jordan et al. 1988; Nottebohm and Arnold 1976; Nottebohm et al. 1982). For example, the "high vocal center" (HVC) and the robust nucleus of the archistriatum (RA) are two sexually dimorphic nuclei within the efferent motor pathway responsible for the production of male bird song. Both regions are three to six times larger in males compared to females (Nottebohm and Arnold 1976; Nottebohm 1991). The neurons of these nuclei were later found to be more numerous, larger in size, and more scattered in males than in females (Arnold and Gorski 1984; Mathews et al. 1988). If females are treated with estradiol early post-hatching and then treated with either testosterone or dihydrotestosterone (DHT) as adults, the HVC and the RA become similar in size to that of the males, and the treated females are then able to produce song (Gurney and Konishi 1980).

Characterization of the sexually dimorphic nuclei controlling song production in birds provided some of the first direct evidence linking hormone-induced changes in brain structure to sexually dimorphic behavior but also provided converging support for the organizational-activational hypothesis of sexual differentiation of the brain. That said, evidence from these studies also suggested that the organizational and activational effects of hormones cannot completely account for the sexual differentiation of these brain nuclei. For example, male birds that are castrated as hatchlings, prior to the secretion of testosterone, sing normally as adults, suggesting the possibility that activation of genes on the avian sex chromosomes may also induce masculinization of the brain in song birds (Arnold 2003; Wade et al. 2001).

2.4 Genetic Sex Differences Impact the Developing Brain

While the evidence above elucidates androgens as an essential mediator of sexual differentiation of the brain, it is also important to keep in mind that each cell's *intrinsic* sex may also have the ability to contribute to the organizational-activational hypothesis, as the presence of either XX or XY chromosomes is maintained within every cell. In other words, organization of sexually dimorphic brain regions may not rely on gonadal sex hormones alone but may also involve sex chromosome complement, such as the direct expression of the *Sry* gene within male brain cells or sex differences in gene expression that result from the process of X inactivation within female brain cells (Arnold and Chen 2009; Carrel and Willard 2005; Dewing et al. 2006). This phenomenon has been investigated using a mouse model described as the "four core genotypes" (FCG) model, in which gonadal sex and chromosomal sex are purposely misaligned to examine the impact of each factor either separately or interdependently (Carruth et al. 2002; De Vries et al. 2002). This phenomenon has also been investigated in a gynandromorphic finch, in which sex chromosome complement was bilaterally split in a bird, but testosterone derived from the one testis (on the "genetic male" side of the body) could circulate throughout the bloodstream and influence the entire organism, including its brain (Agate et al. 2003). Thus, when sex differences were present across both hemispheres of the brain in the gynandromorphic finch, despite the similar hormonal exposure, they could be attributed to sex differences in gene expression that resulted from the avian sex chromosomes. Notably, this finch could still produce male song, indicative of the important role testosterone has in differentiating man regions within male brain and associated behavior. Taken together, these models have been effective in shedding light on how sex chromosome complement affects sex differences in gene expression and phenotypic variation in the brain beyond the scope of sex hormones alone (Carruth et al. 2002; Xu and Disteche 2006).

It is also important to consider the evolutionary basis underlying the mechanisms for sexual differentiation of the brain in general. Despite the fact that males and females each have a unique set of sex chromosomes, the sex chromosomes have an important, but relatively *limited* role in the process of sexual differentiation of the body and brain in many species, including humans (McCarthy and Arnold 2011). Instead, the sex hormones are the primary mechanism by which sexual differentiation of the body and brain occurs. Sexual differentiation requires the coordinated differentiation of a wide variety of cells and tissues throughout the body. These cells and tissues are also mutually dependent upon each other for proper function and coordinated behavior later in life. As a result, sex hormones may have "taken on" the important role of sexual differentiation because of their ability to target specific cells and tissues throughout the entire body and because of their important role in sex-specific physiology and behavior in general. Despite this, there may be many other processes of neural development that are directly affected by the genes of the sex chromosomes or by the sexually dimorphic expression of genes independent of sex hormones that remain to be discovered. Thus, the complexity of sex-specific neurodevelopmental and long-term behavioral outcomes resulting from either chromosomal or hormonal sex, including those that do not apply directly to reproductive or copulatory behavior, continues to be explored.

The organizational-activational hypothesis is supported by sex differences in the development of a number of brain regions including the SDN-POA (described above), the spinal nucleus of the bulbocavernosus (SNB), the bed nucleus of the stria terminalis (BNST), and the anteroventral periventricular nucleus of the hypothalamus (AVPV) (Fig. 3). These brain regions and their associated sex-specific behaviors will be described throughout the remaining chapter. Importantly, we will focus our discussion on the mechanisms by which neurodevelopment is influenced by sex hormones to induce these well-known sex differences in the brain, and in the final conclusions of this chapter, we will discuss important future questions and ideas that arise from our understanding of sex differences in neurodevelopmental processes.

3 Sex Differences in the Size of Brain Nuclei

3.1 The Spinal Nucleus of the Bulbocavernosus

The spinal nucleus of the bulbocavernosus (SNB) was the first region identified in the rodent central nervous system (CNS) to exhibit an androgen-dependent sex difference in cell number. Specifically, males have significantly more neurons within the SNB than females, and treatment of neonatal females with testosterone increases the number of neurons in the female SNB to that seen in males (Sengelaub and Forger 2008). Breedlove and Arnold (1980) identified this collection of motor neurons, located within the fifth and sixth lumbar segments of the rat spinal cord, which they named the SNB after the perineal striated muscles these neurons innervate. They observed that these particular motor neurons were more sensitive to treatment with testosterone than treatment with estradiol, the aromatized product of testosterone. Interestingly, when they examined genetically male rats with a testicular feminization mutation that severely reduced the amount of functional androgen receptors, Breedlove and Arnold (1981) observed a reduction in, or demasculinization of, the total cell number within the SNB of these mutated animals. Based on these observations, they concluded that testosterone signaling at the *androgen* receptor during development plays an important role in determining the sex difference in the number of neurons within this nucleus of the CNS.

Nordeen et al. (1985) subsequently identified an epoch within the perinatal period – between embryonic day 22 (E22) and postnatal day 10 (P10) – during which the SNB cell numbers are decreased via naturally occurring apoptotic processes, and this neurodevelopmental process is *more* robust in female compared to male rats, resulting in the sex difference in cell number described above (Fig. 3). Female rats that were treated with testosterone during the critical period of spinal cord development lost significantly *fewer* neurons in the SNB, thereby exhibiting a masculinized SNB cell number (Nordeen et al. 1985). In concordance with these data, Breedlove and Arnold (1983) demonstrated greater cell death of SNB motor neuron number when treating male rats with flutamide, an androgen receptor antagonist, which blocks the androgen receptor. Together, these data indicate that androgens are effective in preventing programmed cell death within the SNB during perinatal development. Ultimately, this proposed mechanism of sexually dimorphic cell death is involved in the establishment of the sexually dimorphic size of the SNB, which is maintained into adulthood and allows for successful male reproductive behavior, in particular the control of the perineal muscles.

In addition to the impact of neonatal testosterone on naturally occurring apoptosis in the SNB, Lubischer and Arnold (1995) found that treatment of castrated juvenile male rats with testosterone delayed naturally occurring synapse elimination in the spinal cord by increasing the percentage of perineal muscle fibers innervated by multiple axons, thus increasing the size of the associated motor units. The observed changes in synapse elimination persisted for at least 12 months after the completion of testosterone treatment, demonstrating that the effects of testosterone are persistent and have an important role in the establishment of sex differences in the spinal cord that are maintained long-term (Lubischer and Arnold 1995).

These data sparked an investigation into the ontogeny of androgen receptor expression in the muscles innervated by SNB motor neurons; and, in 1997, Jordan and colleagues definitively showed that the rat perineal muscles that are innervated by SNB motor neurons *also* express androgen receptors. More importantly, the emergence of androgen receptors on the perineal muscles actually *precedes* their emergence on the motor neurons that synapse with the SNB (Jordan et al. 1997). In fact, in the mouse SNB circuit, androgen receptors were not expressed on SNB motor neurons until P4, but they were present on the SNB target muscles as early as E15 (Smith et al. 2012). Ultimately, these discoveries have challenged the notion that neonatal androgen exposure is necessary for the sex difference in cell number in the SNB via the direct action of androgens on the motor neurons themselves, particularly given that androgen receptors were not detected on SNB motor neurons until P4, a time point outside of the natural surge in testosterone that occurs in developing males. Instead, androgens' action on the perineal muscles influences the way by which these muscles are subsequently innervated by SNB motor neurons at a later date, and thereby, it is androgens' action on the muscles itself that has a critical role in establishing the male-biased sex difference in SNB cell number that is maintained into adulthood. This is an important point to consider when understanding the various mechanisms by which sex differences in the nervous system can be established. Sex hormones, in this case testosterone, can influence one set of target cells (muscle tissue) to, in turn, influence the innervation, number, and function of motor neurons in the spinal cord, likely via the androgen-induced structural support and growth factors that are present at the neuromuscular junction (Fig. 3). These findings highlight the idea that sexual differentiation of the nervous system can be initiated early, at the time of perinatal testosterone exposure, but that the effects can also be persistent, as testosterone is able to maintain cell survival at a later point in development, as naturally occurring apoptosis continues postnatally.

3.2 The Anteroventral Periventricular Nucleus of the Hypothalamus

Unlike many of the other sexually dimorphic brain regions discussed in this chapter, the anteroventral periventricular nucleus of the hypothalamus (AVPV) is significantly larger in females than in males. This particular nucleus of the hypothalamus plays an important role in gonadotropin release via the robust expression of kisspeptin, found in a subset of AVPV neurons (Kauffman et al. 2007). Specifically, the *Kiss1* neurons in the AVPV control a preovulatory surge in luteinizing hormone (LH) that drives ovulation in females (Hu et al. 2015). Prior to its designation as the AVPV, this collection of cells was originally identified by Bleier et al. (1982) as the medial preoptic nucleus (MPN). In a comparative anatomy study, they discovered that the MPN (or AVPV) of the guinea pig, rat, and hamster contained more cells in adult females compared to adult males (Bleier et al. 1982). The work of Bleier and colleagues established that the sexually dimorphic nature of this specific brain region is, indeed, conserved across a number of species, corresponding with the later discovery of the AVPV's important role in female sex behaviors and the control of ovulation.

In previous sections, we discussed how neonatal androgen receptor signaling mediated the establishment of a sex difference in cell number within the SNB and the SDN-POA via its impact on cell survival – by *preventing* naturally occurring apoptosis. Similar to these two brain regions, sex-specific cell death is also the cause of differences in cell number within the AVPV. In developing rats, the AVPV undergoes neurogenesis between E13 and E18, and treatment of females with testosterone or estradiol does not affect the number of newly born neurons in the AVPV (Sumida et al. 1993). In contrast, treatment with either testosterone or estradiol *increases* subsequent apoptosis in the AVPV, resulting in a sex difference in the size of the AVPV by the time it was examined at E21 (Sumida et al. 1993; Fig. 3). Notably, this pro-apoptotic effect of testosterone in the AVPV is contrary to the role of neonatal testosterone signaling in the SNB, which protects *against* apoptosis (Arai et al. 1996; Sumida et al. 1993). Furthermore, the data indicate that the aromatization of testosterone into estradiol and thus the actions of estradiol, in particular, contribute to increased cell death in the developing male AVPV. These data are corroborated by the high level of estrogen receptor alpha (ERα) expression on the cells of the AVPV (Shughrue et al. 1997). Bax, a protein that promotes apoptosis, is found at higher concentrations in the developing male AVPV; while Bcl-2, a protein that promotes cell survival, is found at lower levels in the developing male AVPV (Tsukahara 2009). Thus, estradiol aromatized from testosterone results in the sexually dimorphic expression of these pro- and anti-apoptotic genes (see Forger 2006 for a comprehensive review), pushing the newly born neurons in the male AVPV toward a fate of cell death during this embryonic period of neurodevelopment (Fig. 3).

It is also important to note that the AVPV controls the LH surge in adult females as a result of its ability to integrate hormonal signals with circadian signals, via the function of kisspeptin neurons localized in this brain region (Simerly 2002). The LH surge is the result of a neuroendocrine positive feedback loop between kisspeptin and estradiol, which is only fully functional *after* puberty (Clarkson et al. 2009). This positive feedback loop of hormones acting on the AVPV results in the addition of even more cells to the female AVPV. Thus, similar to the mechanisms and timeline of sexual differentiation of the SNB, the sex difference in cell number in the AVPV is initially established during the prenatal period via apoptosis; however, there remain additional changes to cell number during puberty that are necessary for the establishment and maintenance of sex differences in the adult AVPV necessary to control female ovulation. Interestingly, Mohr et al. (2016) found that this increase in cell number in the AVPV at puberty not only includes the addition of new neurons but also includes the addition of glial cells, indicating that both neurons and glia appear to be contributing to the establishment and maintenance of sex differences in the size and function of the AVPV (Mohr et al. 2016; Fig. 3). This important idea that various cell types in the developing brain contribute to sex-dependent development and function of the brain is seen in other examples that will be discussed in Sect. 4.

3.3 The Bed Nucleus of the Stria Terminalis

The bed nucleus of the stria terminalis (BNST) is located at one end of the stria terminalis, which is the bundle of axon fibers that connects the BNST to the amygdala. Because of its structure and its role in social and emotional behaviors, the BNST has been described as the "extended amygdala" (Lebow and Chen 2016). Though the BNST forms a continuum with amygdaloid structures, it is now clear that there are at least two distinct regions, including the medial division containing the medial amygdaloid nucleus and the medial BNST, as well as a central division containing projections between the central amygdaloid nucleus and the lateral BNST (reviewed in Crestani et al. 2013). The BNST contains approximately 20 distinct nuclei that are characterized based on the various cell types and their distinct projections to other brain regions. Characterization of the various cell types and the afferent and efferent projections of the BNST (e.g., limbic structures, hypothalamic structures, and brain stem nuclei) suggests that it serves as a critical relay for a variety of neuronal circuits and other systems (e.g., neuroendocrine) to coordinate activity related to the physiology and behavior of the organism. The BNST neurons contain a variety of neurotransmitter and neuropeptides and their respective receptors including vasopressin, glutamate, GABA, noradrenaline, acetylcholine, nitric oxide, cannabinoids, and corticotropin-releasing factor (Crestani et al. 2013).

There is a significant sex difference in the size of the principle nucleus of the BNST (pBNST) in addition to sexually dimorphic innervation of vasopressin neurons into the BNST. Vasopressin has been studied extensively in the BNST for its role in various physiological and behavioral functions (see Sect. 4.3). For this reason, much of our understanding of the functional role of the BNST in males and females throughout brain development has its foundation in the vasopressin system. Males have significantly more vasopressin immunoreactivity, mRNA expression, biosynthetic capacity, and a greater fiber density within the BNST compared to females (Bales et al. 2007; Miller et al. 1989a, b). One of the main vasopressin receptors, V1a, is more abundant in the BNST of males compared to females in the medial posterior BNST, but not in the lateral dorsal or lateral posterior BNST, suggesting that there is considerable heterogeneity within the subnuclei of this region for V1a receptor distribution (Dumais and Veenema 2016). Vasopressin neurons in the medial BNST were identified as the source of the sexually dimorphic vasopressin fiber innervation of the brain in the mid- to late 1980s with the discovery that males have nearly twice as many vasopressin-immunoreactive fibers than females (Fig. 3). Additionally, Miller et al. (1989a, b) determined that males have significantly more vasopressin-expressing cells and more vasopressin grains per cell than females. These data indicate that the sex differences in vasopressin BNST fiber densities are due to males having more vasopressin-expressing neurons as well as increased biosynthetic capacity on a per cell basis compared to females.

Progesterone is also important for establishing the sexually dimorphic expression of vasopressin cells in the BNST. Testosterone, converted to estradiol, can induce the synthesis of progesterone in the brain during postnatal development (Shughrue et al. 1992), but more so, it is thought that progesterone from the pregnant mother may induce sex differences in the developing fetal brain if sex differences are present in the expression of progesterone receptors in the brain (Wagner et al. 1998). In fact, there is a striking sex difference in progestin receptor (PR) expression on vasopressin cells in the BNST (Quadros et al. 2002), and subcutaneous injection of progesterone reduces the number of vasopressin mRNA-expressing cells in the BNST. To determine the role of PRs in the development of sex differences in the vasopressin system, Rood et al. (2008) compared PR^{lacZ} reporter mice to wild-type (WT) mice. PR^{lacZ} reporter mice have a functional disruption in the PR gene that renders them unable to respond to circulating progesterone. Surprisingly, PR^{lacZ} reporter mice showed a similar sex difference in the number of vasopressin-expressing cells in the BNST as WT mice, suggesting that PRs do not play a major role in sexual differentiation of vasopressin cells in the BNST; however, additional data from this study indicated that PRs have a role in establishing the density of projections to brain regions that are innervated by BNST originating vasopressin neurons (Rood et al. 2008). These findings highlight the more "unconventional" mechanisms by which sex differences in the developing brain can be established, in this case via the function of the progesterone receptor. While it is clear that the BNST is sexually dimorphic, with the vasopressin system playing an integral role in this sex difference, it remains to be determined how disruptions in the development of the BNST can affect brain development and sex-specific behavior later in life.

4 Sexual Differentiation of the Brain Requires Communication Across and Involvement of Multiple Cell Types in the Brain

As described in the previous sections, the sex of an organism is determined at the moment of conception, and from that point of determination, every cell in the organism has a sex. The sex chromosomes and, more importantly, sex hormones differentiate the developing brain, body, and behavior of the organism early in development. These early-life organizational and activational effects of sex hormones "program" the function of various cells throughout the nervous system. As a result, it is important that we also understand the *cellular mechanisms* by which basic sex differences in the brain are established, as this may provide additional insight into the cellular mechanisms resulting in disease processes that often occur in a sex-specific manner throughout neurodevelopment.

The brain is arguably one of the most active endocrine organs as it produces and responds to a diverse array of chemical messengers. Chemical communication, such as the kind that occurs in the endocrine system, via hormones and their cognate receptors, is integral to all levels of biological organization, from the mediation of intracellular processes to communication between organs and even communication between individuals. Traditionally, chemical messengers have been classified into separate systems, mainly the nervous system (i.e., neurotransmitters), endocrine system (i.e., hormones), and the immune system (i.e., cytokines). However, it has become increasingly clear that there is substantial overlap and intercellular communication between these systems. In fact, many neurons express receptors for hormones and cytokines; at the same time, many immune cells express receptors for neurotransmitters and hormones. With this knowledge in mind, it is easy to imagine how the sex hormones responsible for sexual differentiation are also capable of affecting neural development via their actions on multiple cell types and various chemical messenger systems to differentiate the brain and determine neural function.

4.1 An Immune Molecule and a Neurotransmitter Coordinate Masculinization of the Preoptic Area

The medial preoptic area (POA) in the rodent brain is a sexually dimorphic cluster of cells, located in the preoptic nucleus (Gorski 1978). Discussed earlier, the sexually dimorphic nucleus (SDN), a small cluster of cells of the POA, is five to seven times larger in males compared to females and is necessary for the appropriate expression of appetitive and consummatory male sex behavior (De Jonge et al. 1989; Houtsmuller et al. 1994; Jarzab et al. 1990; Jeong et al. 2008; Todd et al. 2005). In addition to this macroscopic, structural difference in the size of the SDN-POA, the neurons throughout the POA of males also have two to three times more dendritic spines than females. Within the first 2 weeks of postnatal life, male rats have a higher density of spine synapses along the dendrite on neurons in the POA compared to female rats. Moreover, astrocytes in the male POA have longer processes with a greater amount of branching than females (Amateau and McCarthy 2002a, b). Masculinization of the POA is dependent on testosterone, which is converted to estradiol in the rodent brain during the critical period of sexual differentiation. While it is clear that naturally occurring apoptosis plays a critical role in establishing the SDN-POA, another question remained – *how* does estradiol induce these structural and synaptic changes to the neurons of the male POA? The answer, which was surprising at the time, is that these changes are induced through increases in the synthesis of prostaglandin E2 (PGE2), a pro-inflammatory signaling molecule.

PGE2 is derived from arachidonic acid following cyclinization by the cyclooxygenase enzymes COX-1 and COX-2, and PGE2 is best known for its role in fever production and inflammation. Amateau and McCarthy (2004) determined that activation of estrogen receptors in the POA upregulates the production of COX-1 and

COX-2 leading to a nearly sevenfold increase in PGE2. Notably, PGE2 has been shown to be synthesized by other cells, such as microglia and astrocytes, within the POA via activation of glial adenosine receptors by ATP (Fiebich et al. 2014); however, Amateau and McCarthy (2004) showed that the increase in PGE2 synthesis in the POA was specific to neurons. Neuronal PGE2 then induces the growth of neuronal dendritic spines via the release of glutamate and activation of AMPA receptors (Amateau and McCarthy 2004). Astrocytes are also responsive to estradiol; they release glutamate following stimulation with PGE2, and their morphology is sexually dimorphic in the POA, as described above (Amateau and McCarthy 2002a; Bezzi et al. 1998). Thus, astrocytes play a critical role in the establishment of sexually dimorphic synaptic connectivity within the POA (Fig. 3). Treating females with estradiol or PGE2 produces male-like patterns of dendritic spines, and blocking COX-1 or COX-2 in males leads to female-like patterns of dendritic spines in the POA that are maintained into adulthood (Amateau and McCarthy 2002a, 2004; Wright et al. 2008). Furthermore, PGE2 receptors 2 and 4 (EP2 and EP4) have been shown to be the primary contributors to the masculinizing effects of PGE2 in the POA (Wright et al. 2008). Specifically, activation of EP2 or EP4 leads to increases in protein kinase A (PKA), and if PKA signaling is disrupted during neonatal development, the masculinization of the POA is blocked (Wright and McCarthy, unpublished observation). Taken together, these data highlight a unique multicellular signaling process through which sex steroid hormones can sexually differentiate the developing brain.

4.2 *Sexual Differentiation Across the Synapse in the Ventromedial Nucleus of the Hypothalamus*

A central region of the hypothalamus, the ventromedial nucleus (VMN), is a brain region necessary for female sex behavior (Pfaff and Sakuma 1979). The VMN is characterized by its oval shape and its sparse, thin dendrites (Millhouse 1973). Similar to the POA, the male VMN neurons contain more than three times as many dendritic spines and shaft synapses as those measured on female VMN neurons (Matsumoto and Arai 1983). This sex difference in synaptic connectivity is detected as early as postnatal day 2 and is maintained in the rodent brain until postnatal day 100 (Matsumoto and Arai 1986; Pozzo Miller and Aoki 1991), consistent with the long-term organizational effects of estradiol in the developing brain. This masculinized pattern of dendritic spine morphology in the VMN can be induced in females by treatment with either testosterone or its metabolite, estradiol, within the first few days of life. In adulthood, the female estrous cycle or exogenous estradiol treatment causes changes in the dendritic spine density and patterning of the VMN, which correlate with changes in female sex receptivity (Frankfurt et al. 1990), indicating that changes in the synaptic connectivity of the VMN are a critical mechanism of inducing or constraining female sex behavior.

Glutamate is the primary neurotransmitter in the VMN (Ziegler et al. 2002), and estradiol mediates the changes in neuronal morphology in the developing VMN by directly enhancing glutamate release from these developing neurons (Schwarz et al. 2008). During the critical period of sexual differentiation, estradiol exposure induces the activation of PI3 kinase within only 3 h, which leads to a cascade of events that can induce the defeminization of the synaptic connections in the VMN and the defeminization of sexual behavior in adulthood. Specifically, activation of PI3 kinase in the developing VMN increases the release of glutamate from presynaptic terminals within this region. This, in turn, activates glutamatergic receptors, activating MAP kinase and masculinizing the dendritic spine number measured in the VMN (Schwarz et al. 2008; Fig. 3). Blocking any one of these steps, including PI3 kinase activation, MAP kinase activation, or AMPA/NMDA glutamate receptor activation, completely prevents the increase in dendritic spine proteins seen following estradiol treatment of the female rat pup and ultimately prevents the defeminization of the brain and behavior later in life. Specifically, blocking NMDA receptors during the critical period of sexual differentiation blocks estradiol-induced defeminization of behavior, which results in the production of an adult rat that is capable of expressing both male *and female* sex behavior (Schwarz and McCarthy 2008). These findings describing sexual differentiation of the VMN indicate that the effect of the sex hormone estradiol *is not limited* to the cells that express the estrogen receptor but rather that estradiol-induced masculinization in the VMN initiates the coordination of synaptic connections and entire circuits that control male sex behavior. As stated above, the fact that estradiol activates PI3 kinase in the developing VMN within 3 h after estradiol exposure indicates that the effects of estradiol on modulating neurotransmitter function to initiate the process of sexual differentiation within the VMN occur in a relatively rapid time course. The relatively rapid time course of estradiol's effects suggests that the developing circuitry is extremely sensitive to the organizational effects of estradiol exposure during the critical period of sexual differentiation, and this is likely the case for other brain regions that undergo sexual differentiation during this brief and critical developmental window.

4.3 Development of Neuropeptides Involved in Sex-Specific Behaviors

Vasopressin and oxytocin are neuropeptides involved in the regulation of many sexually dimorphic social behaviors including social recognition, pair bonding, and social cognition in various species (Ross and Young 2009; Veenema and Neumann 2008). These sex differences in behavior are the result of sexually dimorphic innervation of these two neuropeptides within specific brain regions, which are established during brain development. The vasopressin system was first shown to be sexually dimorphic in 1981 by de Vries and colleagues, who discovered that males have more vasopressin-immunoreactive fibers in the lateral septum (LS) and lateral habenula than females and approximately 90% of vasopressin cells in the BNST (see Sect. 3.3) express androgen receptors, the absolute number of which is significantly greater in males compared to females (Bales et al. 2007; de Vries et al. 1981). Previous work has shown that gonadectomy of neonatal male rats significantly reduces vasopressin fiber density later in life, indicating that androgens are necessary for the organization of the sexually dimorphic vasopressin system in the BNST (De Vries et al. 1983). Interestingly, administration of testosterone to gonadectomized males *or intact females* during the first, second, or even third postnatal week can induce masculinization of the vasopressin system. These findings suggest that not only are androgens necessary for the organization of this particular neuropeptide system in males but that vasopressin neurons in the female brain are also sensitive to the effects of androgens. Gonadectomy of *adult* males and ovariectomy of *adult* females cause vasopressin immunostaining in the BNST to disappear almost completely within several weeks post-surgery. This finding suggests that vasopressin expression in the BNST is responsive to circulating steroid hormones even into adulthood and indicates that males and females may respond differently to testosterone (or its metabolites) in order to maintain the sexually dimorphic features of the vasopressin system in adulthood. In fact, later studies investigating this possibility found that when *adult* male and female rats were gonadectomized or ovariectomized, respectively, and then subsequently administered equal doses of testosterone (i.e., via a subcutaneous implant), the sexual dimorphism in the number of vasopressin cells was maintained in the BNST, supporting the hypothesis that males and females show different sensitivities to either testosterone or the metabolites of testosterone (De Vries et al. 1994). It is likely that this sex difference in sensitivity to androgens in the BNST is established early in life by perinatal testosterone exposure. Indeed, the testosterone metabolites estradiol and 5α-dihydrotestosterone (DHT) can affect vasopressin cell number in a sex-dependent manner, such that treatment with estradiol increases the number of vasopressin cells in the BNST in males and females compared to controls and DHT-treated animals; however, this effect was greater in males than females (De Vries et al. 1994). Additionally, males treated with estradiol and DHT combined had significantly more vasopressin cells in the BNST than when treated with estradiol alone, while this same effect was not true for females. These data reiterate the fact that, in addition to the circulating hormones of the neuroendocrine system, every cell has a sex, which means that male and female cells have the potential to respond differently to the same stimulus, thus impacting the brain and behavior in a sexually dimorphic manner.

The finding that the vasopressin system is more prominent in males compared to females suggests that males, more than females, may rely on this neuropeptide for the regulation of social behavior. In support of this notion, injections of vasopressin enhanced social recognition in both males and females, but injection of a vasopressin receptor antagonist prevented social recognition in males only (Bluthe et al. 1994; Dantzer and Bluthe 1993). Additionally, vasopressin mRNA expression in the paraventricular nucleus of the hypothalamus (PVN) was determined to be positively correlated with the amount of social play behavior in juvenile males, but not females (Paul et al. 2014). Indeed, vasopressin has been shown to regulate some behaviors differently in males and females. For example, vasopressin administered to males stimulates aggressive behavior, while this same administration of vasopressin inhibited aggressive behavior in females (Terranova et al. 2016). Furthermore, dominance in males (i.e., more aggressive males) is strongly associated with activation of vasopressin cell bodies in the supraoptic nucleus of the hypothalamus compared to subordinate or control males; however, in females, activation of vasopressin cell bodies was observed in dominant and subordinate females to the same degree, which were both higher than that observed in control females (Terranova et al. 2016). These findings indicate that the vasopressin system exhibits a fundamental sex difference in the regulation of aggressive behavior, whereby the male vasopressin system shows a strong correlation to dominance status, but the female vasopressin system is associated with social interaction more generally. While the exact cellular mechanisms underlying the sexually dimorphic innervation of the vasopressin system established by early androgen exposure are not known, it is clear that this neuropeptide is integral to the expression of a myriad of sexually dimorphic behaviors expressed later in life.

In contrast to vasopressin, oxytocin-synthesizing cells are greater in number in females compared to males across a number of brain regions including the PVN, lateral hypothalamus, lateral septum (LS), POA, and the BNST in several species. Furthermore, the number of oxytocin-immunoreactive fibers in the LS and BNST is significantly greater in female compared to male mice (Haussler et al. 1990) and in the lateral hypothalamus of mandarin voles (Qiao et al. 2014); however, others have found no sex differences in oxytocin mRNA expression (Dumais et al. 2013) nor in the number of oxytocin neurons in humans (Wierda et al. 1991). It has been shown, however, that administration of oxytocin or the oxytocin receptor antagonist can increase oxytocin immunoreactivity in the PVN of female prairie voles, but not male prairie voles (Yamamoto et al. 2004). Furthermore, when given neonatally, oxytocin increased ERα immunoreactivity in the ventromedial hypothalamus, and oxytocin receptor antagonist decreased ERα immunoreactivity in the POA of females, but not males (Yamamoto et al. 2006). Unfortunately, there has been surprisingly little investigation into the development of the oxytocin system in males and females, so there is still much that remains to be elucidated regarding the neuroendocrine regulation of oxytocin in the brain. It is also important to note that while less is known about the oxytocin system in the brain, there is an abundance of evidence demonstrating the sex-specific effects of oxytocin on behavior (for a review see Caldwell 2017). For example, neonatal oxytocin exposure facilitates mate-guarding behavior (a component of pair bonding that increases aggression and reduces social behavior toward other voles) in female, but not male prairie voles (Bales and Carter 2003), and ICV administration of oxytocin was shown to facilitate partner preference formation in females, but not males (Insel and Hulihan 1995; Williams et al. 1994; Winslow et al. 1993).

Despite the evidence indicating that oxytocin may play a more prominent role in females, Yao et al. (2017) found that oxytocin can impact certain social behaviors in adult male, but *not* female, mice. Adult male mice that lack oxytocin signaling did not prefer a female over a male social partner in a 3-chamber-social-investigation paradigm, as is typically observed in wild-type male mice. In contrast, no changes in partner preference were observed in adult female mice with altered oxytocin signaling. Also in contrast with the data described above, vasopressin did *not* impact partner preference in either adult male or female mice within this specific paradigm. Further examination indicated that social recognition in adult male mice appears to be exclusively dependent upon oxytocin signaling in the medial amygdala, as endogenous vasopressin was unsuccessful in deterring the negative effects of altered oxytocin signaling on male partner preference. Notably, the authors indicated that this male-specific effect of oxytocin on social partner preference is likely the result of oxytocin rapidly modulating the sensory responses of aromatase neurons to social cues; therefore, the basis for these contrasting results likely stems from the fact that while oxytocin-synthesizing cells are greater in number in females compared to males in the PVN, lateral hypothalamus, LS, POA, and the BNST, there are a greater number of aromatase-expressing neurons in the medial amygdala of *male* mice, compared to female mice (Wu et al. 2009; Yao et al. 2017). Importantly, these results highlight the idea that sex differences in the development of neural circuits may occur via neuropeptides (hormones or neurotransmitter systems) that are not directly influenced by the traditional testosterone-mediated mechanisms of sexual differentiation.

4.4 Microglia Induce Sex-Specific Neuronal Development

Beyond the "traditional" neural cells and sexually dimorphic circuits that have been investigated for many years, the past decade or so has also uncovered sex differences in the immune cells of the brain. For example, microglia, the resident immune cells of the brain, are critical for many normal neurodevelopmental processes including supporting neurogenesis, pruning spurious synaptic connections, and phagocytosing apoptotic cells (Boulanger 2009; Deverman and Patterson 2009). During early development, males have more microglia than females in certain brain regions, including the hippocampus, the parietal cortex, and the POA (Lenz and McCarthy 2014; Schwarz et al. 2012). Interestingly, this sexually dimorphic profile of microglia shifts dramatically throughout development, with females having more microglia with thick processes or stout phenotype than males as adults (Schwarz et al. 2012). Given the striking sex difference in microglia number across multiple brain regions, and given their importance in neural development, investigators have sought to determine the developmental profile of microglia (Hanamsagar et al. 2017) and whether microglia have an active role in the sexual differentiation of the developing brain. A recent study found that during hormone-induced sexual differentiation of the brain, microglia in the female hippocampus have significantly more phagocytic cups than those in the male hippocampus (Nelson et al. 2017). When females are treated with estradiol, the number of phagocytic cups is reduced to male numbers (Nelson et al. 2017). In the POA, males have twice as many amoeboid microglia as females, and estradiol treatment of female rat pups masculinizes microglia number and morphology in the POA (Lenz et al. 2013). Furthermore, inhibiting microglia prevents the normal estradiol-induced masculinization of dendritic spines in the POA and adult copulatory behavior in males, indicating that microglia are also an important cellular mediator of the process of brain sexual differentiation (Lenz et al. 2013).

As the immune cells of the brain, microglia also respond to immune activation or injury with rapid changes in morphology and the release of cytokines, processes critical to maintaining homeostasis in the brain. Emerging literature in humans suggests that early-life immune dysregulation leads to cognitive and behavioral disorders that persist throughout the lifespan (Bilbo and Schwarz 2009; Frick et al. 2013; Maezawa et al. 2011; Sheridan et al. 2014) and many of these disorders show a distinct male bias (e.g., autism spectrum disorders [ASD], ADHD, learning disabilities, etc.; Schwarz and Bilbo 2012). Animal models investigating the effects of early-life immune activation have found several sex-specific deficits in behavior. For example, prenatal exposure to lipopolysaccharide (LPS; which activates microglia via TLR-4 receptors) produces deficits in social play behavior in juvenile males, but not females (Taylor et al. 2012). Prenatal exposure to stress, which activates the neuroimmune system, induces behavioral impairments including hyperactivity, behavioral despair, and anhedonia in males, but not females (Bronson and Bale 2014; Mueller and Bale 2008). Collectively, the findings presented here highlight the intricate communication that must occur between many cell types in the endocrine system, nervous system, and immune system to allow for the development of a sexually dimorphic brain and sex-specific expression of behavior.

4.5 Sex Differences in Cell Genesis and Fate During the Development of Cognitive Circuits

Many of the brain regions described thus far are essential for controlling a variety of sexually dimorphic sexual and social behaviors. There are many other brain regions whose functions are very important, and yet no *overt* sex differences in these associated behaviors have been found. For example, the hippocampus and the amygdala are both important for the formation of memory and cognition, yet there are no *overt* sex differences in the basic function of these circuits. That is, both males and females are able to perform complex cognitive tasks and remember things to a similar degree. That said, there are sex differences in the *development* of the hippocampus and amygdala that have been identified and that are worth noting in this chapter, mainly because of their potential role in the etiology or the prevalence of neurodevelopmental disorders that are associated with deficits or alterations in learning, cognition, or emotion and that also display strong sex biases.

During postnatal brain development, the male hippocampus is only slightly larger than the female hippocampus, even after controlling for sex differences in body or head size (McCarthy and Konkle 2005), and this is thought to be the result of more neurons and glia in the male compared to the female hippocampus (Conejo et al. 2003; Hilton et al. 2003; Nunez et al. 2003). Consistent with this idea, there is a significant sex difference in the number of newly born cells in the developing rat hippocampus, with the neonatal male rostral CA1 and DG having more proliferating cells than the numbers observed in females. Both androgen and estrogen treatment increased the number of proliferating cells in females to the numbers seen in males underscoring, once again, the sexually dimorphic role of testosterone in neurodevelopment (Zhang et al. 2008). These data highlight the fact that, even though the size of the hippocampus is only slightly larger in males than females, there is a striking and significant sex difference in the proliferation of newly born cells during a time when cell genesis, in particular neurogenesis, in the developing brain and hippocampus is most robust. Notably, this sex difference in cell proliferation is an effect that is driven by prenatal androgen exposure (Fig. 3), and more importantly, it is an effect that may have long-lasting consequences for the function of this structure throughout development.

An adjacent brain region in the temporal lobe is the amygdala, which has an important role in the processing of emotional stimuli and learning. The amygdala also influences a number of sexually dimorphic behaviors including juvenile rough and tumble play (which is greater in males and influenced by prenatal testosterone exposure), and it is also involved in mating, parenting, aggression, and territoriality, all of which are sexually dimorphic. The overall size of the medial amygdala is larger in males than in females (similar to the hippocampus; Cooke 2006); however, in contrast to the hippocampus, the female amygdala has significantly more cell proliferation than males during early postnatal brain development. Interestingly, these cells differentiate into astrocytes as opposed to neurons (Krebs-Kraft et al. 2010; Fig. 3). This sex difference is the result of differential endogenous cannabinoid signaling in males and females that influences the likelihood of ongoing cell proliferation and ultimately the fate of these cells as astrocytes.

To date, the question remains, what is the role of these sex differences in cell number in both the hippocampus and the amygdala? It is likely that they have an important role in generating sex differences in social and perhaps also cognitive behaviors; however, that remains to be fully determined. Importantly, there are times when a sex difference occurs in a brain region in order to eliminate sex differences in behavior and in order to compensate for a sex difference in another coordinating brain region (De Vries 2004). This may be particularly important for neural functions or behaviors that are essential, like cognition (De Vries 2004), and may be one explanation for the sex differences seen here, particularly in the hippocampus. Perhaps more importantly, researchers should consider how these sex differences in neural development may confer risk for deficits in social or cognitive function, particularly following a trauma or a challenge that perturbs cell genesis during early brain development, an effect that may persist throughout the lifespan. In this case, one sex may be more vulnerable to the trauma or challenge, resulting in a sex bias in cognitive, social, or emotional disorders.

5 A Whole Body Perspective on Sex Differences in Neural Development

In accordance with the idea that every cell has a sex, there is accumulating evidence that *all* organs are sexually differentiated and that even peripheral organs may affect brain development (for a recent review, see de Vries and Forger 2015). The classic example of peripheral organs affecting the brain is the basic process of sexual differentiation of the brain via hormones released from the gonads, which act on multiple cell types and receptors in the brain during development, as we described above. The idea that sex differences in the body and brain are an important factor affecting the health of an organism has been somewhat slow to be accepted, but ignoring this could have detrimental effects for understanding disparate health outcomes in men and women. In fact, there are well-known sex differences in vulnerability to disease, and in some cases sex is, perhaps, the most predictive factor contributing to the prevalence of a disorder. Moreover, many of these disorders have their origins in early development; thus an understanding of how other organs may interact with ongoing processes that affect sexual differentiation of the brain is key to elucidating mechanisms governing the neuroendocrine regulation of neural development and behavior. Here we will present just a few examples of the sex differences in peripheral organs and how they may affect brain function (Fig. 4).

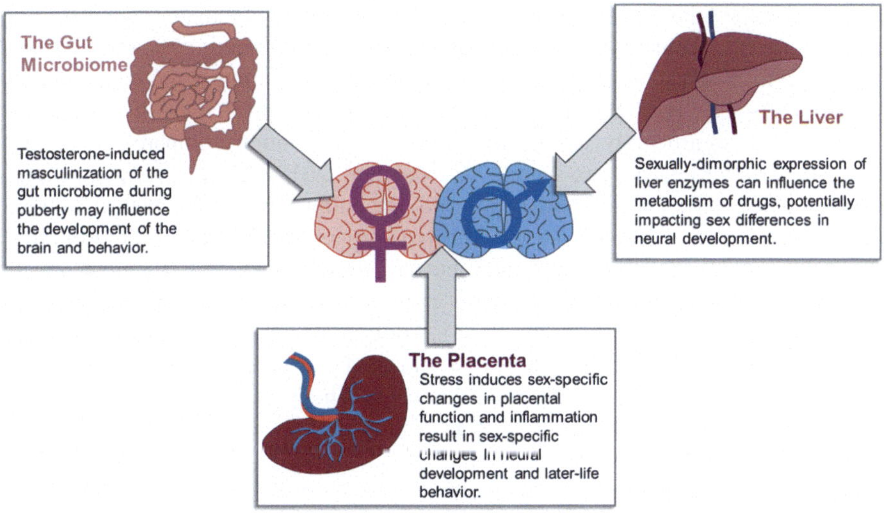

Fig. 4 Examples of how peripheral organs can influence sex-specific development of the brain. Sex differences in various peripheral tissues and organs, including the gut microbiome, the placenta, and the liver, can produce sex-specific influences on the brain that consequently influence neural development in a sex-specific manner

5.1 The Gut Microbiome

Communication between the gut microbiome and the brain is now recognized as an integral factor that impacts brain development and behavior. Following bacterial inoculation of the infant with the maternal vaginal microbiome at birth, communication between these flourishing gut microbes and the brain can function to integrate various signals from the body (e.g., autonomic, neuroendocrine, etc.) that in turn impact the brain. Importantly, the microbiome is highly dynamic, and alterations in its composition occur throughout development, particularly during critical times of sexual differentiation of the brain (Borre et al. 2014). Indeed, sex differences in the gut microbiome emerge during puberty and are actively maintained into adulthood (Yurkovetskiy et al. 2013). Specifically, the female microbiome stays relatively similar to that of the prepubertal male and female microbiome, whereas the male microbiome begins to diverge during puberty and exhibits a distinctive phenotype by adulthood (Yurkovetskiy et al. 2013). Interestingly, if the surge in testosterone during puberty is prevented via gonadectomy in males, this masculinization of the gut microbiota is eliminated in adulthood. Furthermore, transfer of the microbiome of an adult male to a pubertal female completely masculinizes the microbiota composition, metabolomic profile, and produces an increase in testosterone levels that persist into adulthood suggesting that the differentiation of the microbiome during development subsequently impacts neural and physiological function (Markle et al. 2013). If, however, the antiandrogen flutamide is administered with the microbiome transfer, all of these changes are prevented in females demonstrating the important, mechanistic role of testosterone in producing this microbiome-induced masculinization (Fig. 4). In a well-established mouse model of autism spectrum disorder (i.e., BTBR inbred mice), it was found that, while males and females have significantly altered microbiome composition compared to control mice, these alterations are sex-specific (Coretti et al. 2017). These mice also display sex differences in their inflammatory profile in colon tissue, with males showing significantly higher levels of IL-6 and CD11c compared to females and control mice. In another model of ASD, male mice, but not female mice, that are treated with valproic acid on gestational day 11 show reduced social interactions and increased gut inflammation (de Theije et al. 2014a). Additionally, valproic acid significantly alters the composition of the gut microbiota, and the levels of social interaction correlate with metabolites of the microbiota (de Theije et al. 2014b). These data suggest that the gut microbiome is capable of impacting the developing brain and are quite intriguing given the increased prevalence of ASD in males and autism's associated risk with early-life inflammation.

5.2 The Liver

Approximately 72% of genes that have been examined in the liver show sex differences in their expression. Given the important role of the liver and many of these genes in drug metabolism, this striking sex difference can in turn influence the amount of drug that is delivered to the brain in a sex-specific manner (Yang et al. 2006). In fact, the best-known example of sex differences in liver function is drug metabolism, which leads to differences in pharmacodynamics and pharmacokinetics between males and females (Waxman and Holloway 2009). Sex differences in liver metabolism can have profound effects on brain function in adults. For example, the prescription drug zolpidem, which acts as a sleep aid by binding to $GABA_A$ receptors in the brain, was found to have a significantly lower clearance rate in women than men (Greenblatt et al. 2014), making it necessary to decrease the dosage by 50% for women. Despite the fact that 6–7% of drugs that have examined sex as a variable have shown a large difference in pharmacokinetics between men and women, there is still very little known about how sex differences in drug metabolism by the liver influence how drugs affect male and female brains. The liver secretes steroid-binding proteins and enzymes that metabolize circulating gonadal hormones, which influence the available levels of these hormones in the periphery of males and females. Exactly how these liver enzymes alter the amount of gonadal hormones that reach the brain to affect brain development in males and females is not yet known. The maternal liver also synthesizes proteins that are critical for the development of the fetus (Roy and Chatterjee 1983); however, it is not clear exactly how these proteins affect long-term brain development of the fetus.

Furthermore, sexually dimorphic gene expression in the liver is regulated by growth hormone-dependent transcription factors (Legraverend et al. 1992), which is the result of a significant sex difference in the expression of growth hormone itself. Therefore, not only can sexually dimorphic genes of the liver differentially impact brain function, but the brain (via sex differences in growth hormone expression) may also impact the sexually dimorphic expression of genes in the liver, thereby creating an interesting "brain affects liver, liver affects brain" mechanism. Still, sex differences in liver gene expression are also, in part, regulated by *growth hormone-independent* transcription factors, suggesting that the sex differences in liver gene expression extend beyond sex differences in growth hormone expression alone (Yang et al. 2006). Clearly, more work is necessary to provide a complete understanding of the intricate processes by which peripheral tissues can impact sex differences in the brain.

5.3 *The Placenta*

Beyond the developing fetus, the placenta is comprised of both maternal and fetal cells. As a result, the placenta also has a sex (i.e., that of the developing fetus) that has been shown to influence the relative risk or resilience of the fetus to maternal insults in a sex-specific manner (Clifton 2010; Gabory et al. 2013). For example, when pregnant rats are exposed to a stressor during early gestation, male but not female placentas show increases in IL-6 and IL-1β gene expression. Importantly, following this stress, male but not female offspring display anhedonia, adverse stress responsivity, and altered responses to selective serotonin reuptake inhibitors as adults (Mueller and Bale 2008), and blocking the action of these cytokines in the male placenta blunts the stress phenotype in adulthood, suggesting that the changes in immune gene expression in the placenta are the mechanistic source of these changes in the developing brain. The O-glycotransferase enzyme, O-GlcNAc transferase (OGT), is important for regulating gene expression through chromatin remodeling in the placenta and is naturally lower in the male placenta compared to the female placenta. Howerton and Bale (2014) showed that key features of the male offspring that were subjected to early prenatal stress were recapitulated in males that had genetically reduced levels of OGT in the placenta including HPA axis dysregulation, reduced postpubertal growth, and hypothalamic mitochondrial dysfunction. Furthermore, females in this model, instead, had a blunted HPA axis phenotype indicating sex-specific programming of placental OGT and supporting the possible involvement of placental X- and Y-linked genes in determining stress axis programming in adulthood (Howerton and Bale 2014; Fig. 4).

These are just a few known examples by which peripheral tissues can influence the development and function of the brain in a sex-specific manner. Given that there are many sex differences in the function of tissues and organs throughout the body, an important area of research is to better understand how these tissues and organs may also impact sex differences in the development of the brain, during normal brain development, and also following early-life challenges (e.g., stress, infection, exposure to drugs of abuse) or during disease states that affect the function of these organs.

6 Epigenetics and the Sexual Differentiation of the Brain

The mechanisms described thus far that contribute to sexual differentiation of the developing brain involve both the direct actions of sex hormones, as well as the direct actions of the different genes located on the sex chromosomes that are inherent in each cell (Arnold 2009; Gorski et al. 1978). Importantly, gene expression plays a critical role in both of these interdependent mechanisms of brain organization throughout early development. Therefore, it is perhaps not surprising that changes in gene expression, induced via *epigenetic* modifications, also contribute to sexual differentiation of the developing brain.

Epigenetics refers to the mechanisms that change gene expression via the proteins or enzymes that moderate gene expression – mechanisms including modifications to DNA methylation, histones, or DNA-binding proteins. These modifications take place without causing significant changes to the core DNA sequence (Menger et al. 2010; Nugent and McCarthy 2011), but rather these changes determine whether and under what circumstances genes are transcribed or silenced. Epigenetic changes can be induced as a result of the interactions between an organism's genome and changes in the organism's internal or external environment. Notably, gonadal hormones, which are key regulators of our internal environment, can significantly impact epigenetic modifications to the DNA (Auger and Auger 2011; Nugent et al. 2015). Therefore, given the sexually dimorphic nature of gonadal hormone secretion, epigenetic processes are likely contributors in the process of sexual differentiation, thereby determining in which sex and under what circumstances certain genes are expressed in the body and the brain. While we have a strong understanding of how sex differences in the brain are *established*, our understanding of how these sex differences in the brain are *maintained* throughout the lifespan requires further examination (McCarthy et al. 2009). Fortunately, epigenetic modifications of the DNA, which are capable of inducing lifelong and even transgenerational changes in gene expression (Roth et al. 2010), can and have been examined to answer this exciting question of how sex differences may be established and maintained in the developing brain.

We previously discussed the sexually dimorphic nature of the bed nucleus of the stria terminalis (BNST), wherein the number of cells and the overall volume of the primary nucleus of this brain region are significantly greater in the male compared to the female brain. This sex difference in the size of the BNST is a result of the neuroprotective effects of testosterone in the developing male brain, which prevents naturally occurring cell death in this brain region (Guillamon et al. 1988); however, the mechanism by which testosterone protects neurons from this naturally occurring apoptosis was previously not well understood. To address this gap in the research, Murray et al. (2009) examined the role of epigenetic changes in the brain, specifically by manipulating histone acetylation. Histone acetylation, via histone acetyltransferases (HATs), is the process by which an acetyl group is added to the lysine residues at the N-terminus of a histone protein, essentially "loosening" the DNA from a tightly wrapped spool. Loosening the DNA from the histone *increases* the opportunity for gene transcription. Conversely, histone deacetylases (HDACs) *decrease* the opportunity for gene transcription by removing that same acetyl group from the histone protein, keeping the DNA tightly wrapped around the histone. HDAC activity can be blocked by the administration of HDAC inhibitors, like valproic acid (VPA), thus resulting in increased histone acetylation and *increased* gene transcription. When Murray et al. (2009) treated newborn male mice with VPA, not only was histone acetylation increased throughout the brain, but the neuroprotective effect of testosterone on cell number in the BNST was prevented. In other words, the VPA-treated males had reduced BNST volumes. This effect was also observed in newborn female mice that had been treated with testosterone and thus, prior to VPA treatment, had BNST volumes equivalent to developing males (Murray et al. 2009). Therefore, the masculinization of the BNST, via neonatal testosterone exposure, also involves histone deacetylation, or decreased gene expression, within this brain region that persists even after the initial testosterone exposure. These findings highlight the idea that androgens can disrupt an ongoing neurodevelopmental process, in this case apoptosis, by disrupting the expression of genes and thereby inducing a sex difference in the brain. Importantly, when these genes are turned back on by HDAC inhibitors, like VPA, the sex difference can be reversed (Fig. 5). Still, the question remains of whether or not this sex difference can also be modulated outside the critical period of sexual differentiation. One might

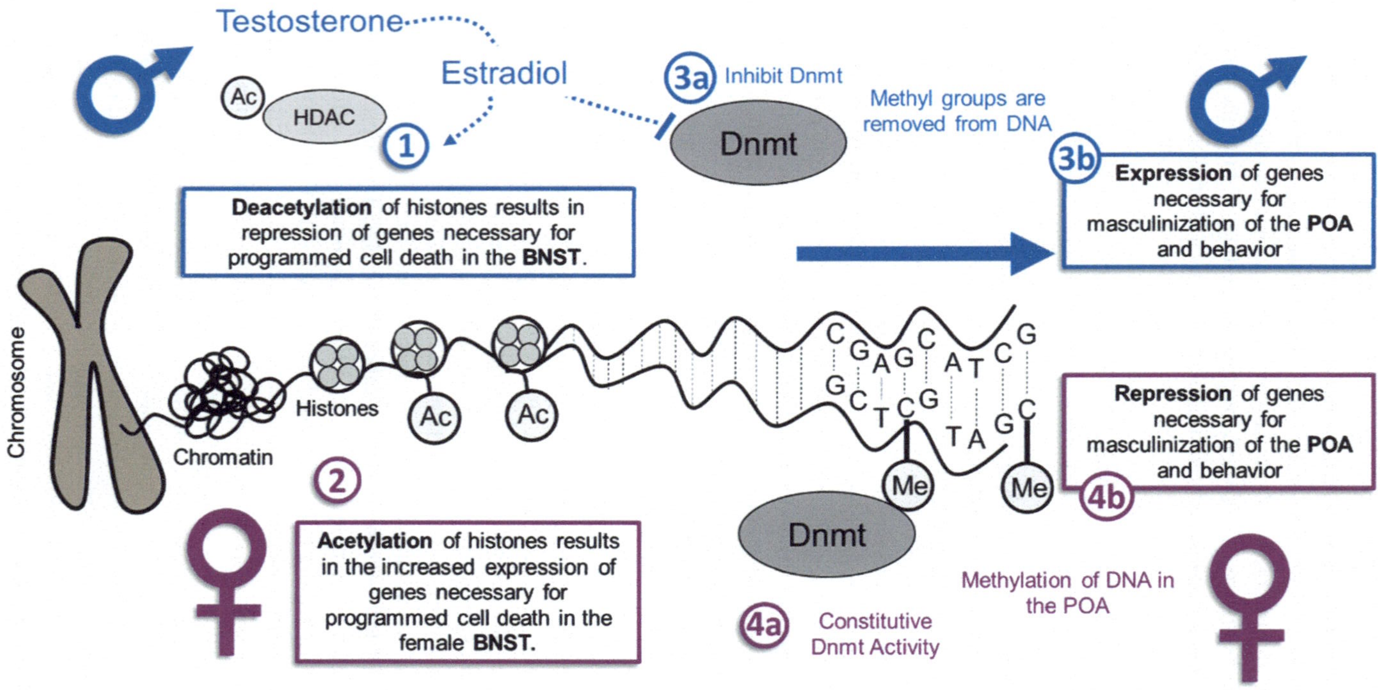

Fig. 5 Epigenetic mechanisms by which sex differences in the brain are established during neural development. **(1)** In males, testosterone converted to estradiol can activate histone deacetylase (HDAC) resulting in the deacetylation of histones in the bed nucleus of the stria terminalis (BNST), which represses gene expression necessary for programmed cell death in this brain region. **(2)** In females, in the absence of testosterone exposure, acetylation of histones increases gene expression necessary for the programmed cell death in the female BNST. **(3a)** In males, testosterone converted to estradiol inhibits DNA methyltransferase (Dnmt) in the preoptic area (POA) which results in the removal of methyl groups from the cytosines (C) of the DNA. **(3b)** In the absence of methylation in the POA, genes necessary for masculinization of the brain and sexual behavior are expressed. **(4a)** In females, Dnmt is constitutively active in the POA, in the absence of testosterone. **(4b)** Methylation of the DNA is maintained by Dnmt activity, which results in the repression of genes necessary for masculinization of the POA and behavior

assume that apoptosis in the BNST that occurs during the period of neurodevelopment occurs *exclusively* during this critical period. However, it is also possible that testosterone *permanently* alters the likelihood that the genes necessary for programed cell death in the BNST are expressed, making it possible for gene transcription to be turned on *outside* of the critical period of sexual differentiation. If so, could cell death again be induced, thereby reversing the sex difference in the size of this nucleus?

This fascinating question may be answered in part by a more recent study that examined the epigenetic mechanisms by which sex differences in the structure and function of the POA are established and maintained. As mentioned previously, the POA exhibits a striking sex difference in the structure and function of its neurons that is established as a result of perinatal testosterone exposure and is necessary for the maintenance and expression of sex differences in sexual behavior later in life (Amateau and McCarthy 2004). A study by Nugent et al. (2015) found that androgens, converted to estradiol in the rodent brain, induce this sex difference in the POA via the inhibition of DNA methyltransferase (Dnmt) activity. Dnmt is an enzyme that methylates DNA, which is thought to silence gene transcription. Thus, androgen exposure during the critical period of sexual differentiation induces the expression of certain genes that are necessary to masculinize the POA by inhibiting the process of DNA methylation. This effect may be maintained as the developing POA matures, particularly given that others have shown that testosterone (or estradiol) can induce a global decrease in DNA methylation that is maintained as cells multiply and mature (Bramble et al. 2016). Furthermore, Nugent et al. (2015) found that when Dnmt3 activity was conditionally knocked out later in life (*after* the critical period of sexual differentiation), in the absence of Dnmt3 and decreases in DNA methylation in the POA, genes necessary for masculinization of the POA could be turned *on*, resulting in masculinization of the brain and displays of male sex behavior. These data highlight a few important points: (1) in the female brain, methylation represses the genes necessary to induce masculinization of the POA; (2) during the critical period of sexual differentiation, testosterone (via estradiol) induces the transcription of genes that are actively epigenetically repressed by Dnmts in the female brain; and (3) masculinization can be induced later in life if sufficient gene transcription is re-induced in the POA, via the knockout of Dnmt3, which was actively maintaining DNA methylation in the female brain. Furthermore, these data suggest that during the critical period of sexual differentiation, Dnmt activity is particularly sensitive to androgen exposure, but outside of this critical period, Dnmt activity is not as responsive to androgen exposure and thus, estrogen exposure as well. Therefore, one might conclude that the ability of Dnmt to be inhibited by testosterone (via estrogen) may be one mechanism by which the critical period of sexual differentiation "closes," and the fate of the male and female brain is finalized.

While the relationship between neonatal testosterone and histone acetylation or DNA methylation within the BNST and POA, respectively, has proven to be fairly straightforward, the same cannot be said for other epigenetic changes that may contribute to the sexual differentiation of the brain. For example, contradictory evidence has emerged for how gonadal hormone exposure and maternal behavior may impact DNA methylation of estrogen receptor alpha (ERα) in the rodent POA. ERα expression in the rodent POA is critical to the organization of typical sex behavior in both male and female rats. In combination with estradiol, which is converted from testosterone by aromatase, ERα initiates defeminization and masculinization of the POA (Kudwa et al. 2006; McCarthy et al. 2008). Interestingly, maternal behavior can impact the methylation of the ERα gene in the neonatal pup (Cameron et al. 2008), thereby impacting masculinization and defeminization of the brain and leading to male sexual behavior or female maternal behavior later in life. The impact of maternal behavior on ERα methylation differed depending on the whether the behavior was simulated as opposed to naturally occurring (Kurian et al. 2010; Lenz et al. 2013); and moreover, sex differences in the methylation of the ERα gene were largely dependent upon the developmental time point at which they were measured (Schwarz et al. 2010). Overall, these divergent results complicate our understanding of the ways in which DNA methylation and possibly other epigenetic modifications may be involved in the maintenance of sex differences in the brain.

There is still so much to be explored in the realm of epigenetic changes and their potential impact on the establishment and maintenance of sex differences in the brain. Importantly, this new field of research has highlighted that sexual differentiation of the brain requires *large, complex networks of gene transcription*. It has long been thought that hormone-induced sexual differentiation of the brain involved the transcription of genes that *exclusively* had a hormone response element (HRE). Rather, we now know that hormones such as androgens (and estrogens metabolized from testosterone) can modulate epigenetic mechanisms such as histone acetylation and DNA methylation, which in turn produce a concert of changes in gene expression – all of which are likely necessary for the masculinization of the brain. These findings also have the potential to increase our understanding of the mechanisms by which the brain is feminized. While many studies have used testosterone treatment of female rodents to induce the process of brain masculinization and examine the expression of genes induced by testosterone during the critical period of sexual differentiation (Armoskus et al. 2014a, b; Nakachi et al. 2015), it has been difficult to identify the genes that are actively involved in the process of brain *feminization*, given that there is no known factor that initiates the process at one specific time. Now that research indicates that feminization of the brain requires active repression of masculinization via DNA methylation (Nugent et al. 2015), it is possible that we may finally be able to identify the genes that are explicitly regulated in the process of brain feminization during this time by exploring the genes that are targeted for methylation by androgens (Shen et al. 2015). From highly recognized mechanisms of gene silencing due to the sexually dimorphic process of X inactivation (Deng et al. 2014) to the understudied impact of sex-specific microRNAs (miRNAs) on post-transcriptional silencing of genes or even steroid hormone-related genes (Cochrane et al. 2011), the exciting field of epigenetic research still has a long way to go, but it may be the key to unlocking a more complete understanding of the processes that differentiate the male and female brain.

7 Conclusions

Sexual differentiation of the brain is an important neurodevelopmental process that results in the sex-specific development of neural structures and circuits that control sexual behavior and associated physiology. Studying the neural mechanisms of sexual differentiation not only informs our understanding of how these sex differences in the brain are established but also increases our understanding of how gender identity and sexual orientation may be influenced. In addition, our understanding of sexual differentiation has informed our *general* understanding of how behavioral circuits develop during critical periods of life. It has also informed our understanding of how sex steroid hormones permanently influence the function of the brain and how various cell types communicate with each other during early brain development to shape the sexually dimorphic brain. There are still many aspects of sexual differentiation that have yet to be determined. In particular, these primary questions remain: what are the mechanisms underlying brain feminization? What are the genes, proteins, or cellular processes that actively shape the development of the female brain? As mentioned earlier, if we could potentially identify genes that are suppressed (e.g., methylated) or downregulated by testosterone exposure, this might help to identify the genes that induce brain feminization.

Studies of sexual differentiation of the developing brain have revealed the mechanisms by which hypothalamic and forebrain structures are differentiated in order to control sexual and social behaviors. An important future area of research will be to understand whether there are also sex differences in the development of other circuits that are important for cognition and emotion. Alternatively, if there are no sex differences in the development of these circuits, perhaps we can understand how events that disrupt or perturb development can disrupt the development of cognitive and emotional circuits in a sex-dependent manner, or disrupt the neurotransmitters systems that control emotion. In this way, perhaps we can understand the sex bias in neurodevelopmental disorders such as autism, schizophrenia, and generalized or pervasive developmental disorders that affect the development and cognitive functioning of young boys more than girls.

Along those lines, we have yet to understand the full ontogeny of sex differences in the brain from prenatal to postnatal stages of life and throughout pubertal and adolescent development. Given that sexual differentiation of the brain permanently organizes neural development in one direction (male) or the other (female), sexual differentiation of the brain can thus determine the fate and function of cells throughout each of these stages of life. To that end, we have yet to understand how the cells maintain a memory of their sex, particularly as it was influenced by sex hormones during early brain development. This "memory" of the cell's "hormonal sex" continuously influences the function of the cells and the expression of genes throughout the lifespan.

In conclusion, as we understand more about the processes underlying sexual differentiation, we can inform our general understanding of neural development and the formation of functional neural circuits. A secondary and important goal of understanding sex differences in neural development is to understand the possible mechanisms by which mental health and neurodevelopmental disorders arise, particularly those that have origins in early development and have a well-known sex bias.

References

Agate RJ, Grisham W, Wade J, Mann S, Wingfield J, Schanen C et al (2003) Neural, not gonadal, origin of brain sex differences in a gynandromorphic finch. Proc Natl Acad Sci U S A 100(8): 4873–4878. https://doi.org/10.1073/pnas.0636925100

Amateau SK, McCarthy MM (2002a) A novel mechanism of dendritic spine plasticity involving estradiol induction of prostaglandin-E2. J Neurosci 22(19):8586–8596

Amateau SK, McCarthy MM (2002b) Sexual differentiation of astrocyte morphology in the developing rat preoptic area. J Neuroendocrinol 14(11):904–910

Amateau SK, McCarthy MM (2004) Induction of PGE2 by estradiol mediates developmental masculinization of sex behavior. Nat Neurosci 7(6):643–650

Arai Y, Sekine Y, Murakami S (1996) Estrogen and apoptosis in the developing sexually dimorphic preoptic area in female rats. Neurosci Res 25(4):403–407

Armoskus C, Moreira D, Bollinger K, Jimenez O, Taniguchi S, Tsai HW (2014a) Identification of sexually dimorphic genes in the neonatal mouse cortex and hippocampus. Brain Res 1562: 23–38. https://doi.org/10.1016/j.brainres.2014.03.017

Armoskus C, Mota T, Moreira D, Tsai HW (2014b) Effects of prenatal testosterone exposure on sexually dimorphic gene expression in the neonatal mouse cortex and hippocampus. J Steroids Horm Sci 5(3):1000139

Arnold AP (2003) The gender of the voice within: the neural origin of sex differences in the brain. Curr Opin Neurobiol 13(6):759–764

Arnold AP (2009) The organizational-activational hypothesis as the foundation for a unified theory of sexual differentiation of all mammalian tissues. Horm Behav 55(5):570–578. https://doi.org/10.1016/j.yhbeh.2009.03.011

Arnold AP, Chen X (2009) What does the "four core genotypes" mouse model tell us about sex differences in the brain and other tissues? Front Neuroendocrinol 30(1):1–9. https://doi.org/10.1016/j.yfrne.2008.11.001

Arnold AP, Gorski RA (1984) Gonadal steroid induction of structural sex differences in the central nervous system. Annu Rev Neurosci 7:413–442

Auger AP, Auger CJ (2011) Epigenetic turn ons and turn offs: chromatin reorganization and brain differentiation. Endocrinology 152(2):349–353. https://doi.org/10.1210/en.2010-0793

Bakker J, DeMees C, Douhard Q, Balthazart J, Gabant P, Szpirer J (2006) Alpha-fetoprotein protects the developing female mouse brain from masculinization and defeminization by estrogens. Nat Neurosci 9(2):20–26

Bales K, Carter CS (2003) Developmental exposure to oxytocin facilitates partner preferences in male prairie voles (microtus ochrogaster). Behav Neurosci 117:854–859

Bales KL, Plotsky PM, Young LJ, Lim MM, Grotte N, Ferrer E, Carter CS (2007) Neonatal oxytocin manipulations have long-lasting, sexually dimorphic effects on vasopressin receptors. Neuroscience 144(1):38–45

Bezzi P, Carmignoto G, Pasti L, Vesce S, Rossi D, Rizzini BL et al (1998) Prostaglandins stimulate calcium-dependent glutamate release in astrocytes. Nature 391(6664):281–285

Bilbo SD, Schwarz JM (2009) Early-life programming of later-life brain and behavior: a critical role for the immune system. Front Behav Neurosci 3:14. https://doi.org/10.3389/neuro.08.014.2009

Bleier R, Byne W, Siggelkow I (1982) Cytoarchitectonic sexual dimorphisms of the medial preoptic and anterior hypothalamic areas in guinea pig, rat, hamster, and mouse. J Comp Neurol 212(2): 118–130

Bluthe RM, Walter V, Parnet P, Laye S, Lestage J, Verrier D et al (1994) Lipopolysaccharide induces sickness behaviour in rats by a vagal mediated mechanism. Comptes Rendus De L'Academie Des Sciences. Serie III Sciences De La Vie 317(6):499–503

Borre YE, O'Keeffe GW, Clarke G, Stanton C, Dinan TG, Cryan JF (2014) Microbiota and neurodevelopmental windows: implications for brain disorders. Trends Mol Med 20(9): 509–518. https://doi.org/10.1016/j.molmed.2014.05.002

Boulanger LM (2009) Immune proteins in brain development and synaptic plasticity. Neuron 64(1): 93–109. https://doi.org/10.1016/j.neuron.2009.09.001

Bramble MS, Roach L, Lipson A, Vashist N, Eskin A, Ngun T et al (2016) Sex-specific effects of testosterone on the sexually dimorphic transcriptome and epigenome of embryonic neural stem/ progenitor cells. Sci Rep 6:36916. https://doi.org/10.1038/srep36916

Breedlove SM, Arnold AP (1980) Hormone accumulation in a sexually dimorphic motor nucleus of the rat spinal cord. Science 210:564–566

Breedlove SM, Arnold AP (1981) Sexually dimorphic motor nucleus in the rat lumbar spinal cord: response to adult hormone manipulation, absence in androgen-insensitive rats. Brain Res 225:297–307

Breedlove SM, Arnold AP (1983) Hormonal control of a developing neuromuscular system. I. Complete demasculinization of the male rat spinal nucleus of the bulbocavernosus using the anti-androgen flutamide. J Neurosci 3:417–423

Bronson SL, Bale TL (2014) Prenatal stress-induced increases in placental inflammation and offspring hyperactivity are male-specific and ameliorated by maternal antiinflammatory treatment. Endocrinology 155(7):2635–2646. https://doi.org/10.1210/en.2014-1040

Caldwell HK (2017) Oxytocin and vasopressin: powerful regulators of social behavior. Neuroscientist 23:517–528. https://doi.org/10.1177/1073858417708284

Cameron NM, Shahrokh D, Del Corpo A, Dhir SK, Szyf M, Champagne FA, Meaney MJ (2008) Epigenetic programming of phenotypic variations in reproductive strategies in the rat through maternal care. J Neuroendocrinol 20(6):795–801. https://doi.org/10.1111/j.1365-2826.2008.01725.x

Carrel L, Willard HF (2005) X-inactivation profile reveals extensive variability in X-linked gene expression in females. Nature 434(7031):400–404

Carruth LL, Reisert I, Arnold AP (2002) Sex chromosome genes directly affect brain sexual differentiation. Nat Neurosci 5(10):933–934. https://doi.org/10.1038/nn922

Chassot AA, Gregoire EP, Magliano M, Lavery R, Chaboissier MC (2008) Genetics of ovarian differentiation: Rspo1, a major player. Sex Dev 2(4–5):219–227. https://doi.org/10.1159/000152038

Clarkson J, Boon WC, Simpson ER, Herbison AE (2009) Postnatal development of an estradiol-kisspeptin positive feedback mechanism implicated in puberty onset. Endocrinology 150(7): 3214–3220. https://doi.org/10.1210/en.2008-1733

Clifton VL (2010) Review: Sex and the human placenta: mediating differential strategies of fetal growth and survival. Placenta 31(Suppl):S33–S39. https://doi.org/10.1016/j.placenta.2009.11.010

Cochrane DR, Cittelly DM, Richer JK (2011) Steroid receptors and microRNAs: relationships revealed. Steroids 76(1–2):1–10. https://doi.org/10.1016/j.steroids.2010.11.003

Conejo NM, Gonzalez-Pardo H, Pedraza C, Navarro FF, Vallejo G, Arias JL (2003) Evidence for sexual difference in astrocytes of adult rat hippocampus. Neurosci Lett 339(2):119–122

Cooke BM (2006) Steroid-dependent plasticity in the medial amygdala. Neuroscience 138(3): 997–1005

Coretti L, Cristiano C, Florio E, Scala G, Lama A, Keller S et al (2017) Sex-related alterations of gut microbiota composition in the BTBR mouse model of autism spectrum disorder. Sci Rep 7: 45356. https://doi.org/10.1038/srep45356

Crestani CC, Alves FH, Gomes FV, Resstel LB, Correa FM, Herman JP (2013) Mechanisms in the bed nucleus of the stria terminalis involved in control of autonomic and neuroendocrine

functions: a review. Curr Neuropharmacol 11(2):141–159. https://doi.org/10.2174/1570159X11311020002
Dantzer R, Bluthe RM (1993) Vasopressin and behavior: from memory to olfaction. Regul Pept 45(1–2):121–125
De Jonge FH, Louwerse AL, Ooms MP, Evers P, Endert E, van de Poll NE (1989) Lesions of the SDN-POA inhibit sexual behavior of male wistar rats. Brain Res Bull 23(6):483–492
de Theije CG, Wopereis H, Ramadan M, van Eijndthoven T, Lambert J, Knol J et al (2014a) Altered gut microbiota and activity in a murine model of autism spectrum disorders. Brain Behav Immun 37:197–206. https://doi.org/10.1016/j.bbi.2013.12.005
de Theije CG, Wu J, Koelink PJ, Korte-Bouws GA, Borre Y, Kas MJ et al (2014b) Autistic-like behavioural and neurochemical changes in a mouse model of food allergy. Behav Brain Res 261:265–274. https://doi.org/10.1016/j.bbr.2013.12.008
De Vries GJ (2004) Minireview: Sex differences in adult and developing brains: compensation, compensation, compensation. Endocrinology 145(3):1063–1068. https://doi.org/10.1210/en.2003-1504
de Vries GJ, Forger NG (2015) Sex differences in the brain: a whole body perspective. Biol Sex Differ 6:15. eCollection 2015. https://doi.org/10.1186/s13293-015-0032-z
de Vries GJ, Buijs RM, Swaab DF (1981) Ontogeny of the vasopressinergic neurons of the suprachiasmatic nucleus and their extrahypothalamic projections in the rat brain--presence of a sex difference in the lateral septum. Brain Res 218(1–2):67–78
De Vries GJ, Best W, Sluiter AA (1983) The influence of androgens on the development of a sex difference in the vasopressinergic innervation of the rat lateral septum. Brain Res 284(2–3):377–380
De Vries GJ, Wang Z, Bullock NA, Numan S (1994) Sex differences in the effects of testosterone and its metabolites on vasopressin messenger RNA levels in the bed nucleus of the stria terminalis of rats. J Neurosci 14(3 Pt 2):1789–1794
De Vries GJ, Rissman EF, Simerly RB, Yang LY, Scordalakes EM, Auger CJ et al (2002) A model system for study of sex chromosome effects on sexually dimorphic neural and behavioral traits. J Neurosci 22(20):9005–9014
Deng X, Berletch JB, Nguyen DK, Disteche CM (2014) X chromosome regulation: diverse patterns in development, tissues and disease. Nat Rev Genet 15(6):367–378. https://doi.org/10.1038/nrg3687
Deverman BE, Patterson PH (2009) Cytokines and CNS development. Neuron 64(1):61–78. https://doi.org/10.1016/j.neuron.2009.09.002
Dewing P, Chiang CW, Sinchak K, Sim H, Fernagut PO, Kelly S et al (2006) Direct regulation of adult brain function by the male-specific factor SRY. Curr Biol 16(4):415–420
Dodson RE, Gorski RA (1993) Testosterone propionate administration prevents the loss of neurons within the central part of the medial preoptic nucleus. J Neurobiol 24(1):80–88. https://doi.org/10.1002/neu.480240107
Dumais KM, Veenema AH (2016) Vasopressin and oxytocin receptor systems in the brain: sex differences and sex-specific regulation of social behavior. Front Neuroendocrinol 40:1–23. https://doi.org/10.1016/j.yfrne.2015.04.003
Dumais KM, Bredewold R, Mayer TE, Veenema AH (2013) Sex differences in oxytocin receptor binding in forebrain regions: correlations with social interest in brain region- and sex-specific ways. Horm Behav 64(4):693–701. https://doi.org/10.1016/j.yhbeh.2013.08.012
Fiebich BL, Akter S, Akundi RS (2014) The two-hit hypothesis for neuroinflammation: role of exogenous ATP in modulating inflammation in the brain. Front Cell Neurosci 8:260
Forger NG (2006) Cell death and sexual differentiation of the nervous system. Neuroscience 138(3):929–938
Frankfurt M, Gould E, Woolley CS, McEwen BS (1990) Gonadal steroids modify dendritic spine density in ventromedial hypothalamic neurons: a golgi study in the adult rat. Neuroendocrinology 51(5):530–535
Frick LR, Williams K, Pittenger C (2013) Microglial dysregulation in psychiatric disease. Clin Dev Immunol 2013:608654. https://doi.org/10.1155/2013/608654

Gabory A, Roseboom TJ, Moore T, Moore LG, Junien C (2013) Placental contribution to the origins of sexual dimorphism in health and diseases: sex chromosomes and epigenetics. Biol Sex Differ 4(1):5. https://doi.org/10.1186/2042-6410-4-5

Gorski RA (1978) Sexual differentiation of the brain. Hosp Pract 13(10):55–62

Gorski RA, Gordon JH, Shryne JE, Southam AM (1978) Evidence for a morphological sex difference within the medial preoptic area of the rat brain. Brain Res 148(2):333–346

Greenblatt DJ, Harmatz JS, Singh NN, Steinberg F, Roth T, Moline ML et al (2014) Gender differences in pharmacokinetics and pharmacodynamics of zolpidem following sublingual administration. J Clin Pharmacol 54(3):282–290. https://doi.org/10.1002/jcph.220

Guillamon A, Segovia S, del Abril A (1988) Early effects of gonadal steroids on the neuron number in the medial posterior region and the lateral division of the bed nucleus of the stria terminalis in the rat. Brain Res 44(2):281–290

Gurney ME, Konishi M (1980) Hormone-induced sexual differentiation of brain and behavior in zebra finches. Science (New York, NY) 208(4450):1380–1383

Hanamsagar R, Alter MD, Block CS, Sullivan H, Bolton JL, Bilbo SD (2017) Generation of a microglial developmental index in mice and in humans reveals a sex difference in maturation and immune reactivity. Glia. https://doi.org/10.1002/glia.23176

Haseltine FP, Ohno S (1981) Mechanisms of gonadal differentiation. Science (New York, NY) 211(4488):1272–1278

Haussler HU, Jirikowski GF, Caldwell JD (1990) Sex differences among oxytocin-immunoreactive neuronal systems in the mouse hypothalamus. J Chem Neuroanat 3(4):271–276

Hilton GD, Nunez JL, McCarthy MM (2003) Sex differences in response to kainic acid and estradiol in the hippocampus of newborn rats. Neuroscience 116(2):383–391

Houtsmuller EJ, Brand T, de Jonge FH, Joosten RN, van de Poll NE, Slob AK (1994) SDN-POA volume, sexual behavior, and partner preference of male rats affected by perinatal treatment with ATD. Physiol Behav 56(3):535–541

Howerton CL, Bale TL (2014) Targeted placental deletion of OGT recapitulates the prenatal stress phenotype including hypothalamic mitochondrial dysfunction. Proc Natl Acad Sci U S A 111(26):9639–9644. https://doi.org/10.1073/pnas.1401203111

Hu MH, Li XF, McCausland B, Li SY, Gresham R, Kinsey-Jones JS et al (2015) Relative importance of the arcuate and anteroventral periventricular kisspeptin neurons in control of puberty and reproductive function in female rats. Endocrinology 156(7):2619–2631. https://doi.org/10.1210/en.2014-1655

Insel TR, Hulihan TJ (1995) A gender-specific mechanism for pair bonding: oxytocin and partner preference formation in monogamous voles. Behav Neurosci 109(4):782–789

Jarzab B, Kaminski M, Gubala E, Achtelik W, Wagiel J, Dohler KD (1990) Postnatal treatment of rats with the beta 2-adrenergic agonist salbutamol influences the volume of the sexually dimorphic nucleus in the preoptic area. Brain Res 516(2):257–262

Jeong JK, Ryu BJ, Choi J, Kim DH, Choi EJ, Park JW et al (2008) NELL2 participates in formation of the sexually dimorphic nucleus of the pre-optic area in rats. J Neurochem 106(4):1604–1613. https://doi.org/10.1111/j.1471-4159.2008.05505.x

Jordan CL, Letinsky MS, Arnold AP (1988) Synapse elimination occurs late in the hormone-sensitive levator ani muscle of the rat. J Neurobiol 19(4):335–356. https://doi.org/10.1002/neu.480190403

Jordan CL, Padgett B, Hershey J, Prins G, Arnold A (1997) Ontogeny of androgen receptor immunoreactivity in lumbar motoneurons and in the sexually dimorphic levator ani muscle of male rats. J Comp Neurol 379(1):88–98. https://doi.org/10.1002/(SICI)1096-9861(19970303)379:13.0.CO;2-E

Josso N, Lamarre I, Picard JY, Berta P, Davies N, Morichon N et al (1993) Anti-mullerian hormone in early human development. Early Hum Dev 33(2):91–99

Jost A (1947) Recherches sur la differenciation sexuelle de l'embryon de lapin. Arch Anat Microsc Morph Exp 36:271–315

Kauffman AS, Gottsch ML, Roa J, Byquist AC, Crown A, Clifton DK et al (2007) Sexual differentiation of Kiss1 gene expression in the brain of the rat. Endocrinology 148(4):1774–1783

Krebs-Kraft DL, Hill MN, Hillard CJ, McCarthy MM (2010) Sex difference in cell proliferation in developing rat amygdala mediated by endocannabinoids has implications for social behavior. Proc Natl Acad Sci U S A 107(47):20535–20540. https://doi.org/10.1073/pnas.1005003107

Kudwa AE, Michopoulos V, Gatewood JD, Rissman EF (2006) Roles of estrogen receptors alpha and beta in differentiation of mouse sexual behavior. Neuroscience 138(3):921–928. https://doi.org/10.1016/j.neuroscience.2005.10.018

Kurian JR, Olesen KM, Auger AP (2010) Sex differences in epigenetic regulation of the estrogen receptor-alpha promoter within the developing preoptic area. Endocrinology 151(5): 2297–2305. https://doi.org/10.1210/en.2009-0649

Lebow MA, Chen A (2016) Overshadowed by the amygdala: the bed nucleus of the stria terminalis emerges as key to psychiatric disorders. Mol Psychiatry 21(4):450–463. https://doi.org/10.1038/mp.2016.1

Legraverend C, Mode A, Westin S, Ström A, Equchi H, Zaphiropoulos PG, Gustafsson JA (1992) Transcriptional regulation of rat P-450 2C gene subfamily members by the sexually dimorphic pattern of growth hormone secretion. Mol Endocrinol 6(2):259–266

Lenz KM, McCarthy MM (2014) A starring role for microglia in brain sex differences. Neuroscientist 21:306–321

Lenz KM, Nugent BM, Haliyur R, McCarthy MM (2013) Microglia are essential to masculinization of brain and behavior. J Neurosci 33:2761–2772

Lubischer JL, Arnold AP (1995) Evidence for target regulation of the development of androgen sensitivity in rat spinal motoneurons. Dev Neurosci 17(2):106–117

Maatouk DM, Capel B (2008) Sexual development of the soma in the mouse. Curr Top Dev Biol 83:151–183. https://doi.org/10.1016/S0070-2153(08)00405-5

Maezawa I, Calafiore M, Wulff H, Jin LW (2011) Does microglial dysfunction play a role in autism and Rett syndrome? Neuron Glia Biol 7(1):85–97. https://doi.org/10.1017/S1740925X1200004X

Markle JG, Frank DN, Mortin-Toth S, Robertson CE, Feazel LM, Rolle-Kampczyk U et al (2013) Sex differences in the gut microbiome drive hormone-dependent regulation of autoimmunity. Science (New York, NY) 339(6123):1084–1088. https://doi.org/10.1126/science.1233521

Mathews GA, Brenowitz EA, Arnold AP (1988) Paradoxical hypermasculinization of the zebra finch song system by an antiestrogen. Horm Behav 22(4):540–551

Matsumoto A, Arai Y (1983) Sex difference in volume of the ventromedial nucleus of the hypothalamus in the rat. Endocrinol Jpn 30(3):277–280

Matsumoto A, Arai Y (1986) Development of sexual dimorphism in synaptic organization in the ventromedial nucleus of the hypothalamus in rats. Neurosci Lett 68(2):165–168

McCarthy MM, Arnold AP (2011) Reframing sexual differentiation of the brain. Nat Neurosci 14(6):677–683. https://doi.org/10.1038/nn.2834

McCarthy MM, Konkle AT (2005) When is a sex difference not a sex difference? Front Neuroendocrinol 26(2):85–102

McCarthy MM, Schwarz JM, Wright CL, Dean SL (2008) Mechanisms mediating oestradiol modulation of the developing brain. J Neuroendocrinol 20(6):777–783. https://doi.org/10.1111/j.1365-2826.2008.01723.x

McCarthy MM, Auger AP, Bale TL, De Vries GJ, Dunn GA, Forger NG et al (2009) The epigenetics of sex differences in the brain. J Neurosci 29(41):12815–12823. https://doi.org/10.1523/JNEUROSCI.3331-09.2009

Menger Y, Bettscheider M, Murgatroyd C, Spengler D (2010) Sex differences in brain epigenetics. Epigenomics 2(6):807–821. https://doi.org/10.2217/epi.10.60

Miller MA, Urban JH, Dorsa DM (1989a) Steroid dependency of vasopressin neurons in the bed nucleus of the stria terminalis by in situ hybridization. Endocrinology 125(5):2335–2340. https://doi.org/10.1210/endo-125-5-2335

Miller MA, Vician L, Clifton DK, Dorsa DM (1989b) Sex differences in vasopressin neurons in the bed nucleus of the stria terminalis by in situ hybridization. Peptides 10(3):615–619

Millhouse OE (1973) The organization of the ventromedial hypothalamic nucleus. Brain Res 55(1): 71–87

Mohr MA, Garcia FL, DonCarlos LL, Sisk CL (2016) Neurons and glial cells are added to the female rat anteroventral periventricular nucleus during puberty. Endocrinology 157(6): 2393–2402. https://doi.org/10.1210/en.2015-2012

Mueller BR, Bale TL (2008) Sex-specific programming of offspring emotionality after stress early in pregnancy. J Neurosci 28(36):9055–9065. https://doi.org/10.1523/JNEUROSCI.1424-08.2008

Murakami S, Arai Y (1989) Neuronal death in the developing sexually dimorphic periventricular nucleus of the preoptic area in the female rat: effect of neonatal androgen treatment. Neurosci Lett 102(2–3):185–190

Murray EK, Hien A, de Vries GJ, Forger NG (2009) Epigenetic control of sexual differentiation of the bed nucleus of the stria terminalis. Endocrinology 150(9):4241–4247. https://doi.org/10.1210/en.2009-0458

Nakachi Y, Iseki M, Yokoo T, Mizuno Y, Okazaki Y (2015) Gene expression profile of the neonatal female mouse brain after administration of testosterone propionate. J Sex Med 12(4):887–896. https://doi.org/10.1111/jsm.12802

Nelson LH, Warden S, Lenz KM (2017) Sex differences in microglial phagocytosis in the neonatal hippocampus. Brain Behav Immun 64:11–22

Nordeen EJ, Nordeen KW, Sengelaub DR, Arnold AP (1985) Androgens prevent normally occurring cell death in a sexually dimorphic spinal nucleus. Science (New York, NY) 229(4714): 671–673

Nottebohm F (1991) Reassessing the mechanisms and origins of vocal learning in birds. Trends Neurosci 14(5):206–211

Nottebohm F, Arnold AP (1976) Sexual dimorphism in vocal control areas of the songbird brain. Science (New York, NY) 194(4261):211–213

Nottebohm F, Kelley DB, Paton JA (1982) Connections of vocal control nuclei in the canary telencephalon. J Comp Neurol 207(4):344–357. https://doi.org/10.1002/cne.902070406

Nugent BM, McCarthy MM (2011) Epigenetic underpinnings of developmental sex differences in the brain. Neuroendocrinology 93(3):150–158. https://doi.org/10.1159/000325264

Nugent BM, Wright CL, Shetty AC, Hodes GE, Lenz KM, Mahurkar A et al (2015) Brain feminization requires active repression of masculinization via DNA methylation. Nat Neurosci 18(5): 690–697. https://doi.org/10.1038/nn.3988

Nunez JL, Alt JJ, McCarthy MM (2003) A new model for prenatal brain damage. I. GABAA receptor activation induces cell death in developing rat hippocampus. Exp Neurol 181(2): 258–269

Pailhoux E, Parma P, Sundstrom J, Vigier B, Servel N, Kuopio T et al (2001) Time course of female-to-male sex reversal in 38,XX fetal and postnatal pigs. Dev Dyn 222(3):328–340. https://doi.org/10.1002/dvdy.1194

Painter TS (1923) Studies in mammalian spermatogenesis. II. The spermatogenesis of man. J Exp Zool 37:291–336. https://doi.org/10.1002/jez.1400370303

Paul MJ, Terranova JI, Probst CK, Murray EK, Ismail NI, de Vries GJ (2014) Sexually dimorphic role for vasopressin in the development of social play. Front Behav Neurosci 8:58. https://doi.org/10.3389/fnbeh.2014.00058

Pfaff DW, Sakuma Y (1979) Deficit in the lordosis reflex of female rats caused by lesions in the ventromedial nucleus of the hypothalamus. J Physiol 288:203–210

Phoenix CH, Goy RW, Gerall AA, Young WC (1959) Organizing action of prenatally administered testosterone propionate on the tissues mediating mating behavior in the female guinea pig. Endocrinology 65:369–382

Pozzo Miller LD, Aoki A (1991) Stereological analysis of the hypothalamic ventromedial nucleus. II. Hormone-induced changes in the synaptogenic pattern. Brain Res Dev Brain Res 61(2): 189–196

Qiao X, Yan Y, Wu R, Tai F, Hao P, Cao Y, Wang J (2014) Sociality and oxytocin and vasopressin in the brain of male and female dominant and subordinate mandarin voles. J Comp Physiol A Neuroethol Sens Neural Behav Physiol 200(2):149–159. https://doi.org/10.1007/s00359-013-0870-2

Quadros PD, Pfau JL, Goldstein AY, de Vries GJ, Wagner CK (2002) Sex differences in progesterone receptor expression: a potential mechanism for estradiol-mediated sexual differentiation. Endocrinology 143(10):3727–3739

Rood BD, Murray EK, Laroche J, Yang MK, Blaustein JD, De Vries GJ (2008) Absence of progestin receptors alters distribution of vasopressin fibers but not sexual differentiation of vasopressin system in mice. Neuroscience 154(3):911–921. https://doi.org/10.1016/j.neuroscience.2008.03.087

Ross HE, Young LJ (2009) Oxytocin and the neural mechanisms regulating social cognition and affiliative behavior. Front Neuroendocrinol 30(4):534–547. https://doi.org/10.1016/j.yfrne.2009.05.004

Roth TL, Roth ED, Sweatt JD (2010) Epigenetic regulation of genes in learning and memory. Essays Biochem 48(1):263–274. https://doi.org/10.1042/bse0480263

Roy AK, Chatterjee B (1983) Sexual dimorphism in the liver. Annu Rev Physiol 45:37–50. https://doi.org/10.1146/annurev.ph.45.030183.000345

Schwarz JM, Bilbo SD (2012) Sex, glia, and development: interactions in health and disease. Horm Behav 62(3):243–253. https://doi.org/10.1016/j.yhbeh.2012.02.018

Schwarz JM, McCarthy MM (2008) The role of neonatal NMDA receptor activation in defeminization and masculinization of sex behavior in the rat. Horm Behav 54(5):662–668. https://doi.org/10.1016/j.yhbeh.2008.07.004

Schwarz JM, Liang SL, Thompson SM, McCarthy MM (2008) Estradiol induces hypothalamic dendritic spines by enhancing glutamate release: a mechanism for organizational sex differences. Neuron 58(4):584–598. https://doi.org/10.1016/j.neuron.2008.03.008

Schwarz JM, Nugent BM, McCarthy MM (2010) Developmental and hormone-induced epigenetic changes to estrogen and progesterone receptor genes in brain are dynamic across the life span. Endocrinology 151(10):4871–4881. https://doi.org/10.1210/en.2010-0142

Schwarz JM, Sholar PW, Bilbo SD (2012) Sex differences in microglial colonization of the developing rat brain. J Neurochem 120(6):948–963. https://doi.org/10.1111/j.1471-4159.2011.07630.x

Sekido R (2014) The potential role of SRY in epigenetic gene regulation during brain sexual differentiation in mammals. Adv Genet 86:135–165. https://doi.org/10.1016/B978-0-12-800222-3.00007-3

Sengelaub DR, Forger NG (2008) The spinal nucleus of the bulbocavernosus: firsts in androgen-dependent neural sex differences. Horm Behav 53(5):596–612. https://doi.org/10.1016/j.yhbeh.2007.11.008

Shen EY, Ahern TH, Cheung I, Straubhaar J, Dincer A, Houston I et al (2015) Epigenetics and sex differences in the brain: a genome-wide comparison of histone-3 lysine-4 trimethylation (H3K4me3) in male and female mice. Exp Neurol 268:21–29. https://doi.org/10.1016/j.expneurol.2014.08.006

Sheridan GK, Wdowicz A, Pickering M, Watters O, Halley P, O'Sullivan NC et al (2014) CX3CL1 is up-regulated in the rat hippocampus during memory-associated synaptic plasticity. Front Cell Neurosci 8:233. https://doi.org/10.3389/fncel.2014.00233

Shughrue PJ, Sar M, Stumpf WE (1992) Progestin target cell distribution in forebrain and midbrain regions of the 8-day postnatal mouse brain. Endocrinology 130(6):3650–3659

Shughrue PJ, Lane MV, Merchenthaler I (1997) Comparative distribution of estrogen receptor-alpha and -beta mRNA in the rat central nervous system. J Comp Neurol 388(4):507–525

Simerly RB (2002) Wired for reproduction: organization and development of sexually dimorphic circuits in the mammalian forebrain. Annu Rev Neurosci 25(1):507–536. https://doi.org/10.1146/annurev.neuro.25.112701.142745

Smith MR, Hamson DK, Poort JE, Jordan CL, Breedlove SM (2012) Ontogeny of androgen receptor expression in spinal nucleus of the bulbocavernosus motoneurons and their target muscles in male mice. Neurosci Lett 513(2):119–123. https://doi.org/10.1016/j.neulet.2012.01.067

Sumida H, Nishizuka M, Kano Y, Arai Y (1993) Sex differences in the anteroventral periventricular nucleus of the preoptic area and in the related effects of androgen in prenatal rats. Neurosci Lett 151(1):41–44

Taketo T, Lee CH, Zhang J, Li Y, Lee CY, Lau YF (2005) Expression of SRY proteins in both normal and sex-reversed XY fetal mouse gonads. Dev Dyn 233(2):612–622. https://doi.org/10.1002/dvdy.20352

Taylor PV, Veenema AH, Paul MJ, Bredewold R, Isaacs S, de Vries GJ (2012) Sexually dimorphic effects of a prenatal immune challenge on social play and vasopressin expression in juvenile rats. Biol Sex Differ 3(1):15. https://doi.org/10.1186/2042-6410-3-15

Terranova JI, Song Z, Larkin TE 2nd, Hardcastle N, Norvelle A, Riaz A, Albers HE (2016) Serotonin and arginine-vasopressin mediate sex differences in the regulation of dominance and aggression by the social brain. Proc Natl Acad Sci U S A 113(46):13233–13238

Todd BJ, Schwarz JM, McCarthy MM (2005) Prostaglandin-E2: a point of divergence in estradiol-mediated sexual differentiation. Horm Behav 48(5):512–521. https://doi.org/10.1016/j.yhbeh.2005.07.011

Tsukahara S (2009) Sex differences and the roles of sex steroids in apoptosis of sexually dimorphic nuclei of the preoptic area in postnatal rats. J Neuroendocrinol 21(4):370–376. https://doi.org/10.1111/j.1365-2826.2009.01855.x

Vainio SJ, Itaranta PV, Perasaari JP, Uusitalo MS (1999) Wnts as kidney tubule inducing factors. Int J Dev Biol 43(5):419–423

Veenema AH, Neumann ID (2008) Central vasopressin and oxytocin release: regulation of complex social behaviours. Prog Brain Res 170:261–276. https://doi.org/10.1016/S0079-6123(08)00422-6

Wade CB, Robinson S, Shapiro RA, Dorsa DM (2001) Estrogen receptor (ER)alpha and ERbeta exhibit unique pharmacologic properties when coupled to activation of the mitogen-activated protein kinase pathway. Endocrinology 142(6):2336–2342

Wagner CK, Nakayama AY, de Vries GJ (1998) Potential role of maternal progesterone in the sexual differentiation of the brain. Endocrinology 139(8):3658–3661

Waxman DJ, Holloway MG (2009) Sex differences in the expression of hepatic drug metabolizing enzymes. Mol Pharmacol 76(2):215–228. https://doi.org/10.1124/mol.109.056705

Wierda M, Goudsmit E, Van der Woude PF, Purba JS, Hofman MA, Bogte H, Swaab DF (1991) Oxytocin cell number in the human paraventricular nucleus remains constant with aging and in Alzheimer's disease. Neurobiol Aging 12(5):511–516

Williams JR, Insel TR, Harbaugh CR, Carter CS (1994) Oxytocin administered centrally facilitates formation of a partner preference in female prairie voles (microtus ochrogaster). J Neuroendocrinol 6(3):247–250

Winslow JT, Hastings N, Carter CS, Harbaugh CR, Insel TR (1993) A role for central vasopressin in pair bonding in monogamous prairie voles. Nature 365(6446):545–548. https://doi.org/10.1038/365545a0

Wright CL, Burks SR, McCarthy MM (2008) Identification of prostaglandin E2 receptors mediating perinatal masculinization of adult sex behavior and neuroanatomical correlates. Dev Neurobiol 68(12):1406–1419. https://doi.org/10.1002/dneu.20665

Wu MV, Manoli DS, Fraser EJ, Coats JK, Tollkuhn J, Honda S, Harada N, Shah NM (2009) Estrogen masculinizes neural pathways and sex-specific behaviors. Cell 139(7):61–72

Xu J, Disteche CM (2006) Sex differences in brain expression of X- and Y-linked genes. Brain Res 1126(1):50–55

Yamamoto Y, Cushing BS, Kramer KM, Epperson PD, Hoffman GE, Carter CS (2004) Neonatal manipulations of oxytocin alter expression of oxytocin and vasopressin immunoreactive cells in

the paraventricular nucleus of the hypothalamus in a gender-specific manner. Neuroscience 125(4):947–955. https://doi.org/10.1016/j.neuroscience.2004.02.028
Yamamoto Y, Carter CS, Cushing BS (2006) Neonatal manipulation of oxytocin affects expression of estrogen receptor alpha. Neuroscience 137(1):157–164
Yang X, Schadt EE, Wang S, Wang H, Arnold AP, Ingram-Drake L et al (2006) Tissue-specific expression and regulation of sexually dimorphic genes in mice. Genome Res 16(8):995–1004
Yao S, Bergan J, Lanjuin A, Dulac C (2017) Oxytocin signaling in the medial amygdala is required for sex discrimination of social cues. elife 6:pii: e31373. https://doi.org/10.7554/eLife.31373
Yurkovetskiy L, Burrows M, Khan AA, Graham L, Volchkov P, Becker L et al (2013) Gender bias in autoimmunity is influenced by microbiota. Immunity 39(2):400–412. https://doi.org/10.1016/j.immuni.2013.08.013
Zhang JM, Konkle AT, Zup SL, McCarthy MM (2008) Impact of sex and hormones on new cells in the developing rat hippocampus: a novel source of sex dimorphism? Eur J Neurosci 27(4):791–800. https://doi.org/10.1111/j.1460-9568.2008.06073.x
Ziegler DR, Cullinan WE, Herman JP (2002) Distribution of vesicular glutamate transporter mRNA in rat hypothalamus. J Comp Neurol 448(3):217–229

Neural Regulation of Paternal Behavior in Mammals: Sensory, Neuroendocrine, and Experiential Influences on the Paternal Brain

Nathan D. Horrell, Peter W. Hickmott, and Wendy Saltzman

Contents

1 Introduction 112
2 Neural Circuitry of Paternal Care 114
  2.1 Overview 115
  2.2 Medial Preoptic Area 116
  2.3 Medial Amygdala 118
  2.4 Mesolimbic Reward Pathway 118
3 Sensory Influences on Paternal Care 120
  3.1 Olfaction 121
  3.2 Audition 123
  3.3 Somatosensation 123
  3.4 Sensory Influences: Conclusions 124
4 Hormonal and Neuropeptide Influences on Paternal Care 125
  4.1 Androgens 129
  4.2 Estrogen 131
  4.3 Progesterone 134
  4.4 Glucocorticoids 135
  4.5 Prolactin 136
  4.6 Vasopressin 137
  4.7 Oxytocin 139
5 Experiential Influences on Paternal Care and Neural Plasticity 140
6 Conclusions and Future Directions 142
References 144

N. D. Horrell and W. Saltzman (✉)
Graduate Program in Neuroscience, University of California, Riverside, Riverside, CA, USA

Department of Evolution, Ecology, and Organismal Biology, University of California, Riverside, Riverside, CA, USA
e-mail: nhorr001@ucr.edu; saltzman@ucr.edu

P. W. Hickmott
Graduate Program in Neuroscience, University of California, Riverside, Riverside, CA, USA

Department of Psychology, University of California, Riverside, Riverside, CA, USA
e-mail: peterh@ucr.edu

Curr Topics Behav Neurosci (2019) 43: 111–160
DOI 10.1007/7854_2018_55
Published Online: 12 September 2018

Abstract Across the animal kingdom, parents in many species devote extraordinary effort toward caring for offspring, often risking their lives and exhausting limited resources. Understanding how the brain orchestrates parental care, biasing effort over the many competing demands, is an important topic in social neuroscience. In mammals, maternal care is necessary for offspring survival and is largely mediated by changes in hormones and neuropeptides that fluctuate massively during pregnancy, parturition, and lactation (e.g., progesterone, estradiol, oxytocin, and prolactin). In the relatively small number of mammalian species in which parental care by fathers enhances offspring survival and development, males also undergo endocrine changes concurrent with birth of their offspring, but on a smaller scale than females. Thus, fathers additionally rely on sensory signals from their mates, environment, and/or offspring to orchestrate paternal behavior. Males can engage in a variety of infant-directed behaviors that range from infanticide to avoidance to care; in many species, males can display all three behaviors in their lifetime. The neural plasticity that underlies such stark changes in behavior is not well understood. In this chapter we summarize current data on the neural circuitry that has been proposed to underlie paternal care in mammals, as well as sensory, neuroendocrine, and experiential influences on paternal behavior and on the underlying circuitry. We highlight some of the gaps in our current knowledge of this system and propose future directions that will enable the development of a more comprehensive understanding of the proximate control of parenting by fathers.

Keywords Alloparental behavior · Medial preoptic area · Neuroendocrinology · Parental behavior · Parental care · Social behavior

1 Introduction

In all 5,000–6,000 extant mammalian species, maternal care is essential for the survival and development of offspring. At a minimum, mothers must undergo the physiological processes of gestation, parturition, and lactation, some of the major hallmarks of mammalian evolution. Typically, however, species-specific behaviors performed by mothers, such as warming, transporting, and protecting young, are also critical for infant survival. Because mothers are uniquely capable of gestating and lactating and can provide all behavioral components of maternal care, female mammals presumably evolved to independently support their own offspring following insemination, a pattern observed in most mammalian species. In contrast, males often provide as much care of offspring as mothers, if not more, in birds, as well as in the relatively small number of fish and amphibian species that engage in parental care (Clutton-Brock 1991; Cockburn 2006; Crump 1996; Goodwin et al. 1998; Gross and Sargent 1985), taxa in which offspring development (and in some cases fertilization) often occur outside of the mother's body and in which both parents are capable of feeding and protecting the young. Nonetheless, fathers provide care for their offspring in approximately 10% of mammalian genera, especially among canids, rodents, and primates (Kleiman and Malcolm 1981; Woodroffe and Vincent 1994).

Both the specific behaviors performed and the amount of resources invested in offspring by mammalian fathers can vary substantially. For example, fathers in biparental canid species, such as the gray wolf (*Canis lupus*), coyote (*C. latrans*), and African wild dog (*Lycaon pictus*), play with, babysit and defend their pups, and provision them with food (Bruin et al. 2016; Kleiman and Malcolm 1981; Mech et al. 1999). In many biparental rodents, such as California mice (*Peromyscus californicus*), prairie voles (*Microtus ochrogaster*), and Campbell's Russian dwarf hamsters (*Phodopus campbelli*), fathers retrieve, huddle, and lick their pups, as well as build nests (Gubernick and Alberts 1987; Wang et al. 1994; Wynne-Edwards 1995). Fathers in some biparental rodent species also commonly perform the kyphotic posture typical of nursing mothers, although males do not lactate, and may even help to deliver pups during parturition and ingest placenta and amniotic fluid (Jones and Wynne-Edwards 2000; Perea-Rodriguez and Saltzman 2014). In the biparental nonhuman primates in which paternal care has been studied extensively, particularly titi monkeys (*Callicebus* spp.), marmosets (*Callithrix* spp.), and tamarins (*Saguinus* spp.), fathers often play a major role in transporting infants, thought to be an energetically expensive behavior. Fathers in these species also groom their infants, as well as play and share food with them (Fernandez-Duque et al. 2009; Spence-Aizenberg et al. 2016). In all of these taxa, the amounts and proportions of parental behaviors performed by fathers, as compared to mothers and alloparents, can differ markedly among species as well as across the offspring's development. Paternal care can be essential for survival of offspring and can have lasting impacts on offspring behavioral, neural, cognitive, affective, and social development (Bales and Saltzman 2016; Braun and Champagne 2014; Kentner et al. 2009).

In accordance with the ubiquity of maternal care among mammals, the neural substrates and endocrine determinants of both physiological and behavioral aspects of maternal care have been studied extensively, especially in rodents. Much less is known about paternal care. The neural circuitry underlying paternal behavior, in particular, remains relatively unexplored. While both the neural substrates and endocrine influences on parental care overlap between mothers and fathers, the extent of this overlap is not yet clear. Similarly, although some of the sensory and experiential influences on male parental behavior resemble those in mothers, these influences are also likely to differ substantially between the sexes; however, the extent of these similarities and differences has received little attention.

In this chapter we review neural, sensory, hormonal, and experiential influences on the expression of paternal behavior in mammals. We focus largely on rodents because they are the best-studied; however, we incorporate findings from primates where data exist. Moreover, we present results from both biparental species (e.g., California mouse, prairie vole, Campbell's Russian dwarf hamster, mandarin vole [*Microtus mandarinus*], Mongolian gerbil [*Meriones unguiculatus*]; common marmoset [*Callithrix jacchus*], cotton-top tamarin [*Saguinus oedipus*], human), in which fathers spontaneously provide care for their offspring, and uniparental mammals, such as laboratory mice (*Mus*) and rats (*Rattus*), in which paternal care may be expressed under captive conditions but does not seem to occur spontaneously in natural environments.

Given that paternal care has evolved multiple times among mammals, even within individual orders (West and Capelllini 2016), we should not assume that it is governed by identical mechanisms in all taxa. Moreover, paternal care can encompass several motivational states (e.g., inhibition of aversion or aggression toward infants, attraction to infants) as well as multiple behaviors, which typically change across the period of infant development. In addition, different mechanisms may underlie the expression of paternal (or allopaternal) behavior in reproductively inexperienced males, new fathers, and experienced fathers. Thus, neural, hormonal, sensory, and experiential influences may differ both among the components of paternal care and over time. Furthermore, many, if not most, studies of paternal care have evaluated males' responses to unrelated young, which might not be expressed and/or mediated identically to responses toward a male's own infants. Therefore, we refer to males' behavior toward their own offspring as "paternal" and behavior toward non-descendant young as "allopaternal." Finally, the vast majority of studies on proximate determinants of paternal care have been conducted under highly artificial laboratory conditions; therefore, we cannot yet determine the precise role of these factors in natural settings.

With these caveats in mind, we summarize current understanding of the neural circuitry underlying parental behavior in mammals, followed by discussions of sensory, endocrine, and experiential influences on this circuitry and on the expression of paternal care. Our review of the neural circuity is based heavily on data from mothers, as limited data are available from fathers. In the remaining sections, however, we focus almost exclusively on males.

2 Neural Circuitry of Paternal Care

Building a complete model of the neurobiological basis of mammalian behavior is a daunting task; a legion of genetic, epigenetic, ontogenetic, and proximate causal factors influence a vast glial-neuronal network with intricate and myriad connectivities and signaling mechanisms. While all circuit diagrams are reductive, they do depict evidence-based theories of connectivity and function of nuclei implicated in behavior and therefore provide useful models to ultimately build a complete picture of a particular behavioral outcome. Furthermore, even incomplete information can shed light on distinct and shared mechanisms underlying behaviors and provide clues for understanding and treating disruptions of these behaviors.

One of the earliest parental-care circuit diagrams was published in 1988 by Michael Numan (1988). Since its inception, the parental-care circuit diagram has evolved substantially, with some versions depicting circuitry starting with sensory inputs and ending in motor outputs (Bridges 2015; Gammie 2005; Lambert and Kinsley 2012; Numan 2014; Olazábal et al. 2013; Zilkha et al. 2017). Contemporary parental-care circuitry diagrams are quite complex, as they integrate decades of studies; a recent iteration depicts 23 brain nuclei involved in regulation of infant-directed behavior in rodents (Dulac et al. 2014). Here we produce another circuit diagram iteration of what we consider the core parental-care circuitry underlying the expression of infant-directed care in mammals.

2.1 Overview

Many circuit diagrams, including the one shown in Fig. 1, depict infant-related sensory information being funneled into the medial preoptic area of the hypothalamus (MPOA) and bed nucleus of the stria terminalis (BNST) via several pathways: sensory cues from several modalities are routed through the amygdala via the stria terminalis and amygdalofugal pathway; through the prefrontal cortex, brainstem, and septum via the medial forebrain bundle; and through the thalamus via projections from at least the intralaminar complex, which conveys suckling information to the MPOA (Berk and Finkelstein 1981; Cservenák et al. 2013, 2017; Chiba and Murata 1985; Hoover and Vertes 2007; Leonard and Scott 1971; Myers et al. 2014; De Olmos and Ingram 1972; Vertes 2004). The MPOA is hypothesized to integrate these inputs, a process influenced by the hormonal milieu, and to facilitate parental care in two ways: (1) by inhibiting nuclei in the aggression/fear circuitry, such as the

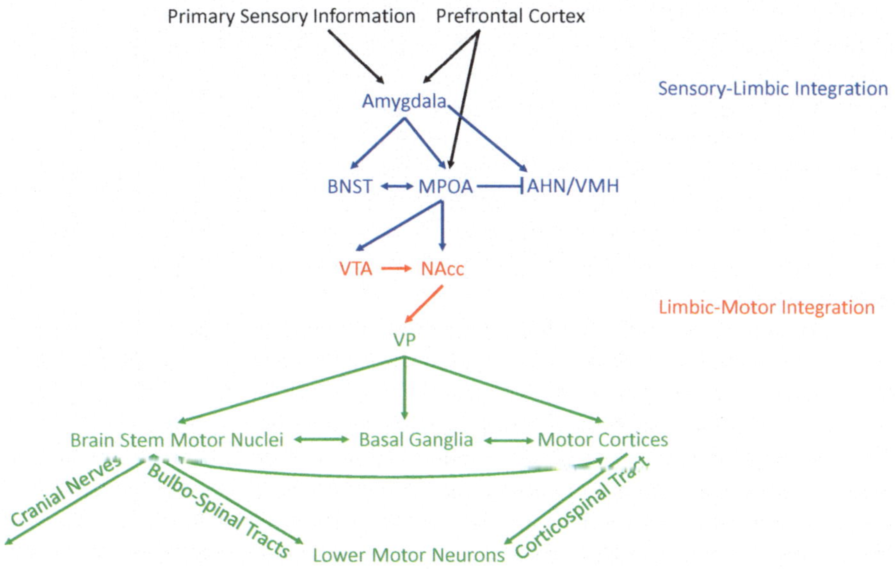

Fig. 1 Simple circuit diagram of parental care in mammals. Blue represents areas associated with sensory-limbic integration, red represents areas associated with limbic-motor integration, and green represents motor areas (see text for details). Abbreviations of brain areas: *BNST* bed nucleus of stria terminalis, *MPOA* medial preoptic area, *AHN* anterior hypothalamic nucleus, *VMH* ventromedial hypothalamus, *VTA* ventral tegmental area, *NAcc* nucleus accumbens, *VP* ventral pallidum

anterior hypothalamic nucleus, ventromedial nucleus of the hypothalamus, and periaqueductal gray, and (2) by exciting the reward circuitry via projections to the ventral tegmental area (VTA) and nucleus accumbens (NAcc). Thus, the MPOA has been conceptualized as a gate for infant-related stimuli to reach the mesolimbic reward pathway and mediate the hedonic tone of infant stimuli (Numan and Stolzenberg 2008).

The NAcc projects to the ventral pallidum. Together these nuclei have been referred to as the "final common pathway" (Mogenson 1987; Smith et al. 2009) for limbic information to influence motor systems, due in part to the ventral pallidum's projection to brainstem motor nuclei such as the pedunculopontine nucleus. The pedunculopontine nucleus sends descending projections to the spinal cord, as well as ascending projections to nuclei heavily implicated in voluntary behavior: the dorsal striatum, globus pallidus, and subthalamic nucleus, as well as supplementary, premotor, and primary motor cortices that contribute to signaling in Betz cells, the upper motor neurons of the descending corticospinal tract (Martinez-Gonzalez et al. 2011; Mena-Segovia et al. 2004; Winn 2006).

2.2 *Medial Preoptic Area*

Many studies implicate the MPOA in parental care in both males and females. In female rodents, the MPOA plays a key role in the expression of maternal behavior, as demonstrated by studies using a variety of techniques (e.g., lesions, stimulation, quantification of immediate-early gene expression, and glucose uptake) (Numan 2014). This brain region expresses receptors for prolactin, oxytocin, estrogen, and progesterone, hormones that fluctuate during the peripartum period in females; correspondingly, these hormones have been found to modulate maternal behavior, at least partly through actions on the MPOA (Ahdieh et al. 1987; Bosch and Neumann 2012; Bridges 1996; Bridges and Freemark 1995; Bridges and Mann 1994; Bridges et al. 1990; Brown et al. 2017; Fahrbach et al. 1986; Pedersen et al. 1994; Ribeiro et al. 2012).

Several studies have also implicated the MPOA in the expression of parental and alloparental behavior by males (Bales and Saltzman 2016), suggesting that paternal care in biparental species may be facilitated by increased activity in "maternal-care circuitry" present in mammals. For example, exposure to pups increases expression of the immediate-early gene protein product fos in the MPOA of fathers and/or allopaternally behaving virgin males in mice, rats, California mice, and prairie voles (De Jong et al. 2009; Horrell et al. 2017; Kirkpatrick et al. 1994a; Lambert et al. 2013). However, one study found no increase in fos expression in the MPOA of male California mice exposed to an unrelated pup compared to those exposed to a novel object (De Jong et al. 2010). In *Mus*, fathers that behave more paternally toward unrelated pups have greater fos expression in the MPOA following pup exposure than males that exhibit less allopaternal behavior (Tsuneoka et al. 2015). Stimulation of the MPOA increases paternal care, while lesions decrease paternal and allo-paternal care, including mate-dependent paternal care, in *Mus* (Akther et al. 2014;

Tsuneoka et al. 2015). Similarly, MPOA lesions in male rats prevent and disrupt pup-induced (i.e., sensitized) allopaternal behavior (Rosenblatt et al. 1996; Sturgis and Bridges 1997). MPOA lesions also disrupt paternal and allopaternal behavior in naturally biparental species such as the California mouse (Lee and Brown 2002, 2007).

Identification of the MPOA cells responsible for the effects observed in these lesion studies has proved more difficult. Colocalization of an immediate-early gene signal (fos, egr-1, or jun protein or mRNA) after parental-care behavior with a cell-type-specific marker can reveal which cell types, categorized by gene expression, are involved in paternal care. Such studies have implicated many cell types in maternal behavior (see Wu et al. 2014), with little evidence of the cell types involved in paternal care (Tsuneoka et al. 2017). Estradiol implants in the MPOA of male rats decrease sensitization times, demonstrating the MPOA to be an action site for estrogenic facilitation of allopaternal care in rats (Rosenblatt and Ceus 1998). Activity in MPOA neurons that express the neuropeptide galanin (Gal+ neurons), which are largely GABAergic, plays a causal role in paternal care in *Mus*: optogenetic activation of Gal+ neurons induced allopaternal behavior, while lesioning them reduced allopaternal care (Wu et al. 2014). These elegant experiments indicate that this Gal+ population is involved in paternal care, but it does not rule out possible roles for other neuronal populations in the MPOA, nor does it explain why only a subset of Gal+ neurons are active during paternal behavior. Recent work in female rats suggests that some Gal+ neurons are responsive to prolactin and receive thalamic projections responsive to nursing stimulation (Cservenák et al. 2017). Because males do not nurse, a role of these inputs in paternal care is questionable, though perhaps ventral tactile stimulation during thermoregulation/huddling may activate this pathway in males. The efferent connectivity of Gal+ MPOA neurons has yet to be determined.

Colocalizing indicators of neural activity can be paired not only with indicators of gene expression to increase our understanding of specific cell types involved in parental care but also with tracers to increase understanding of the functional connectivities of relevant neurons. The cells that express fos after maternal-care behavior in the MPOA of rats show a variety of efferent connectivities, including at least the ventromedial nucleus of the hypothalamus, lateral septum, VTA, and periaqueductal gray (Numan and Numan 1997; Numan and Sheehan 1997). Tsuneoka et al. (2013) suggest that a GABAergic pathway from the central MPOA to the rhomboid nucleus of the BNST regulates infant-directed behavior in male *Mus*. How this pathway relates to the Gal+ findings detailed above is not known. In particular, how distinct populations of MPOA neurons account for the enhancement of activity in reward circuitry and the decrease in activity in aggression circuitry (Numan 2014) in response to a pup stimulus is not understood.

2.3 Medial Amygdala

As with the MPOA, cues from multiple sensory modalities converge on the amygdala, which drives hypothalamic activity directly via the stria terminalis and amygdalofugal pathways. Medial amygdala (MeA) lesions facilitate maternal behavior in rats (Fleming et al. 1980; Numan et al. 1993; Oxley and Fleming 2000). In contrast, lesions of the MeA (or corticomedial amygdala) reduce allopaternal care in virgin male prairie voles (Kirkpatrick et al. 1994a). Lesions of the basolateral amygdala have been reported to decrease allopaternal care in California mice but not in prairie voles (Kirkpatrick et al. 1994a; Lee and Brown 2007).

Limited data are available on the MeA projections to the MPOA. In adult female rats that do not exhibit allomaternal care upon first exposure to pups, indiscriminate stimulation of the MeA produces inhibition or no response (and never excitation) in the MPOA (Gardner and Phillips 1977). The MeA also projects to many other nuclei, including hypothalamic regions associated with defensive and aggressive behaviors such as the anterior hypothalamic nucleus and ventromedial nucleus of the hypothalamus, which likely modulate infant-directed behaviors (Canteras et al. 1995).

The cell types in the MeA or other subnuclei of the amygdala that relay pup-related information to the MPOA or other brain regions have yet to be identified. After identification, investigation into the cues that influence the activity of these cells (e.g., olfactory and auditory cues from pups and mates, hormones, neuropeptides) can be conducted. The identification of these cells, the characterization of their properties, and manipulation of their activity would validate (or weaken) and extend hypotheses about the parental-care circuitry.

2.4 Mesolimbic Reward Pathway

The mesolimbic reward pathway plays a role in parental care, demonstrated mainly by research in female rats. Electrical lesions and $GABA_A$ and $GABA_B$ receptor agonists in the VTA reduce maternal care (Numan and Smith 1984; Numan et al. 2009). The VTA sends dopaminergic projections to the NAcc; however, the role of NAcc in parental care is debated. Some studies report that NAcc lesions profoundly decrease maternal care, while other studies report mixed, minor, or no effects (Lee et al. 1999, 2000; Li and Fleming 2003a, b; Numan et al. 2005; Smith and Holland 1975). The role of dopamine in NAcc is unclear as well. Dopamine levels in this region increase when rat dams interact with their young (Champagne et al. 2004; Hansen et al. 1993). Cis-flupenthixol, a nonspecific dopamine receptor antagonist, decreases retrieval and licking of pups as well as nest building, but so does cocaine, which increases dopamine levels (Keer and Stern 1999; Vernotica et al. 1999). Dopamine-depleting 6-hydroxydopamine lesions of the NAcc increase retrieval latency unless dams are deprived of pups for 3–6 h before retrieval tests (Hansen 1994; Hansen et al. 1991). D1- and D2-type receptors are present in the NAcc: administration of a D1 receptor-specific agonist in the NAcc increases retrieval of pups, while a D1 receptor antagonist into the NAcc shell decreases retrieval (Numan et al. 2005; Stolzenberg et al. 2007). Administration of D2 receptor-specific antagonists in the NAcc shell has been shown to have no effect on or decrease maternal behavior in rats, while administration of a D2 receptor agonist in the NAcc has no effect on maternal behavior (Numan et al. 2005; Silva et al. 2003). These data led Numan and Stolzenberg (2008) to hypothesize that inhibitory D1 receptor signaling in the NAcc disinhibits the ventral pallidum, allowing for the expression of parental behavior, especially the appetitive components.

The preoptic area and ventral BNST (vBNST) are often thought to alter activity in the mesolimbic reward pathway in response to infant-related stimuli via projections to both the VTA and NAcc (Kaufling et al. 2009; Numan and Numan 1997). Both the vBNST and MPOA have glutamatergic and GABAergic projections to the VTA (Geisler et al. 2007; Jennings et al. 2013; Tobiansky et al. 2013, 2016); however, a role of these projections in parental care has yet to be demonstrated. Lesions of the central MPOA (cMPOA) disrupt parental behavior in female *Mus* more strongly than lesions of other regions of the MPOA, and the majority of neurons in the cMPOA that express fos after maternal behavior are GABAergic, while very few are glutamatergic (Tsuneoka et al. 2013). Interestingly, in female rats, the majority of cMPOA neurons that project to the VTA express estrogen receptors (ER), while almost none express progesterone receptors (Fahrbach et al. 1986; Morrell et al. 1984; Tobiansky et al. 2016), which strongly implicates estrogen in gating of pup-related information to the mesolimbic reward pathway through actions on the MPOA. Oxytocin+ and neurotensin+ neurons in the MPOA have also been demonstrated to project to the VTA in female mice and rats, respectively (Geisler and Zahm 2006; Shahrokh et al. 2010; Tsuneoka et al. 2013). Selective manipulation of the activity of any subpopulation of MPOA neurons (e.g., GABAergic, glutamatergic, neurotensin+, oxytocin+) that project to mesolimbic reward circuitry in the context of parental care has not yet been conducted in any species.

As mentioned above, the NAcc projects to the ventral pallidum in what has been called the final common pathway for limbic information to reach the motor system (Mogenson 1987; Smith et al. 2009). Along with the VTA, these areas are responsible for limbic-motor integration and ultimately the control of output (Fig. 1). Excitotoxic lesions or muscimol inactivations of the ventral pallidum decrease maternal behavior in rats (Numan et al. 1988, 2005). NAcc lesions mildly decrease pup retrieval in male California mice but do not affect retrieval in female California mice, partially supporting this view (Lee and Brown 2007). Ventral pallidum lesions decrease paternal care in *Mus* (Akther et al. 2014). Finally, the ventral pallidum projects to brainstem motor nuclei, such as the pedunculopontine nucleus, that project to the basal ganglia and motor cortices, thereby contributing to descending motor pathways (Winn 2006). Limbic areas including the amygdala, MPOA, and BNST are thought to integrate sensory information and facilitate the activity of this part of the circuit to produce the appropriate infant-directed behaviors while suppressing aggressive behaviors (see Fig. 1).

3 Sensory Influences on Paternal Care

The expression of paternal and allopaternal care is influenced by numerous stimuli from both the internal and external environments, detected through multiple sensory modalities (Fig. 2; Brown 1993). Most studies have focused on cues from adult females and offspring. In biparental mammals, fathers typically begin to engage in paternal care after copulation and a period of cohabitation with their pregnant mate. Not surprisingly, therefore, cues from both the mate and pups, including olfactory, auditory, and tactile stimuli, can facilitate the onset and maintenance of paternal care and influence activity in the underlying neural circuitry. Additionally, copulation, specifically ejaculation, activates time-delayed neural mechanisms in male mice and rats, independent of further exposure to the

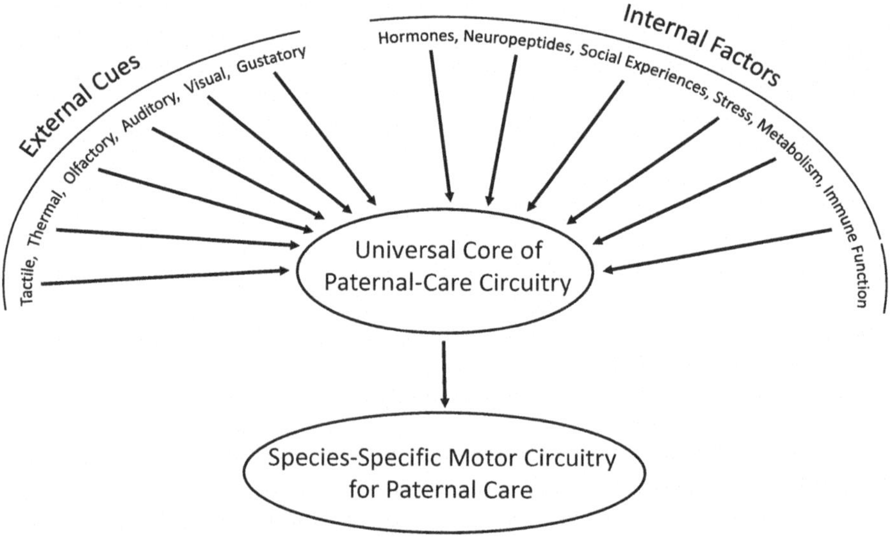

Fig. 2 Overview of external (environmental) and internal factors potentially influencing expression of mammalian paternal behavior via actions on the paternal-care neural circuitry. The specific factors affecting paternal behavior are likely to differ among and potentially within species and may change over time within individual animals based on such variables as age and parental experience. Moreover, these influences are likely to interact in complex ways. For example, social experience (e.g., mating, parental care received) might affect male parental behavior by eliciting changes in hormones (e.g., testosterone, estrogen), which in turn can alter signaling by neuropeptides (e.g., arginine vasopressin, oxytocin). Finally, influences on parental care can occur over different time scales. For example, exposure to specific hormones or neuropeptides during early development, or type of parental behavior received, can affect parental behavior expressed later in life, whereas sensory or metabolic cues might have more acute effects

mate, such that infanticidal behavior is inhibited and paternal behavior is promoted around the time that the males' pups are born (Brown 1993; Mennella and Moltz 1988b; vom Saal 1985). The relative importance of cues from copulation, from females, and from pups in both inhibiting aggression toward pups and facilitating paternal behavior differs markedly among species. Moreover, although olfactory cues play a key role in rodents, the roles of other sensory modalities are largely unknown.

3.1 *Olfaction*

Olfaction is a major sensory modality of rodents. Thus, it is not surprising that olfactory cues from females facilitate paternal care in multiple rodent species. In ICR mice (an outbred strain of *Mus*), olfactory cues from the mate both increase the number of fathers that retrieve pups and elicit fos expression in fathers' MPOA (Liang et al. 2014; Zhong et al. 2014). (Note that we specify strain of mice only when different results have been obtained in other strains.) In the biparental California mouse, fathers engage in high levels of paternal behavior after the birth of their own pups, but paternal responsiveness wanes within the first few days postpartum unless fathers are housed continuously with their mates or exposed to a chemosignal from their mates' urine; exposure to urinary chemosignals from an unfamiliar lactating female, rather than from the familiar mate, does not facilitate paternal care (Gubernick 1990; Gubernick and Alberts 1989). In male prairie voles, in contrast, paternal responsiveness rises following 3 days of housing and mating with a female, but olfactory cues do not appear to play a role in subsequent maintenance of this responsiveness: following removal from their mate, males continue to show high levels of paternal responsiveness whether or not they are continuously exposed to excreta from the female (Jean-Baptiste et al. 2008). Similarly, in Mongolian gerbils, first-time fathers show pronounced inhibition of infanticide around the time of the birth of their first litter, which is dependent upon interactions with the mate. Again, however, olfactory cues do not appear to be important: males removed from their mates during late pregnancy again become infanticidal toward unfamiliar pups, and this effect is not influenced by continuous exposure to excreta or soiled bedding from the mate (Elwood and Ostermeyer 1984).

The olfactory chemical messenger(s) that are emitted from females and influence paternal care have yet to be fully identified, but analyses of the composition of urine have been conducted in multiple species, including house mice and California mice (e.g., Idborg et al. 2005; Jemiolo et al. 1989, 1994; Plumb et al. 2003). Urine and secretions of the preputial gland have been implicated in maternal behavior (Brouette-lahlou et al. 1991b; Londei et al. 1989). In particular, dodecyl propionate, a component of preputial gland secretions, has been implicated in anogenital licking, a component of parental care in rodents. Further analysis of the exact olfactory chemosignals, their effect on olfactory neurons, the connectivity of those neurons, and the modulation by hormones and neuropeptides implicated in parental-care behavior has not been conducted (Brouette-lahlou et al. 1991a).

In reductive parental-care circuitry diagrams, olfactory information is often portrayed as being relayed from the main and accessory olfactory bulbs through the MeA and then conveyed through the stria terminalis to hypothalamic nuclei, which alter activity in the mesolimbic reward circuitry. It is important to remember, though, that the main and accessory olfactory bulbs project to many nuclei, in addition to the MeA, that might influence parental care and, conversely, that hypothalamic nuclei such as the MPOA receive olfactory information from other nuclei besides the MeA (for reviews see Courtiol and Wilson 2014; Pro-Sistiaga et al. 2007; Spehr et al. 2006). Both the main and accessory/vomeronasal (VNO) olfactory systems have been implicated in paternal care in rodents (see below), and at least some main and accessory olfactory bulb signals are integrated in individual MeA neurons (Guthman and Vera 2016; Keshavarzi et al. 2014).

The main olfactory bulb has been implicated in infant-directed behaviors in males of several rodent species. Olfactory bulb lesions increase pup-directed aggression in virgin male prairie voles, which typically display allopaternal care (Kirkpatrick et al. 1994a, b). In adult virgin male rats, the opposite effect is seen: olfactory bulbectomy or zinc sulfate treatment of the nasal epithelium decreases sensitization times (Fleischer et al. 1981; Mayer et al. 1979). In C57BL6 mouse fathers, offspring recognition is at least partly mediated by the integration of newly generated interneurons in the main olfactory bulb and hippocampus (Mak and Weiss 2010).

Lesions or inactivation of the VNO in virgin male mice reduce infanticide and increase allopaternal behavior toward pups (Orikasa et al. 2017; Tachikawa et al. 2013; Wu et al. 2014). Similarly, VNO removal decreases infanticide in male rats (Mennella and Moltz 1988a). Until recently, the cell type(s) in the VNO that process pup-related cues implicated in paternal care were not known. Odorant receptor Olfr692 is highly expressed in the VNO of adult male mice, and correlational studies suggest that VNO sensory neurons expressing Olfr692 may play a crucial role in infant-directed behavior: exposure to pups, to pups encased in wire mesh balls preventing direct contact, or to ligands washed off pups increases expression of the immediate-early gene Egr1 in Olfr692-positive VNO neurons in virgin males, while exposure to odors from predators or adult conspecifics does not (Nakahara et al. 2016). Fathers exposed to unrelated pups show less activity in Olfr692-positive cells than virgin males, although the two groups have the same number of Olfr692-positive cells (Nakahara et al. 2016). Virgin females and mothers also have Olfr692-positive neurons in the VNO, but very few of these neurons express EGR-1 after pup exposure. These correlational studies are intriguing, but studies investigating the causal relationship between activity of Olfr692-expressing neurons and parental behavior remain to be conducted. Moreover, the pup-emitted ligand(s) for Olfr692 has not yet been identified. Other cell types in the VNO have not been investigated with respect to paternal behavior.

Although the bed nucleus of the accessory olfactory tract (BAOT) does not appear in many circuit diagrams to date, BAOT lesions decrease sensitization latencies in male rats (Cruz and Del Cerro 1998; Izquierdo et al. 1992). The particular cell types in the BAOT that respond to pup cues, the properties and connectivities of those cells, and whether they undergo plasticity during the transition into fatherhood have yet to be determined.

3.2 Audition

Infants of virtually all mammalian species cry (i.e., produce age-specific vocalizations when distressed or separated from caregivers) (Newman 2007). Rodent pups produce a variety of vocalizations, some of which are ultrasonic (USVs). USVs may be produced in response to a change in temperature, olfactory stimuli, abnormal tactile stimulation, and/or loss of social contact (e.g., Conely and Bell 1978; Ehret 2005; Geyer 1979; Marchlewska-Koj et al. 1999). In *Mus*, production of USVs in the range of 40–80 kHz raises the pup's body temperature and elicits parental care (Blumberg and Sokoloff 2001; Ehret 2005). Fathers and virgin male *Mus* that have prior experience with pups, but not inexperienced virgin males, show a preference for modeled USVs compared to ultrasonic sounds dissimilar to pup USVs (Ehret 2005; Ehret and Buckenmaier 1994; Ehret and Koch 1989).

Vocalizations from female mates also elicit paternal care in *Mus*. Adult female ICR mice emit a series of 38-kHz USVs when deprived of their pups. When played back to their mates, these vocalizations can increase both retrieval of related pups and fos expression in the MPOA (Liang et al. 2014; Liu et al. 2013; Zhong et al. 2014).

In two biparental primate species, humans and common marmosets, fathers show enhanced behavioral or affective responses to infant distress vocalizations compared to non-fathers, suggesting that infant cries are more likely to elicit nurturing behavior from fathers (Fleming et al. 2002; Ziegler and Sosa 2016). Moreover, responses to infant distress cries are in some cases correlated with or influenced by hormone (estrogen, testosterone, prolactin) or neuropeptide (oxytocin) levels in adult males (Fleming et al. 2002; Li et al. 2017; Ziegler and Sosa 2016).

3.3 Somatosensation

Physical contact can promote paternal care in both rodent and human fathers. In prairie voles, males that have direct physical contact with their mate spend more time in contact with experimentally presented pups, compared to males that receive only distal cues from their mate (Simoncelli et al. 2010). Similarly, in Mongolian gerbils, inhibition of infanticide prior to the birth of a male's first litter appears to be largely dependent upon physical (and possibly visual) contact with the mate (Elwood and Ostermeyer 1984). Skin-to-skin contact between men and their infants has beneficial effects on both infants and fathers, including elevating fathers' feelings of attachment toward their infant (Chen et al. 2017; Cong et al. 2015; Shorey et al. 2016).

The neuroendocrine pathways by which tactile cues (except those from suckling infants) facilitate parental care in either sex are not well known. In primates and rodents, physical contact with an infant can stimulate release of both prolactin and oxytocin in mothers and potentially in fathers (Cong et al. 2015; Dixson and George 1982; Lonstein 2007), which in turn might promote nurturant behavior. At the level of neural circuitry, Seelke et al. (2016) found numerous differences in both extrinsic and intrinsic connections of S1 somatosensory cortex in adult prairie voles that had received different patterns of care, including different amounts of contact, from their own parents, which the authors categorized as "low-contact" and "high-contact" offspring. The two categories of prairie vole offspring express behavioral responses to infants consistent with the parental care that they themselves received (Perkeybile et al. 2015); thus, patterns of S1 connectivity are associated with both parental care received and parental care performed. It is unknown, however, whether these associations reflect causal or simply correlational relationships among parental care, tactile cues, and somatosensory cortex.

In female *Mus*, neurons in the posterior intralaminar complex of the thalamus that express tuberoinfundibular peptide of 39 residues (Tip39) are activated by suckling, project to neurons in the MPOA including galanin+ neurons, and play a role in maternal care (see Cservenák et al. 2017; Dobolyi et al. 2014). The posterior intralaminar complex may also convey auditory information to the MPOA (Campeau and Watson 2000; Dobolyi et al. 2014). The role of these pathways in male parental behavior, if any, has not been investigated.

3.4 Sensory Influences: Conclusions

Clearly, we are far from understanding the neural circuitry of sensory inputs influencing the expression of paternal care. In addition to the olfactory, auditory, and tactile cues discussed above, other types of sensory information from pups and/or mates (i.e., visual, thermal, gustatory), and possibly from other social or nonsocial sources (e.g., male rivals, predators), might converge on central paternal-care circuitry to facilitate or inhibit paternal behavior. Where and how different sensory systems act on core paternal circuitry, and potential interactions among diverse sensory inputs, are not fully understood. In ICR mice, for example, both auditory and olfactory stimuli from dams increase MPOA fos expression and elicit paternal care in their male mates (Zhong et al. 2014). Neither deafening nor anosmia alone decreases the percentage of ICR mouse sires that retrieve pups; however, concurrent deafness and anosmia abolish retrieval, suggesting redundancy between auditory and olfactory cues (Liu et al. 2013). Similarly, in female *Mus*, olfactory and auditory cues from pups presented simultaneously cause more fos expression in the MPOA than either stimulus presented alone (Okabe et al. 2013). In both males and females, it is unknown if olfactory and auditory cues from mates or pups converge on the same MPOA neurons. Compartment analysis of temporal activity by fluorescent in situ hybridization (catFISH), quantifying immediate-early gene expression after sequential presentation of pup-related olfactory and auditory cues, would advance our knowledge of how multimodal sensory cues converge on core parental-care circuitry.

Finally, potential effects of fatherhood and its attendant neuroendocrine changes on plasticity within sensory systems have received little attention. In *Mus* fathers, however, interactions with neonates during the early postpartum period, likely involving physical contact, stimulate neurogenesis in the hippocampus and olfactory bulb, mediated by prolactin (Mak and Weiss 2010). These newly generated neurons respond preferentially to odors from the father's adult offspring and are thought to be involved in kin recognition. Further studies of effects of fatherhood and of so-called "paternal hormones" on sensory plasticity might yield fascinating insights into the proximate control of paternal behavior.

4 Hormonal and Neuropeptide Influences on Paternal Care

Similar to the multiplicity of sensory modalities involved in parental behavior, reviewed above, paternal care involves a confluence of changes in endocrine and neuropeptide signaling acting on multiple sites throughout the brain to orchestrate infant-directed behavior. Many studies in rodents and primates have identified hormonal fluctuations in males during the transition into fatherhood. In most cases, however, these changes have not been shown to have causal effects on the expression of paternal care (reviewed in Saltzman and Ziegler 2014) (Table 1).

Below we briefly review findings on potential neuroendocrine influences on paternal behavior and its underlying neural circuitry. We focus on several categories of steroid hormones (androgens, estrogens, progestagens, glucocorticoids), the peptide hormone prolactin, and the neuropeptides arginine vasopressin (AVP) and oxytocin, all of which have been implicated in paternal behavior in mammals. For each of these signaling systems, we first describe observed changes across reproductive states in fathers and correlations between hormone or neuropeptide levels and paternal behavior. Several recent papers provide detailed reviews of hormonal changes in mammalian fathers (Bales and Saltzman 2016; Gettler 2014; Saltzman and Ziegler 2014; Storey and Ziegler 2016); therefore, we discuss this topic rather briefly. We then review experimental evidence, where available, that these changes influence paternal care. Finally, we discuss known or postulated effects of these hormonal and neuropeptide changes on the neural circuitry underlying paternal care.

Table 1 Changes in circulating or excreted hormone levels of male mammals during the transition into fatherhood

Species	Androgens	Estrogens	Progesterone	Prolactin	Vasopressin	Oxytocin	Glucocorticoids
Norway rat (*Rattus norvegicus*)	nd	nd	nd	nd	nd	nd	nd
House mouse (*Mus musculus*)	nd	nd	nd	nd	nd	nd	nd
California mouse (*Peromyscus californicus*)	No change (Gubernick and Nelson 1989), decrease (Trainor et al. 2003)	nd	Decrease (Trainor et al. 2003)	Increase (Gubernick and Nelson 1989)	nd	Decrease (compared to expectant fathers, Gubernick et al. 1995), no change (compared to virgins, Gubernick et al. 1995)	No change (Chauke et al. 2011; Harris et al. 2013)
Prairie vole (*Microtus ochrogaster*)	nd	nd	nd	nd	nd	nd	No change (Campbell et al. 2009)
Campbell's dwarf hamster (*Phodopus campbelli*)	No change (Schum and Wynne-Edwards 2005), decrease (Reburn and Wynne-Edwards 1999)	No change (Schum and Wynne-Edwards 2005)	Increase (Schum and Wynne-Edwards 2005)	Increase (Reburn and Wynne-Edwards 1999)	nd	nd	Decrease (Reburn and Wynne-Edwards 1999)
Djungarian hamster (*Phodopus sungorus*)	Decrease (Reburn and Wynne-Edwards 1999; Schum and Wynne-Edwards 2005)	Decrease (Schum and Wynne-Edwards 2005)	No change (Schum and Wynne-Edwards 2005)	No change (Reburn and Wynne-Edwards 1999)	nd	nd	No change (Reburn and Wynne-Edwards 1999)
Mongolian gerbil (*Meriones unguiculatus*)	Decrease (Brown et al. 1995)	nd	nd	Increase (Brown et al. 1995)	nd	nd	nd

Volcano mouse (*Neotomodon alstoni*)	No change (Luis et al. 2009)	nd	nd	nd	nd	nd	nd
Striped mouse (*Rhabdomys pumilio*)	Decrease (Schradin and Yuen 2011)	nd	nd	Increase (Schradin 2008)	nd	nd	nd
Mandarin vole (*Microtus mandarinus*)	nd	nd	nd	nd	nd	nd	nd
Common marmoset (*Callithrix jacchus*)	Decrease (Ziegler et al. 2009b)	nd	nd	No change (but increases with infant contact: Mota and Sousa 2000; Mota et al. 2006), increase (Schradin et al. 2003)	nd	nd	nd
Wied's black-tufted-ear marmoset (*Callithrix kuhlii*)	Decrease (Nunes et al. 2000)	Decrease (Nunes et al. 2000)	nd	nd	nd	nd	nd
White-faced marmoset (*Callithrix geoffroyi*)	No change (Cavanaugh and French 2013)	No change (Cavanaugh and French 2013)	nd	nd	nd	nd	No change (Cavanaugh and French 2013)
Cotton-top tamarin (*Saguinus oedipus*)	No change (Ziegler et al. 2000)	nd	nd	No change (Ziegler et al. 2000)	nd	nd	No change (Ziegler et al. 2000)
Titi monkey (*Callicebus cupreus*)	nd	nd	nd	Increase (Schradin et al. 2003)	nd	nd	nd

(continued)

Table 1 (continued)

Species	Androgens	Estrogens	Progesterone	Prolactin	Vasopressin	Oxytocin	Glucocorticoids
Human (*Homo sapiens*)	No change (Magid 2011), decrease (Alvergne et al. 2009; Berg and Wynne-Edwards 2001; Gettler et al. 2011, 2013; Kuzawa et al. 2009; Gray et al. 2002, 2006; Gray and Campbell 2009; Storey et al. 2000; Van Anders and Gray 2007)	nd	Increase (Berg and Wynne-Edwards 2001), decrease (Storey et al. 2000)	nd	No change (Gray et al. 2007)	No change (Gray et al. 2007)	No change (Gray et al. 2007), decrease (Berg and Wynne-Edwards 2001)

Data from both longitudinal and cross-sectional designs are included. "Increase" and "decrease" refer to hormone levels of fathers at any time point after parturition compared to any time point before parturition or in non-fathers (e.g., virgins or mated males that have not produced offspring). *nd* no data

4.1 Androgens

4.1.1 Effects of Fatherhood on Androgen Signaling

Males in many species undergo decreases in circulating or urinary testosterone levels around the time of their mates' parturition. Among rodents, studies of biparental Campbell's dwarf hamsters, Mongolian gerbils, and California mice, as well as uniparental Djungarian hamsters (*Phodopus sungorus*), have found lower plasma testosterone levels in fathers after parturition than during their mates' gestation, before pairing, and/or virgin controls (Brown et al. 1995; Reburn and Wynne-Edwards 1999; Trainor et al. 2003). Other studies, however, have found no differences in plasma testosterone levels of fathers, mated males before parturition, and/or virgin controls in these same biparental species (Gubernick and Nelson 1989; Reburn and Wynne-Edwards 1999; Schum and Wynne-Edwards 2005) as well as in the biparental volcano mouse (*Neotomodon alstoni*) (Luis et al. 2009). Some of these disparities may be due to the lack of daily sampling, circadian effects on androgen levels, and/or the pooling of data from multiple days or weeks for analysis.

Among nonhuman primates, endocrine correlates of and influences on paternal care have been studied primarily in the New World callitrichid monkeys (marmosets and tamarins). Common marmosets and black-tufted-ear marmosets (*C. kuhlii*) have been reported to exhibit decreases in plasma or urinary testosterone levels from before to after their mates' parturition (Ziegler et al. 2009b; Nunes et al. 2000). Consistent with these findings, urinary testosterone levels and infant-carrying are negatively correlated in black-tufted-ear marmoset fathers (Nunes et al. 2001), and exposure to infant scents decreases plasma testosterone in experienced, but not inexperienced, common marmoset fathers (Prudom et al. 2008; Ziegler et al. 2009a). Other studies, however, report no change in testosterone levels in common marmosets, white-faced marmosets (*C. geoffroyi*), and cotton-top tamarins (Cavanaugh and French 2013; Dixson and George 1982; Ziegler and Snowdon 2000), and two studies found elevated urinary testosterone and dihydrotestosterone levels in the final month(s) of their mate's gestation in cotton-top tamarins (Ziegler and Snowdon 2000; Ziegler et al. 2004). Importantly, marmosets and tamarins typically undergo postpartum ovulation (Digby et al. 2007); consequently, the gestation and lactation periods may overlap substantially, complicating the interpretation of changes in males' endocrine function across reproductive stages.

Studies of humans, both cross-sectional and longitudinal, have found reduced salivary or plasma testosterone concentrations in fathers compared to non-fathers in a wide range of countries and cultures (reviewed by Gettler 2014; Storey and Ziegler 2016). Both within and between cultures, testosterone levels often correlate negatively with the amount of time men spend with their children, especially the amount of time in positive interactions. Although causality cannot be inferred from these correlational findings, several experimental studies have found that exposure to infant cues associated with negative affect (e.g., distress cries, sad facial expressions) acutely elevate fathers' testosterone levels, especially when the fathers are not able to comfort the infant (see Storey and Ziegler 2016). Thus, engaging in caretaking behavior, rather than fatherhood per se, may suppress testosterone concentrations in human fathers.

4.1.2 Effects of Androgen Signaling on Paternal Care and the Underlying Neural Circuitry

Androgens can influence the expression of paternal behavior through both long-term organizational effects originating during early stages of development and acute activational effects later in life; however, these effects differ both within and among species. For example, early-life exposure to androgens reduces adult levels of alloparental responsiveness in male rats (McCullough et al. 1974; Rosenberg and Herrenkohl 1976) but increases alloparental responsiveness in adult male prairie voles (Kramer et al. 2009; Lonstein et al. 2002; reviewed in Bales and Saltzman 2016).

Activational effects of androgens on paternal responsiveness, like organizational effects, vary within and among species. In rats, administration of testosterone increases infanticide in males that were castrated in adulthood (Lubin et al. 1972; Rosenberg 1974; Rosenblatt et al. 1996). In California mouse fathers, on the other hand, castration reduces and testosterone replacement restores paternal and/or allopaternal behavior, an effect mediated, in large part, by aromatization of testosterone to estrogen (Trainor and Marler 2002). Similarly, in Mongolian gerbils housed in same-sex groups, testosterone increases allopaternal responsiveness in virgin males castrated in adulthood; however, treatment with either estrogen or dihydrotestosterone (which cannot be aromatized) has the same effect, suggesting that stimulatory effects of androgens in this species do not require aromatization (Martinez et al. 2015). Interestingly, opposite – i.e., inhibitory – effects of testosterone have been found in virgin male gerbils housed with a lactating female (Clark and Galef 1999). Studies of prairie voles have likewise yielded mixed results: castration either reduces (Wang and De Vries 1993) or does not alter (Lonstein and De Vries 1999) responses to unrelated pups in virgin males. Finally, castration did not alter paternal behavior in a study of Campbell's dwarf hamster fathers (Hume and Wynne-Edwards 2005).

Little is known about the neurobiological effects of androgen signaling that mediate paternal care, though substantial research has investigated effects of androgen signaling in limbic circuitry, including regions considered to be part of the parental-care circuitry. Androgen receptors are widely distributed in rat brain, including the limbic areas of canonical parental-care circuitry and throughout the olfactory and auditory sensory systems (Kritzer 2004; Simerly et al. 1990). Cellular effects of testosterone, at least in uniparental rodents, include increasing spine densities (Cunningham et al. 2007; de Castilhos et al. 2008; Garelick and Swann 2014; Zancan et al. 2017); modulating signaling by dopamine, nitric oxide, and substance P (Dees and Kozlowski 1984; Du and Hull 1999; Du et al. 1998; Hadeishi and Wood 1996; Malsbury and McKay 1994; Swann and Newman 1992); and altering Na+/K+-ATPase activity (Guerra et al. 1987). Whether any of these effects contribute to androgenic modulation of paternal care is not known.

Testosterone also exerts organizational effects in the MPOA, often studied in the context of sexual behavior. These effects include modulation of cell death, cell morphology, opiate receptor expression, volumes of subnuclei, and astrocytic development (Arai et al. 1994; Dodson and Gorski 1993; Dohler et al. 1984; Hammer 1985; Jacobson et al. 1981; Mong and McCarthy 1999; Reddy et al. 2015; Roselli et al. 2007, 2015). Androgen levels during development also influence vasopressin levels (see below) in the BNST (Han and De Vries 2003). At least some of these organizational effects of androgens are mediated by intracellular aromatization to estrogen and subsequent binding to estrogen receptors (McCarthy 2010). Again, it is unknown which of these testosterone- or estrogen-dependent processes, if any, affect expression of paternal care later in life.

4.2 *Estrogen*

4.2.1 Effects of Fatherhood on Estrogen Signaling

Fatherhood can influence both circulating estrogen levels and central expression of estrogen receptors in rodents. Plasma estradiol concentrations in California mouse fathers are significantly higher than those of virgin males in the early postpartum period but not the mid- or late postpartum period (Hyer et al. 2017). In the biparental Campbell's dwarf hamster, on the other hand, fathers exhibit no changes in plasma estradiol levels during the transition into fatherhood (Schum and Wynne-Edwards 2005). Surprisingly, uniparental Djungarian hamster fathers undergo systematic fluctuations in plasma estradiol levels across reproductive stages: estradiol rises from before pairing to late gestation, decreases around parturition, and then significantly increases again at day 12 of the mate's lactation period (Schum and Wynne-Edwards 2005). The biological significance of these changes is not clear, since fathers in this species do not provide parental care.

ERα-immunoreactivity in several brain regions studied (MPOA, MeA, BNST) does not change during the transition into fatherhood in either biparental or uniparental hamsters (Timonin et al. 2008). Similarly, ERα mRNA expression in the MPOA, MeA, and BNST does not differ between California mouse fathers and virgin males (Perea-Rodriguez et al. 2015), although fathers in this species have higher activity of aromatase, the enzyme that converts androgens to estrogens, in the MPOA than non-fathers (Trainor et al. 2003). Mandarin vole fathers, in contrast to both hamsters and California mice, have lower ERα-immunoreactivity in the MPOA and BNST and higher ERα-immunoreactivity in the ventromedial nucleus of the hypothalamus, central nucleus of the amygdala, and MeA than virgin males (Song et al. 2010).

Findings to date do not suggest a consistent relationship between estrogen levels and paternal status in nonhuman primates: male common marmosets exhibit no changes in plasma estrogen concentrations from before to after their mate's parturition, while black-tufted-ear marmosets exhibit a decline or strong trend for a decline in urinary estradiol after parturition (Nunes et al. 2000, 2001; Ziegler et al. 2009a, b). White-faced marmoset fathers exhibit no change in urinary estrogen levels 2–8 weeks postpartum; however, no data were collected prepartum (Cavanaugh and French 2013). In paternally experienced, but not inexperienced, cotton-top tamarin fathers, urinary estrogen and estrone levels increase during the mate's late pregnancy (Ziegler et al. 2004), with no data collected after parturition.

Few studies have investigated the relationship between estrogen and fatherhood in men. Fathers were reported to have higher salivary estradiol levels than age-matched non-fathers in a small, cross-sectional study (Berg and Wynne-Edwards 2001); however other studies have found a decrease in men's estradiol levels during the partner's pregnancy or after parturition (Edelstein et al. 2015; Storey et al. 2000). Finally, no changes in estradiol were seen 40 or 70 min after fathers interacted with their toddlers (Gettler et al. 2013).

4.2.2 Effects of Estrogen Signaling on Paternal Care and the Underlying Neural Circuitry

Estrogen signaling has been found to both enhance and inhibit paternal care in rodents. Implantation of 17β-estradiol-releasing capsules in the MPOA of virgin male rats decreases sensitization latencies (Rosenblatt and Ceus 1998), implicating the MPOA as a site of action for estrogenic activation of paternal care. In ICR mouse fathers, aromatase immunoreactivity is stimulated by cues from the female mate and, more potently, suppressed by cues from pups, in brain regions implicated in paternal care, including the MPOA, MeA, NAcc, and ventral pallidum (Akther et al. 2015). Further, suppression of estrogen synthesis by the aromatase inhibitor letrozole inhibits retrieval of related pups by these mouse fathers (Akther et al. 2015). Deletion of either the aromatase gene or the ERα gene increases rates of infanticide and/or decreases expression of paternal behavior in male mice, but whether these effects reflect disruption of estrogen signaling in early life or in adulthood is not clear (Matsumoto et al. 2003; Ogawa et al. 1998). Estrogen also facilitates paternal behavior in California mouse fathers, as stated above (Trainor and Marler 2002). The site of the stimulatory effect of estrogen in this species is not known; however, fathers have higher aromatase activity in the MPOA than males housed with females that have not yet produced a litter (Trainor et al. 2003).

In contrast to uniparental rats and mice, as well as biparental California mice, estrogen inhibits paternal behavior in adult male prairie voles in a site-specific manner: increasing ERα expression in the MeA via viral vector decreases allopaternal behavior (Cushing et al. 2008), while increasing ERα expression in the BNST via viral vector has no effect (Lei et al. 2010). Finally, neither castration nor treatment with an aromatase inhibitor alters paternal behavior in Campbell's dwarf hamster fathers, suggesting that estrogen does not play an important role in the maintenance of paternal care in this biparental species (Hume and Wynne-Edwards 2005, 2006).

The specific cellular mechanisms by which estrogen influences paternal behavior in rats, California mice, prairie voles, and perhaps other species are not known. In general, estrogen can exert both slow effects on target cells by binding to intracellular receptors and inducing changes in gene transcription and rapid effects by binding to membrane receptors and altering cellular activity through nongenomic mechanisms (for reviews see Alexander et al. 2016; Arevalo et al. 2015; Kow and Pfaff 2016; Mani et al. 2012; McEwen et al. 2012; Pfaff et al. 2011). Acting through these mechanisms, estrogen can influence numerous characteristics of preoptic area neurons, including electrical activity, neurogenesis, synaptic plasticity, neurotransmitter release, and cellular morphology (for reviews see Garcia-Galiano et al. 2012; Herbison 1997; Kelly et al. 2013; Ronnekleiv et al. 2012, Zhang et al. 2013). Estrogen modulates many signaling pathways in the preoptic area of rats, including GABAergic (Herbison 1997; Herbison et al. 1989, 1990), glutamatergic (Mahesh and Brann 2005), adrenergic (MacKinnon et al. 1985), noradrenergic (Kelly and Wagner 1999; Szawka et al. 2013), and oxytocinergic pathways (Caldwell et al. 1994; Champagne et al. 2001), among others (see Etgen and Pfaff 2010). In female rats, estrogen decreases excitability of MPOA neurons that project to the MeA (Yoshida et al. 1994). Thus, estrogen has many effects on neuronal signaling in the MPOA in at least some rodent species and therefore might alter activity of MPOA neurons that influence the occurrence of paternal care.

Although no research has been performed on males, intriguing studies have been conducted on the effects of estrogen and motherhood on responses to auditory cues in female rodents (Caras 2013; Liu and Schreiner 2007; Miranda and Liu 2009; Miranda et al. 2014). Estrogen receptors are located throughout the central auditory system in mice (Charitidi and Canlon 2010). In female mice and rats, ERα levels in the cochlea vary over the estrous cycle, as do behavioral responses to infant vocalizations (Charitidi et al. 2012; Ehret and Schmid 2009; Simonoska et al. 2009). Furthermore, estradiol has been implicated in the development of a preference for USVs in female mice (Koch and Ehret 1989). Aromatase and ERs are also expressed in the main and accessory olfactory bulbs of male rats and mice (e.g., Cherian et al. 2014; Dillon et al. 2013; Guo et al. 2001; Hoyk et al. 2014); however, their role in paternal care, if any, is unknown. Effects of hormones on sensory processing in fathers remain a promising area for further research.

4.3 Progesterone

4.3.1 Effects of Fatherhood on Progesterone Signaling

Little is known about associations between paternal care and progesterone signaling. Plasma progesterone concentrations are lower in California mouse fathers 2–3 weeks after the birth of their pups than in virgin males (Trainor et al. 2003). In the same species, however, expression of progesterone receptor mRNA in the BNST is lower in fathers than in virgin males (Perea-Rodriguez et al. 2015). In the biparental Campbell's dwarf hamster, males' circulating progesterone levels are elevated on the day of their mates' parturition, whereas uniparental Djungarian male hamsters do not show any change in plasma progesterone levels throughout their mate's gestation and lactation (Schum and Wynne-Edwards 2005). Progesterone concentrations of human fathers do not change across their partners' pregnancy and have not been studied under baseline conditions during the postpartum period (Edelstein et al. 2015); however, salivary progesterone levels decrease in fathers 40 and 70 min after they play with their infants (Gettler et al. 2013).

4.3.2 Effects of Progesterone Signaling on Paternal Care and the Underlying Neural Circuitry

In female rodents, high levels of progesterone typically inhibit maternal behavior during late pregnancy (Bridges 2015). Similarly, progesterone signaling increases infanticide and reduces paternal and allopaternal behavior in adult male house mice (Schneider et al. 2003, 2009). To our knowledge, however, effects of progesterone on paternal care have not been examined in biparental mammals.

The mechanism by which progesterone reduces paternal care in mice is unknown. Progesterone signaling modulates olfaction in the main and accessory olfactory bulbs, though a connection to parental care has yet to be made (Dey et al. 2015; Kanageswaran et al. 2016). Progesterone can also be metabolized into allopregnanolone, an allosteric modulator of $GABA_A$ receptors, in multiple nuclei, including the MPOA (for review see Henderson 2007). Interestingly, a recent genetic analysis of 32 primate species revealed that the biparental New World monkeys have progesterone response elements in the oxytocin receptor promoter region, implicating progestagenic regulation of oxytocin receptor expression in paternal care in this taxon (Vargas-Pinilla et al. 2015, 2017). Finally, estrogen and progesterone alter volumes of subregions of the MPOA in adult male rats (Bloch and Gorski 1988).

4.4 Glucocorticoids

4.4.1 Effects of Fatherhood on Glucocorticoid Signaling

The glucocorticoid hormones (primarily cortisol and corticosterone) have a multitude of effects on physiology, cognition, affect, and behavior under both baseline and stressful conditions (Sapolsky et al. 2000) and therefore seem likely to influence paternal care. Baseline concentrations of glucocorticoids are notoriously difficult to measure, due to the pronounced sensitivity of glucocorticoid secretion to environmental and organismal influences such as time of day, physical activity, food intake, and stress. Such effects may contribute to differences both within and between studies. Correspondingly, perhaps, findings on associations between glucocorticoid signaling and fatherhood or parental care have been highly variable. Among biparental rodents, for example, male Campbell's dwarf hamsters show a transient elevation of circulating cortisol concentrations during their mate's mid-pregnancy (Reburn and Wynne-Edwards 1999), whereas basal corticosterone levels of male California mice and prairie voles do not differ between fathers and non-fathers (Campbell et al. 2009; Chauke et al. 2011; Harris et al. 2013). Further, repeated exposure of virgin male California mice enhances their paternal responsiveness to unrelated pups but does not alter either baseline plasma corticosterone levels or corticosterone responses to pups (Horrell et al. 2017).

Among nonhuman primates, urinary glucocorticoid levels of male cotton-top tamarins increase around the mate's mid or late pregnancy (Almond et al. 2008; Ziegler and Snowdon 2000; Ziegler et al. 2004) and are lower in experienced fathers living with multiple offspring than in new fathers or unmated males (Ziegler et al. 1996). In black-tufted-ear marmosets, males that carry infants at high rates have lower cortisol than males that carry infants at low rates. White-faced marmoset fathers exhibit no change in urinary cortisol levels 2–8 weeks postpartum, with no data before parturition (Cavanaugh and French 2013).

In humans, salivary cortisol levels of expectant fathers have been reported both to increase during the partner's late pregnancy (Berg and Wynne-Edwards 2001, 2002; Storey et al. 2000) and to remain unchanged throughout pregnancy (Edelstein et al. 2015). Moreover, some studies report no difference in salivary cortisol levels between fathers and non-fathers (Fleming et al. 2002; Gray et al. 2007), while others report lower morning cortisol concentrations in fathers compared to non-fathers (Berg and Wynne-Edwards 2001). Cortisol levels decrease after fathers spend 30 min with their toddlers (Storey et al. 2011).

4.4.2 Effects of Glucocorticoid Signaling on Paternal Care and the Underlying Neural Circuitry

Surprisingly little is known about effects of glucocorticoids on mammalian paternal behavior. In virgin male prairie voles, acute exposure to a swim stressor elevates circulating corticosterone concentrations and enhances paternal responsiveness to an unrelated pup; it is not known, however, whether the endocrine and behavioral consequences of the swim stressor are causally related (Bales et al. 2006). On the other hand, acute treatment of California mouse fathers with supraphysiological doses of corticosterone has little effect on pup-directed behavior (Harris et al. 2011). Moreover, exposure of California mouse fathers to a chronic variable stress paradigm can chronically elevate baseline corticosterone levels but causes only modest reductions in paternal care (Harris et al. 2013).

As there is very limited evidence to date that glucocorticoids affect paternal behavior, the relevant mechanisms mediating such effects remain poorly understood. However, research has begun to characterize effects of glucocorticoid signaling on social circuitry and behavior via genomic and nongenomic mechanisms (for reviews see de Kloet et al. 2008; Groeneweg et al. 2012; Haller et al. 2008).

4.5 *Prolactin*

4.5.1 Effects of Fatherhood on Prolactin Signaling

The anterior pituitary hormone prolactin has received significant interest as a "paternal hormone" (e.g., Hashemian et al. 2016; Schradin and Anzenberger 1999), as prolactin concentrations are elevated in mammalian fathers in numerous biparental species (e.g., rodents: Mongolian gerbils, California mice, and Campbell's dwarf hamsters [Brown et al. 1995; Gubernick and Nelson 1989; Reburn and Wynne-Edwards 1999]; primates: titi monkeys [*Callicebus cupreus*], Goeldi's monkeys [*Callimico goeldii*], common marmosets, and cotton-top tamarins [Ziegler et al. 1996; Dixson and George 1982; Schradin et al. 2003]). Prolactin levels can also be higher in experienced fathers than in first-time fathers (Ziegler et al. 1996, 2000, 2004), and acute changes in prolactin concentrations can occur in association with paternal care, especially tactile cues from infants. For example, common marmoset fathers have significantly higher plasma prolactin concentrations immediately after carrying their infants than at other times (Dixson and George 1982).

In humans, as in other biparental mammals, fathers have been reported to have higher baseline prolactin concentrations than non-fathers (Gettler et al. 2012), and baseline prolactin levels of fathers predict their emotional and behavioral responses to infants. For example, men with higher baseline prolactin concentrations have more positive feelings and more concern in response to infant cries and engage in more coordinated exploratory play with their infants (Fleming et al. 2002; Gordon et al. 2010; Storey et al. 2000). Human fathers also show acute changes in prolactin levels in response to interactions with or cues from infants; however, the direction of these changes is inconsistent: prolactin can either increase or decrease in response to stimuli such as hearing infant cries or holding an infant, effects that can be modulated by the fathers' parity or previous experience with infants (Delahunty et al. 2007; Fleming et al. 2002, 2011).

4.5.2 Effects of Prolactin Signaling on Paternal Care and the Underlying Neural Circuitry

Prolactin promotes the onset of sensitized paternal behavior in virgin male rats (Sakaguchi et al. 1996). Studies in biparental mammals, however, have failed to support a stimulatory effect of prolactin on paternal care. In Campbell's dwarf hamster, treatment of first-time fathers with either bromocriptine, which antagonizes both the D1 and D2 dopamine receptors, or cabergoline, a selective D2 agonist, 1–3 days before the birth of their pups successfully reduces peripheral prolactin concentrations but does not impair paternal behavior (Brooks et al. 2005). Similarly, treatment of common marmoset fathers with cabergoline does not impair the expression of paternal behavior (Almond et al. 2006). Treatment of parentally inexperienced common marmosets with bromocriptine did reduce infant-carrying; however, only two males, including one juvenile and one adult, were tested (Roberts et al. 2001). Thus, in spite of the weight of published evidence documenting a positive association between prolactin and paternal behavior, experiments that pharmacologically reduce prolactin signaling in naturally paternal animal models do not support a causal link.

4.6 *Vasopressin*

4.6.1 Effects of Fatherhood on Arginine Vasopressin Signaling

The neuropeptide AVP acts both peripherally, where it plays a key role in regulation of fluid volume and the stress response, and centrally, where it influences aggressive and affiliative behaviors, affect, and cognition (Caldwell et al. 2008; Zimmermann-Peruzatto et al. 2015). Numerous correlational and experimental studies in several biparental rodents have implicated AVP, especially vasopressinergic projections from the MeA and BNST to the lateral septum and lateral habenular nucleus, in regulating paternal care (reviewed by Bales and Saltzman 2016; Frazier et al. 2006; Perkeybile and Bales 2017; Saltzman et al. 2017). Both within and among species, paternal behavior correlates with expression of AVP and/or AVP V1a receptors. For example, prairie vole fathers have reduced densities of AVP-immunoreactive fibers in the lateral septum and lateral habenular nucleus compared to virgin males and males housed with a pregnant mate, apparently due to increased synthesis and release of AVP from the BNST and MeA following both copulation and parturition (Bamshad et al. 1993, 1994). AVP expression in the paraventricular nucleus of the hypothalamus also correlates with paternal behavior in prairie voles (Perkeybile et al. 2013, 2015), mandarin voles (Wang et al. 2014), and California mice (De Jong et al. 2012; Frazier et al. 2006), generally showing a negative relationship that may be associated with stress, anxiety, and/or activation of the hypothalamic-pituitary-adrenal axis (reviewed by Saltzman et al. 2017).

Few studies have examined the relationship between AVP and fatherhood in either nonhuman or human primates. In common marmosets, fathers have greater expression of AVP V1a receptors in the prefrontal cortex than non-fathers (Kozorovitskiy et al. 2006), whereas in humans, urinary AVP concentrations do not differ between fathers and non-fathers but are negatively correlated with the age of a father's youngest child (Gray et al. 2007). Urinary and circulating AVP levels likely reflect peripheral, rather than central, secretion, and AVP does not readily cross the blood-brain barrier; hence, the neural and behavioral relevance of these levels is not clear.

4.6.2 Effects of AVP Signaling on Paternal Care and the Underlying Neural Circuitry

Experimental manipulations of vasopressinergic signaling support a role of this neuropeptide system in paternal or allopaternal behavior. In both biparental prairie voles (Wang et al. 1994) and facultatively paternal meadow voles (*M. pennsylvanicus*) (Parker et al. 2001), infusion of AVP into the cerebral ventricles or lateral septum increases allopaternal behavior in adult virgin males, whereas similar treatment with AVP antagonists has the opposite effect. Manipulations of AVP (and oxytocin; see below) in nonhuman primates and humans have typically involved intranasal administration of the neuropeptide; however, the physiological relevance of this technique, as well as its efficacy in delivering neuropeptides to the brain, is controversial (Fortuna et al. 2014; Leng and Ludwig 2016). Intranasal AVP treatment of expectant human fathers increased interest in infant-related stimuli (baby-related avatars) in an immersive virtual environment (Cohen-Bendahan et al. 2015); however, intranasal AVP had no effect on fathers' responses to infant-related stimuli in another study of men (Li et al. 2017) and in a study of common marmosets (Taylor and French 2015). Effects of AVP on paternal behavior have not, to our knowledge, been examined directly in other species.

Very little is known about how vasopressinergic signaling modulates transduction or processing of infant-related stimuli and/or activity in the parental-care circuitry. AVP and its receptors are found in many nuclei (Barberis and Tribollet 1996; Kato et al. 1995); however, specific effects of AVP on specific neural substrates of paternal care have not been identified in any species.

4.7 *Oxytocin*

4.7.1 Effects of Fatherhood on Oxytocin Signaling

Like its sister neuropeptide, AVP, oxytocin acts both in the periphery, where it is essential for parturition and milk letdown, and in the brain, where it plays key roles in affiliative and maternal behavior, affect, and social cognition (Caldwell and Albers 2016). Studies of the prosocial effects of oxytocin have traditionally focused on females, whereas those on AVP have focused on males; however, recent evidence has implicated oxytocin in the expression of parental behavior in both sexes.

Oxytocin signaling differs between fathers and virgin males in several biparental and facultatively biparental rodent species, but not in a consistent manner. For example, mandarin vole fathers have higher numbers of oxytocin-immunoreactive fibers in the PVN and SON than virgin males (Song et al. 2010), and several studies of this species have found correlations between paternal behavior and oxytocin-immunoreactivity in the supraoptic nucleus and/or paraventricular nucleus of the hypothalamus (Li et al. 2015; Song et al. 2010; Wang et al. 2014). Paternally behaving meadow vole fathers have higher oxytocin receptor binding in several brain areas, including the BNST and lateral septum, than non-allopaternally behaving virgin males (Parker et al. 2001), whereas California mouse fathers have lower expression of oxytocin receptor mRNA in the BNST compared to virgins (Perea-Rodriguez et al. 2015). Studies of oxytocin-immunoreactivity and oxytocin receptor binding in prairie voles have yielded inconsistent results (Kenkel et al. 2014; Wang et al. 2000). Importantly, where differences between fathers and virgins have been detected, males differed not only in paternal experience but also in sexual experience and/or cohabitation with a female, which appears to contribute to differences in oxytocin signaling. Thus, effects of fatherhood per se on oxytocin signaling are not clear.

In a study examining in vitro hypothalamic release of several neurocrines in common marmosets, cultured hypothalamic explants from fathers released significantly more oxytocin (and significantly less dopamine) than those from virgin males; in contrast, AVP release did not differ between the groups (Woller et al. 2011). Salivary and/or oxytocin concentrations in human fathers rise during the first 6 months of fatherhood (Gordon et al. 2010), can be increased acutely by interactions with infants (Feldman et al. 2010), and correlate with several affiliative components of father-infant interactions (Gordon et al. 2010).

4.7.2 Effects of Oxytocin Signaling on Paternal Care and the Underlying Neural Circuitry

Surprisingly few experimental studies have evaluated effects of oxytocin on paternal or allopaternal behavior in animal models. In a recent, unpublished study of prairie voles, however, the oxytocin receptor antagonist L-368,899 acutely inhibited allopaternal behavior in a dose-dependent manner (see Kenkel et al. 2017), while intranasal oxytocin (but not AVP) treatment increases interest in infant-related stimuli in adult male marmosets. In men, intranasal administration of oxytocin to fathers acutely enhances several components of father-infant interactions, such as positive vocalizations toward and touching of the infant. Several authors have suggested that interactions between oxytocin and paternal behavior may be modulated or mediated by steroid hormones (testosterone, Gordon et al. 2017; progesterone, Vargas-Pinilla et al. 2017).

Very little is known about how oxytocin signaling modulates transduction or processing of infant-related stimuli and/or activity in the paternal-care circuitry in males. The densest collections of oxytocin-containing cell bodies are in the supraoptic nucleus and paraventricular nucleus of the hypothalamus, although cell bodies can also be found in other hypothalamic nuclei and BNST of some species (Sofroniew 1983; Wang et al. 1996). Oxytocin fibers and/or receptors can be found throughout almost the entire brain, including regions critical for parental care such as the main and accessory olfactory bulbs, amygdala, and mesolimbic limbic reward circuitry (Lee et al. 2009; Ross and Young 2009). Oxytocin can be released both synaptically and non-synaptically, which is thought to result in diffuse activation of receptors (Landgraf and Neumann 2004).

One theory of how oxytocin influences parental care is that it reduces the sensitivity of $GABA_A$ receptors to steroids such as allopregnanolone, thus decreasing inhibitory tone in the parental-care circuitry (Koksma et al. 2003). Oxytocin may also act largely in the amygdala to reduce anxiety and neophobia around infants (Bale et al. 2001). Systemic oxytocin administration generally decreases basal- and stress-induced activity of the hypothalamic-pituitary-adrenal axis, which may facilitate parental care (reviewed by Landgraf and Neumann 2004). Oxytocin, perhaps originating from the paraventricular nucleus of the hypothalamus, can also act in the main olfactory bulbs to facilitate maternal care in rats; however, no experiments have been performed in males (Yu et al. 1996a, b).

5 Experiential Influences on Paternal Care and Neural Plasticity

Parental care exhibited by adult males can be influenced by a multitude of experiences occurring during prenatal, early postnatal, and juvenile development and into adulthood. A comprehensive discussion of these factors is beyond the scope of this chapter (but see Saltzman et al. 2017). Briefly, experiential influences on paternal behavior in rodents include intrauterine position (i.e., the number of adjacent brothers during prenatal development), quality and/or amount of parental care received, exposure to younger siblings, stress, cues from the mate and pups, and prior paternal experience. In some cases, these influences are associated with changes in hormonal (e.g., testosterone) or neuropeptide (e.g., AVP, oxytocin) signaling in fathers. Much less is known about experiential effects in primates, but prior experience with infants affects responses of adult males to infant-related stimuli in common marmosets and humans (Storey and Ziegler 2016; Ziegler and Sosa 2016; Ziegler et al. 2009a).

Fatherhood can modulate plasticity in brain regions subserving cognitive, affective, and sensory functions in rodents and primates. Glasper et al. (2011) used the cell-division marker bromodeoxyuridine (BrdU) to evaluate patterns of neurogenesis in male California mice and found evidence that fatherhood inhibits neurogenesis in the hippocampus but not in the subventricular zone. A subsequent study by the same group (Hyer et al. 2016) found that proliferation of new cells in the dentate gyrus did not differ between non-fathers and fathers; however, fathers had reduced 1-week survival but increased 2-week survival of new neurons in the dentate gyrus, as well as a greater proportion of cells with a neuronal phenotype, effects that appear to be estrogen-dependent (Hyer et al. 2017). Franssen et al. (2011), in contrast, found no differences between California mouse fathers and virgin males in indices of hippocampal plasticity, including markers of cell proliferation, number of new neurons, restructuring in mature neurons, or astrocytes.

Fatherhood can also modulate neural plasticity in prairie voles. Lieberwirth et al. (2013) found evidence that fatherhood reduces the survival of new cells in the amygdala, dentate gyrus, and ventromedial hypothalamus, but not in the basolateral amygdala or main olfactory bulbs. In the house mouse, interactions of an adult male with its own pups increase neurogenesis in the father's subventricular zone and dentate gyrus, an effect mediated by prolactin signaling (Mak and Weiss 2010). Some of the new cells mature into olfactory interneurons in the olfactory bulb, where, as described above, they respond preferentially to odors from offspring and appear to function in recognition of mature offspring. Exposure to pups can elicit hippocampal cell proliferation in virgin prairie voles (Ruscio et al. 2008).

In the common marmoset, fatherhood increases spine density on pyramidal neurons in the prefrontal cortex (Kozorovitskiy et al. 2006). Human fathers show increased neural activation in response to pictures of children in several brain regions associated with processing of emotional facial expressions (caudal middle frontal gyrus), mentalizing (temporoparietal junction), and reward processing (medial orbitofrontal cortex), compared to non-fathers (Mascaro et al. 2014).

In sum, these findings from rodents and primates demonstrate that fatherhood can modulate proliferation and survival of neurons, as well as neuronal morphology, in several limbic structures, with pronounced differences among species in both the sites and mechanisms of plasticity. Given the effects of hormones, such as estrogens, on synaptic plasticity in brain areas such as the hippocampus (e.g., Ogiue-Ikeda et al. 2008), it is likely that hormones associated with paternal behavior may also have effects on synapses of the MPOA during the transition to fatherhood.

6 Conclusions and Future Directions

Much work remains to be done to validate, refine, and expand current theories of parental-care circuitry and to elucidate the mechanisms by which activity in this circuitry is modulated by sensory, hormonal, neuropeptide, and experiential factors. Parental-care circuit diagrams have been built mostly by decades of non-cell-type-specific examinations of certain nuclei, including knife cuts, electrolytic lesions, excitotoxic lesions, kindling, electrical stimulation, delivery of various chemicals via intracranial cannula, and quantification of immediate-early gene expression. While these techniques demonstrate that the activity of particular brain regions can affect parental behavior, we know little about which cell types within those regions are involved, what the properties and connectivities of those cell types are, and what plasticity occurs in those particular cells, if any, during the transition into fatherhood.

Future research should identify and characterize specific types of neurons in nuclei that have been implicated in paternal care. For example, optogenetic and chemogenetic techniques can target specific cell types based on gene expression and/or connectivity. To the best of our knowledge, only one study has been conducted using optogenetics to target a specific cell type in terms of either connectivity or gene expression (rather than all the cells in a particular brain region) and examine effects on paternal behavior (Wu et al. 2014). Colocalizing an immunohistochemical or in situ hybridization immediate-early gene signal with another immunohistochemical or in situ hybridization signal in individual animals that have or have not been exposed to infants will also increase our understanding of the cell-type-specific signaling within brain regions implicated in paternal care. New transgenics and gene silencing/inducible knockout techniques can also be used. Localization of immediate-early gene expression after pup exposure in combination with anterograde or retrograde tracers would also be useful in elucidating neural pathways associated with paternal care. With a contemporary toolkit, much will be revealed about the neuroendocrine basis of paternal care that was hidden previously, and paternal-care circuit diagrams will be much refined.

While several brain regions have been implicated strongly in paternal care, very little is known about how specific cell types in these regions change in terms of their gene-expression profiles, morphological properties, and electrophysiological characteristics during the transition into fatherhood. At an even more fundamental level, the MPOA, like many other brain regions, is also poorly characterized in terms of its basic synaptic, intrinsic, and morphological properties. An understanding of these circuit-level characteristics and how they change during the transition to fatherhood is vital for any understanding of larger issues of neural systems or sensory, hormonal, neuropeptide, or experiential influences. Recent preliminary data from California mice demonstrate that such properties can be recorded from MPOA slices and that these circuit properties undergo detectable changes in in fathers (Horrell et al. 2016).

Beyond mammals, a "common core" of paternal-care circuitry may exist across vertebrate taxa. For example, manipulating activity in the preoptic area can affect paternal care in rodents, fish, and birds (Demski and Knigge 1971; Slawski and Buntin 1995; Tsuneoka et al. 2015). If such a "common core" exists, an explanation of how different infant-related sensory information gets funneled into this circuitry and how different parental behaviors emerge from it is needed. These are type-token distinctions: the "types" are infant-related sensory cues and parental-care behavior, and the "tokens" are distinct sensory information and distinct parental-care behaviors of various species. To give a particular example supported by experimental evidence, how does stimulating the preoptic area lead to suppression of infanticide and stimulation of pup grooming in male mice while leading to nest building in male cichlid bluegill (*Lepomis macrochirus*) (Demski and Knigge 1971; Fisher 1956; Wu et al. 2014)? As another example, how does lesioning the preoptic area in male mice decrease pup retrieval while lesioning it in male ring doves (*Streptopelia risoria*) decreases feeding invitations and bouts of feeding (Slawski and Buntin 1995; Tsuneoka et al. 2015; Wu et al. 2014)? The differences in circuitry across taxa that result in the activation of the same hypothalamic nuclei to produce quite different behavioral tokens of the same behavioral type is unknown and is one of the major frontiers in current neuroscience.

Multifunctionality of neurons should be considered as well. Should certain neurons that are active in parental care be considered particular "parental-care" neurons rather than general "prosocial behavior" neurons? For example, while MPOA lesions disrupt paternal care in a number of species, they also disrupt sexual behavior in every species studied thus far (Numan 2014). In fact, recent evidence using compartment analysis of temporal activity by catFISH suggests 20–30% overlap between neurons in the MPOA of mice that express fos during parental care and during sexual behavior (Wu et al. 2014). What are the exact neurons in the MPOA responsible for parental behavior vs. sexual behavior? What causes an MPOA neuron to express fos during parental-care and not during sexual behavior? What causes another MPOA neuron to express fos during both parental care and sexual behavior? Because of the likely multifunctionality of many neurons, analogous to genetic pleiotropy, structures and pathways identified as part of the "parental-care circuitry" may in fact be components of a more general "social circuitry" or other type of circuitry (Goodson and Kabelik 2009; Goodson and Kingsbury 2013). Moreover, within the general "parental-care circuitry" there are more specific circuits for different types of infant-directed behavior. For example, Rosenblatt and Mayer (1995) have posited theories of approach and withdrawal circuitries, and Numan (2014) has proposed theories of appetitive (e.g., retrieval) and consummatory (e.g., thermoregulating and grooming) circuitries in rodents. Thus, we should try not only to differentiate parental-care circuitry from other social circuitries but also to distinguish subtypes of parental-care circuitry. We should also specifically investigate multifunctionality, overlap, and communication between different circuits, as such inquiry may allow us to better understand phenomena like pseudokinship and alloparental care (conflations of kin and non-kin stimuli) and disorders of sociality such as infantile paraphilia and chronophilia (conflations of parental care-inducing and copulation-inducing stimuli).

Acknowledgments We thank the many current and former members of our labs who contributed to our own research described here. Our research was supported by NIH R21HD075021, NIH 1R21MH087806, NSF IOS1256572, and grants from the UC Riverside Academic Senate.

References

Ahdieh HB, Mayer AD, Rosenblatt JS (1987) Effects of brain antiestrogen implants on maternal behavior and on postpartum estrus in pregnant rats. Neuroendocrinology 46:522–531

Akther S, Fakhrul AA, Higashida H (2014) Effects of electrical lesions of the medial preoptic area and the ventral pallidum on mate-dependent paternal behavior in mice. Neurosci Lett 570:21–25

Akther S, Huang Z, Liang M, Zhong J, Fakhrul AA, Yuhi T, Lopatina O, Salmina AB, Yokoyama S, Higashida C, Tsuji T, Matsuo M, Higashida H (2015) Paternal retrieval behavior regulated by brain estrogen synthetase (aromatase) in mouse sires that engage in communicative interactions with pairmates. Front Neurosci 9:450

Alexander A, Irving A, Harvey J (2016) Emerging roles for the novel estrogen-sensing receptor GPER1 in the CNS. Neuropharmacology 113:652–660

Almond RE, Brown GR, Keverne EB (2006) Suppression of prolactin does not reduce infant care by parentally experienced male common marmosets (*Callithrix jacchus*). Horm Behav 49: 673–680

Almond RE, Ziegler TE, Snowdon CT (2008) Changes in prolactin and glucocorticoid levels in cotton-top tamarin fathers during their mate's pregnancy: the effect of infants and paternal experience. Am J Primatol 70:560–565

Alvergne A, Faurie C, Raymond M (2009) Variation in testosterone levels and male reproductive effort: insight from a polygynous human population. Horm Behav 56:491–497

Arai Y, Murakami S, Nishizuka M (1994) Androgen enhances neuronal degeneration in the developing preoptic area: apoptosis in the anteroventral periventricular nucleus (AVPvN-POA). Horm Behav 28:313–319

Arevalo MA, Azcoitia I, Gonzalez-Burgos I, Garcia-Segura LM (2015) Signaling mechanisms mediating the regulation of synaptic plasticity and memory by estradiol. Horm Behav 74:19–27

Bale TL, Davis AM, Auger AP, Dorsa DM, McCarthy MM (2001) CNS region-specific oxytocin receptor expression: importance in regulation of anxiety and sex behavior. J Neurosci 21: 2546–2552

Bales KL, Kramer KM, Lewis-Reese AD, Carter CS (2006) Effects of stress on parental care are sexually dimorphic in prairie voles. Physiol Behav 87:424–429

Bales KL, Saltzman W (2016) Fathering in rodents: neurobiological substrates and consequences for offspring. Horm Behav 77:249–259

Bamshad M, Novak MA, De Vries GJ (1993) Sex and species differences in the vasopressin innervation of sexually naive and parental prairie voles, *Microtus ochrogaster* and meadow voles, *Microtus pennsylvanicus*. J Neuroendocrinol 5:247–255

Bamshad M, Novak MA, de Vries GJ (1994) Cohabitation alters vasopressin innervation and paternal behavior in prairie voles (*Microtus ochrogaster*). Physiol Behav 56:751–758

Barberis C, Tribollet E (1996) Vasopressin and oxytocin receptors in the central nervous system. Crit Rev Neurobiol 10:119–154

Berg SJ, Wynne-Edwards KE (2001) Changes in testosterone, cortisol, and estradiol levels in men becoming fathers. Mayo Clin Proc 76:582–592

Berg SJ, Wynne-Edwards KE (2002) Salivary hormone concentrations in mothers and fathers becoming parents are not correlated. Horm Behav 42:424–436

Berk ML, Finkelstein JA (1981) Afferent projections to the preoptic area and hypothalamic regions in the rat brain. Neuroscience 6:1601–1624

Bloch GJ, Gorski RA (1988) Estrogen/progesterone treatment in adulthood affects the size of several components of the medial preoptic area in the male rat. J Comp Neurol 275:613–622
Blumberg MS, Sokoloff G (2001) Do infant rats cry? Psychol Rev 108:83–95
Bosch OJ, Neumann ID (2012) Both oxytocin and vasopressin are mediators of maternal care and aggression in rodents: from central release to sites of action. Horm Behav 61:293–303
Braun K, Champagne FA (2014) Paternal influences on offspring development: behavioural and epigenetic pathways. J Neuroendocrinol 26:697–706
Bridges RS (1996) Biochemical basis of parental behavior in the rat. Adv Study Behav 25:215–242
Bridges RS (2015) Neuroendocrine regulation of maternal behavior. Front Neuroendocrinol 36: 178–196
Bridges RS, Freemark MS (1995) Human placental lactogen infusions into the medial preoptic area stimulate maternal behavior in steroid-primed, nulliparous female rats. Horm Behav 29: 216–226
Bridges RS, Mann PE (1994) Prolactin-brain interactions in the induction of maternal behavior in rats. Psychoneuroendocrinology 19:611–622
Bridges RS, Numan M, Ronsheim PM, Mann PE, Lupini CE (1990) Central prolactin infusions stimulate maternal behavior in steroid-treated, nulliparous female rats. Proc Natl Acad Sci U S A 87:8003–8007
Brooks PL, Vella ET, Wynne-Edwards KE (2005) Dopamine agonist treatment before and after the birth reduces prolactin concentration but does not impair paternal responsiveness in Djungarian hamsters, *Phodopus campbelli*. Horm Behav 47:358–366
Brouette-lahlou I, Amouroux R, Chastrette F, Cosnier J, Stoffelsma J, Vernet-maury E (1991a) Dodecyl propionate, attractant from rat pup preputial gland: characterization and identification. J Chem Ecol 17:1343–1354
Brouette-lahlou I, Vernet-Maury E, Chanel J (1991b) Is rat-dam licking behavior regulated by pups' preputial gland secretion? Anim Learn Behav 19:177–184
Brown RE (1993) Hormonal and experiential factors influencing paternal behaviour in male rodents: an integrative approach. Behav Processes 30:1–28
Brown RE, Murdoch T, Murphy PR, Moger WH (1995) Hormonal responses of male gerbils to stimuli from their mate and pups. Horm Behav 29:474–491
Brown RE, Aoki M, Ladyman SR, Phillipps HR, Wyatt A, Boehm U, Grattan DR (2017) Prolactin action in the medial preoptic area is necessary for postpartum maternal nursing behavior. Proc Natl Acad Sci U S A 114:10779–10784
Bruin RD, Ganswindt A, Roux AL (2016) From killer to carer: steroid hormones and paternal behaviour. Afr Zool 51:173–182
Caldwell HK, Albers HE (2016) Oxytocin, vasopressin, and the motivational forces that drive social behaviors. In: Behavioral neuroscience of motivation. Springer International Publishing, Basel, pp 51–103
Caldwell JD, Walker CH, Pedersen CA, Baraka AS, Mason GA (1994) Estrogen increases affinity of oxytocin receptors in the medial preoptic area-anterior hypothalamus. Peptides 15:1079–1084
Caldwell HK, Lee HJ, Macbeth AH, Young WS (2008) Vasopressin: behavioral roles of an "original" neuropeptide. Prog Neurobiol 84:1–24
Campbell JC, Laugero KD, Van Westerhuyzen JA, Hostetler CM, Cohen JD, Bales KL (2009) Costs of pair-bonding and paternal care in male prairie voles (*Microtus ochrogaster*). Physiol Behav 98:367–373
Campeau S, Watson SJ (2000) Connections of some auditory-responsive posterior thalamic nuclei putatively involved in activation of the hypothalamo–pituitary–adrenocortical axis in response to audiogenic stress in rats: an anterograde and retrograde tract tracing study combined with fos expression. J Comp Neurol 423:474–491
Canteras NS, Simerly RB, Swanson LW (1995) Organization of projections from the medial nucleus of the amygdala: a PHAL study in the rat. J Comp Neurol 360:213–245
Caras ML (2013) Estrogenic modulation of auditory processing: a vertebrate comparison. Front Neuroendocrinol 34:285–299

Cavanaugh J, French JA (2013) Post-partum variation in the expression of paternal care is unrelated to urinary steroid metabolites in marmoset fathers. Horm Behav 63:551–558

Champagne F, Diorio J, Sharma S, Meaney MJ (2001) Naturally occurring variations in maternal behavior in the rat are associated with differences in estrogen-inducible central oxytocin receptors. Proc Natl Acad Sci U S A 98:12736–12741

Champagne FA, Chretien P, Stevenson CW, Zhang TY, Gratton A, Meaney MJ (2004) Variations in nucleus accumbens dopamine associated with individual differences in maternal behavior in the rat. J Neurosci 24:4113–4123

Charitidi K, Canlon B (2010) Estrogen receptors in the central auditory system of male and female mice. J Neurosci 165:923–933

Charitidi K, Meltser I, Canlon B (2012) Estradiol treatment and hormonal fluctuations during the estrous cycle modulate the expression of estrogen receptors in the auditory system and the prepulse inhibition of acoustic startle response. Endocrinology 153:4412–4421

Chauke M, Malisch JL, Robinson C, de Jong TR, Saltzman W (2011) Effects of reproductive status on behavioral and endocrine responses to acute stress in a biparental rodent, the California mouse (*Peromyscus califor*nicus). Horm Behav 60:128–138

Chen EM, Gau ML, Liu CY, Lee TY (2017) Effects of father-neonate skin-to-skin contact on attachment: a randomized controlled trial. Nurs Res Prect 2017:1–8

Cherian S, Lam YW, McDaniels I, Struziak M, Delay RJ (2014) Estradiol rapidly modulates odor responses in mouse vomeronasal sensory neurons. J Neurosci 269:43–58

Chiba T, Murata Y (1985) Afferent and efferent connections of the medial preoptic area in the rat: a WGA-HRP study. Brain Res Bull 14:261–272

Clark MM, Galef BG Jr (1999) A testosterone-mediated trade-off between parental and sexual effort in male Mongolian gerbils (Meriones unguiculatus). J Comp Psychol 113:388–395

Clutton-Brock TH (1991) The evolution of parental care. Princeton University Press, Princeton, NJ

Cockburn A (2006) Prevalence of different modes of parental care in birds. Proc Biol Sci 273: 1375–1383

Cohen-Bendahan CC, Beijers R, van Doornen LJ, de Weerth C (2015) Explicit and implicit caregiving interests in expectant fathers: do endogenous and exogenous oxytocin and vasopressin matter? Infant Behav Dev 41:26–37

Conely L, Bell RW (1978) Neonatal ultrasounds elicited by odor cues. Dev Psychobiol 11:193–197

Cong X, Ludington-Hoe SM, Hussain N, Cusson RM, Walsh S, Vazquez V, Briere CA, Vittner D (2015) Parental oxytocin responses during skin-to-skin contact in pre-term infants. Early Hum Dev 91:401–406

Courtiol E, Wilson DA (2014) Thalamic olfaction: characterizing odor processing in the mediodorsal thalamus of the rat. J Neurophysiol 111:1274–1285

Crump ML (1996) Parental care among the amphibia. In: Rosenblatt JS, Snowdon CT (eds) Parental care: evolution, mechanisms and adaptive significance, Advances in the study of behavior, vol 25. Academic Press, San Diego, pp 109–144

Cruz R, Del Cerro M (1998) Role of the vomeronasal input in maternal behavior. Psychoneuroendocrinology 23:905–926

Cservenák M, Szabó ÉR, Bodnár I, Lékó A, Palkovits M, Nagy GM, Usdin TB, Dobolyi A (2013) Thalamic neuropeptide mediating the effects of nursing on lactation and maternal motivation. Psychoneuroendocrinology 38:3070–3084

Cservenák M, Kis V, Keller D, Dimén D, Menyhárt L, Oláh S, Szabó ER, Barna J, Renner E, Usdin TB, Dobolyi A (2017) Maternally involved galanin neurons in the preoptic area of the rat. Brain Struct Funct 222:781–798

Cunningham RL, Claiborne BJ, McGinnis MY (2007) Pubertal exposure to anabolic androgenic steroids increases spine densities on neurons in the limbic system of male rats. Neuroscience 150:609–615

Cushing BS, Perry A, Musatov S, Ogawa S, Papademetriou E (2008) Estrogen receptors in the medial amygdala inhibit the expression of male prosocial behavior. J Neurosci 28:10399–10403 de Castilhos J, Claudia DF, Achaval M, Rasia-Filho AA (2008) Dendritic spine density of posterodorsal medial amygdala neurons can be affected by gonadectomy and sex steroid manipulations in adult rats: a Golgi study. Brain Res 1240:73–81

De Jong TR, Chauke M, Harris BN, Saltzman W (2009) From here to paternity: neural correlates of the onset of paternal behavior in California mice (*Peromyscus californicus*). Horm Behav 56(2): 220–231

De Jong TR, Measor KR, Chauke M, Harris BN, Saltzman W (2010) Brief pup exposure induces Fos expression in the lateral habenula and serotonergic caudal dorsal raphe nucleus of paternally experienced male California mice (*Peromyscus californicus*). Neuroscience 169:1094–1104

De Jong TR, Korosi A, Harris BN, Perea-Rodriguez JP, Saltzman W (2012) Individual variation in paternal responses of virgin male California mice (*Peromyscus californicus*): behavioral and physiological correlates. Physiol Biochem Zool 85:740–751

de Kloet ER, Karst H, Joëls M (2008) Corticosteroid hormones in the central stress response: quick-and-slow. Front Neuroendocrinol 29:268–272

De Olmos JS, Ingram WR (1972) The projection field of the stria terminalis in the rat brain. An experimental study. J Comp Neurol 146:303–333

Dees WL, Kozlowski GP (1984) Effects of castration and ethanol on amygdaloid substance P immunoreactivity. Neuroendocrinology 39:231–235

Delahunty KM, McKay DW, Noseworthy DE, Storey AE (2007) Prolactin responses to infant cues in men and women: effects of parental experience and recent infant contact. Horm Behav 51:213–220

Demski LS, Knigge KM (1971) The telencephalon and hypothalamus of the bluegill (*Lepomis macrochirus*): evoked feeding, aggressive and reproductive behavior with representative frontal sections. J Comp Neurol 143:1–16

Dey S, Chamero P, Pru JK, Chien MS, Ibarra-Soria X, Spencer KR, Logan DW, Matsunami H, Peluso JJ, Stowers L (2015) Cyclic regulation of sensory perception by a female hormone alters behavior. Cell 161:1334–1344

Digby LJ, Ferrari SF, Saltzman W (2007) Callitrichines: the role of competition in cooperatively breeding species. In: Campbell JC, Fuentes A, MacKinnon KC, Panger M, Bearder SK (eds) Primates in perspective. Oxford University Press, New York, pp 85–105

Dillon TS, Fox LC, Crystal H, Linster C (2013) 17β-estradiol enhances memory duration in the main olfactory bulb in CD-1 mice. Behav Neurosci 127:923–931

Dixson AF, George L (1982) Prolactin and parental behaviour in a male new world primate. Nature 299:551–553

Dobolyi A, Grattan DR, Stolzenberg DS (2014) Preoptic inputs and mechanisms that regulate maternal responsiveness. J Neuroendocrinol 26:627–640

Dodson RE, Gorski RA (1993) Testosterone propionate administration prevents the loss of neurons within the central part of the medial preoptic nucleus. J Neurobiol 24:80–88

Dohler KD, Coquelin A, Davis F, Hines M, Shryne JE, Gorski RA (1984) Pre- and postnatal influence of testosterone propionate and diethylstilbestrol on differentiation of the sexually dimorphic nucleus of the preoptic area in male and female rats. Brain Res 302:291–295

Du J, Hull EM (1999) Effects of testosterone on neuronal nitric oxide synthase and tyrosine hydroxylase. Brain Res 836:90–98

Du J, Lorrain DS, Hull EM (1998) Castration decreases extracellular, but increases intracellular, dopamine in medial preoptic area of male rats. Brain Res 782:11–17

Dulac C, O'Connell LA, Wu Z (2014) Neural control of maternal and paternal behaviors. Science 345:765–770

Edelstein RS, Wardecker BM, Chopik WJ, Moors AC, Shipman EL, Lin NJ (2015) Prenatal hormones in first-time expectant parents: longitudinal changes and within-couple correlations. Am J Hum Biol 27:317–325

Ehret G (2005) Infant rodent ultrasounds – a gate to the understanding of sound communication. Behav Genet 35:19–29

Ehret G, Buckenmaier J (1994) Estrogen-receptor occurrence in the female mouse brain: effects of maternal experience, ovariectomy, estrogen and anosmia. J Physiol Paris 88:315–329
Ehret G, Koch M (1989) Ultrasound-induced parental behavior in house mouse is controlled by female sex hormones and parental experience. Ethology 80:81–93
Ehret G, Schmid C (2009) Reproductive cycle-dependent plasticity of perception of acoustic meaning in mice. Physiol Behav 96:428–433
Elwood RW, Ostermeyer MC (1984) Infanticide by male and female Mongolian gerbils: ontogeny, causation, and function. In: Hausfater G, Hrdy SB (eds) Infanticide: comparative and evolutionary perspectives. Aldine, New York, pp 367–386
Etgen AM, Pfaff DW (2010) Molecular mechanisms of hormone actions on behavior. Academic Press, Oxford
Fahrbach SE, Morrell JI, Pfaff DW (1986) Identification of medial preoptic neurons that concentrate estradiol and project to the midbrain in the rat. J Comp Neurol 247:364–382
Feldman R, Gordon I, Schneiderman I, Weisman O, Zagoory-Sharon O (2010) Natural variations in maternal and paternal care are associated with systematic changes in oxytocin following parent-infant contact. Psychoneuroendocrinology 35:1133–1141
Fernandez-Duque E, Valeggia CR, Mendoza SP (2009) The biology of paternal care in human and nonhuman primates. Ann Rev Anthropol 38:115–130
Fisher AE (1956) Maternal and sexual behavior induced by intracranial chemical stimulation. Science 123:228–229
Fleischer S, Kordower JH, Kaplan B, Dicker R, Smerling R, Ilgner J (1981) Olfactory bulbectomy and gender differences in maternal behaviors of rats. Physiol Behav 26:957–959
Fleming AS, Vaccarino F, Luebke C (1980) Amygdaloid inhibition of maternal behavior in the nulliparous female rat. Physiol Behav 25:731–743
Fleming AS, Corter C, Stallings J, Steiner M (2002) Testosterone and prolactin are associated with emotional responses to infant cries in new fathers. Horm Behav 42:399–413
Fortuna A, Alves G, Serralheiro A, Sousa J, Falcão A (2014) Intranasal delivery of systemic-acting drugs: small-molecules and biomacromolecules. Eur J Pharm Biopharm 88:8–27
Franssen CL, Bardi M, Shea EA, Hampton JE, Franssen RA, Kinsley CH, Lambert KG (2011) Fatherhood alters behavioural and neural responsiveness in a spatial task. J Neuroendocrinol 23: 1177–1187
Frazier CR, Trainor BC, Cravens CJ, Whitney TK, Marler CA (2006) Paternal behavior influences development of aggression and vasopressin expression in male California mouse offspring. Horm Behav 50:699–707
Gammie SC (2005) Current models and future directions for understanding the neural circuitries of maternal behaviors in rodents. Behav Cog Neurosci Rev 4:119–135
Garcia-Galiano D, Pinilla L, Tena-Sempere M (2012) Sex steroids and the control of the Kiss1 system: developmental roles and major regulatory actions. J Neuroendocrinol 24:22–33
Gardner CR, Phillips SW (1977) The influence of the amygdala on the basal septum and preoptic area of the rat. Exp Brain Res 292:249–263
Garelick T, Swann J (2014) Testosterone regulates the density of dendritic spines in the male preoptic area. Horm Behav 65(3):249–253
Geisler S, Zahm DS (2006) Neurotensin afferents of the ventral tegmental area in the rat: [1] re-examination of their origins and [2] responses to acute psychostimulant and antipsychotic drug administration. Eur J Neurosci 24:116–134
Geisler S, Derst C, Veh RW, Zahm DS (2007) Glutamatergic afferents of the ventral tegmental area in the rat. J Neurosci 27:5730–5743
Gettler LT (2014) Applying socioendocrinology to evolutionary models: fatherhood and physiology. Evol Anthropol 23:146–160
Gettler LT, McDade TW, Feranil AB, Kuzawa CW (2011) Longitudinal evidence that fatherhood decreases testosterone in human males. Proc Natl Acad Sci U S A 108:16194–16199
Gettler LT, McDade TW, Feranil AB, Kuzawa CW (2012) Prolactin, fatherhood, and reproductive behavior in human males. Am J Phys Anthropol 148:362–370

Gettler LT, McDade TW, Agustin SS, Kuzawa CW (2013) Progesterone and estrogen responsiveness to father-toddler interaction. Am J Hum Biol 25:491–498
Geyer LA (1979) Olfactory and thermal influences on ultra-sonic vocalization during development in rodents. Am Zool 19:420–431
Glasper ER, Kozorovitskiy Y, Pavlic A, Gould E (2011) Paternal experience suppresses adult neurogenesis without altering hippocampal function in *Peromyscus californicus*. J Comp Neurol 519:2271–2281
Goodson JL, Kabelik D (2009) Dynamic limbic networks and social diversity in vertebrates: from neural context to neuromodulatory patterning. Front Neuroendocrinol 30:429–441
Goodson JL, Kingsbury MA (2013) What's in a name? Considerations of homologies and nomenclature for vertebrate social behavior networks. Horm Behav 64:103–112
Goodwin NB, Balshine-Earn S, Reynolds JD (1998) Evolutionary transitions in parental care in cichlid fish. Proc Biol Sci 265:2265–2272
Gordon I, Zagoory-Sharon O, Leckman JF, Feldman R (2010) Prolactin, oxytocin, and the development of paternal behavior across the first six months of fatherhood. Horm Behav 58:513–518
Gordon I, Pratt M, Bergunde K, Zagoory-Sharon O, Feldman R (2017) Testosterone, oxytocin, and the development of human parental care. Horm Behav 93:184–192
Gray PB, Campbell BC (2009) Human male testosterone, pair-bonding, and fatherhood. In: Endocrinology of social relationships. In: Ellison PT, Gray PB (eds). Harvard University Press, Cambridge MA, pp 270–293
Gray PB, Kahlenberg SM, Barrett ES, Lipson SF, Ellison PT (2002) Marriage and fatherhood are associated with lower testosterone in males. Evol Hum Behav 23:193–201
Gray PB, Yang CFJ, Pope HG (2006) Fathers have lower salivary testosterone levels than unmarried men and married non-fathers in Beijing, China. Proc Biol Sci 273:333–339
Gray PB, Parkin JC, Samms-Vaughan ME (2007) Hormonal correlates of human paternal interactions: a hospital-based investigation in urban Jamaica. Horm Behav 52:499–507
Groeneweg FL, Karst H, de Kloet ER, Joëls M (2012) Mineralocorticoid and glucocorticoid receptors at the neuronal membrane, regulators of nongenomic corticosteroid signalling. Mol Cell Endocrinol 350:299–309
Gross MR, Sargent RC (1985) The evolution of male and female parental care in fishes. Am Zool 25:807–822
Gubernick DJ (1990) A maternal chemosignal maintains paternal behaviour in the biparental California mouse, *Peromyscus californicus*. Anim Behav 39:936–942
Gubernick DJ, Alberts JR (1987) The biparental care system of the California mouse, *Peromyscus californicus*. J Comp Psychol 101:169–177
Gubernick DJ, Alberts JR (1989) Postpartum maintenance of paternal behaviour in the biparental California mouse, *Peromyscus californicus*. Anim Behav 37:656–664
Gubernick DJ, Nelson RJ (1989) Prolactin and paternal behavior in the biparental California mouse, *Peromyscus californicus*. Horm Behav 23:203–210
Gubernick DJ, Winslow JT, Jensen P, Jeanotte L, Bowen J (1995) Oxytocin changes in males over the reproductive cycle in the monogamous, biparental California mouse, *Peromyscus californicus*. Horm Behav 29:59–73
Guerra M, Del Castillo AR, Battaner E, Mas M (1987) Androgens stimulate preoptic area Na+,K+-ATPase activity in male rats. Neurosci Lett 78:97–100
Guo XZ, Su JD, Sun QW, Jiao BH (2001) Expression of estrogen receptor (ER) -alpha and -beta transcripts in the neonatal and adult rat cerebral cortex, cerebellum, and olfactory bulb. Cell Res 11:321–324
Guthman EM, Vera J (2016) A cellular mechanism for main and accessory olfactory integration at the medial amygdala. J Neurosci 36:2083–2085
Hadeishi Y, Wood RI (1996) Nitric oxide synthase in mating behavior circuitry of male Syrian hamster brain. Dev Neurobiol 30:480–492
Haller J, Mikics É, Makara GB (2008) The effects of non-genomic glucocorticoid mechanisms on bodily functions and the central neural system. A critical evaluation of findings. Front Neuroendocrinol 29:273–291

Hammer RP Jr (1985) The sex hormone-dependent development of opiate receptors in the rat medial preoptic area. Brain Res 360:65–74

Han TM, De Vries GJ (2003) Organizational effects of testosterone, estradiol, and dihydrotestosterone on vasopressin mRNA expression in the bed nucleus of the stria terminalis. J Neurobiol 54:502–510

Hansen S (1994) Maternal behavior of female rats with 6-OHDA lesions in the ventral striatum: characterization of the pup retrieval deficit. Physiol Behav 55:615–620

Hansen S, Harthon C, Wallin E, Lofberg L, Svensson K (1991) The effects of 6-OHDA-induced dopamine depletions in the ventral or dorsal striatum on maternal and sexual behavior in the female rat. Pharmacol Biochem Behav 39:71–77

Hansen S, Bergvall ÅH, Nyiredi S (1993) Interaction with pups enhances dopamine release in the ventral striatum of maternal rats: a microdialysis study. Pharmacol Biochem Behav 45:673–676

Harris BN, Perea-Rodriguez JP, Saltzman W (2011) Acute effects of corticosterone injection on paternal behavior in California mouse (Peromyscus californicus) fathers. Horm Behav 60:666–675

Harris BN, de Jong TR, Yang V, Saltzman W (2013) Chronic variable stress in fathers alters paternal and social behavior but not pup development in the biparental California mouse (*Peromyscus californicus*). Horm Behav 64:799–811

Hashemian F, Shafigh F, Roohi E (2016) Regulatory role of prolactin in paternal behavior in male parents: a narrative review. J Postgrad Med 62:182–187

Henderson LP (2007) Steroid modulation of GABAA receptor-mediated transmission in the hypothalamus: effects on reproductive function. Neuropharmacology 52:1439–1453

Herbison AE (1997) Estrogen regulation of GABA transmission in rat preoptic area. Brain Res Bull 44(4):321–326

Herbison AE, Heavens RP, Dyer RG (1989) Oestrogen and noradrenaline modulate endogenous GABA release from slices of the rat medial preoptic area. Brain Res 486:195–200

Herbison AE, Heavens RP, Dyer RG (1990) Endogenous release of gamma-aminobutyric acid from the medial preoptic area measured by microdialysis in the anaesthetised rat. J Neurochem 55: 1617–1623

Hoover WB, Vertes RP (2007) Anatomical analysis of afferent projections to the medial prefrontal cortex in the rat. Brain Struct Funct 212:149–179

Horrell N, Hickmott P, Luu D, Saltzman W (2016) Effects of fatherhood on synaptic, intrinsic, and morphological characteristics of neurons in the medial preoptic area of male California mice program no. 338.11.2016. Neuroscience meeting planner. Society for Neuroscience, San Diego, CA

Horrell ND, Perea-Rodriguez JP, Harris BN, Saltzman W (2017) Effects of repeated pup exposure on behavioral, neural, and adrenocortical responses to pups in male California mice (*Peromyscus californicus*). Horm Behav 90:56–63

Hoyk Z, Csakvari E, Gyenes A, Siklos L, Harada N, Parducz A (2014) Aromatase and estrogen receptor beta expression in the rat olfactory bulb: neuroestrogen action in the first relay station of the olfactory pathway? Acta Neurobiol Exp 74:1–14

Hume JM, Wynne-Edwards KE (2005) Castration reduces male testosterone, estradiol, and territorial aggression, but not paternal behavior in biparental dwarf hamsters (*Phodopus campbelli*). Horm Behav 48:303–310

Hume JM, Wynne-Edwards KE (2006) Paternal responsiveness in biparental dwarf hamsters (*Phodopus campbelli*) does not require estradiol. Horm Behav 49:538–544

Hyer MM, Hunter TJ, Katakam J, Wolz T, Glasper ER (2016) Neurogenesis and anxiety-like behavior in male California mice during the mate's postpartum period. Eur J Neurosci 43: 703–709

Hyer MM, Khantsis S, Venezia AC, Madison FN, Hallgarth L, Adekola E, Glasper ER (2017) Estrogen-dependent modifications to hippocampal plasticity in paternal California mice (*Peromyscus californicus*). Horm Behav 96:147–155

Idborg H, Zamani L, Edlund P, Schuppe-Koistinen I, Jacobsson SP (2005) Metabolic fingerprinting of rat urine by LC/MS: part 1. Analysis by hydrophilic interaction liquid chromatography–electrospray ionization mass spectrometry. J Chromatogr B 828:9–13

Izquierdo MA, Collado P, Segovia S, Guillamon A, del Cerro MCR (1992) Maternal behavior induced in male rats by bilateral lesions of the bed nucleus of the accessory olfactory tract. Physiol Behav 52:707–712

Jacobson CD, Csernus VJ, Shryne JE, Gorski RA (1981) The influence of gonadectomy, androgen exposure, or a gonadal graft in the neonatal rat on the volume of the sexually dimorphic nucleus of the preoptic area. J Neurosci 1:1142–1147

Jean-Baptiste N, Terleph T, Bamshad M (2008) Changes in paternal responsiveness of prairie voles (*Microtus ochrogaster*) in response to olfactory cues and continuous physical contact with a female. Ethology 114:1239–1246

Jemiolo B, Andreolini F, Xie TM, Wiesler D, Novotny M (1989) Puberty-affecting synthetic analogs of urinary chemosignals in the house mouse, *Mus domesticus*. Physiol Behav 46:293–298

Jemiolo B, Gubernick DJ, Catherine Yoder M, Novotny M (1994) Chemical characterization of urinary volatile compounds of *Peromyscus californicus*, a monogamous biparental rodent. J Chem Ecol 20:2489–2500

Jennings JH, Sparta DR, Stamatakis AM, Ung RL, Pleil KE, Kash TL, Stuber GD (2013) Distinct extended amygdala circuits for divergent motivational states. Nature 496:224–228

Jones JS, Wynne-Edwards KE (2000) Paternal hamsters mechanically assist the delivery, consume amniotic fluid and placenta, remove fetal membranes, and provide parental care during the birth process. Horm Behav 37:116–125

Kanageswaran N, Nagel M, Scholz P, Mohrhardt J, Gisselmann G, Hatt H (2016) Modulatory effects of sex steroids progesterone and estradiol on odorant evoked responses in olfactory receptor neurons. PLoS One 11(8):e0159640

Kato Y, Igarashi N, Hirasawa A, Tsujimoto G, Kobayashi M (1995) Distribution and developmental changes in vasopressin V2 receptor mRNA in rat brain. Differentiation 59:163–169

Kaufling J, Veinante P, Pawlowski SA, Freund-Mercier M-J, Barrot M (2009) Afferents to the GABAergic tail of the ventral tegmental area in the rat. J Comp Neurol 513:597–621

Keer SE, Stern JM (1999) Dopamine receptor blockade in the nucleus accumbens inhibits maternal retrieval and licking, but enhances nursing behavior in lactating rats. Physiol Behav 67:659–669

Kelly MJ, Wagner EJ (1999) Estrogen modulation of G-protein-coupled receptors. Trends Endocrinol Metab 10:369–374

Kelly MJ, Zhang C, Qiu J, Ronnekleiv OK (2013) Pacemaking kisspeptin neurons. Exp Physiol 98: 1535–1543

Kenkel WM, Suboc G, Carter CS (2014) Autonomic, behavioral and neuroendocrine correlates of paternal behavior in male prairie voles. Physiol Behav 128:252–259

Kenkel WM, Perkeybile AM, Carter CS (2017) The neurobiological causes and effects of alloparenting. Dev Neurobiol 77:214–232

Kentner AC, Abizaid A, Bielajew C (2009) Modeling dad: animal models of paternal behavior. Neurosci Biobehav Rev 34:438–451

Keshavarzi S, Sullivan RKP, Ianno DJ, Sah P (2014) Functional properties and projections of neurons in the medial amygdala. J Neurosci 34:8699–8715

Kirkpatrick B, Carter CS, Newman SW, Insel TR (1994a) Axon-sparing lesions of the medial nucleus of the amygdala decrease affiliative behaviors in the prairie vole (*Microtus ochrogaster*): behavioral and anatomical specificity. Behav Neurosci 108:501–513

Kirkpatrick B, Williams JR, Slotnick BM, Carter CS (1994b) Olfactory bulbectomy decreases social behavior in male prairie voles (*M. ochrogaster*). Physiol Behav 55:885–889

Kleiman DG, Malcolm JR (1981) The evolution of male parental investment in mammals. In: Gubernick DJ, Klopfer PH (eds) Parental care in mammals. Springer, Boston, pp 347–387

Koch M, Ehret G (1989) Estradiol and parental experience, but not prolactin are necessary for ultrasound recognition and pup-retrieving in the mouse. Physiol Behav 45:771–776

Koksma JJ, van Kesteren RE, Rosahl TW, Zwart R, Smit AB, Lüddens H, Brussaard AB (2003) Oxytocin regulates neurosteroid modulation of GABAA receptors in supraoptic nucleus around parturition. J Neurosci 23:788–797

Kow LM, Pfaff DW (2016) Rapid estrogen actions on ion channels: a survey in search for mechanisms. Steroids 111:46–53

Kozorovitskiy Y, Hughes M, Lee K, Gould E (2006) Fatherhood affects dendritic spines and vasopressin V1a receptors in the primate prefrontal cortex. Nat Neurosci 9:1094–1095

Kramer KM, Perry AN, Golbin D, Cushing BS (2009) Sex steroid are necessary in the second postnatal week for the expression of male alloparental behavior in prairie voles (*Microtus ochrogaster*). Behav Neurosci 123:958–963

Kritzer M (2004) The distribution of immunoreactivity for intracellular androgen receptors in the cerebral cortex of hormonally intact adult male and female rats: localization in pyramidal neurons making corticocortical connections. Cereb Cortex 14:268–280

Kuzawa CW, Gettler LT, Muller MN, McDade TW, Feranil AB (2009) Fatherhood, pairbonding and testosterone in the Philippines. Horm Behav 56:429–435

Lambert KG, Kinsley CH (2012) Brain and behavioral modifications that accompany the onset of motherhood. Parenting 12:74–88

Lambert KG, Franssen CL, Hampton JE, Rzucidlo AM, Hyer MM, True M, Kaufman C, Bardi M (2013) Modeling paternal attentiveness: distressed pups evoke differential neurobiological and behavioral responses in paternal and nonpaternal mice. Neurosci 234:1–12

Landgraf R, Neumann ID (2004) Vasopressin and oxytocin release within the brain: a dynamic concept of multiple and variable modes of neuropeptide communication. Front Neuroendocrinol 25:150–176

Lee AW, Brown RE (2002) Medial preoptic lesions disrupt parental behavior in both male and female California mice (*Peromyscus californicus*). Behav Neurosci 116:968

Lee AW, Brown RE (2007) Comparison of medial preoptic, amygdala, and nucleus accumbens lesions on parental behavior in California mice (*Peromyscus californicus*). Physiol Behav 92:617–628

Lee A, Li M, Watchus J, Fleming AS (1999) Neuroanatomical basis of maternal memory in postpartum rats: selective role for the nucleus accumbens. Behav Neurosci 113:523–538

Lee A, Clancy S, Fleming AS (2000) Mother rats bar-press for pups: effects of lesions of the mpoa and limbic sites on maternal behavior and operant responding for pup-reinforcement. Behav Brain Res 108:215–231

Lee HJ, Macbeth AH, Pagani JH, Young WS (2009) Oxytocin: the great facilitator of life. Prog Neurobiol 88:127–151

Lei K, Cushing BS, Musatov S, Ogawa S, Kramer KM (2010) Estrogen receptor-α in the bed nucleus of the stria terminalis regulates social affiliation in male prairie voles (*Microtus ochrogaster*). PLoS One 5:e8931

Leng G, Ludwig M (2016) Intranasal oxytocin: myths and delusions. Biol Psychiatry 79:243–250

Leonard CM, Scott JW (1971) Origin and distribution of the amygdalofugal pathways in the rat: an experimental neuronatomical study. J Comp Neurol 141:313–329

Li M, Fleming AS (2003a) Differential involvement of nucleus accumbens shell and core subregions in maternal memory in postpartum female rats. Behav Neurosci 117:426–445

Li M, Fleming AS (2003b) The nucleus accumbens shell is critical for normal expression of pup retrieval in postpartum rats. Behav Brain Res 145:99–111

Li Y, Lian Z, Wang B, Tai F, Wu R, Hao P, Qiao X (2015) Natural variation in paternal behavior is associated with central estrogen receptor alpha and oxytocin levels. J Comp Physiol A 201: 285–293

Li T, Chen X, Mascaro J, Haroon E, Rilling JK (2017) Intranasal oxytocin, but not vasopressin, augments neural responses to toddlers in human fathers. Horm Behav 93:193–202

Liang M, Zhong J, Liu HX, Lopatina O, Nakada R, Yamauchi AM, Higashida H (2014) Pairmate-dependent pup retrieval as parental behavior in male mice. Front Neurosci 8:186

Lieberwirth C, Wang Y, Jia X, Liu Y, Wang Z (2013) Fatherhood reduces the survival of adult-generated cells and affects various types of behavior in the prairie vole (*Microtus ochrogaster*). Behav Neurosci 38:3345–3355
Liu RC, Schreiner CE (2007) Auditory cortical detection and discrimination correlates with communicative significance. PLoS Biol 5:e173
Liu HX, Lopatina O, Higashida C, Fujimoto H, Akther S, Inzhutova A, Liang M, Zhong J, Tsuji T, Yoshihara T, Sumi K, Ishiyama M, Ma WJ, Ozaki M, Yagitani S, Yokoyama S, Mukaida N, Sakurai T, Hori O, Yoshioka K, Hirao A, Kato Y, Ishihara K, Kato I, Okamoto H, Cherepanov SM, Salmina AB, Hirai H, Asano M, Brown DA, Nagano I, Higashida H (2013) Displays of paternal mouse pup retrieval following communicative interaction with maternal mates. Nat Commun 4:1346
Londei T, Segala P, Leone VG (1989) Mouse pup urine as an infant signal. Physiol Behav 45: 579–583
Lonstein JS (2007) Regulation of anxiety during the postpartum period. Front Neuroendocrinol 28: 115–141
Lonstein LS, De Vries GJ (1999) Sex differences in the parental behaviour of adult virgin prairie voles: independence from gonadal hormones and vasopressin. J Neuroendocrinol 11:441–449
Lonstein JS, Rood BD, De Vries GJ (2002) Parental responsiveness is feminized after neonatal castration in virgin male prairie voles, but is not masculinized by perinatal testosterone in virgin females. Horm Behav 41:80–87
Lubin M, Leon M, Moltz H, Numan M (1972) Hormones and maternal behavior in the male rat. Horm Behav 3:369–374
Luis J, Ramíre L, Carmona A, Ortiz G, Delgado J, Cárdenas R (2009) Paternal behavior and testosterone plasma levels in the volcano mouse *Neotomodon alstoni* (Rodentia: Muridae). Rev Biol Trop 57:433–439
MacKinnon P, Clark C, Clement EM, Sheaves R (1985) Changes in concentrations and rate of decline of adrenaline in the medial preoptic area and mediobasal hypothalamus of acutely ovariectomized steroid-treated rats. J Endocrinol 104:129–135
Magid KS (2011) Reproductive ecology and life history of human males: a migrant study of Bangladeshi men. University College London, London
Mahesh VB, Brann DW (2005) Regulatory role of excitatory amino acids in reproduction. Endocrine 28:271–280
Mak GK, Weiss S (2010) Paternal recognition of adult offspring mediated by newly generated CNS neurons. Nat Neurosci 13:753–758
Malsbury CW, McKay K (1994) Neurotrophic effects of testosterone on the medial nucleus of the amygdala in adult male rats. J Neuroendocrinol 6:57–69
Mani SK, Mermelstein PG, Tetel MJ, Anesetti G (2012) Convergence of multiple mechanisms of steroid hormone action. Horm Metab Res 44:569–576
Marchlewska-Koj A, Kapusta J, Olejniczak P (1999) Ultrasonic response of CBA newborn mice to bedding odour. Behaviour 136:269–278
Martinez A, Ramos G, Martinez-Torres M, Nicolás L, Carmona A, Cárdenas M, Luis J (2015) Paternal behavior in the Mongolian gerbil (Meriones unguiculatus): estrogenic and androgenic regulation. Horm Behav 71:91–95
Martinez-Gonzalez C, Bolam JP, Mena-Segovia J (2011) Topographical organization of the pedunculopontine nucleus. Front Neuroanat 5:1–10
Mascaro JS, Hackett PD, Rilling JK (2014) Differential neural responses to child and sexual stimuli in human fathers and non-fathers and their hormonal correlates. Psychoneuroendocrinology 46:153–163
Matsumoto T, Honda S-I, Harada N (2003) Alteration in sex-specific behaviors in male mice lacking the aromatase gene. Neuroendocrinology 77:416–424
Mayer AD, Freeman NCG, Rosenblatt JS (1979) Ontogeny of maternal behavior in the laboratory rat: factors underlying changes in responsiveness from 30 to 90 days. Dev Psychobiol 12:425–439

McCarthy MM (2010) How it's made: organisational effects of hormones on the developing brain. J Neuroendocrinol 22:736–742
McCullough J, Quadagno DM, Goldman BD (1974) Neonatal gonadal hormones: effect on maternal and sexual behavior in the male rat. Physiol Behav 12:183–188
McEwen BS, Akama KT, Spencer-Segal JL, Milner TA, Waters EM (2012) Estrogen effects on the brain: actions beyond the hypothalamus via novel mechanisms. Behav Neurosci 126:4–16
Mech LD, Wolf PC, Packard JM (1999) Regurgitative food transfer among wild wolves. Can J Zool 77:1192–1195
Mena-Segovia J, Bolam JP, Magill PJ (2004) Pedunculopontine nucleus and basal ganglia: distant relatives or part of the same family? Trends Neurosci 27:585–588
Mennella JA, Moltz H (1988a) Infanticide in the male rat: the role of the vomeronasal organ. Physiol Behav 42:303–306
Mennella JA, Moltz H (1988b) Infanticide in rats: male strategy and female counter-strategy. Physiol Behav 42:19–28
Miranda JA, Liu RC (2009) Dissecting natural sensory plasticity: hormones and experience in a maternal context. Hear Res 252:21–28
Miranda JA, Shepard KN, McClintock SK, Liu RC (2014) Adult plasticity in the subcortical auditory pathway of the maternal mouse. PLoS One 9:e101630
Mogenson GJ (1987) Limbic-motor integration. Prog Psychobiol Physiol Psychol 12:117–170
Mong JA, McCarthy MM (1999) Steroid-induced developmental plasticity in hypothalamic astrocytes: implications for synaptic patterning. J Neurobiol 40:602–619
Morrell JI, Schwanzel-Fukuda M, Fahrbach SE, Pfaff DW (1984) Axonal projections and peptide content of steroid hormone concentrating neurons. Peptides 5:227–239
Mota MT, Sousa MBC (2000) Prolactin levels of fathers and helpers related to alloparental care in common marmosets, *Callithrix jacchus*. Folia Primatol 71:22–26
Mota MT, Franci CR, de Sousa MB (2006) Hormonal changes related to paternal and alloparental care in common marmosets (*Callithrix jacchus*). Horm Behav 49:293–302
Myers B, Dolgas CM, Kasckow J, Cullinan WE, Herman JP (2014) Central stress-integrative circuits: forebrain glutamatergic and GABAergic projections to the dorsomedial hypothalamus, medial preoptic area, and bed nucleus of the stria terminalis. Brain Struct 219:1287–1303
Nakahara TS, Cardozo LM, Ibarra-Soria X, Bard AD, Carvalho VMA, Trintinalia GZ, Logan DW, Papes F (2016) Detection of pup odors by non-canonical adult vomeronasal neurons expressing an odorant receptor gene is influenced by sex and parenting status. BMC Biol 14:12
Newman JD (2007) Neural circuits underlying crying and cry responding in mammals. Behav Brain Res 182:155–165
Numan M (1988) Neural basis of maternal behavior in the rat. Psychoneuroendocrinology 13: 47–62
Numan M (2014) Neurobiology of social behavior: toward an understanding of the prosocial and antisocial brain. Academic Press, London
Numan M, Numan MJ (1997) Projection sites of medial preoptic area and ventral bed nucleus of the stria terminalis neurons that express Fos during maternal behavior in female rats. J Neuroendocrinol 9:369–384
Numan M, Sheehan TP (1997) Neuroanatomical circuitry for mammalian maternal behavior. Ann N Y Acad Sci 807:101–125
Numan M, Smith HG (1984) Maternal behavior in rats: evidence for the involvement of preoptic projections to the ventral tegmental area. Behav Neurosci 98:712–727
Numan M, Stolzenberg DS (2008) Hypothalamic interaction with the mesolimbic dopamine system and the regulation of maternal responsiveness. In: Bridges RS (ed) Neurobiology of the parental brain. Academic Press, Burlington, MA, pp 3–22
Numan M, Corodimas KP, Numan MJ, Factor EM, Piers WD (1988) Axon-sparing lesions of the preoptic region and substantia innominata disrupt maternal behavior in rats. Behav Neurosci 102:381–396

Numan M, Numan MJ, English JB (1993) Excitotoxic amino acid injections into the medial amygdala facilitate maternal behavior in virgin female rats. Horm Behav 27:56–81

Numan M, Numan MJ, Schwarz JM, Neuner CM, Flood TF, Smith CD (2005) Medial preoptic area interactions with the nucleus accumbens-ventral pallidum circuit and maternal behavior in rats. Behav Brain Res 158:53–68

Numan M, Stolzenberg DS, Dellevigne AA, Correnti CM, Numan MJ (2009) Temporary inactivation of ventral tegmental area neurons with either muscimol or baclofen reversibly disrupts maternal behavior in rats through different underlying mechanisms. Behav Neurosci 123:740

Nunes S, Fite JE, French JA (2000) Variation in steroid hormones associated with infant care behaviour and experience in male marmosets (*Callithrix kuhlii*). Anim Behav 60:857–865

Nunes S, Fite JE, Patera KJ, French JA (2001) Interactions among paternal behavior, steroid hormones, and parental experience in male marmosets (*Callithrix kuhlii*). Horm Behav 39:70–82

Ogawa S, Washburn TF, Taylor J, Lubahn DB, Korach KS, Pfaff DW (1998) Modifications of testosterone-dependent behaviors by estrogen receptor-α gene disruption in male mice. Endocrinology 139:5058–5069

Ogiue-Ikeda M, Tanabe N, Mukai H, Hojo Y, Murakami G, Tsurugizawa T, Takata N, Kimoto T, Kawato S (2008) Rapid modulation of synaptic plasticity by estrogens as well as endocrine disrupters in hippocampal neurons. Brain Res Rev 57:363–375

Okabe S, Nagasawa M, Kihara T, Kato M, Harada T, Koshida N, Mogi K, Kikusui T (2013) Pup odor and ultrasonic vocalizations synergistically stimulate maternal attention in mice. Behav Neurosci 127:432–438

Olazábal DE, Pereira M, Agrati D, Ferreira A, Fleming AS, González-Mariscal G, Lévy F, Lucion AB, Morrell JI, Numan M, Uriarte N (2013) Flexibility and adaptation of the neural substrate that supports maternal behavior in mammals. Neurosci Biobehav Rev 37:1875–1892

Orikasa C, Kondo Y, Katsumata H, Terada M, Akimoto T, Sakuma Y, Minami S (2017) Vomeronasal signal deficiency enhances parental behavior in socially isolated male mice. Physiol Behav 168:98–102

Oxley G, Fleming AS (2000) The effects of medial preoptic area and amygdala lesions on maternal behavior in the juvenile rat. Dev Psychobiol 37:253–265

Parker KJ, Kinney LF, Phillips KM, Lee TM (2001) Paternal behavior is associated with central neurohormone receptor binding patterns in meadow voles (*Microtus pennsylvanicus*). Behav Neurol 115:1341–1348

Pedersen CA, Caldwell JD, Walker C, Ayers G, Mason GA (1994) Oxyotcin activates the postpartum onset of rat maternal behavior in the ventral tegmental and medial preoptic areas. Behav Neurosci 108:1163–1171

Perea-Rodriguez JP, Saltzman W (2014) Differences in placentophagia in relation to reproductive status in the California mouse (*Peromyscus californicus*). Dev Psychobiol 56:812–820

Perea-Rodriguez JP, Takahashi EY, Amador TM, Hao RC, Saltzman W, Trainor BC (2015) Effects of reproductive experience on central expression of progesterone, oestrogen α, oxytocin and vasopressin receptor mRNA in male California mice (*Peromyscus californicus*). J Neuroendocrinol 27:245–252

Perkeybile AM, Bales KL (2017) Intergenerational transmission of sociality: the role of parents in shaping social behavior in monogamous and non-monogamous species. J Exp Biol 220:114–123

Perkeybile AM, Griffin LL, Bales KL (2013) Natural variation in early parental care correlates with social behaviors in adolescent prairie voles (*Microtus ochrogaster*). Front Behav Neurosci 7:21

Perkeybile AM, Delaney-Busch N, Hartman S, Grimm KJ, Bales KL (2015) Intergenerational transmission of alloparental behavior and oxytocin and vasopressin receptor distribution in the prairie vole. Front Behav Neurosci 9:191

Pfaff D, Waters E, Khan Q, Zhang X, Numan M (2011) Minireview: estrogen receptor-initiated mechanisms causal to mammalian reproductive behaviors. Endocrinology 152:1209–1217

Plumb R, Granger J, Stumpf C, Wilson ID, Evans JA, Lenz EM (2003) Metabonomic analysis of mouse urine by liquid-chromatography-time of flight mass spectrometry (LC-TOFMS): detection of strain, diurnal and gender differences. Analyst 128:819–823

Pro-Sistiaga P, Mohedano-Moriano A, Ubeda-Bañon I, del mar Arroyo-Jimenez M, Marcos P, Artacho-Pérula E, Crespo C, Insausti R, Martinez-Marcos A (2007) Convergence of olfactory and vomeronasal projections in the rat basal telencephalon. J Comp Neurol 504:346–362

Prudom SL, Broz CA, Schultz-Darken NJ, Ferris CT, Snowdon CT, Ziegler TE (2008) Exposure to infant scent lowers serum testosterone in father common marmosets (*Callithrix jacchus*). R Soc Biol Lett 4:603–605

Reburn CJ, Wynne-Edwards KE (1999) Hormonal changes in males of a naturally biparental and a uniparental mammal. Horm Behav 35:163–176

Reddy RC, Amodei R, Estill CT, Stormshak F, Meaker M, Roselli CE (2015) Effect of testosterone on neuronal morphology and neuritic growth of fetal lamb hypothalamus-preoptic area and cerebral cortex in primary culture. PLoS One 10:e0129521

Ribeiro AC, Musatov S, Shteyler A, Simanduyev S, Arrieta-Cruz I, Ogawa S, Pfaff DW (2012) siRNA silencing of estrogen receptor-α expression specifically in medial preoptic area neurons abolishes maternal care in female mice. Proc Natl Acad Sci U S A 109:16324–16329

Roberts RL, Jenkins KT, Lawler T Jr, Wegner FH, Newman JD (2001) Bromocriptine administration lowers serum prolactin and disrupts parental responsiveness in common marmosets (*Callithrix j. jacchus*). Horm Behav 39:106–112

Ronnekleiv OK, Bosch MA, Zhang C (2012) 17beta-oestradiol regulation of gonadotrophin-releasing hormone neuronal excitability. J Neuroendocrinol 24:122–130

Roselli CE, Stadelman H, Reeve R, Bishop CV, Stormshak F (2007) The ovine sexually dimorphic nucleus of the medial preoptic area is organized prenatally by testosterone. Endocrinology 148:4450–4457

Roselli CE, Amodei R, Meaker M, Reddy RC, Estill CT, Stormshak F (2015) Effect of testosterone on neuronal morphology and neuritic growth of fetal lamb hypothalamus-preoptic area and cerebral cortex in primary culture. PLoS One 10:e0129521

Rosenberg KM (1974) Effects of pre- and post-pubertal castration and testosterone on pup-killing behavior in the male rat. Physiol Behav 13:159–161

Rosenberg PA, Herrenkohl LR (1976) Maternal behavior in male rats: critical times for the suppressive actions of androgens. Physiol Behav 16(293):297

Rosenblatt JS, Ceus K (1998) Estrogen implants in the medial preoptic area stimulate maternal behavior in male rats. Horm Behav 33:23–30

Rosenblatt JS, Mayer AD (1995) An analysis of approach/withdrawal processes in the initiation of maternal behavior in the laboratory rat. In: Hood KE, Greenberg G, Tobach E (eds) Handbook of behavioral neurobiology. Plenum Press, New York, pp 229–298

Rosenblatt JS, Hazelwood S, Poole J (1996) Maternal behavior in male rats: effects of medial preoptic area lesions and presence of maternal aggression. Horm Behav 30:201–215

Ross HE, Young LJ (2009) Oxytocin and the neural mechanisms regulating social cognition and affiliative behavior. Front Neuroendocrinol 30(4):534–547

Ruscio MG, Sweeny TD, Hazelton JL, Suppatkul P, Boothe E, Carter CS (2008) Pup exposure elicits hippocampal cell proliferation in the prairie vole. Behav Brain Res 187:9–16

Sakaguchi K, Tanaka M, Ohkubo T, Doh-ura K, Fujikawa T, Sudo S, Nakashima K (1996) Induction of brain prolactin receptor long-form mRNA expression and maternal behavior in pup-contacted male rats: promotion by prolactin administration and suppression by female contact. Neuroendocrinology 63:559–568

Saltzman W, Ziegler TE (2014) Functional significance of hormonal changes in mammalian fathers. Neuroendocrinology 26:685–696

Saltzman W, Harris BN, de Jong TR, Perea-Rodriguez JP, Horrell ND, Zhao M, Andrew JR (2017) Paternal care in biparental rodents: intra- and inter-individual variation. Integr Comp Biol 57:589–602

Sapolsky RM, Romero LM, Munck AU (2000) How do glucocorticoids influence stress responses? Integrating permissive, suppressive, stimulatory and preparative actions. Endocr Rev 21:55–89

Schneider JS, Stone MK, Wynne-Edwards KE, Horton TH, Lydon J, O'Malley B, Levine JE (2003) Progesterone receptors mediate male aggression toward infants. Proc Natl Acad Sci U S A 100: 2951–2956
Schneider JS, Burgess C, Horton TH, Levine JE (2009) Effects of progesterone on male-mediated infant-directed aggression. Behav Brain Res 199:340–344
Schradin C (2008) Differences in prolactin levels between three alternative male reproductive tactics in striped mice (*Rhabdo*mys *pumilio*). Proc Biol Sci 275:1047–1052
Schradin C, Anzenberger G (1999) Prolactin, the hormone of paternity. News Physiol Sci 14: 223–231
Schradin C, Yuen CH (2011) Hormone levels of male African striped mice change as they switch between alternative reproductive tactics. Horm Behav 60:676–680
Schradin C, Reeder DM, Mendoza SP, Anzenberger G (2003) Prolactin and paternal care: comparison of three species of monogamous New World monkeys (*Callicebus cupreus, Callithrix jacchus*, and *Callimico goeldii*). J Comp Psychol 117:166
Schum JE, Wynne-Edwards KE (2005) Estradiol and progesterone in paternal and non-paternal hamsters (*Phodopus*) becoming fathers: conflict with hypothesized roles. Horm Behav 47: 410–418
Seelke AMN, Perkeybile AM, Grunewald R, Bales KL, Krubitzer LA (2016) Individual differences in cortical connections of somatosensory cortex are associated with parental rearing style in prairie voles (*Microtus ochrogaster*). J Comp Neurol 524:564–577
Shahrokh DK, Zhang TY, Diorio J, Gratton A, Meaney MJ (2010) Oxytocin-dopamine interactions mediate variations in maternal behavior in the rat. Endocrinology 151:2276–2286
Shorey S, He HG, Morelius E (2016) Skin-to-skin contact by fathers and the impact on infant and paternal outcomes: an integrative review. Midwifery 40:207–217
Silva MRP, Bernardi MM, Cruz-Casallas PE, Felicio LF (2003) Pimozide injections into the nucleus accumbens disrupt maternal behaviour in lactating rats. Basic Clin Pharmacol Toxicol 93:42–47
Simerly RB, Swanson LW, Chang C, Muramatsu M (1990) Distribution of androgen and estrogen receptor mRNA-containing cells in the rat brain: an in situ hybridization study. J Comp Neurol 294:76–95
Simoncelli LA, Delevan CJ, Al-Naimi OAS, Bamshad M (2010) Female tactile cues maximize paternal behavior in prairie voles. Behav Ecol Sociobiol 64:865–873
Simonoska R, Stenberg A, Masironi B, Sahlin L, Hultcrantz M (2009) Estrogen receptors in the inner ear during different stages of pregnancy and development in the rat. Acta Otolaryngol 129:1175–1181
Slawski BA, Buntin JD (1995) Preoptic area lesions disrupt prolactin-induced parental feeding behavior in ring doves. Horm Behav 29:248–266
Smith MO, Holland RC (1975) Effects of lesions of the nucleus accumbens on lactation and postpartum behavior. Physiol Psychol 3:331–336
Smith KS, Tindell AJ, Aldridge JW, Berridge KC (2009) Ventral pallidum roles in reward and motivation. Behav Brain Res 196:155–167
Sofroniew MV (1983) Vasopressin and oxytocin in the mammalian brain and spinal cord. Trends Neurosci 6:467–472
Song Z, Tai F, Yu C, Wu R, Zhang X, Broders H, Guo R (2010) Sexual or paternal experiences alter alloparental behavior and the central expression of ERα and OT in male mandarin voles (*Microtus mandarinus*). Behav Brain Res 214:290–300
Spehr M, Spehr J, Ukhanov K, Kelliher KR, Leinders-Zufall T, Zufall F (2006) Parallel processing of social signals by the mammalian main and accessory olfactory systems. Cell Mol Life Sci 63:1476–1484
Spence-Aizenberg A, Di Fiore A, Fernandez-Duque E (2016) Social monogamy, male–female relationships, and biparental care in wild titi monkeys (*Callicebus discolor*). Primates 57: 103–112

Stolzenberg DS, McKenna JB, Keough S, Hancock R, Numan MJ, Numan M (2007) Dopamine D_1 receptor stimulation of the nucleus accumbens or the medial preoptic area promotes the onset of maternal behavior in pregnancy-terminated rats. Behav Neurol 121:907–919

Storey AE, Ziegler TE (2016) Primate paternal care: interactions between biology and social experience. Horm Behav 77:260–271

Storey AE, Walsh CJ, Quinton RL, Wynne-Edwards KE (2000) Hormonal correlates of paternal responsiveness in new and expectant fathers. Evol Hum Behav 21:79–95

Storey AE, Noseworthy DE, Delahunty KM, Halfyard SJ, McKay DW (2011) The effects of social context on the hormonal and behavioral responsiveness of human fathers. Horm Behav 60: 353–361

Sturgis JD, Bridges RS (1997) N-methyl-DL-aspartic acid lesions of the medial preoptic area disrupt ongoing parental behavior in male rats. Physiol Behav 62:305–310

Swann JM, Newman SW (1992) Testosterone regulates substance P within neurons of the medial nucleus of the amygdala, the bed nucleus of the stria terminalis and the medial preoptic area of the male golden hamster. Brain Res 590:18–28

Szawka RE, Poletini MO, Leite CM, Bernuci MP, Kalil B, Mendonça LBD, Corlino ROG, Helena CVV, Bertram R, Franci CR, Anselmo-Franci JA (2013) Release of norepinephrine in the preoptic area activates anteroventral periventricular nucleus neurons and stimulates the surge of luteinizing hormone. Endocrinology 154:363–374

Tachikawa KS, Yoshihara Y, Kuroda KO (2013) Behavioral transition from attack to parenting in male mice: a crucial role of the vomeronasal system. J Neurosci 33:5120–5126

Taylor JH, French JA (2015) Oxytocin and vasopressin enhance responsiveness to infant stimuli in adult marmosets. Horm Behav 75:154–159

Timonin ME, Cushing BS, Wynne-Edwards KE (2008) In three brain regions central to maternal behaviour, neither male nor female *Phodopus* dwarf hamsters show changes in oestrogen receptor alpha distribution with mating or parenthood. J Neuroendocrinol 20:1301–1309

Tobiansky DJ, Roma PG, Hattori T, Will RG, Nutsch VL, Dominguez JM (2013) The medial preoptic area modulates cocaine-induced activity in female rats. Behav Neurosci 127:293–302

Tobiansky DJ, Will RG, Lominac KD, Turner JM, Hattori T, Krishnan K, Martz JR, Nutsch VL, Dominguez JM (2016) Estradiol in the preoptic area regulates the dopaminergic response to cocaine in the nucleus accumbens. Neuropsychopharmacology 41:1897–1906

Trainor BC, Marler CA (2002) Testosterone promotes paternal behaviour in a monogamous mammal via conversion to oestrogen. Proc Biol Sci 269:823–829

Trainor BC, Bird IM, Alday NA, Schlinger BA, Marler CA (2003) Variation in aromatase activity in the medial preoptic area and plasma progesterone is associated with the onset of paternal behavior. Neuroendocrinology 78:36–44

Tsuneoka Y, Maruyama T, Yoshida S, Nishimori K, Kato T, Numan M, Kuroda KO (2013) Functional, anatomical, and neurochemical differentiation of medial preoptic area subregions in relation to maternal behavior in the mouse. J Comp Neurol 521:1633–1663

Tsuneoka Y, Tokita K, Yoshihara C, Amano T, Esposito G, Huang AJ, Yu LM, Odaka Y, Shinozuka K, McHugh TJ, Kuroda KO (2015) Distinct preoptic-BST nuclei dissociate paternal and infanticidal behavior in mice. EMBO J 34:2652–2670

Tsuneoka Y, Yoshida S, Takase K, Oda S, Kuroda M, Funato H (2017) Neurotransmitters and neuropeptides in gonadal steroid receptor-expressing cells in medial preoptic area subregions of the male mouse. Sci Rep 7:9809

Vargas-Pinilla P, Paixão-Côrtes VR, Paré P, Tovo-Rodrigues L, Vieira CMDAG, Xavier A, Comas D, Pissinatti A, Sinigaglia M, Rigo MM, Viera GF, Lucion AB, Salzano FM, Vieira GF (2015) Evolutionary pattern in the OXT-OXTR system in primates: coevolution and positive selection footprints. Proc Natl Acad Sci U S A 112:88–93

Vargas-Pinilla P, Babb P, Nunes L, Paré P, Rosa G, Felkl A, Longo Dâae L, Salzano FM, Paixão-Côrtes VR, Gonçalves GL, Bortolini MC (2017) Progesterone response element variation in the OXTR promoter region and paternal care in New World monkeys. Behav Genet 47:77–87

van Anders SM, Gray PB (2007) Hormones and human partnering. Annu Rev Sex Res 18:60–93

Vernotica EM, Rosenblatt JS, Morrell JI (1999) Microinfusion of cocaine into the medial preoptic area or nucleus accumbens transiently impairs maternal behavior in the rat. Behav Neurosci 113: 377–390

Vertes RP (2004) Differential projections of the infralimbic and prelimbic cortex in the rat. Synapse 51:32–58

Vom Saal FS (1985) Time-contingent change in infanticide and parental behavior induced by ejaculation in male mice. Physiol Behav 34:7–15

Wang Z, De Vries GJ (1993) Testosterone effects on paternal behavior and vasopressin immunoreactive projections in prairie voles (Microtus ochrogaster). Brain Res 631:156–160

Wang Z, Ferris CF, De Vries GJ (1994) Role of septal vasopressin innervation in paternal behavior in prairie voles (*Microtus ochrogaster*). Proc Natl Acad Sci U S A 91:400–404

Wang Z, Zhou L, Hulihan TJ, Insel TR (1996) Immunoreactivity of central vasopressin and oxytocin pathways in microtine rodents: a quantitative comparative study. J Comp Neurol 366:726–737

Wang ZX, Liu Y, Young LJ, Insel TR (2000) Hypothalamic vasopressin gene expression increases in both males and females postpartum in a biparental rodent. J Neuroendocrinol 12:111–120

Wang J, Tai F, Lai X (2014) Cocaine withdrawal influences paternal behavior and associated central expression of vasopressin, oxytocin and tyrosine hydroxylase in mandarin voles. Neuropeptides 48:29–35

West HER, Capelllini I (2016) Male care and life history traits in mammals. Nat Commun 7:11854

Winn P (2006) How best to consider the structure and function of the pedunculopontine tegmental nucleus: evidence from animal studies. J Neurol Sci 248:234–250

Woller MJ, Sosa ME, Chiang Y, Prudom SL, Keelty P, Moore JE, Ziegler TE (2011) Differential hypothalamic secretion of neurocrines in male common marmosets: parental experience effects. J Neuroendocrinol 24:413–421

Woodroffe R, Vincent A (1994) Mother's little helpers: patterns of male care in mammals. Trends Ecol Evol 9:294–297

Wu Z, Autry AE, Bergan JF, Watabe-Uchida M, Dulac CG (2014) Galanin neurons in the medial preoptic area govern parental behaviour. Nature 509:325–330

Wynne-Edwards KE (1995) Biparental care in Djungarian but not Siberian dwarf hamsters (*Phodopus*). Anim Behav 50:1571–1585

Yoshida M, Suga S, Sakuma Y (1994) Estrogen reduces the excitability of the female rat medial amygdala afferents from the medial preoptic area but not those from the lateral septum. Exp Brain Res 101:1–7

Yu GZ, Kaba H, Okutani F, Takahashi S, Higuchi T (1996a) The olfactory bulb: a critical site of action for oxytocin in the induction of maternal behaviour in the rat. Neurosci 72:1083–1088

Yu GZ, Kaba H, Okutani F, Takahashi S, Higuchi T, Seto K (1996b) The action of oxytocin originating in the hypothalamic paraventricular nucleus on mitral and granule cells in the rat main olfactory bulb. Neuroscience 72:1073–1082

Zancan M, Dall'Oglio A, Quagliotto E, Rasia-Filho AA (2017) Castration alters the number and structure of dendritic spines in the male posterodorsal medial amygdala. Eur J Neursci 45: 572–580

Zhang C, Tonsfeldt KJ, Qiu J, Bosch MA, Kobayashi K, Steiner RA, Ronnekleiv OK (2013) Molecular mechanisms that drive estradiol-dependent burst firing of Kiss1 neurons in the rostral periventricular preoptic area. Am J Physiol Endocrinol Metab 305:E1384–E1397

Zhong J, Liang M, Akther S, Higashida C, Tsuji T, Higashida H (2014) c-Fos expression in the paternal mouse brain induced by communicative interaction with maternal mates. Mol Brain 7:66

Ziegler TE, Snowdon CT (2000) Preparental hormone levels and parenting experience in male - cotton-top tamarins, *Saguinus oedipus*. Horm Behav 38:159–167

Ziegler TE, Sosa ME (2016) Hormonal stimulation and paternal experience influence responsiveness to infant distress vocalizations by adult male common marmosets, *Callithrix jacchus*. Horm Behav 78:13–19

Ziegler TE, Wegner FH, Snowdon CT (1996) Hormonal responses to parental and nonparental conditions in male cotton-top tamarins, *Saguinus oedipus*, a New World primate. Horm Behav 30:287–297

Ziegler TE, Wegner FH, Carlson AA, Lazaro-Perea C, Snowdon CT (2000) Prolactin levels during the periparturitional period in the biparental cotton-top Tamarin (*Saguinus Oedipus*): interactions with gender, androgen levels, and parenting. Horm Behav 38:111–122

Ziegler TE, Washabaugh KF, Snowdon CT (2004) Responsiveness of expectant male cotton-top tamarins, *Saguinus oedipus*, to mate's pregnancy. Horm Behav 45:84–92

Ziegler TE, Prudom SL, Zahed SR (2009a) Variations in male parenting behavior and physiology in the common marmoset. Am J Hum Biol 21:739–744

Ziegler TE, Prudom SL, Zahed SR, Parlow AF, Wegner F (2009b) Prolactin's mediative role in male parenting in parentally experienced marmosets (*Callithrix jacchus*). Horm Behav 56: 436–443

Zilkha N, Scott N, Kimchi T (2017) Sexual dimorphism of parental care: from genes to behavior. Annu Rev Neurosci 40:273–305

Zimmermann-Peruzatto JM, Lazzari VM, de Moura AC, Almeida S, Giovenardi M (2015) Examining the role of vasopressin in the modulation of parental and sexual behaviors. Front Psych 6:130

Neuroendocrinology and Adaptive Physiology of Maternal Care

Kristina O. Smiley, Sharon R. Ladyman, Papillon Gustafson, David R. Grattan, and Rosemary S. E. Brown

Contents

1 Introduction 162
2 Neural Networks Mediating Parental Behavior 164
2.1 Motivation, Reward, and Reinforcement 166
2.2 Suppression of Avoidance and Aversion 169
2.3 Sensory Processing 172
3 Hormonal Changes Required for the Establishment of Maternal Behavior 173
3.1 Patterns of Hormone Changes During Pregnancy and Lactation 173
3.2 Neuroendocrine Modulation of Young-Directed Maternal Behavior 176
3.3 Neuroendocrine Modulation of Young-Related Behaviors and Physiological Adaptations Supporting Pregnancy and Lactation 183
4 Translation to Human Parenting 191
4.1 Sensory Input 192
4.2 Neuroendocrinology of Maladaptive Maternal Care 193
5 Conclusion 197
References 198

Abstract Parental care is critical for offspring survival in many species. In mammals, parental care is primarily provided through maternal care, due to obligate pregnancy and lactation constraints, although some species also show paternal and alloparental care. These behaviors are driven by specialized neural circuits that receive sensory, cortical, and hormonal input to generate a coordinated and timely

K. O. Smiley and P. Gustafson
Centre for Neuroendocrinology and Department of Anatomy, School of Biomedical Sciences, University of Otago, Dunedin, New Zealand

S. R. Ladyman and D. R. Grattan
Centre for Neuroendocrinology and Department of Anatomy, School of Biomedical Sciences, University of Otago, Dunedin, New Zealand

Maurice Wilkins Centre for Molecular Biodiscovery, Auckland, New Zealand

R. S. E. Brown (✉)
Centre for Neuroendocrinology and Department of Physiology, School of Biomedical Sciences, University of Otago, Dunedin, New Zealand
e-mail: rosemary.brown@otago.ac.nz

Curr Topics Behav Neurosci (2019) 43: 161–210
DOI 10.1007/7854_2019_122
Published Online: 6 December 2019

change in behavior, and sustain that behavior through activation of reward pathways. Importantly, the hormonal changes associated with pregnancy and lactation also act to coordinate a broad range of physiological changes to support the mother and enable her to adapt to the demands of these states. This chapter will review the neural pathways that regulate maternal behavior, the hormonal changes that occur during pregnancy and lactation, and how these two facets merge together to promote both young-directed maternal responses (including nursing and grooming) and young-related responses (including maternal aggression and other physiological adaptions to support the development of and caring for young). We conclude by examining how experimental animal work has translated into knowledge of human parenting, particularly in regards to maternal mental health issues.

Keywords Lactation · Maternal behavior · Maternal care · Metabolism · Postpartum depression · Pregnancy · Reproduction · Reproductive hormones · Stress

1 Introduction

Mammalian offspring require parental support for survival. Parents must protect new-born offspring, keep them warm, and provide nutrition. This requires significant investment in time and resources on behalf of the parents. In particular, the sexually dimorphic nature of pregnancy and the necessity for lactation means that mammalian mothers carry a markedly higher proportion of the cost of reproduction. While both sexes can contribute to parental care after birth, the degree of involvement of males and females varies considerably between species, and between individuals within a species. Because the role of the mother in parenting is ubiquitous in all mammals, at least up until the time of delivery of the offspring at birth, this chapter will focus primarily on maternal behavior. However, we will also examine whether similar or distinct neuroendocrine mechanisms act to facilitate parental behavior in fathers and alloparents (individuals who are not the biological parents but contribute to the care and raising of young).

Maternal behavior can be defined as the collection of behaviors by a postpartum mother that maximizes the likelihood that her offspring will survive. As per Bob Bridges' definition (Bridges 2015), maternal behavior can be broadly classified into two types of responses: young-directed responses and young-related responses. In rodents, young-directed responses include nest-building, retrieval of young, licking/grooming, crouching over the pups, and nursing. These behavioral responses immediately aid in the caring and survival of her pups. Young-related responses include displaying increased maternal aggression to protect offspring, increasing food intake to meet metabolic demands of lactation, and reduced anxiety. These types of behaviors and responses are not directed at the young per se, but are nonetheless required to ensure safety of the young and importantly, to ensure that the mother is in an appropriate physiological state to be able to perform the necessary young-directed responses. For this chapter, we extend Bridges' definition of young-related responses to include the physiological adaptions which occur in the mother during pregnancy and the postpartum period which prepare and enable the mother to express both young-directed and young-related maternal behaviors after birth. The maternal body undergoes extraordinary physiological adaptions during pregnancy to support the growing fetus and these changes have a significant influence on subsequent maternal behavior.

In order to drive maternal behavior, specialized neural circuits receive sensory, cortical, and hormonal input to generate a coordinated change in behavior. The original demonstration that simple electrical stimulation of the hypothalamic preoptic region could induce complex maternal behaviors in virgin female rats (Fisher 1956) provided evidence for an endogenous "maternal circuit" that existed within the brain. Arguably, one of the most important systems supporting parental behavior within this neural circuitry is the midbrain dopaminergic pathways that have evolved to encourage particular offspring-directed behaviors. Compared with other more overt rewards, such as consumption of pleasurable food, learning to complete a motor task, or engaging in sexual activity, parental behavior is a long-term investment that does not provide immediate feedback. Hence, it seems likely that the brain must engage these reward circuits in a different manner to promote the ongoing sacrifices and effort required for parental care of their offspring. The focus of this chapter is on how hormones act on this existing neural circuitry to induce timely adaptations in maternal behavior. Indeed, maternal behavior represents the ultimate neuroendocrine collaboration between neuronal circuits within the brain and hormonal signals arising in the periphery, to unite the different facets of maternal behavior. Insight into the neural circuits underlying maternal behavior, and the hormonal regulation of this behavior, has largely been provided by research conducted in rats and mice. Therefore, in focusing on what is currently understood regarding the neurobiology of maternal behavior, this chapter will predominately utilize research conducted in rodents.

Convincing evidence for the humoral basis of maternal behavior was first provided by parabiotic experiments where blood was transfused from late pregnant animals into virgin female controls (Terkel et al. 1972). This resulted in a rapid induction of maternal care, clearly demonstrating the neuroendocrine interactions underlying this behavior. Hormonal changes coinciding with pregnancy include increased levels in circulating estradiol, progesterone, and lactogenic hormones (prolactin and placental lactogens). These hormones play an integral role in establishing the physiological adaptions in the maternal body both to support the growing fetus during pregnancy and to prepare the mother to be able to display the normal expression of maternal behavior during the postpartum period. These hormonal changes are facilitated by the placenta itself, in addition to maternal changes in the hypothalamic control and release of hormones from the pituitary gland. This change in the pattern of hormone release will be described followed by an explanation of how these hormonal changes affect the maternal body and maternal behavior.

While maternal care is the predominant form of parental behavior in most mammalian species, for approximately 3–5% of mammalian species the father also contributes to parental care, and this is referred to as paternal care. Although this chapter predominantly focuses on maternal care, it is informative to also examine changes that happen during paternal behavior, and to consider whether these changes are meditated through similar or different mechanisms to those described for maternal behavior. With the exception of nursing behavior, male parents typically engage in similar behaviors as females such as nest building, licking and grooming young, and retrieving pups. Males can also play a large role in transporting, playing with, socializing, defending and warming offspring, as well as providing food, shelter, or other resources (Nagaishi et al. 2014). In some cases, such as hamsters, males even assist in the delivery and cleaning of pups during birth (Sorenson and Brelje 1997). Alloparental care also occurs in several cooperatively breeding mammalian species including primates, rodents, and carnivores, where alloparents, or individuals other than the biological parents, contribute to the raising and provisioning of the young. Importantly, humans are an example of a species where both paternal and alloparental care play a role in the safety and development of the offspring, meaning that it is important to consider the neuroendocrinology of alloparents as well. Alloparents are typically juveniles caring for younger siblings, or in some cases, adults which have not dispersed from the natal family group. Studies in both fathers and alloparents (of either sex) reveal that many of the neural regions underlying paternal and alloparental care are similar to maternal care, suggesting that a conserved neural network for parental care exists in most mammals of both sexes. Within this chapter, where appropriate, we will briefly describe the neuroendocrine basis of paternal and alloparental care and discuss the similarities and differences in relation to maternal care.

Finally, we conclude with a description of how the animal literature discussed in this chapter may translate to improved understanding of human parental care. For many human mothers, the transition into maternal care is not easy and upwards to 30% of women may experience postpartum depression (Brummelte and Galea 2016) or other mental illnesses which contribute to maladaptive maternal behaviors – or behaviors that present obstacles in raising and caring for the child. In some cases, mothers will also show atypical maternal behavior or fail to provide maternal care altogether. Here, we will focus on what is known about healthy mothers and how this is informative in understanding maladaptive maternal behavior.

2 Neural Networks Mediating Parental Behavior

Maternal behavior can be considered in the context of two parallel pathways: one promoting offspring approach and parental care, and the other suppressing avoidance of and aversion to offspring (Fig. 1). These pathways have been extensively mapped using markers of recent cellular activity, such as immediate early gene c-fos expression, and by lesion studies characterizing key parts of the network. The advent

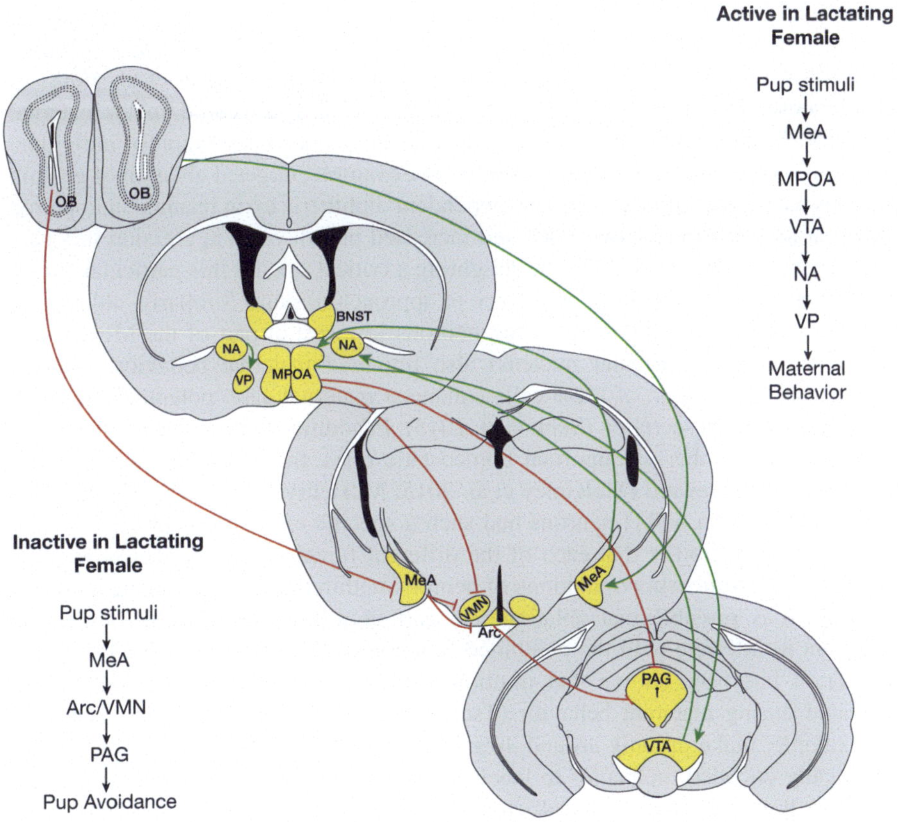

Fig. 1 Dual circuits underlying maternal behavior. Appropriate display of maternal behavior requires both the removal of aversion (shown in red) behavior and activation of pup-directed behavior (shown in green). *OB* olfactory bulb, *NA* nucleus accumbens, *VP* ventral pallidum, *MPOA* medial preoptic area, *BNST* bed nucleus of the stria terminalis, *MeA* medial amygdala, *VMN* ventromedial nucleus, *Arc* arcuate nucleus, *PAG* periaqueductal grey, *VTA* ventral tegmental area

of modern molecular genetic tools has extended our ability to visualize and functionally modulate complex neuronal networks, providing new insights into the nature of these networks. Although the neuroanatomical basis of paternal care has not been as well studied as maternal care, it is clear that several overlapping regions are critical to the expression of both maternal and paternal care, suggesting the presence of a conserved neural network to promote parental care in both sexes. Sensory information from the offspring is important to stimulate either avoidant or approach responses. Therefore, we will also discuss the importance of sensory stimuli processing in relationship with parental care.

The medial preoptic area (MPOA) of the hypothalamus and the ventral bed nucleus of the stria terminalis (vBNST) have long been identified as critical structures in mediating the expression of maternal behavior (Fig. 1). Key initial evidence came from demonstrating that electrolytic lesions of the MPOA blocked the expression of maternal behavior (Numan 1974; Numan et al. 1977). Subsequent refinements showed that even excitotoxic fiber-sparing lesions (Numan et al. 1988) or small focused lesions within the MPOA (Jacobson et al. 1980) could impair maternal behavior. Modern molecular genetic strategies have allowed ablation of individual populations of neurons within the MPOA. For example, targeted ablation of galanin neurons within the MPOA using Cre-dependent diphtheria toxin resulted in impaired maternal responses in mother mice and increased pup-directed aggression in virgin female animals (Wu et al. 2014), highlighting a critical role for this particular subset of neurons in switching from avoidance to approach of pups. Similarly, ablation of tyrosine hydroxylase (TH)-expressing neurons in a rostral part of the MPOA (the anteroventral periventricular nucleus) also impaired maternal behavior, whereas optogenetic stimulation of these cells enhanced maternal care, potentially through enhanced oxytocin secretion (Scott et al. 2015). In addition to maternal behavior, the MPOA has a number of important homeostatic roles, including thermoregulation, sleep, and social reward (McKinley et al. 2015; McHenry et al. 2017). The fact that deletion of galanin or TH neurons had such a specific effect on maternal behavior suggests the possibility that each of the different functions of the MPOA may be mediated by distinct subpopulations of neurons within this complex nucleus. Alternatively, it is possible that ablating one part of a complex network of cells is sufficient to interfere with the combined function of the network. Using c-fos as a marker, it has been shown that multiple different cell types in the MPOA are activated during maternal behavior (Tsuneoka et al. 2013), and galanin neurons, for example, make up only around 40% of activated cells (Wu et al. 2014). Using genetically encoded molecules to label or manipulate specific populations of neurons, we now have the technical capability to tease these possibilities apart. For example, Cre-directed tract tracing has recently been used to map projections of neurotensin neurons in the MPOA to the VTA, and assess their function during social reward (McHenry et al. 2017).

2.1 Motivation, Reward, and Reinforcement

2.1.1 Maternal Care

A critical feature of the maternal behavior network is its ability to engage with reward mechanisms in the brain. Mothers must be motivated to engage in care of their offspring, and must express offspring-directed behavior in preference over other behaviors. Using operant conditioning methods developed for investigation of reward-related learning, nulliparous mice can be trained to press a lever, using pups as reinforcement (van Hemel 1973). Reproductive experience significantly influences this behavior, with postpartum female mice pressing the lever much more frequently than non-pregnant controls (Hauser and Gandelman 1985). Interactions with offspring are highly rewarding for a postpartum female, with a place-preference paradigm showing that postpartum female rats find pups more rewarding than cocaine (Mattson and Morrell 2005). Work by Alison Fleming's group demonstrated that lesions to the MPOA could significantly reduce the reward value of pups (Lee et al. 2000). Importantly, in that study, the MPOA was again highlighted as the only area where greater activation of neurons was seen with a preference for pup-associated versus cocaine-associated cues. It is now apparent that the MPOA provides a key output of the maternal behavior circuit, engaging with the midbrain reward systems through activation of dopamine projections from the ventral tegmental area (VTA) to the nucleus accumbens (NAcc) (Numan et al. 2005; Numan and Stolzenberg 2009; Stolzenberg and Numan 2011) (Fig. 1). This efferent projection from the MPOA to the VTA was observed to be critical, as knife cuts to disrupt this pathway significantly impaired maternal behavior (Terkel et al. 1979; Numan and Smith 1984; Numan et al. 1990). Specific ablation of the dopamine neurons using 6-hydroxydopamine lesions in the VTA or ventral striatum (NAcc, a downstream target of VTA dopaminergic neurons) also impaired maternal behavior (Hansen et al. 1991; Hansen 1994). Moreover, endogenous dopamine release in the NAcc was markedly elevated during maternal behavior (Hansen et al. 1993; Afonso et al. 2013), with dopamine D1, rather than D2, receptors required for the display of offspring-directed maternal behavior (Numan et al. 2005). Dopamine agonist treatment can also induce maternal behavior in estrogen-primed rats (Stolzenberg et al. 2007).

Although lesion and knife-cut studies were consistent with the hypothesis that an MPOA projection to the VTA was critical in maternal behavior, it quickly became apparent that much more widespread connections from the MPOA to other brain regions were also involved in the demonstration of maternal behavior (Numan and Sheehan 1997). To assess further regional activation of neurons during maternal behavior, c-fos labeling was used to identify neurons that were activated during exposure to pups. Consistent with lesion studies, there is a dense activation of neurons throughout the MPOA and vBNST during expression of maternal behavior (Tsuneoka et al. 2013; Numan et al. 1998; Stack and Numan 2000). By combining c-fos studies with classical anterograde and retrograde tract tracing techniques, the distribution of projections from the MPOA and vBNST has been mapped and characterized with respect to maternal behavior (Numan and Sheehan 1997). These data showed extensive projections from the MPOA to the lateral septum and the paraventricular (PVN) and ventromedial hypothalamic nuclei (VMN), as well as the VTA, whereas the vBNST projected more strongly to the VTA, periaqueductal gray matter, and brainstem (Fig. 1). Many of these areas also showed elevated c-fos expression during maternal behavior (Stack et al. 2002; Matsushita et al. 2015; Lonstein and de Vries 2000). Collectively these observations demonstrate that although the MPOA is a critical nexus for the maternal behavior circuit, additional neuronal systems complete a distributed network regulating maternal behavior.

Consistent with the work in rodents, multiple human neuroimaging studies have shown activation of the mesolimbic reward areas during various aspects of parental care. For example, increased fMRI activity in the mesolimbic reward circuit is repeatedly observed when subjects listen to baby cries or viewing baby images, with stronger activation occurring when viewing happy baby faces, relative to neutral or sad faces (Lonstein et al. 2015). Infant features such as wide eyes, large, round heads, and small noses are found attractive and elicit positive emotions in adults when viewing them (Lonstein et al. 2015). Using a paradigm where participants could control the length of time they viewed images of infants, women demonstrated increased motivation to extend the viewing time of infant-related stimuli (Charles et al. 2013). Mothers who show increased synchronicity with their babies' actions during mother–infant interactions showed increased activity in the NAcc while viewing infant-related videos. In contrast, intrusive mothers show higher activation in the amygdala, which plays a role in emotional processing, fear, and arousal (Feldman 2016). In addition to areas such as the amygdala and NAcc, which also show activation patterns that complement their proposed functions in rodents, many higher-order regions such as prefrontal cortex, anterior cingulate cortex, and the orbital frontal cortex (OFC) are also found to be involved with various aspects of human parental care (Rutherford et al. 2011). For example, the OFC, an area known to be involved in hedonic stimuli processing, shows increased activity when viewing image stimuli of one's own infant and is positively correlated with increased mood (Lonstein et al. 2015). Together, these studies point to what is likely a highly conserved network supporting reward and motivation during offspring-directed parental care.

The dopamine (DA) and endogenous opioid (EO) systems are highly involved in the wanting and liking reward processing respectively, which has been strongly supported by animal research. In addition, EO mediates motivation, social reward, and the attenuation of pain and anxiety (Swain 2011). While a direct role for EO in human maternal care is currently unknown, neuroimaging studies show increased activity in central opioid systems. Specifically, PET imaging for μ-opioid neurotransmission during maternal behavior shows an increase in the prefrontal cortex, anterior cingulate cortex, NAcc, and midbrain (Swain 2011). In addition, neuropeptides that are important for social behavior, such as oxytocin, have been shown to regulate reward aspects of parental care in prairie voles (*Microtus ochrogaster*) in the NAcc (Olazabal and Young 2006; Keebaugh et al. 2015), and hence it is possible that oxytocin may have similar effects in humans.

2.1.2 Paternal and Alloparental Care

Similar to females, it is thought that two main neural circuits regulate paternal care: stimulation of one circuit promotes approach responses and motivation to care for young and inhibition of another circuit that regulates the young-avoidance or infanticidal response. Surprisingly, little research has focused on whether fathers find interactions with young rewarding or how this young-directed motivation develops. Since the motivational circuits have been well established in females, one would predict that similar reward pathways are involved in paternal care and that these changes are likely set up in the males during the gestation of his offspring, either by hormones or via female-elicited social cues. While the reward circuitry underlying paternal behaviors are described elsewhere in this book, several behavioral studies suggest that young are indeed found rewarding to fathers. For example, Mandarin vole (*Lasiopodomys mandarinus*) fathers can develop a conditioned place-preference for cues associated with pups throughout the postpartum period (Wang et al. 2012). Paternal voles showed increased dopamine release in the NAcc as a result of pup exposure (Lei et al. 2017), with blockade of dopamine receptors in male prairie voles decreasing pup licking and grooming behavior (Lei et al. 2017). Similar to females, the MPOA has been identified as a key player in regulating male parental behavior, with MPOA lesions disrupting parental care in both male and female California mice (*Peromyscus californicus*) (Lee and Brown 2007). As in females, galanin neurons in the MPOA are also an important modulator of paternal care (Wu et al. 2014). Genetic ablation of these neurons decreased parental behaviors in both males and females, whereas optogenetic activation of these neurons in virgin males decreased pup aggression and increased pup-grooming. Furthermore, Tsuneoka et al. (2015) showed that specific neuronal activity patterns in the central MPOA (cMPOA) are indicative of the degree of parental behavior displayed by male mice. Lesioning the cMPOA increased infanticidal occurrences in fathers, while activation of this area decreased infanticide in virgin males. Together, these studies suggest that the MPOA serves as a "hub" for integrating the neural circuitry of paternal care. Additionally, lesions to the basolateral, corticomedial, and medial nucleus of the amygdala also disrupt paternal care in California mice and prairie voles (Lee and Brown 2007; Kirkpatrick et al. 1994). Other regions such as NAcc and lateral septum, which have been implicated in maternal care and other affiliative social behaviors, have been established as important for paternal care (Lee and Brown 2007; Wang et al. 1994).

To date, the reward mechanisms of alloparental care have not been well studied. Therefore, it is unknown whether alloparents care for offspring due to reward reinforcement from the infant stimulus, whether crying or other stimuli become increasingly aversive, or a combination of both. Although it has been reported that prairie vole alloparents do not show a preference for pups in a conditioned place paradigm (Kenkel et al. 2017), more work is required before definitive conclusions can be drawn.

2.2 Suppression of Avoidance and Aversion

2.2.1 Maternal Care

While the motivation to care for offspring is critical, many studies have shown that there also needs to be a suppression of an alternative pathway that drives non-pregnant animals to avoid young. Non-pregnant female or male rats appear to find pups aversive, and will actively ignore or even attack them. In contrast, primiparous female rats will respond maternally immediately after parturition, suggesting that aversion to pups has been actively suppressed. Hence, there appears to be a second circuit that acts to inhibit the display of certain behaviors (Fig. 1). Both olfactory cues (via the main olfactory bulb) and pheromonal cues from the vomeronasal system (via the accessary olfactory bulb) seem to be important in generating the aversive response (Fleming et al. 1979; Fleming and Luebke 1981). Projections from both of these sources pass into the medial amygdala, a structure long associated with emotionality and fear. Consistent with this nucleus exerting an inhibitory effect on maternal behavior, lesions to the medial amygdala in nulliparous rats facilitate maternal responsiveness. Similarly, lesions to the stria terminalis, the major output from the amygdala into the hypothalamus, are equally effective at stimulating maternal behavior (Fleming et al. 1979). This pathway could function by inhibiting the MPOA maternal behavior circuits. However, most data indicate that the medial amygdala circuit acts in the VMN to promote aversion to pups. A clear inhibitory role for the VMN in maternal behavior was demonstrated with VMN lesions advancing maternal responsiveness in nulliparous rats (Bridges et al. 1999; Mann and Babb 2004).

Nulliparous females will eventually tolerate pups after repeated exposure, and even though they cannot lactate, they begin to show maternal behaviors such as retrieving pups to a nest, grooming, and crouching over them. Known as sensitization, this process in rats has provided an excellent model for investigating the pathways involved in maternal behavior, and has been described as the "approach versus avoidance switch" (Mayer and Rosenblatt 1980; Numan 2006). Evidence suggests that the hormone-induced sensitization process involves both an increase in motivation to approach and a decrease in avoidance, with these two modalities controlled by distinct circuits. Interestingly, in laboratory mice, virgin females are spontaneously maternal and so may not require suppression of the avoidance pathway. However, this loss of aversion may have resulted from selection for laboratory domestication as wild female mice, similar to nulliparous rats, still display aversion towards pups or infanticide (McCarthy and vom Saal 1985). Virgin male laboratory mice, on the other hand, have retained infanticidal tendencies (Tsuneoka et al. 2015), and thus suppression of the avoidance pathway may be of particular interest in male parental behavior in mice and is discussed later.

The key to understanding maternal behavior is in identifying the connectivity of neural circuits that establish the behavior, the complex hormonal changes of pregnancy and lactation, as well as other events associated with the termination of pregnancy and birth of the offspring. These aspects can then be linked into our understanding of the reward system to shed light on how the expression of appropriate behavior is promoted.

2.2.2 Paternal and Alloparental Care

In many rodent species, males are infanticidal outside of the time they may have sired offspring. When these behaviors were first observed in lab settings, they were considered to be effects of poor housing or maladaptive lab environments. However, it is now clear that infanticide can be part of an adaptive reproductive strategy in order for males to increase their own reproductive success. Because fertility is suppressed during lactation (see section on young-related responses), killing offspring sired by other males can reestablish the breeding cycle of the females to allow the male to mate and sire offspring of his own. Through mating, however, infanticidal behavior can be inhibited and males can become paternal. Thus, the trigger to shift behavior from killing to caring for young is the act of mating and ejaculation, and in some cases, cohabitation with the pregnant female. There is considerable inter- and intra-species variation in the stimulus required to decrease infanticidal tendencies and increase paternal behaviors and to whether this extends to their own or foreign pups, which are presumably based on the life history, ecology, and mating system of each species (Saltzman et al. 2017).

In mice, there is a specific time-period following mating during which infanticide is inhibited. The duration of pregnancy in female mice is approximately 20 days, with pups weaned 3–4 weeks postpartum. Interestingly, male infanticide behavior markedly decreases 12 days after mating, and lasts for up to 60 days, after which infanticidal tendencies reemerge (vom Saal 1985). This timeframe corresponds exactly with the period when a male would interact with pups that he sired. Importantly, infanticidal behaviors will only diminish if ejaculation occurs during mating, further supporting the adaptive advantage of this time-locked behavioral change for the males' reproductive success. The precise timing of this dramatic behavioral change suggests that there is an endogenous timing mechanism that controls this switch. To test this hypothesis, Perrigo and colleagues (Perrigo et al. 1990) kept mated male mice on 22-h "fast day" cycles (11:11 h light:dark) or 27-h "slow day" cycles (13.5:13.5 h light:dark). Indeed, they found that the number of light cycles a male experienced since mating and ejaculation, rather than absolute time, determined when the inhibition of infanticide emerged, indicating there is an underlying circadian rhythm which facilitates this behavioral transition.

Similar to females, the amygdala and the BNST play a significant role in the suppression of avoidant tendencies in order to facilitate the expression of paternal care. Lesions to the rhomboid nucleus of the BNST specifically decrease infanticide in virgin males (Tsuneoka et al. 2015). Infanticidal tendencies may also be regulated by progesterone as progesterone receptor knockout mice exhibit reduced aggression towards pups, whereas progesterone treatments increase the occurrences of pup-directed aggression (Schneider et al. 2003).

Aggressive behaviors are triggered by olfactory cues from the pups in virgin rodents. Male mice lacking a functional vomeronasal system do not attack pups and are spontaneously paternal, suggesting that the mating-induced switch in behavior might result from a temporal suppression of vomeronasal activation by pup-related cues (Tachikawa et al. 2013). Similarly, anosmia facilitates maternal care in virgin female rats, which also initially display avoidant responses towards unfamiliar pups (Numan 2007). In male prairie voles (non-fathers), pup exposure increases c-fos activity in the olfactory bulb, with bilateral olfactory bulb lesions increasing the number of attack occurrences towards pups, which is a rare behavior in prairie voles (Bales and Saltzman 2016).

In other species, such as rats, that do not normally cooperatively breed, postweaning-prepubertal males and females will show alloparental behaviors towards pups, with avoidant and infanticidal behaviors not emerging in males until adulthood (Kuroda et al. 2011). Additionally, certain laboratory strains of mice will show alloparental care in the form of communal nesting, while this behavior is rarer in free-living mice (Garner et al. 2016). In contrast to paternal care, alloparents do not require suppression of infanticidal or aggression tendency, in order to care for offspring. Due to alloparental care occurring independently of reproduction, alloparental behaviors do not appear to be tightly linked to hormonal changes as is observed in mothers and fathers.

2.3 *Sensory Processing*

Olfactory, visual, auditory, and tactile stimuli all play a positive role in promoting maternal approach of pups and subsequent maternal interactions. The relative importance of each mode, however, differs depending on context, and may be subject to variation between species. Early work in rats showed that individual manipulations blocking either olfaction, vision, or tactile senses had limited effects on retrieval of pups, but combinations of these manipulations caused significant impairments, suggesting that rats can use redundant systems to achieve the appropriate response (Beach and Jaynes 1956). While anosmic rats will still exhibit maternal behavior, olfactory bulb removal in mice eliminates maternal behavior (Gandelman et al. 1971), indicative of an essential role for the sense of smell in mice. The role of olfaction is complex, however, and may be hormonally modulated. In non-pregnant females, olfactory cues from pups seem to contribute to the aversion and avoidance, but in postpartum females, the olfactory cue is clearly attractive (Levy et al. 2004). There is also a role for auditory cues in stimulating maternal behavior, with retrieval behavior in the mother triggered by the emission of ultrasonic vocalizations from pups isolated outside of a nest (Ehret 2005). Remarkably, responses in the auditory cortex seem to be altered depending on the presence of appropriate olfactory cues from pups (Cohen et al. 2011), suggesting complex interactions between different sensory modalities. Olfactory inputs are filtered through the amygdala, before being integrated in the MPOA along with sensory inputs from other modalities (Fig. 1).

The fact that fathers and alloparents can perform parental behaviors in the absence of the hormonal changes associated with pregnancy and lactation suggests that in most mammals, there is a basal amount of parental activity which can be triggered by infant stimuli alone. For example, the MPOA and BNST also show increased activity in adult virgin male California mice exposed to pups, relative to a novel object (Horrell et al. 2017). The ability of offspring stimuli to activate regions involved in parental care, independent of hormonal stimulation, may play an important role in stimulating alloparental care in juveniles and facilitating care in cooperatively breeding species.

3 Hormonal Changes Required for the Establishment of Maternal Behavior

While maternal behavior circuits can operate independently of hormonal input (Rosenblatt 1967), it is very clear that the hormonal changes associated with pregnancy are essential in leading to timely expression of maternal behavior immediately at parturition. In fact, effects of pregnancy hormones on the brain begin at the start of pregnancy, or arguably even before pregnancy begins. Cyclical changes in ovarian steroids prepare the reproductive tract for conception, and induce adaptive changes in the brain in preparation for pregnancy. Mating-induced changes in hormone secretion provide the first signal that a pregnancy may have been initiated, and these immediate changes in maternal hormones have been shown to influence brain function. The placenta also contributes to the hormonal profile of the maternal circulation during pregnancy and plays a role in maternal adaptation to pregnancy and in establishing maternal behaviors. Subsequent changes in hormone secretion at birth and suckling-induced hormonal release through the postpartum period also continue to modulate maternal behavior. Changes in hormone secretion during pregnancy, parturition, and the postpartum period are outlined below and are presented in Fig. 2, followed by a summary describing what we currently know about how these hormones act on neural circuitry to influence maternal behavior. This description of the neuroendocrine modulations of maternal behavior will include both young-directed behaviors and young-related responses that encompass a broad-range of physiological adaptations that enable a mother to be able to fully respond maternally to her offspring.

3.1 Patterns of Hormone Changes During Pregnancy and Lactation

Before the establishment of pregnancy, the mating stimulus itself induces a change in the pattern of hormone secretion that can exert long-lasting effects on maternal behavior (Fig. 2). In rodents, mating induces twice-daily surges of prolactin release, with these surges persisting into the early stages of pregnancy (Dilley and Adler 1968; Gunnet and Freeman 1983; Erskine 1995; Larsen and Grattan 2010). These prolactin surges act in the ovary to maintain the corpus luteum in rodents, thereby ensuring the continued elevation in progesterone levels, a requirement for the establishment of pregnancy (Bouilly et al. 2012; Stocco et al. 2007). In humans, the prolonged luteal phase sustains progesterone long enough to enable the pregnancy signal to come from the implanted embryo. Human chorionic gonadotropin (hCG) is secreted very early from the syncytiotrophoblast, and is present in high levels in the blood from around 4 weeks of pregnancy, before peaking at around 12 weeks. hCG is luteotrophic, sustaining progesterone secretion from the corpus luteum. Although the hormonal mechanism by which the corpus luteum is

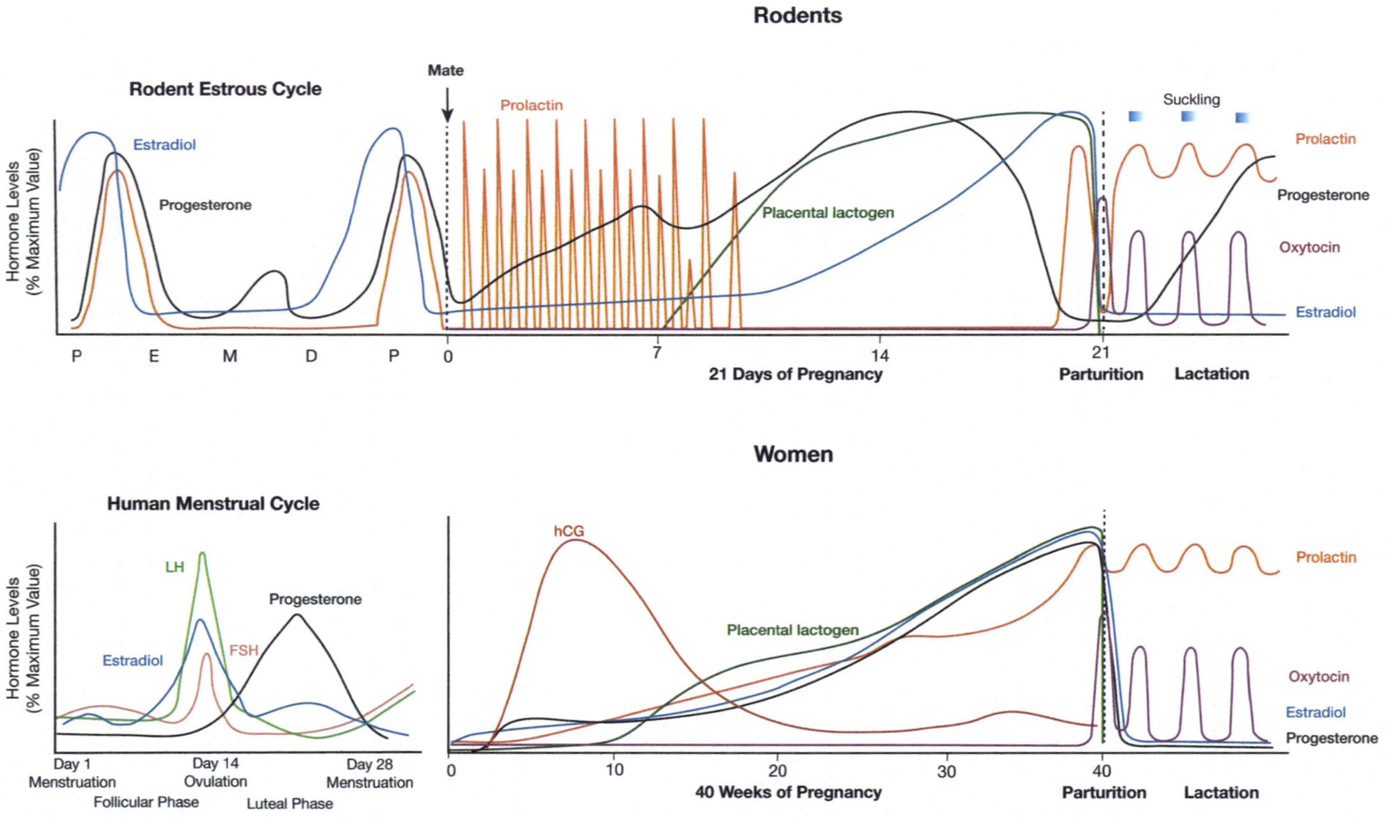

Fig. 2 Hormonal profiles of virgin, pregnant, and lactating rodents and women. In rodents, circulating levels of estradiol, progesterone, and prolactin are shown during the female estrous cycle (*P* proestrus, *E* estrus, *M* metestrus, *D* diestrus), with circulating placental lactogen and oxytocin levels also indicated during 21 days of pregnancy and lactation. In women, profiles of circulating LH, FSH, estradiol, and progesterone are shown during the menstrual cycle, with hCG, placental lactogen, prolactin, and oxytocin levels also indicated during 40 weeks of pregnancy and lactation. Note the key similarities in patterns of lactogens (prolactin and placental lactogen), estradiol, progesterone, and oxytocin secretion during pregnancy and lactation between rodents and women

maintained differs between mammalian species, mating-induced and orgasm-induced prolactin secretion has been shown to occur in other species, including rabbits and women (Pau et al. 2000; Kruger et al. 2002). Rodent studies have also demonstrated that mating induces oxytocin release in both males and females (Waldherr and Neumann 2007; Ross and Young 2009), with this being important for sexual behavior and the formation of partner preference in the socially monogamous prairie vole (Veening et al. 2015).

Throughout pregnancy there is a distinct change in the pattern of steroid hormone secretion away from the cyclical secretion patterns that characterize the estrous or menstrual cycles. These changing levels of steroids play a role in shifting the focus of reproduction from facilitating fertilization and implantation to providing in utero support for a growing fetus and preparing the mother for parturition and the subsequent demands of lactation. Estradiol-17β, the most bioactive form of estrogen, falls after ovulation and remains at low concentrations in the serum throughout the early stages of pregnancy. During the latter half of pregnancy (Day 15 in a rat) the secretion of estradiol begins to rise, peaking on or shortly before the day of parturition (Fig. 2) (Shaik 1971; Garland et al. 1987). Progesterone is secreted by the corpora lutea of the ovaries in rodents and also by the placenta in women, with levels essentially following an inverse pattern of estradiol during pregnancy. Serum concentrations of progesterone begin to rise in early pregnancy (Day 4 in a rat) and reach peak concentration on Day 14. After Day 14, progesterone levels slowly decline, before an abrupt decline late in pregnancy (Day 20) approximately 36 h before parturition (Martin et al. 1977). The requirement of progesterone for implantation and the maintenance of pregnancy have been well documented in numerous species, including rodents and primates. Ovariectomy will terminate a pregnancy at any time during gestation in rodents due to loss of progesterone production (Stocco et al. 2007). While progesterone withdrawal is required for the onset of parturition in rodents, serum levels of progesterone in women remain elevated in late pregnancy in humans. In humans, functional progesterone withdrawal before parturition has been proposed to occur through changes in progesterone receptor activity in the myometrium (Zakar and Hertelendy 2007; Vannuccini et al. 2016).

The placenta serves an important endocrine function during pregnancy and contributes to the altered hormonal profile of a pregnant female. In many mammalian species, the placenta contributes sex steroid hormones, although this is not the case in the rodent. Placental lactogen, the placental analogue of prolactin, is synthesized early in pregnancy and remains elevated until parturition. In the mouse, placental lactogen is detectable in the maternal serum from Day 6 of pregnancy, with concentrations on Day 10 of pregnancy correlating with litter size (Ogren et al. 1989). There are two major forms of placental lactogen produced in rodents, placental lactogens I and II, secreted in a sequential manner such that total placental lactogen levels are chronically elevated throughout the second half of pregnancy until parturition (Robertson and Friesen 1981; Robertson et al. 1982). Elevated placental lactogens act via prolactin receptors in the brain to inhibit maternal prolactin secretion through a negative feedback loop system (Tonkowicz and Voogt 1983; Lee and Voogt 1999), meaning that for most of the second half of pregnancy, pituitary prolactin levels are low while placental lactogens are high (Voogt et al. 1982). This situation continues until a large surge of prolactin occurs during the night preceding parturition (Grattan and Averill 1990). The fact that this surge occurs at a time when placental lactogen is still high is indicative of a change in the short-loop feedback system that normally regulates prolactin secretion, with a loss of dopamine production from the tuberoinfundibular dopamine (TIDA) neurons in the arcuate nucleus of the hypothalamus (Romano et al. 2013). This change in feedback persists throughout lactation, enabling a prolonged period of hyperprolactinemia to be sustained unopposed by negative feedback regulation. Interestingly, human pregnancy is associated with concurrent rises in both prolactin and human placental lactogen (or chorionic somatomammotropin) (Aghaeepour et al. 2018; Romero et al. 2017), suggesting that a change in negative feedback regulation of prolactin also occurs, albeit earlier, during pregnancy in women.

Key endocrine events occurring around the time of parturition include progesterone withdrawal, described above, and activation of both oxytocin and prolactin secretion. Oxytocin is released at parturition from the posterior pituitary, when it stimulates uterine contractions for parturition and has an essential role in postpartum milk ejection, as shown by the complete failure of lactation in oxytocin and oxytocin receptor knockout mice (Gross et al. 1998; Nishimori et al. 1996; Young et al. 1996; Takayanagi et al. 2005). In the postpartum period, suckling induces oxytocin release from the posterior pituitary into the periphery, which in turn is required for milk ejection from the mammary gland. The suckling stimulus simultaneously induces central release of oxytocin from the paraventricular nucleus (PVN) of the hypothalamus. During lactation, mammals are chronically hyperprolactinemic, with elevated circulating prolactin required for milk production. The relative levels of circulating hormones throughout pregnancy and during lactation in both rodents and humans are summarized in Fig. 2.

3.2 Neuroendocrine Modulation of Young-Directed Maternal Behavior

Among the earliest neuroendocrine events of pregnancy in rodents are the mating-induced prolactin surges. In addition to prolactin action on the corpus luteum, prolactin acts in the maternal brain to increase neurogenesis in the sub-ventricular zone (Shingo et al. 2003). Pharmacological suppression of the prolactin surges and the subsequent neurogenesis during early pregnancy resulted in increased postpartum anxiety and impaired maternal behavior (Larsen and Grattan 2010). A role for these surges in adaptive changes in the brain during early pregnancy may explain why these mating-induced prolactin surges persist in species that do not require the luteotrophic actions of prolactin.

The role of the sex steroids in modulating young-directed maternal behavior has been the subject of extensive research. Estradiol was first demonstrated to stimulate the onset of maternal behavior, with estradiol administration reducing the latency to demonstrate young-directed maternal behavior in both virgin and pregnancy-terminated primigravid rats (Siegel and Rosenblatt 1975a, b). These actions of estradiol are mediated through the estrogen receptor alpha (ERα), with ERα knock-out virgin female mice displaying deficits in pup retrieval behavior (Ogawa et al. 1998). Within the neural circuit regulating maternal behavior, the MPOA is a key site of estradiol action, with estradiol benzoate implants into the MPOA-inducing maternal behavior (Numan et al. 1977). Furthermore, knockdown of ERα expression specifically in the MPOA using adeno-associated viral (AAV) vector delivery of a short hairpin RNA (siRNA), suppressed postpartum retrieval and nursing behavior (Ribeiro et al. 2012). In contrast to the actions of estradiol, administration of progesterone on its own is ineffective in reducing the latency of virgin female rats to respond maternally to pups (Bridges 1984). Instead, progesterone acts to modulate the actions of other hormones on maternal behavior. When progesterone is chronically coadministered with estradiol, followed by acute progesterone removal, progesterone further increases the capacity of estradiol to reduce the latency of rats to demonstrate maternal behavior (Bridges 1984), indicating that progesterone sensitizes the female to the subsequent effects of estradiol. Because progesterone's actions on maternal behavior are indirect, the neural sites of progesterone action have been more difficult to define. Nonetheless, the MPOA also appears to be an important target for progesterone in modulating estradiol actions on maternal behavior. Sustained progesterone and estradiol administration in pregnancy-terminated rats blocks both estrogen-stimulated c-fos immunoreactivity in the MPOA and the display of maternal behavior in these animals (Sheehan and Numan 2002). Progesterone administration in virgin female rats also suppresses prolactin receptor mRNA expression in the MPOA (Bridges and Hays 2005), the significance of which is discussed below. From these studies in rodents, progesterone has emerged as a negative regulator of maternal behavior, highlighting the necessity of progesterone withdrawal prior to parturition for appropriately timed expression of maternal behavior.

Like prolactin, placental lactogens readily cross the blood–brain barrier and activate prolactin receptors throughout the brain. Placental lactogen was observed to bind to the choroid plexus and hypothalamus in the maternal brain (Pihoker et al. 1993), and is likely to be responsible for high levels of phosphorylated signal transducer and activator of transcription 5 (pSTAT5) expression (activated downstream of the prolactin receptor (Brown et al. 2010)) detected throughout the brain during pregnancy (Salais Lopez et al. 2017) compared with non-pregnant females. Evidence for a role for the lactogenic hormones in maternal behavior was first suggested in 1935, with repeated injections of prolactin-inducing maternal behavior in virgin female rats (Riddle et al. 1935). The failure of estradiol and progesterone administration alone to induce maternal behavior in hypophysectomized virgin rats (Bridges et al. 1985) indicated the requirement of a pituitary-derived factor for the expression of maternal behavior. The addition of either prolactin or an ectopic prolactin-secreting pituitary graft was sufficient to significantly reduce the latency of inexperienced virgin female rats to respond maternally, thus confirming the importance of prolactin in maternal behavior (Bridges et al. 1985). Subsequently, it was shown that central infusions of prolactin in virgin mice were able to replicate these findings at concentrations that were ineffective when administered peripherally (Bridges et al. 1990). With placental lactogen and prolactin showing a similar potency for inducing maternal behavior in virgin rats (Bridges et al. 1997), placental lactogen is likely to replace the role of prolactin in modulating prolactin-sensitive neural circuits from early pregnancy when circulating prolactin is suppressed, until maternal prolactin levels rise in late pregnancy. Generation of prolactin receptor knockout mice confirmed that actions of prolactin on pup-induced maternal care are mediated through the prolactin receptor (Lucas et al. 1998). However, due to the obligate role of prolactin and placental lactogen in sustaining ovarian progesterone production during pregnancy in rodents (Stocco et al. 2007), the role of prolactin or placental lactogen in the context of a normal pregnancy was not able to be investigated until recently. Specific removal of prolactin receptors from the MPOA of adult female mice using stereotaxic delivery of an AAV-Cre, resulted in substantial deficits in postpartum maternal behavior and the subsequent death of all pups (Brown et al. 2017). With prolactin-receptor signaling intact elsewhere in the maternal neural circuit, this data confirms a key role for the MPOA in mediating the action of lactogenic hormones on maternal behavior. The source of the lactogenic hormone that is required for induction of maternal care remains to be clarified, with both maternal prolactin and placental lactogens likely contributing. During lactation, the suckling stimulus provides a major drive for prolactin secretion and prolactin responsiveness remains significantly heightened in the MPOA and throughout the maternal neural network (Brown et al. 2011). However, the significance of this on the maintenance of maternal behavior is unclear. While an essential role in the onset of maternal behavior has been clearly demonstrated (Brown et al. 2017), it has not been shown that prolactin plays an equally important role in maintenance of young-directed maternal behavior throughout the postpartum period. The prolonged state of hyperprolactinemia is, however, likely to contribute to other young-related maternal adaptations and this will be discussed below.

Circulating oxytocin poorly penetrates the blood–brain barrier (Leng and Russell 2016) and does not appear to contribute to maternal behavior (Rosenblatt 1969; Herrenkohl and Rosenberg 1974). However, magnocellular oxytocin neurons in the supraoptic nucleus (SON) and PVN that project to the posterior pituitary to release oxytocin in the periphery also have collateral branches that project centrally to release oxytocin within the brain (Son et al. 2013), and central release of oxytocin from these collaterals may be important for maternal care. Within the PVN, parvocellular oxytocin neurons that project centrally are activated during parturition and in response to suckling (Ludwig and Stern 2015). Moreover, there is also evidence that oxytocin is secreted within the SON, PVN, and in surrounding regions from the dendrites of the oxytocin neurons (Ludwig and Stern 2015; Lambert and Kinsley 1993; Rossoni et al. 2008). Dendritically released oxytocin may diffuse over long distances to activate oxytocin receptors in regions of the brain that lack projections from oxytocin neurons. Therefore, collectively there are multiple mechanisms to enable central release of oxytocin around the time of parturition and throughout lactation. Administration of central oxytocin to estrogen-primed virgin rats rapidly induced the onset of maternal behavior (Pedersen and Prange 1979; Pedersen et al. 1982). It appears that oxytocin is primarily involved in facilitating the onset of maternal behavior at parturition, rather than being important for the maintenance of these behaviors through the postpartum period (Yoshihara et al. 2017). Lesion studies provided clarity into the time-window when oxytocin primarily acts to modulate maternal behavior, with lesions to the PVN during late pregnancy severely impacting maternal behavior, whereas lesions in the early postpartum period had no effect (Insel and Harbaugh 1989; Numan et al. 1985). During parturition there are also high levels of oxytocin binding in the VTA and MPOA (Pedersen et al. 1994), and infusion of an oxytocin antagonist into either of these sites during late pregnancy blocks subsequent maternal behavior in postpartum rats (Pedersen et al. 1994). The preoptic area is also involved in regulating the release of oxytocin, with ablation of tyrosine hydroxylase-expressing neurons in the preoptic area decreasing circulating oxytocin levels and impairing maternal behavior. Similarly, optogenetic stimulation of these neurons increased oxytocin levels (Scott et al. 2015), indicating that this key region within the maternal neural network is involved in modulating release of neurochemicals that influence maternal behavior. During the postpartum period, oxytocin has more subtle modulatory effects, with central administration of an oxytocin antagonist reducing the time dams spend licking pups and in the kyphotic nursing posture (Pedersen and Boccia 2003). Furthermore, it has been proposed that oxytocin's role in postpartum maternal behavior may be dependent on the stressed state of the animal, with oxytocin administration only inducing maternal behavior under certain conditions of stress (Pedersen et al. 1992). In environments of minimal stress, oxytocin does not appear to play an essential role in the continued display of maternal behavior through the postpartum period (Yoshihara et al. 2017).

In humans, the surge of oxytocin released during parturition and breast-feeding-induced oxytocin release is thought to be associated with central release of oxytocin, which is important for bonding and increasing maternal sensitivity (Rilling 2013). In humans, certain polymorphisms in the oxytocin receptor gene are associated with lower plasma OT and less parental touch towards infants, while other polymorphisms in the oxytocin receptor gene are related to greater sensitivity during infant interactions, and increased infant-directed vocalizations (Rodrigues et al. 2009). Although there are many positive associations with oxytocin and parental care, which are consistent with many animal findings, oxytocin is also associated with prosocial behavior in humans and other animals generally, and thus, it is difficult to tease apart prosocial effects versus specific parental care effects.

The concurrent activation of prolactin and oxytocin secretion raises some interesting questions about interactions of these two hormones in the regulation of maternal behavior. In rats, prolactin receptors are expressed on oxytocin neurons, and prolactin acutely suppresses the firing of these neurons (Townsend et al. 2005; Kokay et al. 2006; Sirzen-Zelenskaya et al. 2011). However, this inhibitory effect is lost in late pregnancy, allowing activation of these neurons in the presence of elevated prolactin (Augustine et al. 2017). As well as prolactin regulating oxytocin secretion, there is significant evidence that oxytocin regulates prolactin secretion (Kennett and McKee 2012). Predominantly, oxytocin has been shown to stimulate prolactin release (Egli et al. 2006), although it can also activate tuberoinfundibular dopamine neurons (Briffaud et al. 2015). While such an action might be expected to inhibit prolactin secretion (Grattan 2015), we have shown in mice that TIDA neurons do not release significant dopamine during lactation (Romano et al. 2013). The precise role of this interdependent, mutual regulation of prolactin and oxytocin in maternal behavior is unclear, but it is an important consideration when evaluating experiments that have manipulated one or other of these hormones.

3.2.1 Paternal and Alloparental Care

Parental males appear to undergo systematic neuroendocrine changes across their mate's pregnancy and postpartum period, in many cases mirroring the hormonal changes of the pregnant mate. For example, in humans, men and women partners show similar, correlated stage-specific hormonal changes across the females' pregnancy, including higher circulating prolactin and cortisol and lower testosterone and estradiol concentrations just before the birth of their child (Storey et al. 2000). However, most of the studies on the neuroendocrine changes in fathers are correlative in nature and preclude causal conclusions. Hence, there is a deep lack of understanding of the functional relationships between hormones and paternal behavior. A more detailed description of the hormonal changes can be found elsewhere in this book; therefore, the neuroendocrine influences on paternal care will only be briefly summarized here.

Paternal care is often characterized by reduced levels of androgens and progesterone, and higher levels of prolactin, estrogens, and the nonapeptides oxytocin and vasopressin. For instance, high levels of testosterone are often thought to be incompatible with paternal behavior, as proposed in the challenge hypothesis (Wingfield et al. 1990) although it does not appear to be as straightforward as this suggests. In Djungarian hamsters and prairie voles, castration had no detectable effect on paternal behavior (Saltzman and Ziegler 2014). However, in the California mouse, testosterone is critical for the expression of paternal care, with castration disrupting paternal care, and testosterone replacement being able to rescue paternal care (Saltzman and Ziegler 2014). While men across cultures show lower testosterone levels as fathers, compared to non-fathers, fathers have been shown to respond to frightened baby cries with an increase in testosterone, presumably for preparation for a potential aggressive encounter with a threatening stimulus (van Anders et al. 2012). Similar to androgens, a high level of progesterone is often cited as being incompatible with paternal care, with progesterone inhibiting paternal care and stimulating infanticidal tendencies in male mice (Schneider et al. 2009). However, the pattern of progesterone does not always seem to match this assumption, with progesterone higher in Djungarian hamsters but lower in California mice fathers towards the end of their mate's pregnancies (Saltzman and Ziegler 2014).

Similar to females, fathers tend to show increased estrogen levels near the end of their mate's gestation period and for some species, fathers also show declines in estradiol concentrations after birth, similar to mothers. Additionally, in almost every paternal species measured to date, circulating prolactin levels are higher in fathers relative to non-fathers, yet causal studies do not support a role for prolactin in paternal care thus far. Although injections of prolactin can induce paternal care behavior in the uniparental rat, blocking prolactin with a dopamine D_2 receptor agonist (a suppressor of prolactin) had no effect on paternal care in Djungarian hamsters or common marmoset fathers (Saltzman and Ziegler 2014).

Some of the strongest evidence for the neuroendocrine regulation of paternal behavior comes from work in prairie voles looking at vasopressin action. Paternal care tightly correlates with neuronal vasopressin-immunoreactivity and central binding patterns at the v1a receptor. Central infusions of vasopressin increase paternal care behavior while vasopressin antagonists inhibit paternal care (Wang and Insel 1996). Oxytocin also tends to be higher in fathers, compared to non-fathers, yet there is less experimental evidence to support a causal role in paternal care. In prairie voles, paternal care can be disrupted when both oxytocin and vasopressin receptors are antagonized, but paternal care is not affected if either receptor is antagonized alone (Bales and Saltzman 2016). Intranasal administration of oxytocin in human fathers has been shown to positively modulate several aspects of father–child interactions; however, causal studies beyond this manipulation are not permitted.

In males, sensory cues and experimental effects may play a stronger role in promoting paternal care than hormones. Exposure to pregnant female sensory and social cues can lead to an increase in paternal tendencies in males. In particular, access to olfactory cues from the pregnant female are important for priming paternal responsiveness in many rodents. In fact, simply housing males with soiled bedding from the pregnant female is often sufficient to stimulate paternal care in prairie voles and California mice (Brown 1993). This effect is specific to the male partner as urine from novel lactating females or virgin females does not facilitate paternal care in males separated from the litter (Jean-Baptiste et al. 2008). Female odors may also stimulate neuroendocrine changes in males that facilitate paternal care. Male gerbils and male California mice that have cohabited with their pregnant mate and pups had higher prolactin than males that did not cohabit with a female (Brown 1993). In addition, cohabitation led to lower testosterone in male gerbils during the postpartum period. These hormonal changes are suspected to influence paternal care, though fewer causal studies have been conducted in males, relative to females, to support this hypothesis.

In the absence of mating and cohabitation with a female, repeated exposure to pups in male rodents can eventually induce paternal care over the course of several days, as in females, a phenomenon known as sensitization. Injections of hormones associated with pregnancy such as estradiol, progesterone, and prolactin, can stimulate a quicker onset of paternal behaviors in animals that are not normally paternal, such as rats. However, blocking prolactin secretion and castration failed to disrupt paternal care in hamsters, making the role of hormones less clear (Wynne-Edwards and Timonin 2007). Neuroendocrine mechanisms may play a role in mediating the impact of previous parental experience on the expression of paternal care. For example, experienced male cotton top tamarins, but not inexperienced males, show an increase in estradiol during their mate's pregnancy. On the other hand, suppressing prolactin secretion in inexperienced males disrupted paternal care, while this treatment had no effect in experienced males (Wynne-Edwards and Timonin 2007).

Hormonal manipulation in alloparents has not been as successful as equivalent studies on normal maternally and paternally behaving animals. The relative ineffectiveness of hormone treatment has given rise to the notion that alloparental care is less hormonally regulated than maternal or paternal care, and instead, is primarily stimulated by offspring stimuli alone. For example, in prairie voles, both castration and testosterone treatment in juveniles reduced alloparental care when these subjects became adults (Lonstein et al. 2002). However, castration in adulthood had no effect on alloparental behavior (Lonstein and De Vries 1999). Similarly, treating juveniles with estrogen receptor agonists decreased pup-directed behaviors, yet blocking the estrogen receptor also decreased alloparental behaviors (Kenkel et al. 2017). Similar to parental care, prolactin has been thought to play a role in alloparental care. Male meerkats who remain at the nest to care or "babysit" the young while others leave to forage have higher plasma prolactin. However, these higher levels may result from spending increased time in contact with young, rather than causing babysitting behavior (Carlson et al. 2006).

Oxytocin and vasopressin, acting as neurotransmitters, may play a role in regulating alloparental care in both males and females. In adult female prairie voles, the amount of "spontaneous" or alloparental care that is exhibited towards pups is positively correlated with oxytocin receptor (OTR) expression in the NAcc (Kenkel et al. 2017). Infusions of an OTR antagonist into the NAcc blocked the expression of alloparental care, whereas viral vector infusions of the OTR gene into the NAcc, which increased OTR expression, facilitated alloparental care in adult females (Olazabal 2014). In male prairie voles, however, both OTR and vasopressin receptors needed to be blocked in order to reduce alloparental care (Bales et al. 2004). OTR density in the NAcc is also related to species differences in the alloparental care observed in juveniles. Whereas juvenile prairie voles will immediately show alloparental care, these behaviors are low or virtually nonexistent in juvenile mice and meadow voles. Accordingly, prairie voles show the highest expression of OTR in the NAcc, whereas mice and meadow voles show the lowest expression of OTR in the NAcc. In rats, only juveniles from ages 20 to 22 days will display alloparental tendencies. Around postnatal day 24–27, rats develop adult-like neophobic responses towards pups that results in rejection and avoidance of pups (Kinsley and Bridges 1988). Interestingly, OTR expression begins to decline in juvenile rats after 20 days postnatal, suggesting that juvenile rats are only alloparental when OTR expression is highest. Together, these data suggest that the OTR density may predispose animals to differing propensities to display alloparental care. This propensity can be altered by early life manipulations of the oxytocin system. Injecting neonates (postnatal day 1) with oxytocin increased alloparental care behaviors in adult female prairie voles, whereas injecting oxytocin antagonists decreased alloparental care and increased attacks toward pups in males (Kenkel et al. 2017). Further exploration into the organizational effects of oxytocin and AVP on later adult alloparental behavior would be informative.

3.3 Neuroendocrine Modulation of Young-Related Behaviors and Physiological Adaptations Supporting Pregnancy and Lactation

While hormones directly influence young-directed maternal behavior (nesting, retrieval, grooming, nursing), they also have indirect actions by promoting other adaptive changes in the mother. These have been collectively described by Bridges (2015) as "young-related behaviors," as distinct from young-directed behaviors described above. These young-related behaviors have typically included increased maternal aggression towards intruders, suppression of the stress response, and increased food intake. We propose that this definition be broadened to include young-related adaptations that incorporate the multiple physiological adaptations that occur in a mother to facilitate and support the behavioral changes. Such changes include, but are not limited to, changes in the cardiovascular system, respiratory system, skeletal muscles, sleep patterns, glucose homeostasis, and gut physiology (Fig. 3). For the purposes of this chapter, we will focus in detail on a few of these young-related behaviors and adaptations in which significant progress has been made. Those topics are suppression of fertility, suppression of the stress responses, and metabolic adaptations during pregnancy. Similar to the young-directed behaviors described above, neuroendocrine changes during pregnancy and the postpartum period play a key role in modulating these adaptations.

3.3.1 Maternal Aggression

During the postpartum period, females show increased levels of aggression towards intruders as a way of protecting their offspring. Research into the neurocircuitry governing maternal aggression has highlighted two regions that play a significant role in modulating postpartum aggressive behavior; the PVN and VMN. Interestingly, these two hypothalamic regions appear to have opposing roles in modulating aggression levels, with the PVN inhibiting and the VMN promoting aggression. Integration of outputs of these two regions is likely to occur in the midbrain, where both the PVN and VMN are known to project. A role for the VMN in promoting maternal aggression was revealed when electrolytic lesions of the mediobasal hypothalamus or more specific knife cuts along the border of the VMN, attenuated maternal aggression (Hansen 1989). Due to higher levels of aggression in males compared to females in many species, work into the neurobiology of aggression has largely focused on males. These studies have identified the ventrolateral VMN

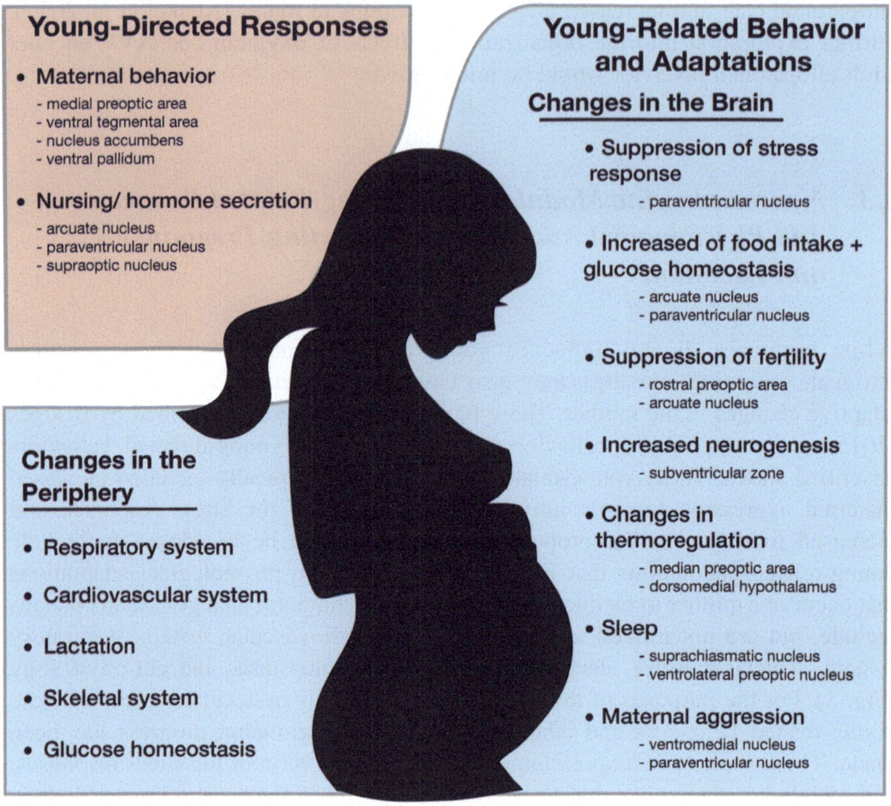

Fig. 3 Summary of adaptive changes in the maternal body and key brain regions known to be involved in facilitating these changes. Young-directed responses (shown in orange) include maternal behavior and changes in hormone secretion. Young-related behaviors and adaptations (shown in blue) include key changes in both the periphery and the brain which enable the mother to display appropriate maternal care

(VMNvl) in particular, as being important in controlling aggressive behavior, with neural stimulation activating aggression and neural inhibition inhibiting aggression in males (Lin et al. 2011; Lee et al. 2014; Falkner et al. 2016). Recent work has investigated whether similar circuitry may also be involved in female aggression. There are some clear sex differences, with ablation of progesterone receptor-expressing cells in the VMNvl reducing aggressive behavior in males, but failing to reduce postpartum aggression in females (Yang et al. 2013). In males, there is a high degree of overlap in progesterone receptor and estrogen receptor expression in the VMNvl. However, in females, estrogen receptor-expressing neurons, rather than progesterone receptor-expressing neurons, are the key player in mediating postpartum aggression, with a significant increase in expression of c-fos observed in the VMNvl of lactating females following aggressive behavior (Hashikawa et al. 2017).

These estrogen receptor-expressing cells in the VMNvl show increased in vivo activity during aggression, and chemogenetic inhibition of these cells reduces both the duration and frequency of attacks on intruders by a postpartum dam (Hashikawa et al. 2017). Although these studies show that estrogen receptor-expressing cells in the VMN are important for maternal aggression, a role for the female sex steroids themselves in modulating aggression has yet to be confirmed. Furthermore, it is not currently known how this circuit is modulated during pregnancy and the postpartum period to lead to the onset of aggression specifically in the early postpartum period.

In addition to the VMN, the PVN also plays an important role in modulating maternal aggression. The impact of PVN lesions on this behavior was dependent on the extent and localization of damage. For example, electrolytic lesions, causing widespread damage to the PVN, resulted in reduced aggression towards intruders (Consiglio and Lucion 1996), whereas restricting damage to the parvocellular region of PVN caused an increase in aggressive behavior (Giovenardi et al. 1998). Activation of parvocellular neurons in the PVN, therefore, is important for reducing the display of aggressive behavior. Within this region, oxytocin has emerged as an important neuroendocrine modulator, with administration of an oxytocin antisense increasing aggression (Giovenardi et al. 1998). Oxytocin may be acting locally, with local oxytocin release stimulated by the addition of an intruder into the home cage of a lactating female rat (Bosch et al. 2004), or acting on other components of the neural circuitry regulating aggression, such as the amygdala. Blockade of oxytocin actions by bilateral infusion of an oxytocin antagonist into the central amygdala increased maternal aggression in postpartum rats (Lubin et al. 2003). In addition to oxytocin, there is evidence suggesting vasopressin is also involved in modulating levels of maternal aggression, although these effects have been different between species. In mice, vasopressin 1b receptor, but not 1a, knockout animals displayed reduced levels of maternal aggression (Wersinger et al. 2007a, b). Whereas in rats, central administration of an arginine vasopressin receptor 1a antagonist led to increased maternal aggression (Nephew and Bridges 2008). Interestingly, work in rats has also shown that the impact of vasopressin on maternal aggression is dependent on the anxiety-state of the female (Bosch 2013; Bosch and Neumann 2010), indicating that the importance of vasopressin as a neuromodulator of aggression is influenced by other factors. A role for other hormones in modulating maternal aggression throughout the neural circuitry regulating aggression has yet to be demonstrated.

3.3.2 Metabolism

In addition to appropriate young-directed maternal behavior following parturition, the mother must also supply energy to the pups for growth and development by way of milk. Indeed, supply of energy begins during pregnancy with the maternal body providing all the energy requirements for the growing fetus via the placenta. In addition, fat deposition is increased in the mother during pregnancy as an energy reserve for late pregnancy and lactation, when energy demands can dramatically exceed that which the mother can consume. For mammals, the same hormonal events that lead to the development of maternal behavior are also proposed to drive adaptations in the regulation of energy homeostasis to allow the mother to adequately supply the offspring with energy, both of which are required for the long-term success of reproductive efforts.

Food consumption usually increases during both pregnancy and lactation; however, the mechanisms driving this increase are likely different given the physiological differences between pregnancy and lactation. It should be noted, however, that not all mammals increase food intake during pregnancy; for example, some hamster species maintain their pre-pregnancy food intake levels while utilizing their own pre-pregnancy fat stores to supply the energy required during gestation (Quek and Trayhurn 1990; Wade et al. 1986). Nevertheless, in general, pregnancy represents a state of positive energy balance, with increasing maternal body weight gain accompanying the increasing mass of the growing conceptus. In addition to the energy demands of the growing fetus, other maternal adaptations also require energy, such as increased deposition of fat mass and the development of the mammary glands. Work examining both the satiety responses and the activation of intracellular signaling pathways has shown that during pregnancy the brain becomes insensitive to many factors that normally suppress food intake such as leptin, alpha-MSH, CCK, and insulin (Ladyman and Grattan 2004, 2005, 2016, 2017; Ladyman et al. 2009, 2010, 2011, 2012, 2016). Pregnant rats are also less responsive to the satiety actions of central oxytocin administration (Douglas et al. 2007). These data have led to the hypothesis that insensitivity to satiety factors during pregnancy may facilitate the extra drive to increase food intake beyond the immediate demands of the fetus, establishing the positive energy balance (Ladyman et al. 2010). The mechanism underlying this insensitivity is likely to involve hormonal changes in the brain, with chronic prolactin receptor activation, presumably due to increased placental lactogen in pregnancy, a likely contributor of central leptin insensitivity. The hormonal profile of pseudopregnant rats mimics early pregnancy, but these rats do not become leptin insensitive unless the pseudopregnancy is extended by chronic central infusion of prolactin to mimic high placental lactogen levels (Augustine and Grattan 2008). Similarly, chronic central infusion of prolactin in virgin female rats, albeit for a longer duration than that required in pseudopregnant rats, also leads to a state of insensitivity to the satiety actions of leptin (Naef and Woodside 2007). How prolactin action can generate this insensitivity to leptin during pregnancy remains unknown, but given the lack of overlap in the distribution of leptin-sensitive neurons and prolactin-sensitive neurons (Ladyman et al. 2010; Nagaishi et al. 2014) it is unlikely that there is a direct action of prolactin on leptin-responsive neurons.

One of the key metabolic adaptations of pregnancy is the change in the regulation of glucose that allows a constant flow of glucose for the developing fetus despite the intermittent intake of food by the mother. Glucose is transferred to the fetus from the maternal circulation in a passive manner, relying heavily on the gradient of glucose concentration between the fetus and the mother. As pregnancy advances, in order to maintain a higher level of glucose on the maternal side, insulin-sensitive peripheral tissues become relatively insulin resistant. During this time, in response to a glucose load, such as a meal, increasing levels of insulin are required to counterbalance this insulin resistance and protect the fetus from excessive glucose. To cope with these demands, during pregnancy there is an increase in the size and number of beta-cells and increased insulin secretion in response to glucose (Sorenson and Brelje 1997; Baeyens et al. 2016). These adaptations are vital in maintaining a healthy pregnancy, and incomplete or incorrect adaptations to this system can lead to the development of gestational diabetes, and numerous negative consequences for both mother and offspring both in the short and long term. Placental lactogens play a key role in the expansion of the beta-cell and the changes in glucose homeostasis during pregnancy (Sorenson and Brelje 1997). While the increase in insulin is necessary for peripheral effects during pregnancy, increased central actions of insulin such as suppression of food intake (Strubbe and Mein 1977; McGowan et al. 1992; Obici et al. 2002a) or suppression of hepatic glucose production (Obici et al. 2002b; Inoue 2016) are less desirable during pregnancy. Recent data indicates that central sensitivity to insulin, at least in the hypothalamus, is attenuated during pregnancy (Ladyman and Grattan 2017), allowing for increased peripheral actions without an increase in central actions.

Lactation is highly metabolically demanding and there are a range of strategies that mammals can utilize to provide for the increased energy demands of lactation. For example, mice and rats greatly increase food intake during lactation, in addition to utilizing the fat mass accumulated during pregnancy to supply energy, especially if food is scarce or restricted. For rodents, the pups grow rapidly during the first few weeks until they are weaned. Lactating rats and mice can consume up to 300% of the food intake of virgin control rodents and yet the dams can still lose weight, suggesting that they are in a state of negative energy balance despite high food intake levels (Woodside 2007). The metabolic demand of milk production likely accounts for the large drive to eat. When lactating dams can't deliver milk but continue receiving the suckling stimulus, food intake still increases across lactation, although to a lesser level than if they were producing milk, indicating that hormonal and/or physical (suckling) cues drive food intake over and above the metabolic demand of milk production (Woodside et al. 2000). Furthermore, extremely low leptin levels during lactation may also contribute to greatly increased food intake at this time in rats and mice. Another strategy is to rely on increased fat stores accumulated during pregnancy as the primary source of energy for lactation. Large animals that require high-fat, low sugar milk for offspring are more likely to rely predominately on this strategy than smaller animals (Oftedal 2000). For example, the Weddell seal accumulates a large amount of body fat during pregnancy and primarily uses this resource during lactation to provide the required energy for milk production. Although more recent studies have suggested that lactating Weddell seals do participate in some food foraging, this accumulated fat mass is the primary source of energy and the mother experiences huge weight loss during lactation. Regardless of the strategy employed, the mother must successfully navigate the trade-off between lactation and foraging. It is perhaps not surprising that reproductively experienced rodents become more efficient at foraging for food (Kinsley and Lambert 2008), with exposure to pups contributing to this change (Lambert et al. 2005).

The hypothalamus, particularly the arcuate nucleus, plays a key role in the regulation of energy balance. Within the arcuate nucleus two of the most widely studied neuronal populations involved in appetite regulation are the anorectic proopiomelanocortin (POMC) neurons and the orexigenic neuropeptide Y (NPY)/agouti-related peptide (AgRP) neurons. During lactation, as demonstrated in rodent models, key peptides involved in driving an increase in food intake such as NPY and AgRP are elevated (Smith 1993; Chen et al. 1999), while the anorectic precursor peptide of alpha-MSH, POMC, is reduced (Smith 1993). Furthermore, there is a suckling-induced increase in expression of NPY mRNA in the dorsal medial hypothalamus (DMH) during lactation that is likely to be driven by suckling-induced secretion of prolactin (Li et al. 1998). The DMH only appears to express NPY in the metabolically demanding state of lactation and some incidences of obesity in rodents (Bi et al. 2012). However, the role of this increased NPY in the DMH in driving increased food intake during lactation is yet to be confirmed with empirical evidence. In the arcuate nucleus, prolactin does not appear to be directly responsible for the increased NPY mRNA during lactation as these neurons are not responsive to prolactin as demonstrated by a lack of prolactin receptor expression in both rats (Li et al. 1998) and mice (Ladyman and Grattan 2017).

3.3.3 Suppression of Fertility

In many species, including woman, maternal fertility is suppressed during the lactational period. This important adaptation enables the focusing of maternal resources towards supporting newly born offspring, rather than supporting a further pregnancy (Valeggia and Ellison 2009). The suckling stimulus is a key factor in suppressing pulsatile luteinizing hormone (LH) secretion from the pituitary gland during lactation, which in turn fails to stimulate the ovarian production of sex steroids and ovulation (Fox and Smith 1984; Tsukamura and Maeda 2001). In women, the duration of lactational amenorrhea correlates with the strength of the suckling stimulus (McNeilly 2001). Loss of LH pulsatile secretion indicates that release of gonadotropin-releasing hormone secretion (GnRH), the key driver of LH secretion, from the hypothalamus is suppressed. Studies from mice have revealed that during lactation there is suppression in the rostral hypothalamic production and release of kisspeptin, the potent stimulator of GnRH neuronal activity (Liu et al. 2014). The mechanism underlying the suppression of kisspeptin synthesis during lactation is yet to be determined, and may include either a direct neural component of suckling or endocrine or metabolic changes that are induced by suckling or milk production. Suckling-induced elevation in circulating prolactin is a potential candidate, with hyperprolactinemia being a well-recognized cause of infertility, and the length of lactational amenorrhea in woman positively correlating with high prolactin levels (Campino et al. 1999). There may be species differences in the mechanism underlying lactational infertility with prolactin appearing to not play a significant role in number of species including pigs, cows, and some monkeys (McNeilly 2001).

3.3.4 Stress Responses

Pregnancy and lactation are both periods of "stress hypo-responsiveness." This state is brought about by adaptive changes to hypothalamic-pituitary-adrenal (HPA) axis activity (da Costa et al. 1996; Douglas et al. 2003; Johnstone et al. 2000; Lightman and Young 1989; Neumann et al. 1998; Shanks et al. 1999) which ultimately serves to protect both the mother and her offspring from the potentially negative consequences associated with stress (Altemus et al. 1995; Brunton and Russell 2008). Failure of this adaptation to be induced or maintained is associated with loss of pregnancy, developmental changes to the systems regulating stress, anxiety and metabolism in the offspring, and development of postpartum mood disorders in the mother (Brunton et al. 2014; Hillerer et al. 2011; Meaney et al. 2007; Parker and Douglas 2010; Sandman et al. 2006). In addition to the HPA axis, other maternal stress-responsive systems are also hypo-responsive. For example, the sympathetic-medullary response to stress is attenuated, with restraint stress-induced adrenaline and noradrenaline secretion significantly reduced in lactating rats (Higuchi et al. 1989). Stress-induced oxytocin and prolactin secretion are also significantly reduced (Lightman and Young 1989; Altemus et al. 1995; Higuchi et al. 1989; Neumann et al. 1995, 2000a). Collectively, reduced activation of these pathways is considered to promote successful lactation through energy conservation and maintenance of stores of the prolactin and oxytocin peptides (Higuchi et al. 1989).

The PVN serves as the main central regulator of the stress response, receiving and integrating inputs from a variety of stress-responsive regions in the brainstem, hypothalamus, and limbic system (Ulrich-Lai and Herman 2009). Activation of the corticotropin-releasing hormone (CRH) and arginine vasopressin (AVP) neurons in the PVN stimulates the secretion of CRH at the median eminence. CRH acts on anterior pituitary corticotrophs to stimulate the synthesis and secretion of adrenocorticotropic hormone (ACTH), an action potentiated by AVP (Gillies et al. 1982). In turn, ACTH stimulates the synthesis and secretion of glucocorticoids from the adrenal cortex. Glucocorticoids act at the level of the anterior pituitary, hypothalamus, and extra-hypothalamic regions such as the hippocampus (Knigge and Hays 1963) and medial prefrontal cortex (Diorio et al. 1993) to limit further activation of the HPA axis. During late pregnancy and lactation, there is an attenuation of the extent to which physical and psychological stressors are able to induce an increase in CRH mRNA expression in the rat PVN (Lightman and Young 1989; Shanks et al. 1999; da Costa et al. 2001), with reduced stress-induced c-fos, indicating reduced neuronal activation in a lactating rat compared to a virgin control (da Costa et al. 1996; Shanks et al. 1999; Woodside and Amir 1997). Consistent with these hypothalamic changes, stress-induced secretion of ACTH and CORT is also attenuated during late pregnancy and lactation (Johnstone et al. 2000; Neumann et al. 1998; Shanks et al. 1999; da Costa et al. 2001; Brunton et al. 2005). During pregnancy, the reduction in stress-induced ACTH has been attributed to reduced hypothalamic AVP secretion (Ma et al. 2005) while during lactation increased AVP drives the secretion of ACTH and CORT (Toufexis et al. 1999).

The diurnal rhythms of ACTH and CORT secretion are flattened during pregnancy and lactation, while basal levels of both hormones are elevated during lactation (Atkinson and Waddell 1995). The suckling stimulus plays a key role in the elevated basal activity of the HPA axis, with ACTH and CORT secretion restored to virgin levels within 24 h following removal of pups (Walker et al. 1992). Additionally, stress-induced increases in c-fos are restored 48 h following the removal of the pups as is the CRH mRNA response to stress (Lightman and Young 1989; Woodside and Amir 1997), demonstrating the importance of the suckling stimulus for the regulation of HPA axis activity under both basal and stress conditions.

Considering that the activation of the HPA axis is regulated by negative feedback, it has been hypothesized that increased basal circulating CORT during pregnancy and lactation may mediate the reduced activation of the HPA axis (Lightman et al. 2001). In support of this, levels of 11-β-hydroxysteroid dehydrogenase-1, an enzyme which reactivates metabolized CORT, are increased in the PVN and anterior pituitary gland during late pregnancy, and in lactation, removal of CORT negative feedback by adrenalectomy increases expression of CRH mRNA in the PVN to virgin levels (Johnstone et al. 2000; Lightman and Young 1989). However, adrenalectomy does not reverse the attenuated stress-induced ACTH secretion in lactation, suggesting that enhanced glucocorticoid feedback alone cannot explain the stress hypo-responsiveness (Lightman et al. 2001; Walker et al. 1995). Rather, reduced forward drive of HPA axis activation (resulting from decreased CRH and/or AVP neuronal activation) seems to underlie the attenuated responsiveness of the axis during pregnancy and lactation (Johnstone et al. 2000; Shanks et al. 1999; da Costa et al. 2001).

The mechanism underlying reduced HPA axis activation may involve the altered hormonal milieu of pregnancy and lactation. During pregnancy, the progesterone metabolite allopregnanolone plays an important role in regulating the response to immune stress. In parallel with changes in progesterone in blood, allopregnanolone levels in the brain are elevated during pregnancy (Concas et al. 1998), and inhibition of allopregnanolone synthesis significantly increases HPA axis activation in response to an immune challenge (Brunton et al. 2014). Chronically elevated prolactin receptor-mediated signaling during pregnancy and lactation may also play a role in the regulation of the maternal HPA axis. In virgin rats, chronic central prolactin infusion significantly reduces stress-induced activation of the HPA axis, while the converse is observed when central prolactin receptors are downregulated (Donner et al. 2007; Torner et al. 2001, 2002). During lactation, blocking central prolactin receptors increases stress-induced ACTH secretion (Torner et al. 2002). A reduction in stress-induced c-fos mRNA in the PVN following central prolactin treatment suggests prolactin may act by modulating neuronal activation (Donner et al. 2007). As the CRH neurons do not express prolactin receptors (Gustafson et al. 2017), the mechanism by which prolactin modulates these neurons is currently unknown. Oxytocin has also been implicated in reducing stress responses. Central infusion of oxytocin attenuated noise and restraint-stress-induced ACTH and CORT release in virgin female rats and induction of both CRH and c-fos mRNAs in the

PVN in response to restraint stress is abolished following central oxytocin treatment (Windle et al. 1997, 2004). Conversely, central administration of an oxytocin antagonist significantly increased both basal and stress-induced (elevated plus maze and forced-swimming) ACTH and CORT secretion in both male and virgin female rats (Neumann et al. 2000a, b). Interestingly, this suppressive effect of oxytocin upon the HPA axis is dependent on the reproductive state, thus inhibiting oxytocin in either pregnant or lactating rats is not sufficient to alleviate the hypo-responsive nature of the HPA axis (Neumann et al. 2000a).

3.3.5 Physiological Adaptations in Males

Although males do not undergo pregnancy or lactation, some fathers appear to undergo similar metabolic changes in preparation for energy-demanding paternal care. Male cotton top tamarins and California mice, for example, show significant weight gain during the end of their mates' pregnancy, as well as significant weight loss during the offspring-rearing period (Saltzman and Ziegler 2014). This effect may be mediated by prolactin as a dopamine D_2 receptor agonist treatment (to suppress prolactin) caused increased weight loss in males. There has also been some investigation as to whether the stress response is altered in fathers as it is in mothers. Behavioral evidence supports the notion that California mice are less neophobic during paternal care, yet changes in plasma corticosterone or c-fos expression in the PVN, central amygdala, or BNST do not support changes in the CRH system (Saltzman and Ziegler 2014). In contrast, prairie vole fathers actually show increased anxiety- and depressive-like behaviors during paternal care (Lieberwirth et al. 2013). Therefore, while the modulation of the stress response may indeed be different in males, it is not clear whether it is to the same extent as it is in females.

4 Translation to Human Parenting

The hormonal changes that drive behavioral changes and maternal adaptations also occur in humans. Where examined, many of the neural regions that have been implicated in rodent parental care have been found to be active during human parental care. Furthermore, there is a great deal of overlap in the neural regions which respond to infant cues in non-pregnant women, mothers, and fathers. Collectively, these studies point to what is likely a highly conserved parental brain network and the study of maternal care has proven to be one of the most translatable fields of neurobiological research. Importantly for humans, many higher-order cortical regions, especially those involved in decision making, cognition, and attention, are highly recruited during parental care. The neuroendocrine state of the mother can influence her attention and response to infant cues, such as cries or facial expressions, while simultaneously, infant cues can also elicit certain neuroendocrine states within the mother, which may later affect her behavior. Attraction to infant cues is important as increased exposure to infant stimuli and attention increases mother–infant bonding. The amount of attention given to cues is important as well; too little attention results in low sensitivity or neglect, whereas too much attention can induce anxiety and pathological mothering behaviors (Lonstein et al. 2015; Pawluski et al. 2017). These regions must also be highly regulated to produce optimal level of parental care that is healthy for both mother and offspring. We will briefly review the role of sensory stimuli in eliciting parental care in humans, then discuss how the neuroendocrine findings in both animals and humans contribute to our understanding of how atypical maternal behavior arises during postpartum depression.

4.1 Sensory Input

A mother is honed into her infant's sensory cues immediately following birth. Mothers can discriminate between her own and other infant's odor within 1 h after giving birth, even after a caesarean section delivery (Lonstein et al. 2015). Mothers with increased CORT postpartum have better recognition of their infant's smell, and are more attracted to their newborn's odor (Lonstein et al. 2015). While the neurobiology of interacting with infant odor is limited, one study found that the PFC was only activated in mothers, and not non-mothers, when exposed to infant odor in an fMRI (Nishitani et al. 2014). Mothers can also recognize the tactile characteristics of their newborn baby's skin within 1 h after giving birth (Lonstein et al. 2015). The neurobiology of touch is difficult to assess as procedures such as fMRI preclude real-time interactions with babies while in the scanner. However, circulating oxytocin concentrations collected in mothers after interacting with their baby positively correlate to ventral striatum brain responses when viewing images of their infant inside the scanner, suggesting that oxytocin-mediated reward circuitry, which has been implicated in rodent maternal care, may also apply to humans (Rilling 2013). Increased touching- and nursing-induced oxytocin can promote increased sensitivity to infant cues and mother–infant bonding (Lonstein et al. 2015).

Humans are extremely reliant on visual cues for appropriate parental behavior. There are different neural activation patterns in response to looking at one's own versus another baby, especially in the frontal and thalamocortical circuits (Swain et al. 2014a). Mothers show increased activity in the midbrain, periaqueductal grey and striatum, OFC, medial prefrontal cortex, ACC, anterior insula, and precuneus when viewing pictures or videos of their own baby, compared to other babies. In addition, happy faces from their own baby reduced amygdala activity in the mother more than non-related happy faces (Barrett et al. 2012), indicating that mothers are particularly sensitive to the mood and state of their infant. The selectivity of the response to own versus other babies is present in some biparental animals, such as mandarin voles, but other rodents, such as mice, show nonselective responses.

For humans, crying is one of the most effective cues to elicit parental care. Baby cries activate similar neural regions in non-pregnant women, mothers, and fathers.

Using fMRI, Lorberbaum et al. (2002) have identified a specific thalamocingulate circuit, which is unique in its response to baby cries. Swain (2011) extended these findings to show another hypothalamic-midbrain-limbic-paralimbic-cortical brain circuit which is selectively activated during times of baby crying. Cries also activate the cingulate and thalamus, MPOA, and BNST (Swain et al. 2014b). Most often, brain imaging and hormonal studies in humans are performed in response to standard prerecorded infant cries and are not from one owns infant. However, Swain and colleagues have found that there are different responses to one's own versus other's cries. First time mothers listening to their own infant cries elicited stronger fMRI activity in the OFC, midbrain, striatum, cingulate, amygdala, and insula then when listening to non-related infants (Swain 2011). Furthermore, mothers have been reported to be able to discriminate between different types of crying (e.g., hunger versus pain), which is consistent with studies showing different patterns of brain activity while listening to different types of cries (Swain et al. 2014b). Fathers are also able to discriminate between different types of baby cries and show differential neuroendocrine responses, accordingly. Whereas cries requiring sympathetic responses result in higher prolactin, cries signaling threat, or the need for aggression actions (i.e., protection of infant) result in higher testosterone (van Anders et al. 2012; Delahunty et al. 2007).

4.2 *Neuroendocrinology of Maladaptive Maternal Care*

4.2.1 Hormonal Correlates of Postpartum Depression and Anxiety

Aside from the unique timing, postpartum depression (PPD) shares many of the same behavioral and psychological characteristics as major depressive disorder. These include sad mood, anhedonia (decreased capacity to experience pleasure), restlessness and/or agitation, difficulty with concentration, and alterations in appetite, sleep, and/or motor activity (Pawluski et al. 2017; Gonzalez-Mariscal and Melo 2013), all of which have consequences for the growing infant. Mothers with PPD are less contingent, less responsive, and are less sensitive to baby cues (Gonzalez-Mariscal and Melo 2013). They are also less accurate at interpreting the infant's emotions and show less rhythmicity and synchrony during parent–infant interactions (Gonzalez-Mariscal and Melo 2013).

The changing hormonal profile of pregnancy suggests that ovarian hormones may play a role in PPD. The hormonal profile during human pregnancy is relatively similar to that observed in rodents (Fig. 2), and these hormonal changes are likely to be driving appropriate changes in mood and behavior, as seen in animal models. However, potential imbalances in these endocrine adaptations could lead to instances of maladaptive maternal mood and behavior. Studies in primates suggest that high levels of estradiol during pregnancy are important for stimulating maternal motivation and responsiveness to infant stimuli, and specifically, the ratio of estrogen:progesterone at the end of pregnancy plays a critical role in the onset of maternal care (Saltzman and Maestripieri 2011). In human mothers, a low estrogen:progesterone ratio correlates with positive mood, increased attachment, and increased sensitivity towards their infants, while high estrogen:progesterone ratios are associated with depressive and anxiety symptoms (Bridges 2015; Brummelte and Galea 2016). In addition, increased androgen and estrogen exposure during the last 4 weeks of pregnancy is related to depressed mood, which is consistent with more mothers reporting increased postpartum "blues" after giving birth to boys, relative to girls (Brummelte and Galea 2016). Since levels of both estrogen and progesterone immediately drop after expulsion of the placenta, a "estradiol withdrawal" hypothesis has been proposed, which has received support for contributing to PPD in animal work. In humans, fluctuating estrogen levels are associated with depressed mood and increased serotonin transporter in the neocortex of women. This likely results in decreased central serotonin levels, which for some time has been implicated in depression (Brummelte and Galea 2016). Furthermore, DNA methylation patterns have been found on genes involved in estrogen signaling, supporting a role for altered estrogenic activity in women with PPD (Brummelte and Galea 2016).

Increased glucocorticoids are strongly associated with increased stress; however, they are also responsible for increasing energy to meet new metabolic demands, and the relationship between cortisol (CORT), the primary glucocorticoid in humans, and maternal behavior varies throughout the postpartum period. For instance, high salivary CORT during early postpartum is associated with high positivity and high levels of responsiveness to infants, and greater sympathy toward infant cries (Levy 2016). There is evidence for a role of CORT in the enhancement of attention, which may influence responsiveness to infant stimuli (Levy 2016) and hence be beneficial immediately postpartum. In contrast, high CORT concentrations at 6 months postpartum is inversely related to maternal sensitivity and maternal behavior (Swain 2011), and is associated with decreased responses to cries (Swain et al. 2014b; Swain and Ho 2017). It has been documented that exposure to repeated or chronic stress during pregnancy or the postpartum period can lead to impairments in maternal care, a finding confirmed by numerous animal studies (Rutherford et al. 2011), and may explain why high CORT at 6 months is associated with poor mothering. Long-term overactive HPA responses and stress have been implicated in depression and, thus, are a prime candidate to study for PPD. PPD is associated with heightened HPA activity and increased CORT, which has been linked with delayed fetal brain development (Field et al. 2006). The transition from pregnancy to motherhood is inherently stressful, yet adaptive mechanisms are in place to attenuate the stress response during this time, as described earlier in this chapter. HPA feedback is suppressed in non-depressed mothers and studies in rats show that there is an increase in the amount of free CORT during the postpartum period (Brummelte and Galea 2016). Understanding how HPA activity is disrupted in PPD mothers will be beneficial, as currently most antidepressant medications used to treat PPD and anxiety are aimed at regulating HPA activity.

Other neurotransmitters involved in regulating the stress response are also altered during the postpartum period such as serotonin, dopamine, norepinephrine, vasopressin, oxytocin, and CRH (Rutherford et al. 2011), all of which may play a role in the adaptive stress response during motherhood. Animal studies have shown that high oxytocin has anxiolytic effects during stressful situations. In women, oxytocin released during breastfeeding can inhibit ACTH secretion, which may lead to decreased CORT levels in human mothers in response to stressful stimuli, such as baby cries (Swain et al. 2011). Women with low oxytocin during the last trimester are more at risk for PPD than women with higher oxytocin (Brummelte and Galea 2016). Oxytocin plays a significant role in shifting the perception of pups from being aversive to rewarding, which may explain why lower oxytocin mothers have more trouble bonding with their infants. Consistent with this hypothesis, PPD mothers treated with oxytocin show an increase in the perception of the bond with their infant, but not in their mood (Mah et al. 2013). However, recent studies have shown that synthetic oxytocin, which is used to induce labor, is positively correlated with higher depressive, anxious, and somatization symptoms 2 months postpartum. Furthermore, synthetic oxytocin is associated with increased risk of postpartum depressive and anxiety disorders in mothers both with and without previous histories of pregnancy depressive or anxiety disorders (Kroll-Desrosiers et al. 2017). Oxytocin has a complex relationship with maternal care and the dysregulation of oxytocin in PPD mothers is likely related to the dysfunction of the HPA response in mothers. Like oxytocin, prolactin is also thought to be a protective factor during pregnancy and lactation against stress-induced biological modifications such as gastric ulcers or hypothermia (Drago et al. 1989). Interestingly, in rodents, there appears to be a greater reduction in stress and anxiety in multiparous rats, relative to virgin rats, suggesting there are permanent effects on the stress system following motherhood (Lambert 2012).

While concerns about the baby's wellbeing, vulnerability, and safety peak after birth and are beneficial to the health and care of the infant, too much anxiety can lead to maladaptive behaviors and negative consequences for the baby. Postpartum anxiety does not have a unique diagnostic criteria, but shares common characteristics of generalized anxiety disorder such as excessive concern or worry that cannot be controlled, and with obsessive-compulsive disorder, such as intrusive thoughts, impulses, or behavior (Pawluski et al. 2017). Highly anxious mothers show less positive, less warm, and less relationship-building behaviors (Gonzalez-Mariscal and Melo 2013). In addition, women with anxiety disorders are more intrusive and show more fearful responses to the baby and less support for baby exploration (Gonzalez-Mariscal and Melo 2013). Anxiety is often present simultaneously with PPD, and as such, they share many similar symptoms and consequences for infant care and development. Postpartum anxiety and depression are both associated with decreased breastfeeding, reduced emotional sensitivity to baby cues, decreased bonding, and negative infant temperament, which has been linked with atypical neurodevelopment, and emotional and behavior problems later in the child's life (Pawluski et al. 2017).

4.2.2 Neural Correlates of Postpartum Depression and Anxiety

It is important to note that although PPD shares many of the characteristics of major depressive disorder, PPD women present unique neuroendocrine signatures which coincide with the birth of their child (Brummelte and Galea 2016; Pawluski et al. 2017). One of the difficulties in studying PPD is that although it shares many behavioral symptoms with major depressive disorder, the neurobiological signatures of PPD are distinctly different, but overlap with many structures that have been identified for maternal care behavior (Pawluski et al. 2017). Furthermore, there is a significant lack of studies in women with depression or anxiety before birth, which is the largest predictor of PPD (Pawluski et al. 2017). Animal research has begun to study PPD in rodent models, which often involves using paradigms such as applying repeated stressors, administering high levels of exogenous glucocorticoids, separating mothers from her pups, or administrating and quickly withdrawing ovarian hormones, in order to mimic PPD-like behaviors (Pawluski et al. 2017). Results from studies using these methods indicate that rodents subjected to these treatments show modified synaptic plasticity in areas such as the PFC, NAcc, amygdala, and hippocampus, all of which also show altered fMRI activity in PPD mothers relative to healthy mothers (Pawluski et al. 2017). Along with these findings, Chase et al. (2014) showed decreased connectivity in the amygdala and the post cingulate cortex in depressed mothers, indicative of limited neural plasticity, which is widely seen in new mothers' brains, especially in animals (for a review, see Lambert 2012). The response of mothers with PPD to baby cry cues is eliminated in the paralimbic and prefrontal regions, which are normally responsive to baby cries in healthy mothers, while striatal and medial thalamic activities are increased (Swain 2011). Decreased activity in the reward, empathy, and emotional regulation circuits are observed in PPD mothers which may play a large role in the lowered response to infant cues (Feldman 2016). Overall, the disturbances in these networks may help explain deficits in maternal care of mothers with PPD.

4.2.3 Substance Abuse

Substance abuse in mothers is associated with preterm birth and decreased birth weight and is later associated with increased aggression and the development of neurological and behavior problems in children (Gonzalez-Mariscal and Melo 2013). A plethora of rodent work shows that cocaine use before, during, or after pregnancy disrupts maternal care, but does not abolish it, suggesting drug use creates deficits in the reward saliency of pups (Gonzalez-Mariscal and Melo 2013). Substance-abusing parents are less attentive and interact less with their babies (Gonzalez-Mariscal and Melo 2013), consistent with research that has found decreased activity in the parahippocampus, which is associated with vigilance, and the PFC, which is involved in cognitive control (Feldman 2016). Maternal care involves unique adaptions of both the reward and stress systems, and both of these systems are altered in drug addicts. Therefore, based on human and animal work, Rutherford and colleagues (Rutherford et al. 2011) have proposed the hypothesis that substance-abusing parents find parenting cues stressful, rather than rewarding. The complex system between dopamine, endogenous opioids, and the reward system has made it difficult to develop a treatment of addiction during motherhood. However, a study of women undergoing controlled treatment of methadone, for example, may prove useful into understanding the neurobiology of maladaptive maternal care in substance-abusing mothers.

5 Conclusion

Parental care is critical to the reproductive strategy of mammals, and without it, the young are incapable of survival. This function is driven by specialized neural circuits that receive sensory, cortical, and hormonal input to generate a coordinated and timely change in behavior, and sustain that behavior through activation of reward pathways. Predominantly, parental care is delivered by mothers, although data suggests that similar circuits exist in the male brain, enabling several species to adopt a strategy of paternal care in support of the maternal behavior. While the neural circuitry can largely function in the absence of hormonal inputs, the role of hormones is crucial to promote the behavior at the appropriate time. Importantly, the hormonal changes associated with pregnancy and lactation also act to coordinate a broad range of physiological changes to support the mother and help her adapt to the demands of these states, including changes in metabolism, reproductive system, and responses to stress. Hence, alongside the well-studied young-directed behaviors, such as retrieval, grooming, and nursing, and the related young-associated behaviors, such as maternal aggression, the physiological adaptations enabling successful pregnancy and lactation (Fig. 3) are a significant part of the overall process of maternal care.

The neural and hormonal basis of human parental care is remarkably similar to that of the various animal models that have been investigated in order to understand the mechanisms controlling this essential behavior. This makes the study of maternal care a highly translatable field of research. Importantly, no single animal model is going to be the same as the human, highlighting the value of using a broad comparative approach, including use of non-mammalian models, to study different aspects of parental care that might be prevalent in the different species. While the hormones of pregnancy are arguably not as critical for the onset of maternal care in woman as they are in some animal models, they are nonetheless present in human mothers and are having an impact. Hormones increase maternal sensitivity and promote mother–infant interactions and bonding, and abnormalities in these hormones or their actions may be contributing to the etiology of pregnancy complications. Complications such as maternal mental health issues are somewhat under-recognized and certainly under-treated. The methodology used in humans often precludes establishment of direct causality, and thus insights from both human and animal studies are necessary in order to create a complete picture of maternal care. Such knowledge will continue to be of benefit in enhancing our understanding of parenting, both to celebrate the enjoyment and pleasure that parenting can bring, and also to help diagnose and treat disorders that can occur.

References

Afonso VM, Shams WM, Jin D, Fleming AS (2013) Distal pup cues evoke dopamine responses in hormonally primed rats in the absence of pup experience or ongoing maternal behavior. J Neurosci 33(6):2305–2312

Aghaeepour N, Lehallier B, Baca Q, Ganio EA, Wong RJ, Ghaemi MS, Culos A, El-Sayed YY, Blumenfeld YJ, Druzin ML, Winn VD, Gibbs RS, Tibshirani R, Shaw GM, Stevenson DK, Gaudilliere B, Angst MS (2018) A proteomic clock of human pregnancy. Am J Obstet Gynecol 218(3):347.e1–347e14

Altemus M, Deuster PA, Galliven E, Carter C, Gold PW (1995) Suppression of hypothalmic-pituitary-adrenal axis responses to stress in lactating women. J Clin Endocrinol Metab 80 (10):2954–2959

Atkinson HC, Waddell BJ (1995) The hypothalamic-pituitary-adrenal axis in rat pregnancy and lactation: circadian variation and interrelationship of plasma adrenocorticotropin and corticosterone. Endocrinology 136(2):512–520

Augustine RA, Grattan DR (2008) Induction of central leptin resistance in hyperphagic pseudopregnant rats by chronic prolactin infusion. Endocrinology 149(3):1049–1055

Augustine RA, Ladyman SR, Bouwer GT, Alyousif Y, Sapsford TJ, Scott V, Kokay IC, Grattan DR, Brown CH (2017) Prolactin regulation of oxytocin neurone activity in pregnancy and lactation. J Physiol Lond 595(11):3591–3605

Baeyens L, Hindi S, Sorenson RL, German MS (2016) beta-Cell adaptation in pregnancy. Diabetes Obes Metab 18(Suppl 1):63–70

Bales KL, Saltzman W (2016) Fathering in rodents: neurobiological substrates and consequences for offspring. Horm Behav 77:249–259

Bales KL, Kim AJ, Lewis-Reese AD, Sue Carter C (2004) Both oxytocin and vasopressin may influence alloparental behavior in male prairie voles. Horm Behav 45(5):354–361

Barrett J, Wonch KE, Gonzalez A, Ali N, Steiner M, Hall GB, Fleming AS (2012) Maternal affect and quality of parenting experiences are related to amygdala response to infant faces. Soc Neurosci 7(3):252–268

Beach FA, Jaynes J (1956) Studies of maternal retrieving in rats III: sensory cues involved in the lactating female's response to her young. Behavior 10:104–125

Bi S, Kim YJ, Zheng F (2012) Dorsomedial hypothalamic NPY and energy balance control. Neuropeptides 46(6):309–314

Bosch OJ (2013) Maternal aggression in rodents: brain oxytocin and vasopressin mediate pup defence. Philos Trans R Soc Lond Ser B Biol Sci 368(1631):20130085

Bosch OJ, Neumann ID (2010) Vasopressin released within the central amygdala promotes maternal aggression. Eur J Neurosci 31(5):883–891

Bosch OJ, Kromer SA, Brunton PJ, Neumann ID (2004) Release of oxytocin in the hypothalamic paraventricular nucleus, but not central amygdala or lateral septum in lactating residents and virgin intruders during maternal defence. Neuroscience 124(2):439–448

Bouilly J, Sonigo C, Auffret J, Gibori G, Binart N (2012) Prolactin signaling mechanisms in ovary. Mol Cell Endocrinol 356(1–2):80–87

Bridges RS (1984) A quantitative analysis of the roles of dosage, sequence, and duration of estradiol and progesterone exposure in the regulation of maternal behavior in the rat. Endocrinology 114 (3):930–940

Bridges RS (2015) Neuroendocrine regulation of maternal behavior. Front Neuroendocrinol 36:178–196

Bridges RS, Hays LE (2005) Steroid-induced alterations in mRNA expression of the long form of the prolactin receptor in the medial preoptic area of female rats: effects of exposure to a pregnancy-like regimen of progesterone and estradiol. Brain Res Mol Brain Res 140 (1–2):10–16

Bridges RS, DiBiase R, Loundes DD, Doherty PC (1985) Prolactin stimulation of maternal behavior in female rats. Science 227(4688):782–784

Bridges RS, Numan M, Ronsheim PM, Mann PE, Lupini CE (1990) Central prolactin infusions stimulate maternal behavior in steroid-treated, nulliparous female rats. Proc Soc Natl Acad Sci U S A 87(20):8003–8007

Bridges RS, Robertson MC, Shiu RP, Sturgis JD, Henriquez BM, Mann PE (1997) Central lactogenic regulation of maternal behavior in rats: steroid dependence, hormone specificity, and behavioral potencies of rat prolactin and rat placental lactogen I. Endocrinology 138 (2):756–763

Bridges RS, Mann PE, Coppeta JS (1999) Hypothalamic involvement in the regulation of maternal behaviour in the rat: inhibitory roles for the ventromedial hypothalamus and the dorsal/anterior hypothalamic areas. J Neuroendocrinol 11(4):259–266

Briffaud V, Williams P, Courty J, Broberger C (2015) Excitation of tuberoinfundibular dopamine neurons by oxytocin: crosstalk in the control of lactation. J Neurosci 35(10):4229–4237

Brown RE (1993) Hormonal and experiential factors influencing parental behaviour in male rodents: an integrative approach. Behav Process 30(1):1–27

Brown RS, Kokay IC, Herbison AE, Grattan DR (2010) Distribution of prolactin-responsive neurons in the mouse forebrain. J Comp Neurol 518(1):92–102

Brown RS, Herbison AE, Grattan DR (2011) Differential changes in responses of hypothalamic and brainstem neuronal populations to prolactin during lactation in the mouse. Biol Reprod 84 (4):826–836

Brown RSE, Aoki M, Ladyman SR, Phillipps HR, Wyatt A, Boehm U, Grattan DR (2017) Prolactin action in the medial preoptic area is necessary for postpartum maternal nursing behavior. Proc Natl Acad Sci U S A 114(40):10779–10784

Brummelte S, Galea LA (2016) Postpartum depression: etiology, treatment and consequences for maternal care. Horm Behav 77:153–166

Brunton PJ, Russell JA (2008) The expectant brain: adapting for motherhood. Nat Rev Neurosci 9 (1):11–25

Brunton PJ, Meddle SL, Ma S, Ochedalski T, Douglas AJ, Russell JA (2005) Endogenous opioids and attenuated hypothalamic-pituitary-adrenal axis responses to immune challenge in pregnant rats. J Neurosci 25(21):5117–5126

Brunton PJ, Russell JA, Hirst JJ (2014) Allopregnanolone in the brain: protecting pregnancy and birth outcomes. Prog Neurobiol 113:106–136

Campino C, Torres C, Ampuero S, Diaz S, Gonzalez GB, Seron-Ferre M (1999) Bioactivity of prolactin isoforms: lactation and recovery of menses in nursing women. Hum Reprod 14 (4):898–905

Carlson AA, Russell AF, Young AJ, Jordan NR, McNeilly AS, Parlow AF, Clutton-Brock T (2006) Elevated prolactin levels immediately precede decisions to babysit by male meerkat helpers. Horm Behav 50(1):94–100

Charles NE, Alexander GM, Saenz J (2013) Motivational value and salience of images of infants. Evol Hum Behav 34(5):373–381

Chase HW, Moses-Kolko EL, Zevallos C, Wisner KL, Phillips ML (2014) Disrupted posterior cingulate-amygdala connectivity in postpartum depressed women as measured with resting BOLD fMRI. Soc Cogn Affect Neurosci 9(8):1069–1075

Chen P, Li C, Haskell-Luevano C, Cone RD, Smith MS (1999) Altered expression of agouti-related protein and its colocalization with neuropeptide Y in the arcuate nucleus of the hypothalamus during lactation. Endocrinology 140(6):2645–2650

Cohen L, Rothschild G, Mizrahi A (2011) Multisensory integration of natural odors and sounds in the auditory cortex. Neuron 72(2):357–369

Concas A, Mostallino MC, Porcu P, Follesa P, Barbaccia ML, Trabucchi M, Purdy RH, Grisenti P, Biggio G (1998) Role of brain allopregnanolone in the plasticity of gamma-aminobutyric acid type A receptor in rat brain during pregnancy and after delivery. Proc Natl Acad Sci U S A 95 (22):13284–13289

Consiglio AR, Lucion AB (1996) Lesion of hypothalamic paraventricular nucleus and maternal aggressive behavior in female rats. Physiol Behav 59(4–5):591–596

da Costa AP, Wood S, Ingram CD, Lightman SL (1996) Region-specific reduction in stress-induced c-fos mRNA expression during pregnancy and lactation. Brain Res 742(1–2):177–184

da Costa AP, Ma X, Ingram CD, Lightman SL, Aguilera G (2001) Hypothalamic and amygdaloid corticotropin-releasing hormone (CRH) and CRH receptor-1 mRNA expression in the stress-hyporesponsive late pregnant and early lactating rat. Brain Res Mol Brain Res 91(1–2):119–130

Delahunty KM, McKay DW, Noseworthy DE, Storey AE (2007) Prolactin responses to infant cues in men and women: effects of parental experience and recent infant contact. Horm Behav 51 (2):213–220

Dilley WG, Adler NT (1968) Postcopulatory mammary gland secretion in rats. Proc Soc Exp Biol Med 129(3):964–966

Diorio D, Viau V, Meaney MJ (1993) The role of the medial prefrontal cortex (cingulate gyrus) in the regulation of hypothalamic-pituitary-adrenal responses to stress. J Neurosci 13 (9):3839–3847

Donner N, Bredewold R, Maloumby R, Neumann ID (2007) Chronic intracerebral prolactin attenuates neuronal stress circuitries in virgin rats. Eur J Neurosci 25(6):1804–1814

Douglas AJ, Brunton PJ, Bosch OJ, Russell JA, Neumann ID (2003) Neuroendocrine responses to stress in mice: hyporesponsiveness in pregnancy and parturition. Endocrinology 144 (12):5268–5276

Douglas AJ, Johnstone LE, Leng G (2007) Neuroendocrine mechanisms of change in food intake during pregnancy: a potential role for brain oxytocin. Physiol Behav 91(4):352–365

Drago F, D'Agata V, Iacona T, Spadaro F, Grassi M, Valerio C, Raffaele R, Astuto C, Lauria N, Vitetta M (1989) Prolactin as a protective factor in stress-induced biological changes. J Clin Lab Anal 3(6):340–344

Egli M, Bertram R, Toporikova N, Sellix MT, Blanco W, Freeman ME (2006) Prolactin secretory rhythm of mated rats induced by a single injection of oxytocin. Am J Physiol Endocrinol Metab 290(3):E566–E572

Ehret G (2005) Infant rodent ultrasounds – a gate to the understanding of sound communication. Behav Genet 35(1):19–29

Erskine MS (1995) Prolactin release after mating and genitosensory stimulation in females. Endocr Rev 16(4):508–528

Falkner AL, Grosenick L, Davidson TJ, Deisseroth K, Lin D (2016) Hypothalamic control of male aggression-seeking behavior. Nat Neurosci 19(4):596–604

Feldman R (2016) The neurobiology of mammalian parenting and the biosocial context of human caregiving. Horm Behav 77:3–17

Field T, Diego M, Hernandez-Reif M (2006) Prenatal depression effects on the fetus and newborn: a review. Infant Behav Dev 29(3):445–455

Fisher AE (1956) Maternal and sexual behavior induced by intracranial chemical stimulation. Science 124(3214):228–229

Fleming AS, Luebke C (1981) Timidity prevents the virgin female rat from being a good mother: emotionality differences between nulliparous and parturient females. Physiol Behav 27 (5):863–868

Fleming A, Vaccarino F, Tambosso L, Chee P (1979) Vomeronasal and olfactory system modulation of maternal behavior in the rat. Science 203(4378):372–374

Fox SR, Smith MS (1984) The suppression of pulsatile luteinizing hormone secretion during lactation in the rat. Endocrinology 115(6):2045–2051

Gandelman R, Zarrow MX, Denenberg VH, Myers M (1971) Olfactory bulb removal eliminates maternal behavior in the mouse. Science 171(3967):210–211

Garland HO, Atherton JC, Baylis C, Morgan MR, Milne CM (1987) Hormone profiles for progesterone, oestradiol, prolactin, plasma renin activity, aldosterone and corticosterone during pregnancy and pseudopregnancy in two strains of rat: correlation with renal studies. J Endocrinol 113(3):435–444

Garner JP, Gaskill BN, Pritchett-Corning KR (2016) Two of a kind or a full house? Reproductive suppression and alloparenting in laboratory mice. PLoS One 11(5):e0154966

Gillies GE, Linton EA, Lowry PJ (1982) Corticotropin releasing activity of the new CRF is potentiated several times by vasopressin. Nature 299(5881):355–357

Giovenardi M, Padoin MJ, Cadore LP, Lucion AB (1998) Hypothalamic paraventricular nucleus modulates maternal aggression in rats: effects of ibotenic acid lesion and oxytocin antisense. Physiol Behav 63(3):351–359

Gonzalez-Mariscal G, Melo AI (2013) Parental behavior. In: Pfaff DW (ed) Neuroscience in the 21st century. Springer, New York, pp 2069–2100

Grattan DR (2015) 60 years of neuroendocrinology: the hypothalamo-prolactin axis. J Endocrinol 226(2):T101–T122

Grattan DR, Averill RL (1990) Effect of ovarian steroids on a nocturnal surge of prolactin secretion that precedes parturition in the rat. Endocrinology 126(2):1199–1205

Gross GA, Imamura T, Luedke C, Vogt SK, Olson LM, Nelson DM, Sadovsky Y, Muglia LJ (1998) Opposing actions of prostaglandins and oxytocin determine the onset of murine labor. Proc Natl Acad Sci U S A 95(20):11875–11879

Gunnet JW, Freeman ME (1983) The mating-induced release of prolactin: a unique neuroendocrine response. Endocr Rev 4(1):44–61

Gustafson P, Kokay I, Sapsford T, Bunn S, Grattan D (2017) Prolactin regulation of the HPA axis is not mediated by a direct action upon CRH neurons: evidence from the rat and mouse. Brain Struct Funct 222(7):3191–3204

Hansen S (1989) Medial hypothalamic involvement in maternal aggression of rats. Behav Neurosci 103(5):1035–1046

Hansen S (1994) Maternal behavior of female rats with 6-OHDA lesions in the ventral striatum: characterization of the pup retrieval deficit. Physiol Behav 55(4):615–620

Hansen S, Harthon C, Wallin E, Lofberg L, Svensson K (1991) The effects of 6-OHDA-induced dopamine depletions in the ventral or dorsal striatum on maternal and sexual behavior in the female rat. Pharmacol Biochem Behav 39(1):71–77

Hansen S, Bergvall AH, Nyiredi S (1993) Interaction with pups enhances dopamine release in the ventral striatum of maternal rats: a microdialysis study. Pharmacol Biochem Behav 45 (3):673–676

Hashikawa K, Hashikawa Y, Tremblay R, Zhang J, Feng JE, Sabol A, Piper WT, Lee H, Rudy B, Lin D (2017) Esr1(+) cells in the ventromedial hypothalamus control female aggression. Nat Neurosci 20(11):1580–1590

Hauser H, Gandelman R (1985) Lever pressing for pups: evidence for hormonal influence upon maternal behavior of mice. Horm Behav 19(4):454–468

Herrenkohl LR, Rosenberg PA (1974) Effects of hypothalamic deafferentation late in gestation on lactation and nursing behavior in the rat. Horm Behav 5(1):33–41

Higuchi T, Negoro H, Arita J (1989) Reduced responses of prolactin and catecholamine to stress in the lactating rat. J Endocrinol 122(2):495–498

Hillerer KM, Reber SO, Neumann ID, Slattery DA (2011) Exposure to chronic pregnancy stress reverses peripartum-associated adaptations: implications for postpartum anxiety and mood disorders. Endocrinology 152(10):3930–3940

Horrell ND, Perea-Rodriguez JP, Harris BN, Saltzman W (2017) Effects of repeated pup exposure on behavioral, neural, and adrenocortical responses to pups in male California mice (Peromyscus californicus). Horm Behav 90:56–63

Inoue H (2016) Central insulin-mediated regulation of hepatic glucose production [Review]. Endocr J 63(1):1–7

Insel TR, Harbaugh CR (1989) Lesions of the hypothalamic paraventricular nucleus disrupt the initiation of maternal behavior. Physiol Behav 45(5):1033–1041

Jacobson CD, Terkel J, Gorski RA, Sawyer CH (1980) Effects of small medial preoptic area lesions on maternal behavior: retrieving and nest building in the rat. Brain Res 194(2):471–478

Jean-Baptiste N, Terleph TA, Bamshad M (2008) Changes in paternal responsiveness of prairie voles (Microtus ochrogaster) in response to olfactory cues and continuous physical contact with a female. Ethology 114:1239–1246

Johnstone HA, Wigger A, Douglas AJ, Neumann ID, Landgraf R, Seckl JR, Russell JA (2000) Attenuation of hypothalamic-pituitary-adrenal axis stress responses in late pregnancy: changes in feedforward and feedback mechanisms. J Neuroendocrinol 12(8):811–822

Keebaugh AC, Barrett CE, Laprairie JL, Jenkins JJ, Young LJ (2015) RNAi knockdown of oxytocin receptor in the nucleus accumbens inhibits social attachment and parental care in monogamous female prairie voles. Soc Neurosci 10(5):561–570

Kenkel WM, Perkeybile AM, Carter CS (2017) The neurobiological causes and effects of alloparenting. Dev Neurobiol 77(2):214–232

Kennett JE, McKee DT (2012) Oxytocin: an emerging regulator of prolactin secretion in the female rat. J Neuroendocrinol 24(3):403–412

Kinsley CH, Bridges RS (1988) Prolactin modulation of the maternal-like behavior displayed by juvenile rats. Horm Behav 22(1):49–65

Kinsley CH, Lambert KG (2008) Reproduction-induced neuroplasticity: natural behavioural and neuronal alterations associated with the production and care of offspring. J Neuroendocrinol 20 (4):515–525

Kirkpatrick B, Carter CS, Newman SW, Insel TR (1994) Axon-sparing lesions of the medial nucleus of the amygdala decrease affiliative behaviors in the prairie vole (Microtus ochrogaster): behavioral and anatomical specificity. Behav Neurosci 108(3):501–513

Knigge KM, Hays M (1963) Evidence of inhibitive role of hippocampus in neural regulation of ACTH release. Proc Soc Exp Biol Med 114:67–69

Kokay IC, Bull PM, Davis RL, Ludwig M, Grattan DR (2006) Expression of the long form of the prolactin receptor in magnocellular oxytocin neurons is associated with specific prolactin regulation of oxytocin neurons. Am J Physiol Regul Integr Comp Physiol 290(5):R1216–R1225

Kroll-Desrosiers AR, Nephew BC, Babb JA, Guilarte-Walker Y, Moore Simas TA, Deligiannidis KM (2017) Association of peripartum synthetic oxytocin administration and depressive and anxiety disorders within the first postpartum year. Depress Anxiety 34(2):137–146

Kruger TH, Haake P, Hartmann U, Schedlowski M, Exton MS (2002) Orgasm-induced prolactin secretion: feedback control of sexual drive? Neurosci Biobehav Rev 26(1):31–44

Kuroda KO, Tachikawa K, Yoshida S, Tsuneoka Y, Numan M (2011) Neuromolecular basis of parental behavior in laboratory mice and rats: with special emphasis on technical issues of using mouse genetics. Prog Neuro-Psychopharmacol Biol Psychiatry 35(5):1205–1231

Ladyman SR, Grattan DR (2004) Region-specific reduction in leptin-induced phosphorylation of signal transducer and activator of transcription-3 (STAT3) in the rat hypothalamus is associated with leptin resistance during pregnancy. Endocrinology 145(8):3704–3711

Ladyman SR, Grattan DR (2005) Suppression of leptin receptor messenger ribonucleic acid and leptin responsiveness in the ventromedial nucleus of the hypothalamus during pregnancy in the rat. Endocrinology 146(9):3868–3874

Ladyman SR, Grattan DR (2017) Region-specific suppression of hypothalamic responses to insulin to adapt to elevated maternal insulin secretion during pregnancy. Endocrinology 158:4257–4269

Ladyman SR, Grattan DR (2016) Central effects of leptin on glucose homeostasis are modified during pregnancy in the rat. J Neuroendocrinol 28(10). https://doi.org/10.1111/jne.12431

Ladyman SR, Tups A, Augustine RA, Swahn-Azavedo A, Kokay IC, Grattan DR (2009) Loss of hypothalamic response to leptin during pregnancy associated with development of melanocortin resistance. J Neuroendocrinol 21(5):449–456

Ladyman SR, Augustine RA, Grattan DR (2010) Hormone interactions regulating energy balance during pregnancy. J Neuroendocrinol 22(7):805–817

Ladyman SR, Sapsford TJ, Grattan DR (2011) Loss of acute satiety response to cholecystokinin in pregnant rats. J Neuroendocrinol 23(11):1091–1098

Ladyman SR, Fieldwick DM, Grattan DR (2012) Suppression of leptin-induced hypothalamic JAK/STAT signalling and feeding response during pregnancy in the mouse. Reproduction 144(1):83–90

Ladyman SR, Augustine RA, Scherf E, Phillipps HR, Brown CH, Grattan DR (2016) Attenuated hypothalamic responses to alpha-melanocyte stimulating hormone during pregnancy in the rat. J Physiol Lond 594(4):1087–1101

Lambert KG (2012) The parental brain: transformations and adaptations. Physiol Behav 107 (5):792–800

Lambert KG, Kinsley CH (1993) Sex differences and gonadal hormones influence susceptibility to the activity-stress paradigm. Physiol Behav 53(6):1085–1090

Lambert KG, Berry AE, Griffins G, Amory-Meyers E, Madonia-Lomas L, Love G, Kinsley CH (2005) Pup exposure differentially enhances foraging ability in primiparous and nulliparous rats. Physiol Behav 84(5):799–806

Larsen CM, Grattan DR (2010) Prolactin-induced mitogenesis in the subventricular zone of the maternal brain during early pregnancy is essential for normal postpartum behavioral responses in the mother. Endocrinology 151(8):3805–3814

Lee AW, Brown RE (2007) Comparison of medial preoptic, amygdala, and nucleus accumbens lesions on parental behavior in California mice (Peromyscus californicus). Physiol Behav 92 (4):617–628

Lee Y, Voogt J (1999) Feedback effects of placental lactogens on prolactin levels and Fos-related antigen immunoreactivity of tuberoinfundibular dopaminergic neurons in the arcuate nucleus during pregnancy in the rat. Endocrinology 140(5):2159–2166

Lee A, Clancy S, Fleming AS (2000) Mother rats bar-press for pups: effects of lesions of the mpoa and limbic sites on maternal behavior and operant responding for pup-reinforcement. Behav Brain Res 108(2):215–231

Lee H, Kim DW, Remedios R, Anthony TE, Chang A, Madisen L, Zeng H, Anderson DJ (2014) Scalable control of mounting and attack by Esr1+ neurons in the ventromedial hypothalamus. Nature 509(7502):627–632

Lei K, Liu Y, Smith AS, Lonstein JS, Wang Z (2017) Effects of pair bonding on parental behavior and dopamine activity in the nucleus accumbens in male prairie voles. Eur J Neurosci 46 (7):2276–2284

Leng G, Russell JA (2016) The peptide oxytocin antagonist F-792, when given systemically, does not act centrally in lactating rats. J Neuroendocrinol 28(4). https://doi.org/10.1111/jne.12331

Levy F (2016) Neuroendocrine control of maternal behavior in non-human and human mammals. Ann Endocrinol (Paris) 77(2):114–125

Levy F, Keller M, Poindron P (2004) Olfactory regulation of maternal behavior in mammals. Horm Behav 46(3):284–302

Li C, Chen P, Smith MS (1998) The acute suckling stimulus induces expression of neuropeptide Y (NPY) in cells in the dorsomedial hypothalamus and increases NPY expression in the arcuate nucleus. Endocrinology 139(4):1645–1652

Lieberwirth C, Wang Y, Jia X, Liu Y, Wang Z (2013) Fatherhood reduces the survival of adult-generated cells and affects various types of behavior in the prairie vole (Microtus ochrogaster). Eur J Neurosci 38(9):3345–3355

Lightman SL, Young WS 3rd. (1989) Lactation inhibits stress-mediated secretion of corticosterone and oxytocin and hypothalamic accumulation of corticotropin-releasing factor and enkephalin messenger ribonucleic acids. Endocrinology 124(5):2358–2364

Lightman SL, Windle RJ, Wood SA, Kershaw YM, Shanks N, Ingram CD (2001) Peripartum plasticity within the hypothalamo-pituitary-adrenal axis. Prog Brain Res 133:111–129

Lin D, Boyle MP, Dollar P, Lee H, Lein ES, Perona P, Anderson DJ (2011) Functional identification of an aggression locus in the mouse hypothalamus. Nature 470(7333):221–226

Liu X, Brown RS, Herbison AE, Grattan DR (2014) Lactational anovulation in mice results from a selective loss of kisspeptin input to GnRH neurons. Endocrinology 155(1):193–203

Lonstein JS, De Vries GJ (1999) Sex differences in the parental behaviour of adult virgin prairie voles: independence from gonadal hormones and vasopressin. J Neuroendocrinol 11 (6):441–449

Lonstein JS, de Vries GJ (2000) Sex differences in the parental behavior of rodents. Neurosci Biobehav Rev 24(6):669–686

Lonstein JS, Rood BD, De Vries GJ (2002) Parental responsiveness is feminized after neonatal castration in virgin male prairie voles, but is not masculinized by perinatal testosterone in virgin females. Horm Behav 41(1):80–87

Lonstein JS, Levy F, Fleming AS (2015) Common and divergent psychobiological mechanisms underlying maternal behaviors in non-human and human mammals. Horm Behav 73:156–185

Lorberbaum JP, Newman JD, Horwitz AR, Dubno JR, Lydiard RB, Hamner MB, Bohning DE, George MS (2002) A potential role for thalamocingulate circuitry in human maternal behavior. Biol Psychiatry 51(6):431–445

Lubin DA, Elliott JC, Black MC, Johns JM (2003) An oxytocin antagonist infused into the central nucleus of the amygdala increases maternal aggressive behavior. Behav Neurosci 117 (2):195–201

Lucas BK, Ormandy CJ, Binart N, Bridges RS, Kelly PA (1998) Null mutation of the prolactin receptor gene produces a defect in maternal behavior. Endocrinology 139(10):4102–4107

Ludwig M, Stern J (2015) Multiple signalling modalities mediated by dendritic exocytosis of oxytocin and vasopressin. Philos Trans R Soc Lond Ser B Biol Sci 370(1672). https://doi.org/10.1098/rstb.2014.0182

Ma S, Shipston MJ, Morilak D, Russell JA (2005) Reduced hypothalamic vasopressin secretion underlies attenuated adrenocorticotropin stress responses in pregnant rats. Endocrinology 146 (3):1626–1637

Mah BL, Van Ijzendoorn MH, Smith R, Bakermans-Kranenburg MJ (2013) Oxytocin in postnatally depressed mothers: its influence on mood and expressed emotion. Prog Neuro-Psychopharmacol Biol Psychiatry 40:267–272

Mann PE, Babb JA (2004) Disinhibition of maternal behavior following neurotoxic lesions of the hypothalamus in primigravid rats. Brain Res 1025(1–2):51–58

Martin CE, Cake MH, Hartmann PE, Cook IF (1977) Relationship between foetal corticosteroids, maternal progesterone and parturition in the rat. Acta Endocrinol 84(1):167–176

Matsushita N, Muroi Y, Kinoshita K, Ishii T (2015) Comparison of c-Fos expression in brain regions involved in maternal behavior of virgin and lactating female mice. Neurosci Lett 590:166–171

Mattson BJ, Morrell JI (2005) Preference for cocaine- versus pup-associated cues differentially activates neurons expressing either Fos or cocaine- and amphetamine-regulated transcript in lactating, maternal rodents. Neuroscience 135(2):315–328

Mayer AD, Rosenblatt JS (1980) Hormonal interaction with stimulus and situational factors in the initiation of maternal behavior in nonpregnant rats. J Comp Physiol Psychol 94(6):1040–1059

McCarthy MM, vom Saal FS (1985) The influence of reproductive state on infanticide by wild female house mice (Mus musculus). Physiol Behav 35(6):843–849

McGowan MK, Andrews KM, Grossman SP (1992) Chronic intrahypothalamic infusions of insulin or insulin antibodies alter body weight and food intake in the rat. Physiol Behav 51(4):753–766

McHenry JA, Otis JM, Rossi MA, Robinson JE, Kosyk O, Miller NW, McElligott ZA, Budygin EA, Rubinow DR, Stuber GD (2017) Hormonal gain control of a medial preoptic area social reward circuit. Nat Neurosci 20(3):449–458

McKinley MJ, Yao ST, Uschakov A, McAllen RM, Rundgren M, Martelli D (2015) The median preoptic nucleus: front and centre for the regulation of body fluid, sodium, temperature, sleep and cardiovascular homeostasis. Acta Physiol (Oxf) 214(1):8–32

McNeilly AS (2001) Lactational control of reproduction. Reprod Fertil Dev 13(7–8):583–590

Meaney MJ, Szyf M, Seckl JR (2007) Epigenetic mechanisms of perinatal programming of hypothalamic-pituitary-adrenal function and health. Trends Mol Med 13(7):269–277

Naef L, Woodside B (2007) Prolactin/leptin interactions in the control of food intake in rats. Endocrinology 148:5977–5983

Nagaishi VS, Cardinali LI, Zampieri TT, Furigo IC, Metzger M, Donato J Jr (2014) Possible crosstalk between leptin and prolactin during pregnancy. Neuroscience 259:71–83

Nephew BC, Bridges RS (2008) Central actions of arginine vasopressin and a V1a receptor antagonist on maternal aggression, maternal behavior, and grooming in lactating rats. Pharmacol Biochem Behav 91(1):77–83

Neumann I, Landgraf R, Bauce L, Pittman QJ (1995) Osmotic responsiveness and cross talk involving oxytocin, but not vasopressin or amino acids, between the supraoptic nuclei in virgin and lactating rats. J Neurosci 15(5 Pt 1):3408–3417

Neumann ID, Johnstone HA, Hatzinger M, Liebsch G, Shipston M, Russell JA, Landgraf R, Douglas AJ (1998) Attenuated neuroendocrine responses to emotional and physical stressors in pregnant rats involve adenohypophysial changes. J Physiol Lond 508(Pt 1):289–300

Neumann ID, Torner L, Wigger A (2000a) Brain oxytocin: differential inhibition of neuroendocrine stress responses and anxiety-related behaviour in virgin, pregnant and lactating rats. Neuroscience 95(2):567–575

Neumann ID, Wigger A, Torner L, Holsboer F, Landgraf R (2000b) Brain oxytocin inhibits basal and stress-induced activity of the hypothalamo-pituitary-adrenal axis in male and female rats: partial action within the paraventricular nucleus. J Neuroendocrinol 12(3):235–243

Nishimori K, Young LJ, Guo Q, Wang Z, Insel TR, Matzuk MM (1996) Oxytocin is required for nursing but is not essential for parturition or reproductive behavior. Proc Natl Acad Sci U S A 93 (21):11699–11704

Nishitani S, Kuwamoto S, Takahira A, Miyamura T, Shinohara K (2014) Maternal prefrontal cortex activation by newborn infant odors. Chem Senses 39(3):195–202

Numan M (1974) Medial preoptic area and maternal behavior in the female rat. J Comp Physiol Psychol 87(4):746–759

Numan M (2006) Hypothalamic neural circuits regulating maternal responsiveness toward infants. Behav Cogn Neurosci Rev 5(4):163–190

Numan M (2007) Motivational systems and the neural circuitry of maternal behavior in the rat. Dev Psychobiol 49(1):12–21

Numan M, Sheehan TP (1997) Neuroanatomical circuitry for mammalian maternal behavior. Ann N Y Acad Sci 807:101–125

Numan M, Smith HG (1984) Maternal behavior in rats: evidence for the involvement of preoptic projections to the ventral tegmental area. Behav Neurosci 98(4):712–727

Numan M, Stolzenberg DS (2009) Medial preoptic area interactions with dopamine neural systems in the control of the onset and maintenance of maternal behavior in rats. Front Neuroendocrinol 30(1):46–64

Numan M, Rosenblatt JS, Komisaruk BR (1977) Medial preoptic area and onset of maternal behavior in the rat. J Comp Physiol Psychol 91(1):146–164

Numan M, Morrell JI, Pfaff DW (1985) Anatomical identification of neurons in selected brain regions associated with maternal behavior deficits induced by knife cuts of the lateral hypothalamus in rats. J Comp Neurol 237(4):552–564

Numan M, Corodimas KP, Numan MJ, Factor EM, Piers WD (1988) Axon-sparing lesions of the preoptic region and substantia innominata disrupt maternal behavior in rats. Behav Neurosci 102 (3):381–396

Numan M, McSparren J, Numan MJ (1990) Dorsolateral connections of the medial preoptic area and maternal behavior in rats. Behav Neurosci 104(6):964–979

Numan M, Numan MJ, Marzella SR, Palumbo A (1998) Expression of c-fos, fos B, and egr-1 in the medial preoptic area and bed nucleus of the stria terminalis during maternal behavior in rats. Brain Res 792(2):348–352

Numan M, Numan MJ, Pliakou N, Stolzenberg DS, Mullins OJ, Murphy JM, Smith CD (2005) The effects of D1 or D2 dopamine receptor antagonism in the medial preoptic area, ventral pallidum, or nucleus accumbens on the maternal retrieval response and other aspects of maternal behavior in rats. Behav Neurosci 119(6):1588–1604

Obici S, Feng Z, Karkanias G, Baskin DG, Rossetti L (2002a) Decreasing hypothalamic insulin receptors causes hyperphagia and insulin resistance in rats. Nat Neurosci 5(6):566–572

Obici S, Zhang BB, Karkanias G, Rossetti L (2002b) Hypothalamic insulin signaling is required for inhibition of glucose production. Nat Med 8(12):1376–1382

Oftedal OT (2000) Use of maternal reserves as a lactation strategy in large mammals. Proc Nutr Soc 59(1):99–106

Ogawa S, Eng V, Taylor J, Lubahn DB, Korach KS, Pfaff DW (1998) Roles of estrogen receptor-alpha gene expression in reproduction-related behaviors in female mice. Endocrinology 139 (12):5070–5081

Ogren L, Southard JN, Colosi P, Linzer DI, Talamantes F (1989) Mouse placental lactogen-I: RIA and gestational profile in maternal serum. Endocrinology 125(5):2253–2257

Olazabal DE (2014) Comparative analysis of oxytocin receptor density in the nucleus accumbens: an adaptation for female and male alloparental care? J Physiol Paris 108(2–3):213–220

Olazabal DE, Young LJ (2006) Oxytocin receptors in the nucleus accumbens facilitate "spontaneous" maternal behavior in adult female prairie voles. Neuroscience 141(2):559–568

Parker VJ, Douglas AJ (2010) Stress in early pregnancy: maternal neuro-endocrine-immune responses and effects. J Reprod Immunol 85(1):86–92

Pau CY, Pau KY, Berria M, Spies HG (2000) Ovarian influence on gonadotropin and prolactin release in mated rabbits. Endocrine 13(1):25–35

Pawluski JL, Lonstein JS, Fleming AS (2017) The neurobiology of postpartum anxiety and depression. Trends Neurosci 40(2):106–120

Pedersen CA, Boccia ML (2003) Oxytocin antagonism alters rat dams' oral grooming and upright posturing over pups. Physiol Behav 80(2–3):233–241

Pedersen CA, Prange AJ Jr (1979) Induction of maternal behavior in virgin rats after intracerebroventricular administration of oxytocin. Proc Natl Acad Sci U S A 76 (12):6661–6665

Pedersen CA, Ascher JA, Monroe YL, Prange AJ Jr (1982) Oxytocin induces maternal behavior in virgin female rats. Science 216(4546):648–650

Pedersen CA, Caldwell JD, Peterson G, Walker CH, Mason GA (1992) Oxytocin activation of maternal behavior in the rat. Ann N Y Acad Sci 652:58–69

Pedersen CA, Caldwell JD, Walker C, Ayers G, Mason GA (1994) Oxytocin activates the postpartum onset of rat maternal behavior in the ventral tegmental and medial preoptic areas. Behav Neurosci 108(6):1163–1171

Perrigo G, Bryant WC, vom Saal FS (1990) A unique timing system prevents male mice from harming their own offspring. Anim Behav 39:535–539

Pihoker C, Robertson MC, Freemark M (1993) Rat placental lactogen-I binds to the choroid plexus and hypothalamus of the pregnant rat. J Endocrinol 139(2):235–242

Quek VS, Trayhurn P (1990) Calorimetric study of the energetics of pregnancy in golden hamsters. Am J Phys 259(4 Pt 2):R807–R812

Ribeiro AC, Musatov S, Shteyler A, Simanduyev S, Arrieta-Cruz I, Ogawa S, Pfaff DW (2012) siRNA silencing of estrogen receptor-alpha expression specifically in medial preoptic area neurons abolishes maternal care in female mice. Proc Natl Acad Sci U S A 109 (40):16324–16329

Riddle O, Lahr EL, Bates RW (1935) Maternal behavior induced in virgin rats by prolactin. Exp Biol Med 32(5):730–734

Rilling JK (2013) The neural and hormonal bases of human parental care. Neuropsychologia 51 (4):731–747
Robertson MC, Friesen HG (1981) Two forms of rat placental lactogen revealed by radioimmunoassay. Endocrinology 108(6):2388–2390
Robertson MC, Gillespie B, Friesen HG (1982) Characterization of the two forms of rat placental lactogen (rPL): rPL-I and rPL-II. Endocrinology 111(6):1862–1866
Rodrigues SM, Saslow LR, Garcia N, John OP, Keltner D (2009) Oxytocin receptor genetic variation relates to empathy and stress reactivity in humans. Proc Natl Acad Sci U S A 106 (50):21437–21441
Romano N, Yip SH, Hodson DJ, Guillou A, Parnaudeau S, Kirk S, Tronche F, Bonnefont X, Le Tissier P, Bunn SJ, Grattan DR, Mollard P, Martin AO (2013) Plasticity of hypothalamic dopamine neurons during lactation results in dissociation of electrical activity and release. J Neurosci 33(10):4424–4433
Romero R, Erez O, Maymon E, Chaemsaithong P, Xu Z, Pacora P, Chaiworapongsa T, Done B, Hassan SS, Tarca AL (2017) The maternal plasma proteome changes as a function of gestational age in normal pregnancy: a longitudinal study. Am J Obstet Gynecol 217(1):67.e1–67e21
Rosenblatt JS (1967) Nonhormonal basis of maternal behavior in the rat. Science 156 (3781):1512–1514
Rosenblatt JS (1969) The development of maternal responsiveness in the rat. Am J Orthopsychiatry 39(1):36–56
Ross HE, Young LJ (2009) Oxytocin and the neural mechanisms regulating social cognition and affiliative behavior. Front Neuroendocrinol 30(4):534–547
Rossoni E, Feng J, Tirozzi B, Brown D, Leng G, Moos F (2008) Emergent synchronous bursting of oxytocin neuronal network. PLoS Comput Biol 4(7):e1000123
Rutherford HJ, Williams SK, Moy S, Mayes LC, Johns JM (2011) Disruption of maternal parenting circuitry by addictive process: rewiring of reward and stress systems. Front Psych 2:37
Salais-Lopez H, Lanuza E, Agustin-Pavon C, Martinez-Garcia F (2017) Tuning the brain for motherhood: prolactin-like central signalling in virgin, pregnant, and lactating female mice. Brain Struct Funct 222(2):895–921
Saltzman W, Maestripieri D (2011) The neuroendocrinology of primate maternal behavior. Prog Neuro-Psychopharmacol Biol Psychiatry 35(5):1192–1204
Saltzman W, Ziegler TE (2014) Functional significance of hormonal changes in mammalian fathers. J Neuroendocrinol 26(10):685–696
Saltzman W, Harris BN, de Jong TR, Perea-Rodriguez JP, Horrell ND, Zhao M, Andrew JR (2017) Paternal care in biparental rodents: intra- and inter-individual variation. Integr Comp Biol 57 (3):589–602
Sandman CA, Glynn L, Schetter CD, Wadhwa P, Garite T, Chicz-DeMet A, Hobel C (2006) Elevated maternal cortisol early in pregnancy predicts third trimester levels of placental corticotropin releasing hormone (CRH): priming the placental clock. Peptides 27(6):1457–1463
Schneider JS, Stone MK, Wynne-Edwards KE, Horton TH, Lydon J, O'Malley B, Levine JE (2003) Progesterone receptors mediate male aggression toward infants. Proc Natl Acad Sci U S A 100 (5):2951–2956
Schneider JS, Burgess C, Horton TH, Levine JE (2009) Effects of progesterone on male-mediated infant-directed aggression. Behav Brain Res 199(2):340–344
Scott N, Prigge M, Yizhar O, Kimchi T (2015) A sexually dimorphic hypothalamic circuit controls maternal care and oxytocin secretion. Nature 525(7570):519–522
Shaik AA (1971) Estrone and estradiol levels in the ovarian venous blood from rats during estrous cycle and pregnancy. Biol Reprod 5:297–307
Shanks N, Windle RJ, Perks P, Wood S, Ingram CD, Lightman SL (1999) The hypothalamic-pituitary-adrenal axis response to endotoxin is attenuated during lactation. [Erratum appears in J Neuroendocrinol 2000 May;12(5):471]. J Neuroendocrinol 11(11):857–865
Sheehan T, Numan M (2002) Estrogen, progesterone, and pregnancy termination alter neural activity in brain regions that control maternal behavior in rats. Neuroendocrinology 75(1):12–23

Shingo T, Gregg C, Enwere E, Fujikawa H, Hassam R, Geary C, Cross JC, Weiss S (2003) Pregnancy-stimulated neurogenesis in the adult female forebrain mediated by prolactin. Science 299(5603):117–120
Siegel HI, Rosenblatt JS (1975a) Estrogen-induced maternal behavior in hysterectomized-overiectomized virgin rats. Physiol Behav 14(04):465–471
Siegel HI, Rosenblatt JS (1975b) Hormonal basis of hysterectomy-induced maternal behavior during pregnancy in the rat. Horm Behav 6(3):211–222
Sirzen-Zelenskaya A, Gonzalez-Iglesias AE, Boutet de Monvel J, Bertram R, Freeman ME, Gerber U, Egli M (2011) Prolactin induces a hyperpolarising current in rat paraventricular oxytocinergic neurones. J Neuroendocrinol 23(10):883–893
Smith MS (1993) Lactation alters neuropeptide-Y and proopiomelanocortin gene expression in the arcuate nucleus of the rat. Endocrinology 133(3):1258–1265
Son SJ, Filosa JA, Potapenko ES, Biancardi VC, Zheng H, Patel KP, Tobin VA, Ludwig M, Stern JE (2013) Dendritic peptide release mediates interpopulation crosstalk between neurosecretory and preautonomic networks. Neuron 78(6):1036–1049
Sorenson RL, Brelje TC (1997) Adaptation of islets of Langerhans to pregnancy: beta-cell growth, enhanced insulin secretion and the role of lactogenic hormones. Horm Metab Res 29 (6):301–307
Stack E, Numan M (2000) The temporal course of expression of c-Fos and Fos B within the medial preoptic area and other brain regions of postpartum female rats during prolonged mother-young interactions. Behav Neurosci 114:609–622
Stack EC, Balakrishnan R, Numan MJ, Numan M (2002) A functional neuroanatomical investigation of the role of the medial preoptic area in neural circuits regulating maternal behavior. Behav Brain Res 131(1–2):17–36
Stocco C, Telleria C, Gibori G (2007) The molecular control of corpus luteum formation, function, and regression. Endocr Rev 28(1):117–149
Stolzenberg DS, Numan M (2011) Hypothalamic interaction with the mesolimbic DA system in the control of the maternal and sexual behaviors in rats. Neurosci Biobehav Rev 35(3):826–847
Stolzenberg DS, McKenna JB, Keough S, Hancock R, Numan MJ, Numan M (2007) Dopamine D1 receptor stimulation of the nucleus accumbens or the medial preoptic area promotes the onset of maternal behavior in pregnancy-terminated rats. Behav Neurosci 121(5):907–919
Storey AE, Walsh CJ, Quinton RL, Wynne-Edwards KE (2000) Hormonal correlates of paternal responsiveness in new and expectant fathers. Evol Hum Behav 21(2):79–95
Strubbe JH, Mein CG (1977) Increased feeding in response to bilateral injection of insulin antibodies in the VMH. Physiol Behav 19(2):309–313
Swain JE (2011) The human parental brain: in vivo neuroimaging. Prog Neuro-Psychopharmacol Biol Psychiatry 35(5):1242–1254
Swain JE, Ho SS (2017) Neuroendocrine mechanisms for parental sensitivity: overview, recent advances and future directions. Curr Opin Psychol 15:105–110
Swain JE, Kim P, Ho SS (2011) Neuroendocrinology of parental response to baby-cry. J Neuroendocrinol 23(11):1036–1041
Swain JE, Dayton CJ, Kim P, Tolman RM, Volling BL (2014a) Progress on the paternal brain: theory, animal models, human brain research, and mental health implications. Infant Ment Health J 35(5):394–408
Swain JE, Kim P, Spicer J, Ho SS, Dayton CJ, Elmadih A, Abel KM (2014b) Approaching the biology of human parental attachment: brain imaging, oxytocin and coordinated assessments of mothers and fathers. Brain Res 1580:78–101
Tachikawa KS, Yoshihara Y, Kuroda KO (2013) Behavioral transition from attack to parenting in male mice: a crucial role of the vomeronasal system. J Neurosci 33(12):5120–5126
Takayanagi Y, Yoshida M, Bielsky IF, Ross HE, Kawamata M, Onaka T, Yanagisawa T, Kimura T, Matzuk MM, Young LJ, Nishimori K (2005) Pervasive social deficits, but normal parturition, in oxytocin receptor-deficient mice. Proc Natl Acad Sci U S A 102(44):16096–16101

Terkel J, Blake CA, Sawyer CH (1972) Serum prolactin levels in lactating rats after suckling or exposure to ether. Endocrinology 91(1):49–53
Terkel J, Bridges RS, Sawyer CH (1979) Effects of transecting lateral neural connections of the medial preoptic area on maternal behavior in the rat: nest building, pup retrieval and prolactin secretion. Brain Res 169(2):369–380
Tonkowicz PA, Voogt JL (1983) Termination of prolactin surges with development of placental lactogen secretion in the pregnant rat. Endocrinology 113(4):1314–1318
Torner L, Toschi N, Pohlinger A, Landgraf R, Neumann ID (2001) Anxiolytic and anti-stress effects of brain prolactin: improved efficacy of antisense targeting of the prolactin receptor by molecular modeling. J Neurosci 21(9):3207–3214
Torner L, Toschi N, Nava G, Clapp C, Neumann ID (2002) Increased hypothalamic expression of prolactin in lactation: involvement in behavioural and neuroendocrine stress responses. Eur J Neurosci 15(8):1381–1389
Toufexis DJ, Tesolin S, Huang N, Walker C (1999) Altered pituitary sensitivity to corticotropin-releasing factor and arginine vasopressin participates in the stress hyporesponsiveness of lactation in the rat. J Neuroendocrinol 11(10):757–764
Townsend J, Cave BJ, Norman MR, Flynn A, Uney JB, Tortonese DJ, Wakerley JB (2005) Effects of prolactin on hypothalamic supraoptic neurones: evidence for modulation of STAT5 expression and electrical activity. Neuro Endocrinol Lett 26(2):125–130
Tsukamura H, Maeda K (2001) Non-metabolic and metabolic factors causing lactational anestrus: rat models uncovering the neuroendocrine mechanism underlying the suckling-induced changes in the mother. Prog Brain Res 133:187–205
Tsuneoka Y, Maruyama T, Yoshida S, Nishimori K, Kato T, Numan M, Kuroda KO (2013) Functional, anatomical, and neurochemical differentiation of medial preoptic area subregions in relation to maternal behavior in the mouse. J Comp Neurol 521(7):1633–1663
Tsuneoka Y, Tokita K, Yoshihara C, Amano T, Esposito G, Huang AJ, Yu LM, Odaka Y, Shinozuka K, McHugh TJ, Kuroda KO (2015) Distinct preoptic-BST nuclei dissociate paternal and infanticidal behavior in mice. EMBO J 34(21):2652–2670
Ulrich-Lai YM, Herman JP (2009) Neural regulation of endocrine and autonomic stress responses. Nat Rev Neurosci 10(6):397–409
Valeggia C, Ellison PT (2009) Interactions between metabolic and reproductive functions in the resumption of postpartum fecundity. Am J Hum Biol 21(4):559–566
van Anders SM, Tolman RM, Volling BL (2012) Baby cries and nurturance affect testosterone in men. Horm Behav 61(1):31–36
van Hemel SB (1973) Pup retrieving as a reinforcer in nulliparous mice. J Exp Anal Behav 19 (2):233–238
Vannuccini S, Bocchi C, Severi FM, Challis JR, Petraglia F (2016) Endocrinology of human parturition. Ann Endocrinol (Paris) 77(2):105–113
Veening JG, de Jong TR, Waldinger MD, Korte SM, Olivier B (2015) The role of oxytocin in male and female reproductive behavior. Eur J Pharmacol 753:209–228
vom Saal FS (1985) Time-contingent change in infanticide and parental behavior induced by ejaculation in male mice. Physiol Behav 34(1):7–15
Voogt J, Robertson M, Friesen H (1982) Inverse relationship of prolactin and rat placental lactogen during pregnancy. Biol Reprod 26(5):800–805
Wade GN, Jennings G, Trayhurn P (1986) Energy balance and brown adipose tissue thermogenesis during pregnancy in Syrian hamsters. Am J Phys 250(5 Pt 2):R845–R850
Waldherr M, Neumann ID (2007) Centrally released oxytocin mediates mating-induced anxiolysis in male rats. Proc Natl Acad Sci U S A 104(42):16681–16684
Walker CD, Lightman SL, Steele MK, Dallman MF (1992) Suckling is a persistent stimulus to the adrenocortical system of the rat. Endocrinology 130(1):115–125
Walker CD, Trottier G, Rochford J, Lavallee D (1995) Dissociation between behavioral and hormonal responses to the forced swim stress in lactating rats. J Neuroendocrinol 7(8):615–622

Wang Z, Insel TR (1996) Parental behavior in voles. In: Rosenblatt JS, Snowden CT (eds) Advances in the study of behavior: parental care: evolution, mechanism, and adaptive significance. Academic, Cambridge, pp 361–384

Wang Z, Ferris CF, De Vries GJ (1994) Role of septal vasopressin innervation in paternal behavior in prairie voles (Microtus ochrogaster). Proc Natl Acad Sci U S A 91(1):400–404

Wang J, Tai F, Yu P, Wu R (2012) Reinforcing properties of pups versus cocaine for fathers and associated central expression of Fos and tyrosine hydroxylase in mandarin voles (Microtus mandarinus). Behav Brain Res 230(1):149–157

Wersinger SR, Caldwell HK, Christiansen M, Young WS 3rd. (2007a) Disruption of the vasopressin 1b receptor gene impairs the attack component of aggressive behavior in mice. Genes Brain Behav 6(7):653–660

Wersinger SR, Caldwell HK, Martinez L, Gold P, Hu SB, Young WS 3rd. (2007b) Vasopressin 1a receptor knockout mice have a subtle olfactory deficit but normal aggression. Genes Brain Behav 6(6):540–551

Windle RJ, Shanks N, Lightman SL, Ingram CD (1997) Central oxytocin administration reduces stress-induced corticosterone release and anxiety behavior in rats. Endocrinology 138 (7):2829–2834

Windle RJ, Kershaw YM, Shanks N, Wood SA, Lightman SL, Ingram CD (2004) Oxytocin attenuates stress-induced c-fos mRNA expression in specific forebrain regions associated with modulation of hypothalamo-pituitary-adrenal activity. J Neurosci 24(12):2974–2982

Wingfield JC, Hegner RE, Dufty AMJ, Ball GF (1990) The "challenge hypothesis": theoretical implications for patterns of testosterone secretion, mating systems, and breeding strategies. Am Nat 136:829–846

Woodside B (2007) Prolactin and the hyperphagia of lactation. Physiol Behav 91(4):375–382

Woodside B, Amir S (1997) Lactation reduces Fos induction in the paraventricular and supraoptic nuclei of the hypothalamus after urethane administration in rats. Brain Res 752(1–2):319–323

Woodside B, Abizaid A, Walker C (2000) Changes in leptin levels during lactation: implications for lactational hyperphagia and anovulation. Horm Behav 37(4):353–365

Wu Z, Autry AE, Bergan JF, Watabe-Uchida M, Dulac CG (2014) Galanin neurons in the medial preoptic area govern parental behaviour. Nature 509(7500):325–330

Wynne-Edwards KE, Timonin ME (2007) Paternal care in rodents: weakening support for hormonal regulation of the transition to behavioral fatherhood in rodent animal models of biparental care. Horm Behav 52(1):114–121

Yang CF, Chiang MC, Gray DC, Prabhakaran M, Alvarado M, Juntti SA, Unger EK, Wells JA, Shah NM (2013) Sexually dimorphic neurons in the ventromedial hypothalamus govern mating in both sexes and aggression in males. Cell 153(4):896–909

Yoshihara C, Numan M, Kuroda KO (2017) Oxytocin and parental behaviors. Curr Top Behav Neurosci 35:119–153

Young WS 3rd, Shepard E, Amico J, Hennighausen L, Wagner KU, LaMarca ME, McKinney C, Ginns EI (1996) Deficiency in mouse oxytocin prevents milk ejection, but not fertility or parturition. J Neuroendocrinol 8(11):847–853

Zakar T, Hertelendy F (2007) Progesterone withdrawal: key to parturition. Am J Obstet Gynecol 196(4):289–296

Neural Circuits Underlying Rodent Sociality: A Comparative Approach

Nicole S. Lee and Annaliese K. Beery

Contents

1 Introduction 212
2 Circuits Underlying Social Behaviors and Processes 213
  2.1 Social Recognition and Memory 214
  2.2 Social Reward 216
  2.3 Social Approach/Avoidance 217
3 Studying Rodent Sociality Across Social Structures 218
  3.1 Synthesizing Social Processes to Understand Selective Relationships in Voles 219
4 Sociality in Free-Living Rodents 221
5 New Techniques and Future Directions 226
  5.1 Genetic Techniques 226
  5.2 New Conceptual Directions 227
References 227

Abstract All mammals begin life in social groups, but for some species, social relationships persist and develop throughout the course of an individual's life. Research in multiple rodent species provides evidence of relatively conserved circuitry underlying social behaviors and processes such as social recognition and memory, social reward, and social approach/avoidance. Species exhibiting different complex social behaviors and social systems (such as social monogamy or familiarity preferences) can be characterized in part by when and how they display specific social behaviors. Prairie and meadow voles are closely related species that exhibit similarly selective peer preferences but different mating systems, aiding direct comparison of the mechanisms underlying affiliative behavior. This chapter draws

N. S. Lee (✉)
Neuroscience and Behavior Program, University of Massachusetts, Amherst, MA, USA
e-mail: nlee71@smith.edu

A. K. Beery (✉)
Neuroscience and Behavior Program, University of Massachusetts, Amherst, MA, USA

Department of Psychology, Smith College, Northampton, MA, USA

Neuroscience Program, Smith College, Northampton, MA, USA
e-mail: abeery@smith.edu

Curr Topics Behav Neurosci (2019) 43: 211–238
DOI 10.1007/7854_2018_77
Published Online: 2 February 2019

on research in voles as well as other rodents to explore the mechanisms involved in individual social behavior processes, as well as specific complex social patterns. Contrasts between vole species exemplify how the laboratory study of diverse species improves our understanding of the mechanisms underlying social behavior. We identify several additional rodent species whose interesting social structures and available ecological and behavioral field data make them good candidates for study. New techniques and integration across laboratory and field settings will provide exciting opportunities for future mechanistic work in non-model species.

Keywords Group living · Meadow vole · Model species · Neural circuits · Neuroendocrinology · Prairie vole · Social behavior · Sociality

1 Introduction

All mammals exhibit some degree of social behavior, but the extent to which they are social varies widely across species. Social behavior is associated with costs, including increased risk of disease transmission and competition for resources. Nonetheless, benefits from increased predator detection, defense, and, in some species, cooperative breeding, can lead to the evolution of sociality (Clutton-Brock and Lukas 2012; Lukas and Clutton-Brock 2012a, b, 2013; reviewed in Lee 1994). Social relationships are best studied in species that display specific traits of interest. For example, social monogamy is rare among rodents ($<5\%$) but has been studied in prairie voles and California mice. Group living has been investigated in meadow voles, naked mole rats, social tuco-tucos, striped mice, and other colonial rodents (reviewed in Anacker and Beery 2013; Beery et al. 2016, 2018). In this chapter we discuss mechanisms underlying behaviors supporting life in social groups, and how they may vary between species. We focus on prairie and meadow voles, two closely related species that provide an ideal opportunity to investigate diversity of mechanism and social system.

Despite the advantages of examining species that exhibit particular characteristics of interest, mammalian research is dominated by studies of mice and rats, which in 2009 made up approximately 90% of mammalian physiology studies (Beery and Zucker 2011 supplementary material). Mice and rats have provided important insights into social behavior – intense focus on a small number of model organisms allows for great depth of study, and the development of cutting-edge technologies for these species makes them well situated for mechanistic work. However, both species are gregarious and do not exhibit selective affiliation, making them inappropriate models for studies of adult social bonds and social preferences (Beery et al. 2018; Schweinfurth et al. 2017). Furthermore, laboratory rodents are often highly inbred and far removed from the ecological contexts in which the social traits of interest evolved (but see Chalfin et al. 2014), making it difficult to determine links between behavior and natural history.

It is also possible that the mechanisms that are relevant for one species may not apply to another species with a different social organization. Even among species with similar social organization, the mechanisms that support sociality may be different (e.g., socially monogamous rodents and socially monogamous nonhuman primates and humans likely exhibit some key differences in mechanism). By examining the shared and unique basis of behaviors across species, we can hope to effectively determine how and when we can translate research across species, and potentially to humans. Laboratory studies of diverse species, for whom detailed studies of their field ecology and behavior are available, will improve our understanding of the species-specificity and generality of mechanisms supporting different social behaviors (Johnson and Young 2018; Taborsky et al. 2015).

Although ultimate/evolutionary explanations are not the focus of this chapter, they may inform our understanding of proximate mechanisms of social behavior by suggesting hypotheses concerning whether mechanisms are likely to vary between ecological, phylogenetic, and behavioral contexts. For example, because reward pathways play an essential role in assessing salience of external stimuli such as a potential mate or high-calorie foods, and thereby in motivating appetitive behaviors, we would expect them to be highly conserved across vertebrate taxa (O'Connell and Hofmann 2011b). Thus, we approach the question of what circuits underlie social behavior with evolution in mind, by considering how species with varying social structures derive different benefits from life in groups. We focus on rodents throughout, as they are the most widely used animal model for many fields including neurobiology, and include discussion of both field and laboratory experiments. We describe the neural circuitry underlying specific social behaviors and processes before synthesizing them into complex social behaviors and social systems. We summarize relevant work in rats and mice in service of a more in-depth discussion of work in voles, as these classic models have significantly informed work in voles. We then highlight some of the most socially distinctive rodents in the wild in order to illustrate potential candidates for future study, as well as describe the natural behaviors of already well-studied animals. The chapter concludes with remarks about new techniques that may help advance comparative work, as well as future directions in the study of neural mechanisms of social behavior.

2 Circuits Underlying Social Behaviors and Processes

Complex social behaviors, such as prairie vole pair bonding and rat maternal behavior, rely on specific social behaviors and processes. Here, we discuss social recognition and memory, social reward, and social approach/avoidance. These processes are often interrelated (Fig. 1); for example, social recognition is a form of short-term social memory. Social memory is necessary to form long-lasting social relationships, as individuals must recognize and remember their partner. In order for

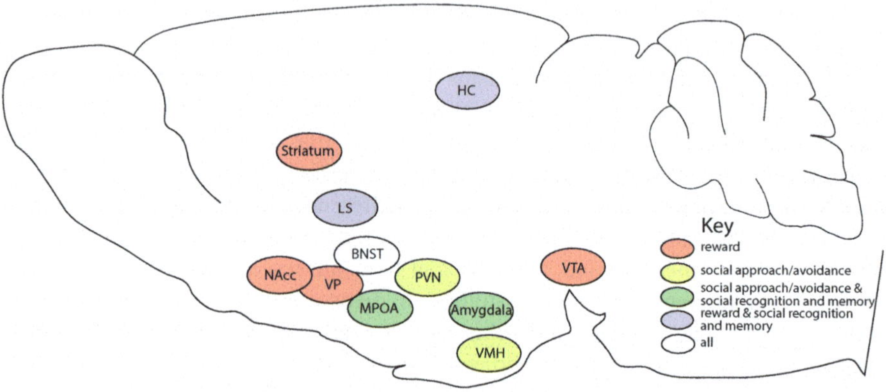

Fig. 1 Key brain regions associated with social behavior. *BNST* bed nucleus of the stria terminalis, *HC* hippocampus, *LS* lateral septum, *MPOA* medial preoptic area, *NAcc* nucleus accumbens, *PVN* paraventricular nucleus of the hypothalamus, *VMH* ventromedial nucleus of the hypothalamus, *VP* ventral pallidum, *VTA* ventral tegmental area. Brain regions explicitly discussed in the text are depicted, as well as: BNST (Bielsky and Young 2004; Lebow and Chen 2016; O'Connell and Hofmann 2011b; Walker et al. 2003), hippocampus (Broadbent et al. 2004; Brown and Aggleton 2001; Hölscher et al. 2003; Kogan et al. 2000; O'Connell and Hofmann 2011b), ventral pallidum (Smith et al. 2009), VMH (Colpaert 1975; Grossman 1972). The sagittal section was adapted from the mouse brain atlas of Paxinos and Franklin (2012)

social recognition and social memory to lead to the formation of a social relationship, an individual must exhibit decreased fear and anxiety toward a prospective partner, thereby allowing for social approach rather than avoidance. Social reward may mediate motivation to approach and may also reinforce social preferences. Although aggression is often thought to directly oppose sociality (and will not be discussed here), aggression can also play an important role in mediating social preferences. For example, if a relationship has been formed that is highly selective and/or motivating, an individual may exhibit aggression toward unfamiliar individuals.

2.1 Social Recognition and Memory

Social recognition and social memory are closely related and are important for life in some social groups, as animals may need to recognize and remember specific individuals in order to assess how to behave toward these individuals (e.g., in cases where strangers need to be quickly identified, or in cases where familiarity is a proxy for kinship recognition and degree of relatedness determines how an individual behaves toward a conspecific). The importance of social recognition and social memory to social structure will be discussed in Sect. 3.

Social recognition in rodents is measured via behavioral tests such as the habituation/dishabituation test, which measures time spent investigating a conspecific stimulus animal after repeated exposure, followed by a novel animal (reviewed in Ferguson et al. 2002; Gheusi et al. 1994). Behavioral tests such as these are based on the tendency of rats and mice to investigate unfamiliar individuals more than they would familiar individuals.

Many laboratory studies on social recognition have demonstrated the importance of two neuropeptides: oxytocin and vasopressin. Male vasopressin 1a receptor (V1aR) knockout mice showed impaired social recognition (Bielsky et al. 2004), and reexpression of V1aR in the lateral septum restored the behavior (Bielsky et al. 2005). Similarly, greater V1aR expression and higher vasopressin activity in the lateral septum are associated with better social recognition (Everts and Koolhaas 1999; Landgraf et al. 1995, 2003) and social investigation in male rats and prairie voles (Ophir et al. 2009). Male mice with a mutant form of the oxytocin gene did not exhibit social recognition (Ferguson et al. 2000), and oxytocin infusion in the medial amygdala rescued the behavior (Ferguson et al. 2001). Studies have mostly focused on males, but evidence suggests that oxytocin and potentially vasopressin are also important for social recognition in females (Clipperton-Allen et al. 2012, but see Bluthe et al. 1990). Furthermore, evidence suggests an important interplay with sex steroid hormones (Bluthe et al. 1990).

Unsurprisingly, social recognition is mediated through more than just neuropeptide action, involving important connections to and from other systems. Dopaminergic, noradrenergic, and glutamatergic systems have all been implicated in social recognition, as well as muscarinic acetylcholine receptor activation, with neuropeptides likely serving as neuromodulators of neurotransmitter release (Bielsky and Young 2004; Dluzen et al. 2000; Ferguson et al. 2002; Winslow and Insel 2004). The vomeronasal system has also been found to play an essential role in social recognition, with the medial amygdala, and oxytocin in the medial amygdala, mediating relevant information processing in a sex-specific manner (Bergan et al. 2014; Li and Dulac 2018; Li et al. 2017; Yao et al. 2017). Due to the complexity and overlap of the neural circuits underlying all social behaviors and processes (Fig. 1), those relevant to social recognition may be difficult to isolate.

Beyond social recognition, some animals maintain long-term social bonds that require social memory. In this chapter, social memory will refer to the set of behaviors routinely cited in the literature as distinct from short-term social recognition, including kin recognition, pair-bond maintenance, selective pregnancy termination, territoriality, and maintenance of stable dominance hierarchies (Gheusi et al. 1994; Winslow and Insel 2004). For example, pair-bond maintenance in prairie voles persists even after long-term separation (1–4 weeks) from a partner; at reunion, prairie voles show a preference for a partner (DeVries and Carter 1999; Sun et al. 2014). The dopamine and opioid systems play important roles in mediating the longevity of these mate bonds (Aragona et al. 2006; Resendez et al. 2012). Social memory is also necessary for the maintenance of territoriality and dominance hierarchies, and the agonistic interactions between individuals that are involved in this maintenance are often modulated by social stress (reviewed in Tamashiro et al. 2005; van der Kooij and Sandi 2012). Rats and mice in particular have been studied for their ability to spontaneously form stable dominance hierarchies. For example, glucocorticoids (Timmer and Sandi 2010) and oxytocin receptor density (OTR) in the medial amygdala and lateral septum (Timmer et al. 2011) play important roles in the formation of dominance hierarchy-related social memory in male rats. Similarly, higher brain gene expression of gonadotropin-releasing hormone (GnRH) in the medial preoptic area (MPOA) of the hypothalamus is associated with opportunity for social ascent in male mice (Williamson et al. 2017). Higher corticotropin releasing factor (CRF) mRNA in the medial and central nuclei of the amygdala and the MPOA, and glucocorticoid receptor mRNA in the hippocampus, are associated with higher dominance status (So et al. 2015).

2.2 Social Reward

In some species, social reward plays an important role in stable social groups by motivating initial social contact and later reinforcing this social contact. The neurobiology of reward has been studied extensively in the context of sexual behavior and appetitive behaviors such as drug addiction (reviewed in Beloate and Coolen 2017; Young et al. 2011). Reward has also been well studied in maternal behavior and social play in rodents, especially rats and mice (Trezza et al. 2011). These studies have focused on dopamine, opioid, and serotonin neurotransmission as targets for manipulation. Furthermore, mounting evidence suggests that reward mechanisms associated with social behavior – specifically, social behavior for which an individual is highly motivated, including parental behavior and pair bonding – are very similar to those associated with sexual behavior and drug addiction. Some have even suggested these kinds of social attachments are themselves addictive (Insel 2003; Young et al. 2011). From a fitness perspective, it is not difficult to understand why it would be beneficial for sexual behavior, parental behavior, and pair bonding to be highly rewarding and therefore highly motivating. Drugs of abuse co-opt conserved reward mechanisms to manipulate behavior.

Investigators seeking to illuminate the reward mechanisms associated with social behavior have likewise designed experiments that target the mesolimbic dopamine reward pathway. For example, activation of dopamine D2-type receptors – and concurrent interaction with OTR – is necessary for opposite-sex pair-bond formation in female prairie voles (Liu and Wang 2003; Wang et al. 1999). Specifically, nucleus accumbens dopamine is critical for pair bonding in both male and female prairie voles (Aragona et al. 2003, 2006; Gingrich et al. 2000; Liu and Wang 2003). D1-like activation in the rostral shell of the nucleus accumbens prevented pair-bond formation in male prairie voles, whereas D2-like activation facilitated it (Aragona et al. 2006). Furthermore, upregulation of D1-type receptors in the nucleus accumbens is associated with pair-bond maintenance. These dopamine-manipulated prairie voles show differences in partner preference but not in the number of mating bouts, distinguishing between sexual and social behavior. Consistent with this finding, prairie and meadow voles show similar increases in extracellular dopamine in the striatum after mating (Curtis et al. 2003). Thus, the necessity of dopamine in prairie vole pair-bond formation is specific to the partner preference, and not to any effects on mating.

It is likely that the mesolimbic dopamine reward pathway's role in reinforcing certain social behaviors is conserved across vertebrates (Bruce and Braford 2009; O'Connell and Hofmann 2011a). It has also been suggested that significant overlap exists between the reward system and the social behavior network, and that these circuits were present even in early vertebrates (O'Connell and Hofmann 2011b). However, despite the importance of dopamine in prairie vole pair bonding, blocking it does not impair peer affiliation in female meadow voles (Beery and Zucker 2010) or prairie voles (Lee et al. unpublished data). Thus, dopamine appears to be more important for selective relationships with mates than with peers.

2.3 Social Approach/Avoidance

Many animals find novel stimuli, including novel social stimuli, fear- and anxiety-inducing. This fear and anxiety must be countered to shift from an initial avoidance response to social approach and thereby social behavior. For example, nulliparous rats are fearful of pups, avoiding them and sometimes acting aggressively toward them, sometimes to the point of infanticide (Fleming and Anderson 1987). The onset of maternal behavior involves changes in approach/avoidance, whereby a maternal rat has greater tendency to approach pup stimuli than to avoid it (Rosenblatt and Mayer 1995). The MPOA has been pinpointed as a significant region for this shift, depressing antagonistic neural systems related to avoidance behaviors and activating appetitive neural systems related to approach behaviors (Numan 2007). Furthermore, maternal memory may involve maternal experience-induced synaptic plasticity within relevant neural circuits, such that pup stimuli can more effectively activate these maternal circuits (Numan and Stolzenberg 2009). These authors suggest such synaptic plasticity may include downregulation of the female rat's withdrawal/avoidance system toward pups, so that pup stimuli are less likely to activate avoidance behavior. This approach-avoidance model applies to non-maternal social behavior as well (reviewed in O'Connell and Hofmann 2011a).

In female meadow voles, a similar shift from avoidance to approach behaviors occurs with seasonal changes in day length. In the summer – or under long day-length conditions in a laboratory setting – meadow voles are aggressive and territorial (Madison 1980; Madison and McShea 1987; McShea and Madison 1984). In the winter – under short day-length conditions – they become more tolerant of conspecifics and live in communal groups. Day-length dependent changes in social behavior in the laboratory are concomitant with changes in brain and peripheral hormone circulation that may facilitate this behavioral shift. For example, both OTR and CRF receptors (1 and 2) vary with day length and with individual huddling in female meadow voles (Beery and Zucker 2010; Beery et al. 2014). Glucocorticoid secretion also varies seasonally, and stress and glucocorticoid exposure alter the formation of both same- and opposite-sex partner preferences in voles (Anacker et al. 2016b; DeVries et al. 1996; Pyter et al. 2005).

Social approach can be measured in multiple ways: latency to approach a conspecific, amount of social contact with a conspecific, and social preference (preference for a social stimulus over a nonsocial stimulus). Social approach is not equivalent to affiliative behavior; it may refer only to the tolerance of an individual for a conspecific, or the amount an individual investigates and interacts with a conspecific. Sociability tests in rats and mice have found that oxytocin facilitates social approach and prevents social avoidance in these animals (reviewed in Lukas and Neumann 2013). Although vasopressin – specifically, AVPR1a antagonist – did not have a clear effect on social approach (Lukas et al. 2011), AVPR1b antagonist reduced social avoidance in mice after social defeat (Litvin et al. 2011).

Social experience modulates the balance between approach and avoidance. For example, social contact itself may decrease fear and anxiety (reviewed in Hostinar et al. 2014). Rats, prairie voles, guinea pigs, mice, California mice, and Siberian hamsters all show social buffering, whereby animals exhibit modulated stress responses with social interaction (reviewed in Beery and Kaufer 2015). Unsurprisingly, oxytocin and opioids have been implicated in social buffering (reviewed in Kikusui et al. 2006). For example, social buffering in prairie voles is oxytocin-dependent (Burkett et al. 2016) – specifically, it involves oxytocin in the paraventricular nucleus (PVN) of the hypothalamus (Smith and Wang 2014). Conversely, social isolation in highly social animals may cause fear and anxiety. Disruption of both peer and mate relationships in prairie voles has been used as a model for affective disorders such as depression and anxiety. Social isolation induces behavioral, cardiac, autonomic, and neuroendocrine changes relevant to anxiety and depression in male and female prairie voles (Grippo et al. 2007a, b, c, 2008). Pair-bond disruption in prairie voles causes depression-related behaviors and an increase in adrenocorticotropic hormone (ACTH) and corticosterone (McNeal et al. 2014). Changes in CRF are also associated with passive stress-coping after pair-bond disruption (Bosch et al. 2009), and isolation in juvenile prairie voles (Ruscio et al. 2007). These studies illustrate the intimate relationship between social approach, social avoidance, and the initiation or loss of a social bond.

3 Studying Rodent Sociality Across Social Structures

The nature of sociality varies widely across rodents, such that group-living species may differ in group size, composition (e.g., kin, nonrelatives, subadults, adult peers, mates), and the role of specific, selective relationships. Many laboratory studies focus on the mechanisms supporting individual social behaviors and processes. In moving from specific behaviors to social structure, rodent groups with inter- and intraspecific variation in sociality are particularly useful for comparison across different social structures. For example, the African mole-rat family is the only mammalian clade to contain eusocial species. Naked mole-rats are the most highly specialized of these species, studied in the lab for mechanisms underlying their social behavior, with a focus on effects of sex and social status (Hathaway et al. 2016; Mooney et al. 2014, 2015; Rosen et al. 2007), as well as for comparison with solitary Cape mole-rats (Coen et al. 2015; Kalamatianos et al. 2010). Similarly, South American tuco-tucos are members of a genus consisting of social and nonsocial species, studied in the lab for oxytocin and vasopressin receptor binding differences (Beery et al. 2008a). Examination of additional species with diverse social behaviors will aid our understanding of the specificity and generality of mechanisms underlying different aspects of life in groups (see Table 1 for examples of some promising species). The majority of research on mechanisms supporting peer sociality is in voles, which provide an excellent opportunity to examine inter- and intra-specific variation in social behavior.

3.1 Synthesizing Social Processes to Understand Selective Relationships in Voles

Prairie voles and meadow voles are closely related but behaviorally disparate species that have been well studied in both the field and lab. By identifying which social behavior processes are important in specific contexts, such as prairie vole pair bonds and meadow vole peer relationships, we can predict important mechanistic differences in prairie and meadow vole social behavior.

Socially monogamous prairie voles are an increasingly popular rodent model of social behavior, and research on this species has informed our understanding of oxytocin and its importance in selective social bonds. Oxytocin, vasopressin, and dopamine are important molecules in the formation and maintenance of pair bonds in both males and females (Aragona et al. 2003; Cho et al. 1999; Wang et al. 1999). Specifically, oxytocin has been found to be particularly important for females (Insel and Hulihan 1995; Williams et al. 1992, 1994), and vasopressin for males (Donaldson et al. 2010; Lim and Young 2004; Liu et al. 2001; Winslow et al. 1993). Prairie voles have also been studied for their biparental behavior (Thomas and Birney 1979) and selective same-sex social bonds (Beery et al. 2018; DeVries et al. 1997; Lee et al. in review).

Sexually promiscuous meadow voles have been studied as contrasts to socially monogamous prairie voles, with some studies assessing whether specific manipulations cause meadow voles to behave more like prairie voles. For example, meadow vole males were successfully manipulated to show more partner preference for an opposite-sex partner using viral vector *V1aR* gene transfer into the ventral forebrain

(Lim et al. 2004). However, a similar experiment using viral vector *OXTR* gene transfer into the nucleus accumbens of females failed to enhance partner preference for opposite-sex partners (Ross et al. 2009). Meadow voles have also been studied for their ability to form selective same-sex relationships (e.g., Beery et al. 2008b; Ondrasek et al. 2015; Parker and Lee 2003). Perhaps unsurprisingly, oxytocin mediates prairie and meadow vole social behavior in different ways. While prairie vole mate bonds rely on oxytocin, oxytocin is not necessary for meadow vole peer relationships, although oxytocin can both enhance and eliminate peer partner preferences, acting in different brain regions (Anacker et al. 2016a; Beery and Zucker 2010). This and other differences in mechanism may be predicted by identifying the social behavior processes relevant to specific complex social behaviors and social systems. For example, the relative importance of social recognition, reward, and social approach shift according to context. Pair bonding, parental care, and group living all provide variations on context that lead to variations in mechanism.

In reproductive pair bonds, prairie voles strongly exhibit all social behaviors and processes discussed previously. Social recognition, as well as social investigation and memory, are essential to pair-bond formation (Young et al. 2005). Social recognition and decreased fear and anxiety are necessary for voles to approach potential mates and form a pair bond. Paired voles show partner preference even after prolonged separation, indicating long-term social memory. Following pair-bond formation, both sexes display increased aggression toward conspecifics of either sex (Carter and Getz 1993). Aggression, especially inter-male aggression, allows prairie voles to maintain the exclusivity of their pair bond by mate guarding and defending their territory, and is also important for parental care via defense of young.

Meadow vole peer relationships share several behavioral elements in common with reproductive pair-bonds: they are also selective, enduring, and depend on long-term social memory. In contrast to prairie vole pair bonds, meadow vole peer relationships do not appear to be dependent on dopamine signaling (Beery and Zucker 2010) and do not result in aggression toward extra-pair conspecifics. It is possible that meadow vole peer relationships involve increased tolerance toward all conspecifics rather than increased affiliation toward individuals, and that these bonds are neither as highly motivating nor as highly selective as those of prairie voles. Ongoing work is investigating the role of reward in prairie vole peer relationships, to elucidate whether the mechanisms of prairie vole pair-bond formation and maintenance are specific to the species (prairie vole) or specific to the behavior (pair bonding). Prairie voles can be socially conditioned to show place preference for mates but not long-term cagemates (Goodwin et al. 2018), and preliminary data suggests that prairie voles exhibit greater lever pressing for mates compared to peers (Lopez et al., Beery Lab, unpublished data). It seems likely that prairie vole mate bonds and meadow vole peer relationships are so different because the former are reproductive in nature and the latter are not. Indeed, preliminary data suggests that prairie vole peer relationships, like meadow vole peer relationships, do not rely on the dopaminergic reward pathway (Lee et al., Beery Lab, unpublished data).

This example of two closely related species exhibiting different social behaviors and underlying mechanisms highlights the advantage of the comparative perspective. Understanding affiliation will require study of both peer and mate relationships in multiple species.

4 Sociality in Free-Living Rodents

Rodents are the most diverse, numerous, and widespread of all mammals (Wolff and Sherman 2007). Sociality is widespread in rodents, with at least 70 documented social species in 39 genera (Lacey and Sherman 2007). Social rodents are found in a vast range of environments and exhibit distinct social behaviors, even within the same genus. These features make them particularly valuable models for comparative work. Rodents are also favorite lab subjects, due to their generally small size, fast life history, ease of attainability, and availability. Thus, rodents are ideal subjects for understanding social behavior mechanisms in the context of ecological relevance and phylogenetic relatedness. Table 1 summarizes the social behaviors of some of the most socially distinctive rodents in the wild. For example, beavers, prairie voles, and California mice all provide an opportunity to study social monogamy, a rare behavior among mammals. Naked mole-rats and social tuco-tucos are examples of opportunities to study sociality in direct comparison to closely related nonsocial species. Although some of these social rodents have been used in laboratory studies of affiliative behavior, the mechanisms of social behavior for many more remain to be elucidated.

The most studied rodents in the lab – rats and lab mice – are not often studied in their natural environments. Special strains of rats and mice have been bred to study specific behaviors or physiological characteristics, and these specific strains do not occur naturally in the wild. While some rodent researchers use wild-caught animals or animals bred from wild-caught as their subjects, or outbreed with wild-caught animals to maintain genetic diversity, these studies are not the norm (but see: research on wild rats, Ibe et al. 2010; Ruan and Zhang 2016). However, inbred strains of mice and rats have allowed for the development of genetic tools in these animals before any others (e.g., knock-out mice have been used in research for many years). Furthermore, a variety of behavioral tests have been standardized in laboratory rodents, including tests of anxiety-like and depressive-like behavior (open field test, light-dark box test, elevated plus maze, forced swim test), spatial memory (Morris water maze), aggression (resident-intruder test), and affiliation (partner preference test). Although these laboratory techniques have allowed for incredible depth in mechanistic research, breadth is also necessary to inform translatability of mechanistic findings. Thus, much may be gained by increasing efforts to research the neurobiology of species with field data, and to research species with neurobiology data in the field.

Table 1 Natural (field) behavior of a small sampling of well-studied rodents that exhibit diverse and interesting social behaviors

Genus/ species	Group size/composition	Mating system/social relationships	Home range/dispersal	Other features	References
Naked mole-rats	• Dominant breeding female and up to a few breeding males • Closely related conspecifics that help care for young, forage, maintain nest/tunnels, defend against predators • Specific caste/body size associated with each worker role • Live in large colonies of 200–300 individuals, with a mean group size of ~80 • Litters of up to 28 offspring	• Eusociality • Monogamy/polyandry • Highly inbred • Nonbreeding adults reproductively suppressed • Reproductive suppression reversible and maintained by aggression/threats from dominant female	• Successful dispersal infrequent • Subordinate adults usually remain in natal nest	• Arguably most social of all rodents • Related species with varying degrees of sociality	Faulkes and Bennett (2007), Jarvis (1981), Lacey and Sherman (1991), Nowak (1999), and Sherman et al. (1991)
Black-tailed prairie dogs	• Live in large colonies (towns) made up of coteries (family groups) • Coteries generally consist of 1–2 adult males, 1–6 adult females, and offspring • One colony may contain 15–26 coteries	• Polygyny • Little or no multiple mating by females, even if other males available • Communal nesting with close female relatives • Mothers nurse communally after juveniles from different litters begin to share burrows	• Delay in dispersal and reproduction compared to more solitary prairie dog species • Coterie territory boundaries well defined and defended • Females generally remain in natal coterie • Males disperse to other coteries within the same colony • Males and females sometimes disperse outside of colony	• Known for alarm calls, specific to species and approach of predators	Foltz and Hoogland (1981), Hoogland (1982, 1983, 1985, 2007, 2013), Michener and Murie (1983), and Nowak (1999)
Beavers	• Pair-bonded male and female breeding pair and their offspring • Older family members may live with family group	• Social and sexual monogamy • Biparental care • Long-lasting pair bonds • Degree of social interaction inversely related to age	• Territory defense • Territory marking (scent mounds) • Older family members may be juveniles delaying dispersal	• Construct lodges and dams from mud and sticks	Busher (2007), Godin (1977), Jenkins and Busher (1979), Nowak (1999), and Wilsson (1971)

Belding's ground squirrels	• Female and her offspring	• Nondefense polygyny • Multiple paternity of litters common (promiscuity rather than true polygyny) • Dominance hierarchies determine male reproductive success • No mate guarding (clear first male advantage) • Mothers and daughters cooperate to defend against infanticide • Antipredator warning calls – females display nepotistic behavior	• All males disperse from natal nest • Mothers and daughters set up territories near each other	• Most comprehensive data on rodent sociality exists for ground-dwelling squirrels (which include prairie dogs, marmots, chipmunks, and ground squirrels) • Kin/social recognition important • Long social memory	Hanken and Sherman (1981), Hare and Murie (2007), Mateo (2010), Nowak (1999), Sherman (1981, 1985), and Sherman and Morton (1984)
Prairie voles	• Male–female pair and their offspring • Shift toward communal groups in fall/winter, but less pronounced change than in meadow voles	• Social but not sexual monogamy • Biparental care • Mate-guarding by males • Alloparental care by older siblings • Exhibits specific preferences for known peers and mates	• Male–female pairs have overlapping home ranges that they defend	• Most well studied of vole species, and extensively studied in the lab	Carter and Getz (1993), Getz et al. (1993), Getz (1962, 1972), Getz and Carter (1996), Ophir et al. (2007), Solomon and Jacquot (2002), and Thomas and Birney (1979)
Meadow voles	• Breeding female dyads in spring • Solitary in summer • Maternal family group, and some adult and subadult males in fall • Mixed-lineage social group in winter	• Promiscuous • Intolerant toward conspecifics during breeding season • Nests in groups and shares territories during nonbreeding season • Exhibits specific preferences for known peers	• Females defend territories during breeding season • Home ranges overlap increasingly through fall/winter • Males disperse from natal nest	• Seasonal sociality	Boonstra et al. (1993), Dark et al. (1983), Ferkin (1988), Madison (1980), Madison et al. (1984), Madison and McShea (1987), McShea and Madison (1984), and Turner et al. (1983)

(continued)

Table 1 (continued)

Genus/ species	Group size/composition	Mating system/social relationships	Home range/dispersal	Other features	References
Mice (Lab)	• Large hierarchical group • Territorial groups (demes) founded by one male and multiple females	• Social hierarchy • Polygyny • Interactions between males relatively rare and usually aggressive	• Less complex burrows than rats (often contain a single cavity occupied by a single male) • Males territorial • Highly variable home ranges (and population densities) based on food availability and other related factors	• One of the most well-studied lab animals • Many genetic techniques exist for mice that do not exist for most (if not all) other species • Spontaneously form stable dominance hierarchies in the lab • Environmental richness is an important factor in determining features of their group living	Bronson (1979), Lidicker (1976), Lloyd (1975), MacKintosh (1973), Nowak (1999), Poole and Morgan (1973, 1976), Schmid-Holmes et al. (2001), and Zegeren (1979)
Rats (Lab)	• Large hierarchical group (colony)	• Social hierarchy, but less absolute than in mice • Polygynandry, with all males mating but dominant male mating the most	• Less territorial than mice • Highly variable home ranges (and population densities) based on food availability, etc.	• One of the most well-studied lab animals • Ideal for studying mechanisms of mothering due to well-defined suite of maternal behaviors • Also studied for their gregariousness and pro-sociality • Environmental richness is an important factor in determining features of their group living	Berdoy and Drickamer (2007), Calhoun (1948, 1962), MacDonald et al. (1999), McClintock and Anisko (1982), McClintock et al. (1982), and Nowak (1999)
California mice	• Small, semi-permanent family groups	• Social and sexual monogamy • Biparental care • Fathers display all parental behaviors beside nursing as much as mothers	• Mated pairs have largely overlapping home ranges	• Studied for their paternal care and winner effect	Bedford and Hoekstra (2015), Gubernick and Alberts (1987), Gubernick and Nordby (1993), Nowak (1999), Ribble (1992), and Ribble and Salvioni (1990)

Social tuco-tuco	• Social species in a clade with many solitary and a few other social species for comparisons • Variation in group size and composition • Up to 6 closely related adult females, at most one immigrant adult male, and dependent young	• All females reproduce • Communal nesting • Male and females contribute to care of young	• 2/3 of females remain in natal burrows • All males disperse after becoming adults and after each breeding season	• Allonursing observed in lab • Cooperative behaviors: digging, alarm calling	Lacey (2004), Lacey et al. (1997, 1998), Lacey and Wieczorek (2004), and Tammone et al. (2012)
Degus	• Colonial • 2–5 females and dependent young • Number of adult males unclear (single or multiple)	• All females reproduce • Communal nesting • Male(s) and females contribute to care of young	• High turnover rate of group members • No sex bias in dispersal	• Allonursing observed in lab • Cooperative behaviors: digging, alarm calling	Ardiles et al. (2013), Colonnello et al. (2011), Ebensperger et al. (2004, 2011a, b), and Fulk (1976)
Striped mice	• Facultative sociality • Social populations: Up to 30 adult individuals (2–4 breeding females, 1 breeding male, adult offspring)	• Social populations: Communal breeding, philopatric young • Solitary populations: Males and females only interact to mate	• Social populations: Territorial social groups, males mostly patrol borders • Solitary populations: Females inhabit exclusively female home ranges and disperse as juveniles; males inhabit home ranges that overlap with many females	• Social flexibility allows for intraspecific comparison between social and solitary states • Reproductive competition in group-living individuals may lead to dispersal and solitary-living • High population density may lead to group-living • Fitness of each tactic dependent on environment • Captive animals show biparental care	Schoepf and Schradin (2012), Schradin (2005, 2006), Schradin et al. (2010, 2012)

5 New Techniques and Future Directions

The studies discussed above use a variety of neural techniques including lesions and stimulation of different brain regions, pharmacological manipulations, and neurochemical measurement. Additional new methods promise both to enhance the specificity of control of particular circuits and to bring genetic manipulations and measurements to species beyond the laboratory mouse. This will help increase the diversity of laboratory models.

5.1 Genetic Techniques

Until recently, lab mice dominated genetic manipulation experiments, with transgenic mice employed extensively to study the roles of specific genes, receptors, and other players. Since the birth of genetic technology, the genomes of many more species have been sequenced, and transgenic prairie voles (Donaldson et al. 2009), zebra finches (Agate et al. 2009), sticklebacks (Kingsley et al. 2004), and marmosets (Kishi et al. 2014; Miller et al. 2016) now exist. CRISPR/Cas9 is one of many new genetic manipulation techniques that is fine-scale, reversible, and can be performed in vivo (in mice, Swiech et al. 2015). It has recently been successfully applied to rhesus macaques (Kang et al. 2015; reviewed in Luo et al. 2016; Niu et al. 2014; Wan et al. 2015). This technique is also being developed in prairie and meadow voles, as well as many other species. Zinc finger nucleases (ZFN) and transcription activator-like effector nucleases (TALENs) can also now be used to make precise, targeted genome modifications in model species that include zebrafish, rats, and mice (Gaj et al. 2013).

However, most organisms still lack fully sequenced genomes. Transcriptome analysis provides a high-throughput, low-cost means of genomic sequencing by which refined differences or changes in gene expression are identified. Because this next-generation sequencing (e.g., RNA-seq) can be used in animals with few to no genomic resources, it is useful for studying non-model systems (Blumstein et al. 2010; Ekblom and Galindo 2011; Ockendon et al. 2016; Taborsky et al. 2015). De novo transcriptome assembly allows for the identification of novel transcripts and does not require a reference genome. In fact, many new techniques, including optogenetics and chemogenetics, can be applied in both traditional and nontraditional systems and are equally effective in each. Optogenetics and chemogenetics, like CRISPR/Cas9, can activate and suppress neural action at a very fine scale; they are also reversible and can be performed in vivo.

Despite the recent advances in technology that can be applied to non-model species, tools currently available primarily in mice provide a level of specificity and refinement that cannot yet be rivaled. Combined with techniques that can be used in both model and non-model species, such as optogenetics and chemogenetics, these methods have much to contribute to understanding the mechanisms underlying social behavior (e.g., Beloate et al. 2016; Burgos-Robles et al. 2017; McHenry et al. 2017).

As new techniques become available for a larger number of species, they should be adopted when possible. We are only now emerging from a bottleneck created by the beginnings of genetic technology, which had been a "genetic revolution" that consisted of sequences and tools for only a few species (Brenowitz and Zakon 2015). Now that genetic techniques are available for more than just a few species, more comparative, mechanistic work will be made possible.

5.2 *New Conceptual Directions*

The idea that no single neurochemical or brain region controls a behavior is now well accepted. For social behaviors, a complex, interconnected circuit is thought to play an important role in diverse social behaviors: the so-called social behavior network (Newman 1999) or social decision-making network (O'Connell and Hofmann 2012). A growing movement has recently called for an integrative and comparative approach, arguing for the importance of both an ultimate and proximate perspective, both field and lab work, and the study of non-model organisms across a wide range of taxa (Blumstein et al. 2010; Hofmann et al. 2014; Rubenstein and Hofmann 2015; Taborsky et al. 2015). Field work alone does not uncover mechanism, and lab work alone does not consider the context of an animal's behavior – its functional significance in a specific ecological environment, and its evolutionary history. Furthermore, there is such incredible diversity in behavior and the mechanisms that underlie it, even within a single genus, that both basic biological processes and the human brain and disease can only hope to be understood by comparative, wide-ranging research.

While the availability of new technology has proven essential to dissecting functional circuits, it has sometimes promoted a focus on technique rather than question. Many researchers argue neuroscientists have forgotten brains belong to behaving animals – that "neuroscience needs behavior" (Krakauer et al. 2017) and that "nothing in neuroscience makes sense except in light of behavior" (Hofmann et al. 2016). Researchers may find themselves inundated with vast amounts of genetic data, but it is important to return to context and life history (Kültz et al. 2013).

Ultimately, we must not seek to understand a specific behavior in a specific species alone, but to understand the social brain, how it functions, and how it evolved.

References

Agate RJ, Scott BB, Haripal B, Lois C, Nottebohm F (2009) Transgenic songbirds offer an opportunity to develop a genetic model for vocal learning. Proc Natl Acad Sci 106 (42):17963–17967

Anacker AMJ, Beery AK (2013) Life in groups: the roles of oxytocin in mammalian sociality. Front Behav Neurosci 7. https://doi.org/10.3389/fnbeh.2013.00185

Anacker AMJ, Christensen JD, LaFlamme EM, Grunberg DM, Beery AK (2016a) Septal oxytocin administration impairs peer affiliation via V1a receptors in female meadow voles. Psychoneuroendocrinology 68:156–162. https://doi.org/10.1016/j.psyneuen.2016.02.025

Anacker AMJ, Reitz KM, Goodwin NL, Beery AK (2016b) Stress impairs new but not established relationships in seasonally social voles. Horm Behav 79:52–57. https://doi.org/10.1016/j.yhbeh.2016.01.004

Aragona BJ, Liu Y, Curtis JT, Stephan FK, Wang Z (2003) A critical role for nucleus accumbens dopamine in partner-preference formation in male prairie voles. J Neurosci 23(8):3483–3490

Aragona BJ, Liu Y, Yu YJ, Curtis JT, Detwiler JM, Insel TR, Wang Z (2006) Nucleus accumbens dopamine differentially mediates the formation and maintenance of monogamous pair bonds. Nat Neurosci 9(1):133–139. https://doi.org/10.1038/nn1613

Ardiles A, Ewer J, Acosta ML, Kirkwood A, Martinez A, Ebensperger L et al (2013) *Octodon degus* (Molina 1782): a model in comparative biology and biomedicine. Cold Spring Harb Protoc 2013(4):312–318. https://doi.org/10.1101/pdb.emo071357

Bedford NL, Hoekstra HE (2015) *Peromyscus* mice as a model for studying natural variation. eLife 4:e06813. https://doi.org/10.7554/eLife.06813

Beery AK, Kaufer D (2015) Stress, social behavior, and resilience: insights from rodents. Neurobiol Stress 1:116–127. https://doi.org/10.1016/j.ynstr.2014.10.004

Beery AK, Zucker I (2010) Oxytocin and same-sex social behavior in female meadow voles. Neuroscience 169(2):665–673. https://doi.org/10.1016/j.neuroscience.2010.05.023

Beery AK, Zucker I (2011) Sex bias in neuroscience and biomedical research. Neurosci Biobehav Rev 35(3):565–572. https://doi.org/10.1016/j.neubiorev.2010.07.002

Beery AK, Lacey EA, Francis DD (2008a) Oxytocin and vasopressin receptor distributions in a solitary and a social species of tuco-tuco (*Ctenomys haigi* and *Ctenomys sociabilis*). J Comp Neurol 507(6):1847–1859. https://doi.org/10.1002/cne.21638

Beery AK, Loo TJ, Zucker I (2008b) Day length and estradiol affect same-sex affiliative behavior in the female meadow vole. Horm Behav 54(1):153–159. https://doi.org/10.1016/j.yhbeh.2008.02.007

Beery AK, Vahaba DM, Grunberg DM (2014) Corticotropin-releasing factor receptor densities vary with photoperiod and sociality. Horm Behav 66:779–786. https://doi.org/10.1016/j.yhbeh.2014.08.014

Beery AK, Kamal Y, Sobrero R, Hayes LD (2016) Comparative neurobiology and genetics of mammalian social behavior. In: Sociobiology of caviomorph rodents: an integrated view. Wiley Blackwell, Hoboken

Beery AK, Christensen JD, Lee NS, Blandino KL (2018) Specificity in sociality: mice and prairie voles exhibit different patterns of peer affiliation. Front Behav Neurosci 12. https://doi.org/10.3389/fnbeh.2018.00050

Beloate LN, Coolen LM (2017) Influences of social reward experience on behavioral responses to drugs of abuse: review of shared and divergent neural plasticity mechanisms for sexual reward and drugs of abuse. Neurosci Biobehav Rev 83:356–372. https://doi.org/10.1016/j.neubiorev.2017.10.024

Beloate LN, Omrani A, Adan RA, Webb IC, Coolen LM (2016) Ventral tegmental area dopamine cell activation during male rat sexual behavior regulates neuroplasticity and d-amphetamine cross-sensitization following sex abstinence. J Neurosci 36(38):9949–9961. https://doi.org/10.1523/JNEUROSCI.0937-16.2016

Berdoy M, Drickamer LC (2007) Comparative social organization and life history of *Rattus* and *Mus*. In: Wolff JO, Sherman PW (eds) Rodent societies: an ecological & evolutionary perspective. The University of Chicago Press, Chicago

Bergan JF, Ben-Shaul Y, Dulac C (2014) Sex-specific processing of social cues in the medial amygdala. eLife 3. https://doi.org/10.7554/eLife.02743

Bielsky IF, Young LJ (2004) Oxytocin, vasopressin, and social recognition in mammals. Peptides 25(9):1565–1574. https://doi.org/10.1016/j.peptides.2004.05.019
Bielsky IF, Hu S-B, Szegda KL, Westphal H, Young LJ (2004) Profound impairment in social recognition and reduction in anxiety-like behavior in vasopressin V1a receptor knockout mice. Neuropsychopharmacology 29(3):483–493. https://doi.org/10.1038/sj.npp.1300360
Bielsky IF, Hu S-B, Ren X, Terwilliger EF, Young LJ (2005) The V1a vasopressin receptor is necessary and sufficient for normal social recognition: a gene replacement study. Neuron 47 (4):503–513. https://doi.org/10.1016/j.neuron.2005.06.031
Blumstein DT, Ebensperger LA, Hayes LD, Vásquez RA, Ahern TH, Burger JR et al (2010) Toward an integrative understanding of social behavior: new models and new opportunities. Front Behav Neurosci 4. https://doi.org/10.3389/fnbeh.2010.00034
Bluthe R-M, Schoenen J, Dantzer R (1990) Androgen-dependent vasopressinergic neurons are involved in social recognition in rats. Brain Res 519(1–2):150–157. https://doi.org/10.1016/0006-8993(90)90073-K
Boonstra R, Xia X, Pavone L (1993) Mating system of the meadow vole, *Microtus pennsylvanicus*. Behav Ecol 4(1):83–89. https://doi.org/10.1093/beheco/4.1.83
Bosch OJ, Nair HP, Ahern TH, Neumann ID, Young LJ (2009) The CRF system mediates increased passive stress-coping behavior following the loss of a bonded partner in a monogamous rodent. Neuropsychopharmacology 34(6):1406–1415. https://doi.org/10.1038/npp.2008.154
Brenowitz EA, Zakon HH (2015) Emerging from the bottleneck: benefits of the comparative approach to modern neuroscience. Trends Neurosci 38(5):273–278. https://doi.org/10.1016/j.tins.2015.02.008
Broadbent NJ, Squire LR, Clark RE (2004) Spatial memory, recognition memory, and the hippocampus. Proc Natl Acad Sci 101(40):14515–14520. https://doi.org/10.1073/pnas.0406344101
Bronson FH (1979) The reproductive ecology of the house mouse. Q Rev Biol 54(3):265–299
Brown MW, Aggleton JP (2001) Recognition memory: what are the roles of the perirhinal cortex and hippocampus? Nat Rev Neurosci 2(1):51–61. https://doi.org/10.1038/35049064
Bruce LL, Braford MR (2009) Evolution of the limbic system. In: Squire LR (ed) Encyclopedia of neuroscience, vol 4. Academic Press, Oxford
Burgos-Robles A, Kimchi EY, Izadmehr EM, Porzenheim MJ, Ramos-Guasp WA, Nieh EH et al (2017) Amygdala inputs to prefrontal cortex guide behavior amid conflicting cues of reward and punishment. Nat Neurosci 20(6):824–835. https://doi.org/10.1038/nn.4553
Burkett JP, Andari E, Johnson ZV, Curry DC, de Waal FBM, Young LJ (2016) Oxytocin-dependent consolation behavior in rodents. Science 351(6271):375–378. https://doi.org/10.1126/science.aac4785
Busher P (2007) Social organization and monogamy in the beaver. In: Wolff JO, Sherman PW (eds) Rodent societies: an ecological & evolutionary perspective. The University of Chicago Press, Chicago
Calhoun JB (1948) Mortality and movement of brown rats (*Rattus norvegicus*) in artificially supersaturated populations. J Wildl Manag 12(2):167–172. https://doi.org/10.2307/3796412
Calhoun JB (1962) The ecology and sociology of the Norway rat (US Public Health Service Publication no. 1008). US Government Printing Office, Washington
Carter CS, Getz LL (1993) Monogamy and the prairie vole. Sci Am 268(6):100–106
Chalfin L, Dayan M, Levy DR, Austad SN, Miller RA, Iraqi FA et al (2014) Mapping ecologically relevant social behaviours by gene knockout in wild mice. Nat Commun 5:4569. https://doi.org/10.1038/ncomms5569
Cho MM, Courtney A, Williams JR, Carter CS (1999) The effects of oxytocin and vasopressin on partner preferences in male and female prairie voles (*Microtus ochrogaster*). Behav Neurosci 113(5):1071–1079. https://doi.org/10.1037/0735-7044.113.5.1071
Clipperton-Allen AE, Lee AW, Reyes A, Devidze N, Phan A, Pfaff DW, Choleris E (2012) Oxytocin, vasopressin and estrogen receptor gene expression in relation to social recognition in female mice. Physiol Behav 105(4):915–924. https://doi.org/10.1016/j.physbeh.2011.10.025

Clutton-Brock TH, Lukas D (2012) The evolution of social philopatry and dispersal in female mammals. Mol Ecol 21(3):472–492. https://doi.org/10.1111/j.1365-294X.2011.05232.x

Coen CW, Kalamatianos T, Oosthuizen MK, Poorun R, Faulkes CG, Bennett NC (2015) Sociality and the telencephalic distribution of corticotrophin-releasing factor, urocortin 3, and binding sites for CRF type 1 and type 2 receptors: a comparative study of eusocial naked mole-rats and solitary Cape mole-rats. J Comp Neurol 523(16):2344–2371. https://doi.org/10.1002/cne.23796

Colonnello V, Iacobucci P, Fuchs T, Newberry RC, Panksepp J (2011) *Octodon degus*. A useful animal model for social-affective neuroscience research: basic description of separation distress, social attachments and play. Neurosci Biobehav Rev 35(9):1854–1863. https://doi.org/10.1016/j.neubiorev.2011.03.014

Colpaert FC (1975) The ventromedial hypothalamus and the control of avoidance behavior and aggression: fear hypothesis versus response-suppression theory of limbic system function. Behav Biol 15(1):27–44. https://doi.org/10.1016/S0091-6773(75)92053-2

Curtis J, Stowe J, Wang Z (2003) Differential effects of intraspecific interactions on the striatal dopamine system in social and non-social voles. Neuroscience 118(4):1165–1173. https://doi.org/10.1016/S0306-4522(03)00032-0

Dark J, Zucker I, Wade GN (1983) Photoperiodic regulation of body mass, food intake, and reproduction in meadow voles. Am J Phys Regul Integr Comp Phys 245(3):R334–R338

DeVries AC, Carter CS (1999) Sex differences in temporal parameters of partner preference in prairie voles *(Microtus ochrogaster)*. Can J Zool 77(6):885–889. https://doi.org/10.1139/z99-054

DeVries AC, DeVries MB, Taymans SE, Carter CS (1996) The effects of stress on social preferences are sexually dimorphic in prairie voles. Proc Natl Acad Sci 93:11980–11984

DeVries AC, Johnson CL, Carter CS (1997) Familiarity and gender influence social preferences in prairie voles (*Microtus ochrogaster*). Can J Zool 75(2):295–301. https://doi.org/10.1139/z97-037

Dluzen DE, Muraoka S, Engelmann M, Ebner K, Landgraf R (2000) Oxytocin induces preservation of social recognition in male rats by activating α-adrenoceptors of the olfactory bulb. Eur J Neurosci 12(2):760–766. https://doi.org/10.1046/j.1460-9568.2000.00952.x

Donaldson ZR, Yang S-H, Chan AWS, Young LJ (2009) Production of germline transgenic prairie voles (*Microtus ochrogaster*) using lentiviral vectors. Biol Reprod 81(6):1189–1195. https://doi.org/10.1095/biolreprod.109.077529

Donaldson ZR, Spiegel L, Young LJ (2010) Central vasopressin V1a receptor activation is independently necessary for both partner preference formation and expression in socially monogamous male prairie voles. Behav Neurosci 124(1):159–163. https://doi.org/10.1037/a0018094

Ebensperger LA, Hurtado MJ, Soto-Gamboa M, Lacey EA, Chang AT (2004) Communal nesting and kinship in degus (*Octodon degus*). Naturwissenschaften 91(8):391–395. https://doi.org/10.1007/s00114-004-0545-5

Ebensperger LA, Chesh AS, Castro RA, Tolhuysen LO, Quirici V, Burger JR et al (2011a) Burrow limitations and group living in the communally rearing rodent, *Octodon degus*. J Mammal 92(1):21–30. https://doi.org/10.1644/09-MAMM-S-383.1

Ebensperger LA, Ramírez-Estrada J, León C, Castro RA, Tolhuysen LO, Sobrero R et al (2011b) Sociality, glucocorticoids and direct fitness in the communally rearing rodent, *Octodon degus*. Horm Behav 60(4):346–352. https://doi.org/10.1016/j.yhbeh.2011.07.002

Ekblom R, Galindo J (2011) Applications of next generation sequencing in molecular ecology of non-model organisms. Heredity 107(1):1–15. https://doi.org/10.1038/hdy.2010.152

Everts HG, Koolhaas J (1999) Differential modulation of lateral septal vasopressin receptor blockade in spatial learning, social recognition, and anxiety-related behaviors in rats. Behav Brain Res 99(1):7–16. https://doi.org/10.1016/S0166-4328(98)00004-7

Faulkes CG, Bennett NC (2007) African mole-rats: social and ecological diversity. In: Wolff JO, Sherman PW (eds) Rodent societies: an ecological & evolutionary perspective. The University of Chicago Press, Chicago

Ferguson JN, Young LJ, Hearn EF, Matzuk MM, Insel TR, Winslow JT (2000) Social amnesia in mice lacking the oxytocin gene. Nat Genet 25(3):284–288. https://doi.org/10.1038/77040

Ferguson JN, Aldag JM, Insel TR, Young LJ (2001) Oxytocin in the medial amygdala is essential for social recognition in the mouse. J Neurosci 21(20):8278–8285

Ferguson JN, Young LJ, Insel TR (2002) The neuroendocrine basis of social recognition. Front Neuroendocrinol 23(2):200–224. https://doi.org/10.1006/frne.2002.0229

Ferkin MH (1988) The effect of familiarity on social interactions in meadow voles, *Microtus pennsylvanicus*: a laboratory and field study. Anim Behav 36(6):1816–1822. https://doi.org/10.1016/S0003-3472(88)80121-0

Fleming AS, Anderson V (1987) Affect and nurturance: mechanisms mediating maternal behavior in two female mammals. Prog Neuro-Psychopharmacol Biol Psychiatry 11(2–3):121–127. https://doi.org/10.1016/0278-5846(87)90049-2

Foltz DW, Hoogland JL (1981) Analysis of the mating system in the black-tailed prairie dog (*Cynomys ludovicianus*) by likelihood of paternity. J Mammal 62(4):706–712. https://doi.org/10.2307/1380592

Fulk GW (1976) Notes on the activity, reproduction, and social behavior of *Octodon degus*. J Mammal 57(3):495–505. https://doi.org/10.2307/1379298

Gaj T, Gersbach CA, Barbas CF (2013) ZFN, TALEN and CRISPR/Cas-based methods for genome engineering. Trends Biotechnol 31(7):397–405. https://doi.org/10.1016/j.tibtech.2013.04.004

Getz LL (1962) Aggressive behavior of the meadow and prairie voles. J Mammal 43(3):351–358. https://doi.org/10.2307/1376942

Getz LL (1972) Social structure and aggressive behavior in a population of *Microtus pennsylvanicus*. J Mammal 53(2):310–317. https://doi.org/10.2307/1379167

Getz LL, Carter CS (1996) Prairie-vole partnerships. Am Sci 84(1):56–62

Getz LL, McGuire B, Pizzuto T, Hofmann JE, Frase B (1993) Social organization of the prairie vole (*Microtus ochrogaster*). J Mammal 74(1):44–58. https://doi.org/10.2307/1381904

Gheusi G, Bluthé R-M, Goodall G, Dantzer R (1994) Social and individual recognition in rodents: methodological aspects and neurobiological bases. Behav Process 33(1–2):59–87

Gingrich B, Liu Y, Cascio C, Wang Z, Insel TR (2000) Dopamine D2 receptors in the nucleus accumbens are important for social attachment in female prairie voles (*Microtus ochrogaster*). Behav Neurosci 114(1):173–183. https://doi.org/10.1037//0735-7044.114.1.173

Godin AJ (1977) Wild mammals of New England. Johns Hopkins University Press, Baltimore

Goodwin NL, Lopez SA, Lee NS, Beery AK (2018) Comparative role of reward in long-term peer and mate relationships in voles. Horm Behav. https://doi.org/10.1016/j.yhbeh.2018.10.012

Grippo AJ, Cushing BS, Carter CS (2007a) Depression-like behavior and stressor-induced neuroendocrine activation in female prairie voles exposed to chronic social isolation. Psychosom Med 69(2):149–157. https://doi.org/10.1097/PSY.0b013e31802f054b

Grippo AJ, Gerena D, Huang J, Kumar N, Shah M, Ughreja R, Carter CS (2007b) Social isolation induces behavioral and neuroendocrine disturbances relevant to depression in female and male prairie voles. Psychoneuroendocrinology 32(8–10):966–980. https://doi.org/10.1016/j.psyneuen.2007.07.004

Grippo AJ, Lamb DG, Carter CS, Porges SW (2007c) Social isolation disrupts autonomic regulation of the heart and influences negative affective behaviors. Biol Psychiatry 62(10):1162–1170. https://doi.org/10.1016/j.biopsych.2007.04.011

Grippo AJ, Wu KD, Hassan I, Carter CS (2008) Social isolation in prairie voles induces behaviors relevant to negative affect: toward the development of a rodent model focused on co-occurring depression and anxiety. Depress Anxiety 25(6):E17–E26. https://doi.org/10.1002/da.20375

Grossman SP (1972) Aggression, avoidance, and reaction to novel environments in female rats with ventromedial hypothalamic lesions. J Comp Physiol Psychol 78(2):274–283. https://doi.org/10.1037/h0032284

Gubernick DJ, Alberts JR (1987) The biparental care system of the California mouse, *Peromyscus californicus*. J Comp Psychol 101(2):169–177. https://doi.org/10.1037//0735-7036.101.2.169

Gubernick DJ, Nordby JC (1993) Mechanisms of sexual fidelity in the monogamous California mouse, *Peromyscus californicus*. Behav Ecol Sociobiol 32(3):211–219

Hanken J, Sherman PW (1981) Multiple paternity in Belding's ground squirrel litters. Science 212 (4492):351–353. https://doi.org/10.1126/science.7209536

Hare JF, Murie JO (2007) Ecology, kinship, and ground squirrel sociality: insights from comparative analyses. In: Wolff JO, Sherman PW (eds) Rodent societies: an ecological & evolutionary perspective. The University of Chicago Press, Chicago

Hathaway GA, Faykoo-Martinez M, Peragine DE, Mooney SJ, Holmes MM (2016) Subcaste differences in neural activation suggest a prosocial role for oxytocin in eusocial naked mole-rats. Horm Behav 79:1–7. https://doi.org/10.1016/j.yhbeh.2015.12.001

Hofmann HA, Beery AK, Blumstein DT, Couzin ID, Earley RL, Hayes LD et al (2014) An evolutionary framework for studying mechanisms of social behavior. Trends Ecol Evol 29 (10):581–589. https://doi.org/10.1016/j.tree.2014.07.008

Hofmann HA, Renn SCP, Rubenstein DR (2016) Introduction to symposium: new frontiers in the integrative study of animal behavior: nothing in neuroscience makes sense except in light of behavior. Integr Comp Biol 56(6):1192–1196. https://doi.org/10.1093/icb/icw127

Hölscher C, Jacob W, Mallot HA (2003) Reward modulates neuronal activity in the hippocampus of the rat. Behav Brain Res 142(1):181–191. https://doi.org/10.1016/S0166-4328(02)00422-9

Hoogland JL (1982) Prairie dogs avoid extreme inbreeding. Science 215(4540):1639–1641. https://doi.org/10.1126/science.215.4540.1639

Hoogland JL (1983) Nepotism and alarm calling in the black-tailed prairie dog (*Cynomys ludovicianus*). Anim Behav 31(2):472–479

Hoogland JL (1985) Infanticide in prairie dogs: lactating females kill offspring of close kin. Science 230(4729):1037–1040. https://doi.org/10.1126/science.230.4729.1037

Hoogland JL (2007) Alarm calling, multiple mating, and infanticide among Black-tailed, Gunnison's, and Utah prairie dogs. In: Wolff JO, Sherman PW (eds) Rodent societies: an ecological & evolutionary perspective. The University of Chicago Press, Chicago

Hoogland JL (2013) Prairie dogs disperse when all close kin have disappeared. Science 339 (6124):1205–1207. https://doi.org/10.1126/science.1231689

Hostinar CE, Sullivan RM, Gunnar MR (2014) Psychobiological mechanisms underlying the social buffering of the HPA axis: a review of animal models and human studies across development. Psychol Bull 140(1):256

Ibe CS, Onyeanusi BI, Hambolu JO, Ayo JO (2010) Sexual dimorphism in the whole brain and brainstem morphometry in the African giant pouched rat (*Cricetomys gambianus*, Waterhouse 1840). Folia Morphol (Warsz) 69(2):69–74

Insel TR (2003) Is social attachment an addictive disorder? Physiol Behav 79(3):351–357. https://doi.org/10.1016/S0031-9384(03)00148-3

Insel TR, Hulihan TJ (1995) A gender-specific mechanism for pair bonding: oxytocin and partner preference formation in monogamous voles. Behav Neurosci 109(4):782–789. https://doi.org/10.1037/0735-7044.109.4.782

Jarvis JUM (1981) Eusociality in a mammal: cooperative breeding in naked mole-rat colonies. Science 212(4494):571–573

Jenkins SH, Busher PE (1979) Castor canadensis. Mamm Species 120:1–8

Johnson ZV, Young LJ (2018) Evolutionary diversity as a catalyst for biological discovery. Integr Zool 13:616. https://doi.org/10.1111/1749-4877.12339. Accepted Author Manuscript

Kang Y, Zheng B, Shen B, Chen Y, Wang L, Wang J et al (2015) CRISPR/Cas9-mediated *Dax1* knockout in the monkey recapitulates human AHC-HH. Hum Mol Genet 24(25):7255–7264. https://doi.org/10.1093/hmg/ddv425

Kalamatianos T, Faulkes CG, Oosthuizen MK, Poorun R, Bennett NC, Coen CW (2010) Telencephalic binding sites for oxytocin and social organization: a comparative study of eusocial naked mole-rats and solitary cape mole-rats. J Comp Neurol 518(10):1792–1813. https://doi.org/10.1002/cne.22302

Kikusui T, Winslow JT, Mori Y (2006) Social buffering: relief from stress and anxiety. Philos Trans Biol Sci 361(1476):2215–2228
Kingsley DM, Zhu B, Osoegawa K, De Jong PJ, Schein J, Marra M et al (2004) New genomic tools for molecular studies of evolutionary change in threespine sticklebacks. Behaviour 141 (11):1331–1344
Kishi N, Sato K, Sasaki E, Okano H (2014) Common marmoset as a new model animal for neuroscience research and genome editing technology. Develop Growth Differ 56(1):53–62. https://doi.org/10.1111/dgd.12109
Kogan JH, Frankland PW, Silva AJ (2000) Long-term memory underlying hippocampus-dependent social recognition in mice. Hippocampus 10(1):47–56
Krakauer JW, Ghazanfar AA, Gomez-Marin A, MacIver MA, Poeppel D (2017) Neuroscience needs behavior: correcting a reductionist bias. Neuron 93(3):480–490. https://doi.org/10.1016/j.neuron.2016.12.041
Kültz D, Clayton DF, Robinson GE, Albertson C, Carey HV, Cummings ME et al (2013) New frontiers for organismal biology. Bioscience 63(6):464–471. https://doi.org/10.1525/bio.2013.63.6.8
Lacey EA (2004) Sociality reduces individual direct fitness in a communally breeding rodent, the colonial tuco-tuco (*Ctenomys sociabilis*). Behav Ecol Sociobiol 56(5):449–457
Lacey EA, Sherman PW (1991) Social organization of naked mole-rat colonies: evidence for divisions of labor. In: The biology of the naked mole-rat, vol 275. Princeton University Press, Princeton, p 336
Lacey EA, Sherman PW (2007) The ecology of sociality in rodents. In: Wolff JO, Sherman PW (eds) Rodent societies: an ecological and evolutionary perspective. The University of Chicago Press, Chicago
Lacey EA, Wieczorek JR (2004) Kinship in colonial tuco-tucos: evidence from group composition and population structure. Behav Ecol 15(6):988–996. https://doi.org/10.1093/beheco/arh104
Lacey EA, Braude SH, Wieczorek JR (1997) Burrow sharing by colonial tuco-tucos (*Ctenomys sociabilis*). J Mammal 78(2):556–562
Lacey EA, Braude SH, Wieczorek JR (1998) Solitary burrow use by adult Patagonian tuco-tucos (*Ctenomys haigi*). J Mammal 79(3):986–991
Landgraf R, Gerstberger R, Montkowski A, Probst JC, Wotjak CT, Holsboer F, Engelmann M (1995) V1 vasopressin receptor antisense oligodeoxynucleotide into septum reduces vasopressin binding, social discrimination abilities, and anxiety-related behavior in rats. J Neurosci 15 (6):4250–4258
Landgraf R, Frank E, Aldag JM, Neumann ID, Sharer CA, Ren X et al (2003) Viral vector-mediated gene transfer of the vole V1a vasopressin receptor in the rat septum: improved social discrimination and active social behaviour. Eur J Neurosci 18(2):403–411. https://doi.org/10.1046/j.1460-9568.2003.02750.x
Lebow MA, Chen A (2016) Overshadowed by the amygdala: the bed nucleus of the stria terminalis emerges as key to psychiatric disorders. Mol Psychiatry 21(4):450–463. https://doi.org/10.1038/mp.2016.1
Lee PC (1994) Social structure and evolution. In: Slater PJB, Halliday TR (eds) Behaviour and evolution. Cambridge University Press, Cambridge
Li Y, Dulac C (2018) Neural coding of sex-specific social information in the mouse brain. Curr Opin Neurobiol 53:120–130. https://doi.org/10.1016/j.conb.2018.07.005
Li Y, Mathis A, Grewe BF, Osterhout JA, Ahanonu B, Schnitzer MJ et al (2017) Neuronal representation of social information in the medial amygdala of awake behaving mice. Cell 171(5):1176–1190.e17. https://doi.org/10.1016/j.cell.2017.10.015
Lidicker WZ (1976) Social behaviour and density regulation in house mice living in large enclosures. J Anim Ecol 45(3):677–697. https://doi.org/10.2307/3575

Lim M, Young L (2004) Vasopressin-dependent neural circuits underlying pair bond formation in the monogamous prairie vole. Neuroscience 125(1):35–45. https://doi.org/10.1016/j.neuroscience.2003.12.008

Lim MM, Wang Z, Olazábal DE, Ren X, Terwilliger EF, Young LJ (2004) Enhanced partner preference in a promiscuous species by manipulating the expression of a single gene. Nature 429 (6993):754–757. https://doi.org/10.1038/nature02539

Litvin Y, Murakami G, Pfaff DW (2011) Effects of chronic social defeat on behavioral and neural correlates of sociality: vasopressin, oxytocin and the vasopressinergic V1b receptor. Physiol Behav 103(3–4):393–403. https://doi.org/10.1016/j.physbeh.2011.03.007

Liu Y, Wang Z (2003) Nucleus accumbens oxytocin and dopamine interact to regulate pair bond formation in female prairie voles. Neuroscience 121(3):537–544. https://doi.org/10.1016/S0306-4522(03)00555-4

Liu Y, Curtis JT, Wang Z (2001) Vasopressin in the lateral septum regulates pair bond formation in male prairie voles (*Microtus ochrogaster*). Behav Neurosci 115(4):910–919. https://doi.org/10.1037//0735-7044.115.4.910

Lloyd JA (1975) Social structure and reproduction in two freely growing populations of house mice (*Mus musculus L.*). Anim Behav 23:413–424. https://doi.org/10.1016/0003-3472(75)90089-5

Lukas D, Clutton-Brock T (2012a) Cooperative breeding and monogamy in mammalian societies. Proc R Soc B Biol Sci 279(1736):2151–2156. https://doi.org/10.1098/rspb.2011.2468

Lukas D, Clutton-Brock T (2012b) Life histories and the evolution of cooperative breeding in mammals. Proc R Soc B Biol Sci 279(1744):4065–4070. https://doi.org/10.1098/rspb.2012.1433

Lukas D, Clutton-Brock TH (2013) The evolution of social monogamy in mammals. Science 341 (6145):526–530. https://doi.org/10.1126/science.1238677

Lukas M, Neumann ID (2013) Oxytocin and vasopressin in rodent behaviors related to social dysfunctions in autism spectrum disorders. Behav Brain Res 251:85–94. https://doi.org/10.1016/j.bbr.2012.08.011

Lukas M, Toth I, Reber SO, Slattery DA, Veenema AH, Neumann ID (2011) The neuropeptide oxytocin facilitates pro-social behavior and prevents social avoidance in rats and mice. Neuropsychopharmacology 36(11):2159–2168. https://doi.org/10.1038/npp.2011.95

Luo X, Li M, Su B (2016) Application of the genome editing tool CRISPR/Cas9 in non-human primates. Zool Res 37(4):241–219. https://doi.org/10.13918/j.issn.2095-8137.2016.4.214

MacDonald DW, Mathews F, Berdoy M (1999) The behaviour and ecology of *Rattus norvegicus*: from opportunism to kamikaze tendencies. In: Singleton GR, Hinds LA, Leirsand H, Zhang Z (eds) Ecologically-based management of rodent pests, ACIAR Monograph No. 59. Australian Centre for International Agricultural Research, Canberra, pp 49–80

MacKintosh JH (1973) Factors affecting the recognition of territory boundaries by mice (*Mus musculus*). Anim Behav 21(3):464–470. https://doi.org/10.1016/S0003-3472(73)80006-5

Madison DM (1980) Space use and social structure in meadow voles, *Microtus pennsylvanicus*. Behav Ecol Sociobiol 7(1):65–71

Madison DM, McShea WJ (1987) Seasonal changes in reproductive tolerance, spacing, and social organization in meadow voles: a microtine model. Am Zool 27(3):899–908. https://doi.org/10.1093/icb/27.3.899

Madison DM, FitzGerald RW, McShea WJ (1984) Dynamics of social nesting in overwintering meadow voles (*Microtus pennsylvanicus*): possible consequences for population cycling. Behav Ecol Sociobiol 15(1):9–17

Mateo JM (2010) Self-referent phenotype matching and long-term maintenance of kin recognition. Anim Behav 80(5):929–935. https://doi.org/10.1016/j.anbehav.2010.08.019

McClintock MK, Anisko JJ (1982) Group mating among Norway rats I. Sex differences in the pattern and neuroendocrine consequences of copulation. Anim Behav 30(2):398–409. https://doi.org/10.1016/S0003-3472(82)80051-1

McClintock MK, Anisko JJ, Adler NT (1982) Group mating among Norway rats II. The social dynamics of copulation: competition, cooperation, and mate choice. Anim Behav 30 (2):410–425. https://doi.org/10.1016/S0003-3472(82)80052-3

McHenry JA, Otis JM, Rossi MA, Robinson JE, Kosyk O, Miller NW et al (2017) Hormonal gain control of a medial preoptic area social reward circuit. Nat Neurosci 20(3):449–462

McNeal N, Scotti M-AL, Wardwell J, Chandler DL, Bates SL, LaRocca M et al (2014) Disruption of social bonds induces behavioral and physiological dysregulation in male and female prairie voles. Auton Neurosci 180:9–16. https://doi.org/10.1016/j.autneu.2013.10.001

McShea WJ, Madison DM (1984) Communal nesting between reproductively active females in a spring population of *Microtus pennsylvanicus*. Can J Zool 62(3):344–346

Michener GR, Murie JO (1983) Black-tailed prairie dog coteries: are they cooperatively breeding units? Am Nat 121(2):266–274

Miller CT, Freiwald WA, Leopold DA, Mitchell JF, Silva AC, Wang X (2016) Marmosets: a neuroscientific model of human social behavior. Neuron 90(2):219–233. https://doi.org/10.1016/j.neuron.2016.03.018

Mooney SJ, Douglas NR, Holmes MM (2014) Peripheral administration of oxytocin increases social affiliation in the naked mole-rat (*Heterocephalus glaber*). Horm Behav 65(4):380–385. https://doi.org/10.1016/j.yhbeh.2014.02.003

Mooney SJ, Coen CW, Holmes MM, Beery AK (2015) Region-specific associations between sex, social status, and oxytocin receptor density in the brains of eusocial rodents. Neuroscience 303:261–269. https://doi.org/10.1016/j.neuroscience.2015.06.043

Newman SW (1999) The medial extended amygdala in male reproductive behavior: a node in the mammalian social behavior network. Ann N Y Acad Sci 877(1):242–257. https://doi.org/10.1111/j.1749-6632.1999.tb09271.x

Niu Y, Shen B, Cui Y, Chen Y, Wang J, Wang L et al (2014) Generation of gene-modified *Cynomolgus* monkey via Cas9/RNA-mediated gene targeting in one-cell embryos. Cell 156 (4):836–843. https://doi.org/10.1016/j.cell.2014.01.027

Nowak RM (1999) Walker's mammals of the world, 6th edn. The Johns Hopkins University Press, Baltimore

Numan M (2007) Motivational systems and the neural circuitry of maternal behavior in the rat. Dev Psychobiol 49(1):12–21. https://doi.org/10.1002/dev.20198

Numan M, Stolzenberg DS (2009) Medial preoptic area interactions with dopamine neural systems in the control of the onset and maintenance of maternal behavior in rats. Front Neuroendocrinol 30(1):46–64. https://doi.org/10.1016/j.yfrne.2008.10.002

O'Connell LA, Hofmann HA (2011a) Genes, hormones, and circuits: an integrative approach to study the evolution of social behavior. Front Neuroendocrinol 32(3):320–335. https://doi.org/10.1016/j.yfrne.2010.12.004

O'Connell LA, Hofmann HA (2011b) The vertebrate mesolimbic reward system and social behavior network: a comparative synthesis. J Comp Neurol 519(18):3599–3639. https://doi.org/10.1002/cne.22735

O'Connell LA, Hofmann HA (2012) Evolution of a vertebrate social decision-making network. Science 336(6085):1154–1157. https://doi.org/10.1126/science.1218889

Ockendon NF, O'Connell LA, Bush SJ, Monzón-Sandoval J, Barnes H, Székely T et al (2016) Optimization of next-generation sequencing transcriptome annotation for species lacking sequenced genomes. Mol Ecol Resour 16(2):446–458. https://doi.org/10.1111/1755-0998.12465

Ondrasek NR, Wade A, Burkhard T, Hsu K, Nguyen T, Post J, Zucker I (2015) Environmental modulation of same-sex affiliative behavior in female meadow voles (*Microtus pennsylvanicus*). Physiol Behav 140:118–126. https://doi.org/10.1016/j.physbeh.2014.12.021

Ophir AG, Phelps SM, Sorin AB, Wolff JO (2007) Morphological, genetic, and behavioral comparisons of two prairie vole populations in the field and laboratory. J Mammal 88 (4):989–999. https://doi.org/10.1644/06-MAMM-A-250R.1

Ophir AG, Zheng D-J, Eans S, Phelps SM (2009) Social investigation in a memory task relates to natural variation in septal expression of oxytocin receptor and vasopressin receptor 1a in prairie voles (*Microtus ochrogaster*). Behav Neurosci 123(5):979

Parker KJ, Lee TM (2003) Female meadow voles (*Microtus pennsylvanicus*) demonstrate same-sex partner preferences. J Comp Psychol 117(3):283–289. https://doi.org/10.1037/0735-7036.117.3.283

Paxinos G, Franklin KB (2012) The mouse brain in stereotaxic coordinates. Elsevier Academic Press, Cambridge

Poole TB, Morgan HDR (1973) Differences in aggressive behaviour between male mice (*Mus musculus L.*) in colonies of different sizes. Anim Behav 21(4):788–795. https://doi.org/10.1016/S0003-3472(73)80105-8

Poole TB, Morgan HDR (1976) Social and territorial behaviour of laboratory mice (*Mus musculus L.*) in small complex areas. Anim Behav 24(2):476–480. https://doi.org/10.1016/S0003-3472(76)80056-5

Pyter LM, Weil ZM, Nelson RJ (2005) Latitude affects photoperiod-induced changes in immune response in meadow voles (*Microtus pennsylvanicus*). Can J Zool 83:1271–1278. https://doi.org/10.1139/z05-121

Resendez SL, Kuhnmuench M, Krzywosinski T, Aragona BJ (2012) κ-opioid receptors within the nucleus accumbens shell mediate pair bond maintenance. J Neurosci 32(20):6771–6784. https://doi.org/10.1523/JNEUROSCI.5779-11.2012

Ribble DO (1992) Dispersal in a monogamous rodent, *Peromyscus californicus*. Ecology 73 (3):859–866

Ribble DO, Salvioni M (1990) Social organization and nest co-occupancy in *Peromyscus californicus*, a monogamous rodent. Behav Ecol Sociobiol 26(1):9–15. https://doi.org/10.1007/BF00174020

Rosen GJ, Vries GJD, Goldman SL, Goldman BD, Forger NG (2007) Distribution of vasopressin in the brain of the eusocial naked mole-rat. J Comp Neurol 500(6):1093–1105. https://doi.org/10.1002/cne.21215

Rosenblatt JS, Mayer AD (1995) An analysis of approach/withdrawal processes in the initiation of maternal behavior in the laboratory rat. In: Hood KE, Greenberg G, Tobach E (eds) Research in developmental and comparative psychology, vol. 1. Behavioral development: concepts of approach/withdrawal and integrative levels. Garland Publishing, New York, pp 177–230

Ross HE, Freeman SM, Spiegel LL, Ren X, Terwilliger EF, Young LJ (2009) Variation in oxytocin receptor density in the nucleus accumbens has differential effects on affiliative behaviors in monogamous and polygamous voles. J Neurosci 29:1312–1318

Ruan C, Zhang Z (2016) Laboratory domestication changed the expression patterns of oxytocin and vasopressin in brains of rats and mice. Anat Sci Int 91(4):358–370. https://doi.org/10.1007/s12565-015-0311-0

Rubenstein DR, Hofmann HA (2015) Editorial overview: the integrative study of animal behavior. Curr Opin Behav Sci 6:v–viii. https://doi.org/10.1016/j.cobeha.2015.11.009

Ruscio MG, Sweeny T, Hazelton J, Suppatkul P, Carter CS (2007) Social environment regulates corticotropin releasing factor, corticosterone and vasopressin in juvenile prairie voles. Horm Behav 51(1):54–61. https://doi.org/10.1016/j.yhbeh.2006.08.004

Schmid-Holmes S, Drickamer LC, Robinson AS, Gillie LL (2001) Burrows and burrow-cleaning behavior of house mice (*Mus musculus domesticus*). Am Midl Nat 146(1):53–62

Schoepf I, Schradin C (2012) Better off alone! Reproductive competition and ecological constraints determine sociality in the African striped mouse (*Rhabdomys pumilio*). J Anim Ecol 81 (3):649–656. https://doi.org/10.1111/j.1365-2656.2011.01939.x

Schradin C (2005) When to live alone and when to live in groups: ecological determinants of sociality in the African striped mouse (*Rhabdomys pumilio*, Sparrman, 1784). Belg J Zool 135 (Suppl. 1):77–82

Schradin C (2006) Whole-day follows of striped mice (*Rhabdomys pumilio*), a diurnal murid rodent. J Ethol 24(1):37–43. https://doi.org/10.1007/s10164-005-0158-2

Schradin C, König B, Pillay N (2010) Reproductive competition favours solitary living while ecological constraints impose group-living in African striped mice. J Anim Ecol 79(3):515–521. https://doi.org/10.1111/j.1365-2656.2009.01651.x

Schradin C, Lindholm AK, Johannesen J, Schoepf I, Yuen C-H, König B, Pillay N (2012) Social flexibility and social evolution in mammals: a case study of the African striped mouse (*Rhabdomys pumilio*). Mol Ecol 21(3):541–553. https://doi.org/10.1111/j.1365-294X.2011.05256.x

Schweinfurth MK, Neuenschwander J, Engqvist L, Schneeberger K, Rentsch AK, Gygax M et al (2017) Do female Norway rats form social bonds? Behav Ecol Sociobiol 71:98

Sherman PW (1981) Kinship, demography, and Belding's ground squirrel nepotism. Behav Ecol Sociobiol 8(4):251–259

Sherman PW (1985) Alarm calls of Belding's ground squirrels to aerial predators: nepotism or self-preservation? Behav Ecol Sociobiol 17(4):313–323

Sherman PW, Morton ML (1984) Demography of Belding's ground squirrels. Ecology 65 (5):1617–1628. https://doi.org/10.2307/1939140

Sherman PW, Jarvis JUM, Alexander RD (eds) (1991) The biology of the naked mole-rat. Princeton University Press, Princeton

Smith AS, Wang Z (2014) Hypothalamic oxytocin mediates social buffering of the stress response. Biol Psychiatry 76(4):281–288. https://doi.org/10.1016/j.biopsych.2013.09.017

Smith KS, Tindell AJ, Aldridge JW, Berridge KC (2009) Ventral pallidum roles in reward and motivation. Behav Brain Res 196(2):155–167. https://doi.org/10.1016/j.bbr.2008.09.038

So N, Franks B, Lim S, Curley JP (2015) A social network approach reveals associations between mouse social dominance and brain gene expression. PLoS One 10(7):e0134509. https://doi.org/10.1371/journal.pone.0134509

Solomon NG, Jacquot JJ (2002) Characteristics of resident and wandering prairie voles, *Microtus ochrogaster*. Can J Zool 80(5):951–955. https://doi.org/10.1139/z02-053

Sun P, Smith AS, Lei K, Liu Y, Wang Z (2014) Breaking bonds in male prairie vole: long-term effects on emotional and social behavior, physiology, and neurochemistry. Behav Brain Res 265:22–31. https://doi.org/10.1016/j.bbr.2014.02.016

Swiech L, Heidenreich M, Banerjee A, Habib N, Li Y, Trombetta J et al (2015) In vivo interrogation of gene function in the mammalian brain using CRISPR-Cas9. Nat Biotechnol 33(1):102–106. https://doi.org/10.1038/nbt.3055

Taborsky M, Hofmann HA, Beery AK, Blumstein DT, Hayes LD, Lacey EA et al (2015) Taxon matters: promoting integrative studies of social behavior. Trends Neurosci 38(4):189–191. https://doi.org/10.1016/j.tins.2015.01.004

Tamashiro KLK, Nguyen MMN, Sakai RR (2005) Social stress: from rodents to primates. Front Neuroendocrinol 26(1):27–40. https://doi.org/10.1016/j.yfrne.2005.03.001

Tammone MN, Lacey EA, Relva MA (2012) Habitat use by colonial tuco-tucos (*Ctenomys sociabilis*): specialization, variation, and sociality. J Mammal 93(6):1409–1419. https://doi.org/10.1644/11-MAMM-A-266.1

Thomas JA, Birney EC (1979) Parental care and mating system of the prairie vole, *Microtus ochrogaster*. Behav Ecol Sociobiol 5(2):171–186

Timmer M, Sandi C (2010) A role for glucocorticoids in the long-term establishment of a social hierarchy. Psychoneuroendocrinology 35(10):1543–1552. https://doi.org/10.1016/j.psyneuen.2010.05.011

Timmer M, Cordero MI, Sevelinges Y, Sandi C (2011) Evidence for a role of oxytocin receptors in the long-term establishment of dominance hierarchies. Neuropsychopharmacology 36 (11):2349–2356. https://doi.org/10.1038/npp.2011.125

Trezza V, Campolongo P, Vanderschuren LJMJ (2011) Evaluating the rewarding nature of social interactions in laboratory animals. Dev Cogn Neurosci 1(4):444–458. https://doi.org/10.1016/j.dcn.2011.05.007

Turner BN, Iverson SL, Severson KL (1983) Seasonal changes in open-field behavior in wild male meadow voles (*Microtus pennsylvanicus*). Behav Neural Biol 39(1):60–77. https://doi.org/10.1016/S0163-1047(83)90637-4

van der Kooij MA, Sandi C (2012) Social memories in rodents: methods, mechanisms and modulation by stress. Neurosci Biobehav Rev 36(7):1763–1772. https://doi.org/10.1016/j.neubiorev.2011.10.006

Walker DL, Toufexis DJ, Davis M (2003) Role of the bed nucleus of the stria terminalis versus the amygdala in fear, stress, and anxiety. Eur J Pharmacol 463(1):199–216. https://doi.org/10.1016/S0014-2999(03)01282-2

Wan H, Feng C, Teng F, Yang S, Hu B, Niu Y et al (2015) One-step generation of p53 gene biallelic mutant *Cynomolgus* monkey via the CRISPR/Cas system. Cell Res 25(2):258

Wang Z, Yu G, Cascio C, Liu Y, Gingrich B, Insel TR (1999) Dopamine D2 receptor-mediated regulation of partner preferences in female prairie voles (*Microtus ochrogaster*): a mechanism for pair bonding? Behav Neurosci 113(3):602

Williams JR, Carter CS, Insel T (1992) Partner preference development in female prairie voles is facilitated by mating or the central infusion of oxytocin. Ann N Y Acad Sci 652(1):487–489. https://doi.org/10.1111/j.1749-6632.1992.tb34393.x

Williams JR, Insel TR, Harbaugh CR, Carter CS (1994) Oxytocin administered centrally facilitates formation of a partner preference in female prairie voles (*Microtus ochrogaster*). J Neuroendocrinol 6 (3):247–250. https://doi.org/10.1111/j.1365-2826.1994.tb00579.x

Williamson CM, Romeo RD, Curley JP (2017) Dynamic changes in social dominance and mPOA GnRH expression in male mice following social opportunity. Horm Behav 87:80–88. https://doi.org/10.1016/j.yhbeh.2016.11.001

Wilsson L (1971) Observations and experiments on the ethology of the European beaver (*Castor fiber L.*). Viltrevy 8(3):115–266

Winslow JT, Insel TR (2004) Neuroendocrine basis of social recognition. Curr Opin Neurobiol 14 (2):248–253. https://doi.org/10.1016/j.conb.2004.03.009

Winslow JT, Hastings N, Carter CS, Harbaugh CR, Insel TR (1993) A role for central vasopressin in pair bonding in monogamous prairie voles. Nature 365(6446):545–548

Wolff JO, Sherman PW (2007) Rodent societies: an ecological and evolutionary perspective. The University of Chicago Press, Chicago

Yao S, Bergan J, Lanjuin A, Dulac C (2017) Oxytocin signaling in the medial amygdala is required for sex discrimination of social cues. eLife 6:e31373. https://doi.org/10.7554/eLife.31373

Young LJ, Murphy Young AZ, Hammock EAD (2005) Anatomy and neurochemistry of the pair bond. J Comp Neurol 493(1):51–57. https://doi.org/10.1002/cne.20771

Young KA, Gobrogge KL, Wang Z (2011) The role of mesocorticolimbic dopamine in regulating interactions between drugs of abuse and social behavior. Neurosci Biobehav Rev 35(3):498–515. https://doi.org/10.1016/j.neubiorev.2010.06.004

Zegeren KV (1979) Variation in aggressiveness and the regulation of numbers in house mouse populations. Neth J Zool 30(4):635–770. https://doi.org/10.1163/002829679X00241

Hypothalamic Integration of the Endocrine Signaling Related to Food Intake

Anica Klockars, Allen S. Levine, and Pawel K. Olszewski

Contents

1 Introduction 240
2 Hypothalamic Regulation of Food Intake: Key Integrative Processes 241
2.1 Peripheral Input into the Hypothalamus 241
2.2 Hypothalamic Circuit 248
3 Failure to Control Appetite in an Obesogenic Environment: Is Neuronal Plasticity the Culprit? 252
3.1 Peripheral Signals: A Brief Overview 252
3.2 Neuronal Plasticity in the Hypothalamus: A Path Forward in Deciphering Causative Factors Underlying Hyperphagia? 254
4 Concluding Remarks 260
References 260

Abstract Hypothalamic integration of gastrointestinal and adipose tissue-derived hormones serves as a key element of neuroendocrine control of food intake. Leptin, adiponectin, oleoylethanolamide, cholecystokinin, and ghrelin, to name a few, are in a constant "cross talk" with the feeding-related brain circuits that encompass hypothalamic populations synthesizing anorexigens (melanocortins, CART, oxytocin) and orexigens (Agouti-related protein, neuropeptide Y, orexins). While this integrated neuroendocrine circuit successfully ensures that enough energy is acquired, it does not seem to be equally efficient in preventing excessive energy intake, especially in the obesogenic environment in which highly caloric and palatable food is constantly available. The current review presents an overview of intricate

A. Klockars
Department of Biological Sciences, University of Waikato, Hamilton, New Zealand

A. S. Levine
Department of Food Science and Nutrition, University of Minnesota, Saint Paul, MN, USA

P. K. Olszewski (✉)
Department of Biological Sciences, University of Waikato, Hamilton, New Zealand

Department of Food Science and Nutrition, University of Minnesota, Saint Paul, MN, USA
e-mail: pawel@waikato.ac.nz

Curr Topics Behav Neurosci (2019) 43: 239–270
DOI 10.1007/7854_2018_54
Published Online: 10 June 2018

mechanisms underlying hypothalamic integration of energy balance-related peripheral endocrine input. We discuss vulnerabilities and maladaptive neuroregulatory processes, including changes in hypothalamic neuronal plasticity that propel overeating despite negative consequences.

Keywords Brain · Hunger · Hypothalamus · Obesity · Plasticity · Satiety

1 Introduction

All living forms need energy to survive and maintain their basic functions; animals acquire it with food. Throughout the evolutionary history, energy sources have not been overabundant but rather a scarce "commodity." Consequently, the successful genotypes have been those that facilitate rapid acquisition of safe (i.e., nontoxic) energy, maximize the efficiency of energy utilization, and promote its storage. Genes that underpin such processes would be well conserved. It is not surprising therefore that the evolutionary predecessors of many molecular systems that govern energy balance in mammals, such as melanocortins, arose early in the vertebrate evolution (Scherag et al. 2010; Schioth et al. 2005).

The intra- and intercellular signaling molecules that indicate the need to obtain energy facilitate a complex set of physiological and behavioral responses that allow the animal to search for, acquire, and process high-energy foods. The state in which these responses occur has been dubbed, somewhat in a simplistic manner, as "hunger." As clearly evidenced by the obesity "epidemic," "hunger" is a much broader phenomenon than a mere reflection of balancing the energy intake-expenditure equation. In fact, a drive to eat or to discontinue consumption stems from a complex set of processes and mechanisms that include environmental cues and endocrine, metabolic, homeostatic, and reward-related signals (Olszewski and Levine 2007; Alsio et al. 2012).

Paradoxically, constant overabundance of highly caloric and palatable food (alone or in combination with additional environmental cues) appears to propel consumption, oftentimes to the level that jeopardizes internal milieu (Olszewski and Levine 2007). It is counterintuitive considering the existence of seemingly effective peripheral hormonal mechanisms originating in the gut and in the adipose tissue, as well as the fact that the hypothalamus appears to be able to integrate peripheral signaling and promote satiety-driven termination of consumption. Therefore, in the current review, not only do we delineate intricate mechanisms that underpin hypothalamic integration of energy balance-related peripheral endocrine input, but we also ponder on whether the hypothalamus is in fact molecularly and functionally ill equipped to curb consumption and keep it at safe levels when palatable and highly caloric food is overabundant in an obesogenic environment.

2 Hypothalamic Regulation of Food Intake: Key Integrative Processes

The early approach to characterizing how the hypothalamus controls energy balance was focused on examining whether discrete hypothalamic nuclei play unique roles in feeding. Already in the 1960s, it was determined that ablation of the ventromedial hypothalamic nucleus (VMH) leads to excessive food intake and body weight, whereas lateral hypothalamic (LH) lesions cause hypophagia and weight loss in rodents (Mayer and Thomas 1967). Early postnatal treatment with monosodium glutamate in mice was found to disrupt the development of the hypothalamus, particularly its arcuate nucleus (ARC), as well as the proper development of multiple facets of endocrine functioning; not surprisingly, those animals exhibited dysregulation of energy balance and obesity (Olney 1969). In subsequent studies, several research groups found that also ablation of the paraventricular nucleus of the hypothalamus (PVN) as well as disruption of the hypothalamic-brain stem pathways promoted excessive appetite and weight gain in laboratory animals (Leibowitz et al. 1981; Sims and Lorden 1986; Shor-Posner et al. 1985; Kirchgessner and Sclafani 1988).

As a consequence of those findings, specific hypothalamic nuclei were often referred to as hunger or satiety "centers." However, as the ability to identify the molecular cytoarchitecture of the brain progressed, and so did the capacity to measure and modify site-specific gene expression and pharmacologically target select subtypes of receptors, it became apparent that such broad definition of the roles of specific hypothalamic areas is a gross oversimplification. Instead, it was proposed that feeding is shaped by the combination of neural and chemical signals integrated by the network of hypothalamic regions that communicate with each other, with extrahypothalamic areas and (directly or indirectly) with peripheral tissues.

2.1 Peripheral Input into the Hypothalamus

2.1.1 Integration of Peripheral Endocrine Signaling: CNS Entry Points

The basis of effective food intake control is the ability of the circulating signaling molecules to reach the hypothalamus. The entry of relevant endocrine system-mediated information into specific subregions of the hypothalamus is facilitated by discrete mechanisms. Those molecules that penetrate the BBB can act directly at their CNS receptors, thereby being able to affect multiple hypothalamic and extrahypothalamic sites. On the other hand, BBB non-penetrant factors have to rely on the entry point in those areas where the BBB is weak, especially the median eminence and – in the case of activating wider networks that incorporate the hypothalamus as a downstream component – the area postrema (AP) in the brain stem (Morita et al. 2016; Miyata 2015). Alternatively, the vagus nerve serves as an additional path through which periphery-CNS communication can be achieved (Schwartz 2010). Cell bodies of afferent fibers of the vagal abdominal branch are located in the nodose ganglia, and they project onto the brain stem. Therein, the dorsal vagal complex, which encompasses the nucleus of the solitary tract (NTS), the dorsal motor nucleus of the vagus (DMNV), and AP, contains cells that innervate hypothalamic and other CNS regions, and this innervation is reciprocal (Schwartz 2010). Vagal afferent brain stem neurons express a variety of receptors for adipose tissue and GI tract hormones, including leptin and insulin receptors (Burdyga et al. 2002; Ellacott et al. 2006; Blake and Smith 2012), Y2 receptor (Koda et al. 2005), GLP1/2 receptors (Nakagawa et al. 2004; Nelson et al. 2007), CCK1/2 receptors (Moriarty et al. 1997), and GHS receptor (Date 2012) (Fig. 1).

It should be noted that endocrine-based signaling received by the CNS is intertwined with information conveyed by vagal afferents (e.g., stemming from mechanoreceptor stimulation due to gastric distension) and by circumventricular organs (integrating, e.g., osmolality/toxicity status). In some circumstances, this complexity of signals can promote a synergistic effect, strengthening hormone-mediated neural and feeding behavioral responses; in others, these independent signals interfere with each other (Schwartz 2000, 2010; Schwartz et al. 1999).

2.1.2 Adiposity-Relevant Signals

The hypothalamus and the peripheral endocrine system are in a constant "cross talk" relevant to appetite control. One of the most spectacular discoveries showing that the peripheral tissues affect hypothalamic regulatory processes related to energy balance was the identification of the adipocyte hormone, leptin (Lep), a product of the ob gene (Friedman and Halaas 1998) (it should be noted that since then, also non-adipose tissues have been found to contribute to the circulating leptin pool, as the ob gene is expressed, e.g., in the gastrointestinal (GI) tract and placenta (Sobhani et al. 2000; Challier et al. 2003)).

The amount of secreted leptin parallels adiposity (Murphy and Bloom 2004). Friedman's group was the first to report that Lep-null mice were deficient in leptin due to the loss-of-function mutation (Friedman and Halaas 1998; Zhang et al. 1994). On the other hand, db-deficient animals synthesized this satiety peptide in excess but were resistant to it, and this effect was due to mutations in the gene encoding a class 1 cytokine receptor specific for leptin (Lepr) (Farooqi et al. 2007; Maffei et al. 1995). Administration of leptin rescues Lep-null mice from the obese/hyperphagic phenotype (Friedman and Halaas 1998). Peripheral and central injections of the endogenous agonist of the Lepr reduce food intake in normal animals (Friedman and Halaas 1998; Ahima et al. 1996). Starvation of wild-type rats and mice suppresses Lep mRNA expression, and this change in the expression profile is reversed by refeeding or by administration of insulin (Frederich et al. 1995). Not surprisingly, changes within the Lep or Lepr gene have been associated with obesity in many populations in humans (Farooqi et al. 2007; Mazen et al. 2011; Andiran et al. 2011). Importantly,

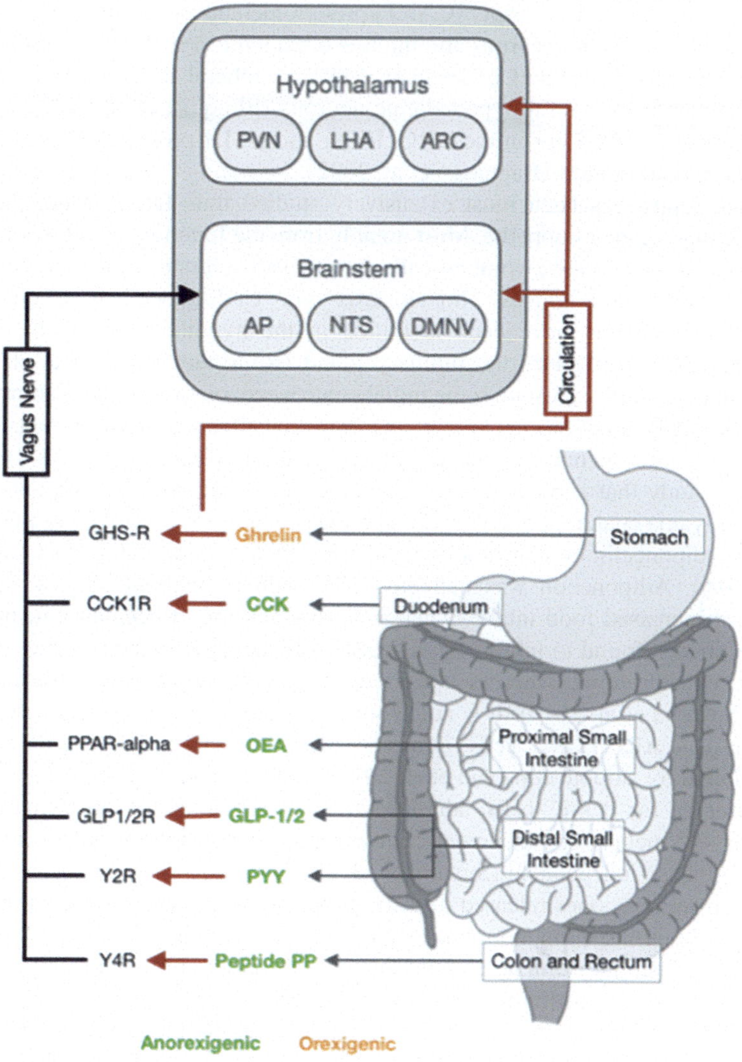

Fig. 1 Schematic diagram of gut signal secretion

obesity caused by congenital leptin deficiency can be treated by recombinant leptin injections.

It has been established that leptin can cross the blood-brain barrier (BBB), and it engages the JAK2-STAT3 and PI3K-PDE3B-cAMP pathways in target neurons relevant to energy balance control (Mori et al. 2004). The Lepr is highly expressed in the hypothalamus, especially in the ARC, VMN, DMH, and LH. In the ARC, the Lepr can be found on neurons synthesizing orexigens, neuropeptide Y (NPY)/ Agouti-related protein (AgRP), as well as anorexigens, and cocaine- and amphetamine-related transcript (CART)/proopiomelanocortin (POMC). Leptin directly activates POMC neurons and inhibits orexigenic AgRP/NPY neurons leading to a decrease in appetite (Cowley et al. 2001). Peripheral administration of leptin affects activation of ventral hypothalamic neurons (Elmquist et al. 1997). Finally, overexpression of the Lepr in the ARC, PVN, and VMH promotes hypophagia in transgenic rodent models (Bagnasco et al. 2002).

Though leptin has been most extensively studied thus far, it is not the only adipokine that regulates appetite. Most notably from the functional (and, potentially, therapeutic) point of view, white fat cells secrete also adiponectin which influences energy balance and glucose homeostasis via CNS-dependent mechanisms. Adiponectin receptors have been identified in the hypothalamus, including the PVN and ARC, as well as the pituitary gland (Rodriguez-Pacheco et al. 2007; Wilkinson et al. 2007). Despite some initial controversy on the ability of adiponectin to cross the BBB, more recent experiments show that after peripheral administration, the peptide levels increase in the cerebrospinal fluid (Kubota et al. 2007). In an interesting study that explained some discrepancies in the effect of adiponectin on feeding, Suyama et al. reported that glucose levels determine excitatory or inhibitory effects of adiponectin on consumption and ARC POMC neuronal activity (Suyama et al. 2016). Adiponectin was delivered ICV with or without glucose: at high glucose, it increased food intake, whereas at low glucose, it decreased it. In vitro, adiponectin was found to inhibit activity of POMC neurons in the presence of high glucose concentrations, whereas it caused an opposite effect when glucose was absent (Suyama et al. 2016). It is noteworthy that leptin and adiponectin synergistically activate ARC POMC cells (Sun et al. 2016). Activity of NPY/AgRP neurons was inhibited by adiponectin (Sun et al. 2016), and – importantly – unlike what has been observed for POMC cells, adiponectin enhanced inhibitory postsynaptic current onto NPY neurons to attenuate action potential firing in these cells in a manner that was glucose independent (Suyama et al. 2017).

Aside from hormones released directly by the adipose tissue, also insulin levels serve as a brain-targeting hormonal signal relevant to adiposity. Human studies revealed that fasting insulin levels correlate with the amount of white adipose tissue (Rocha et al. 2011). Insulin receptor knockout (KO) experiments have shown that female KOs exhibit hyperphagia, and both sexes develop diet-induced obesity with increases in leptin and insulin secretion and insulin resistance (Bruning et al. 2000). Insulin receptors are highly expressed in various hypothalamic nuclei, including the PVN, ARC, and DMH (Marks et al. 1990). Intraventricular and hypothalamic site-specific administration of this hormone causes early termination of food intake in rodents subjected to deprivation as well as in non-deprived animals during the nighttime feeding (McGowan et al. 1992a, b). One mechanism through which insulin reduces appetite is by engaging the ARC POMC pathway. Benoit and colleagues reported that third ventricular insulin infusions in fasted rats reduced food intake and this effect was abolished by subthreshold doses of the melanocortin receptor antagonist, SHU-9119. ICV insulin also increased ARC POMC mRNA levels (Benoit et al. 2002). Obici et al. pointed to the importance of the functional relationship with NPY (Obici et al. 2002). They injected ICV antisense oligodeoxynucleotide against the insulin receptor precursor in rats and found the rapid onset of hyperphagia and obesity, as well as upregulation of ARC NPY.

2.1.3 Gut Signaling

While adiposity-related hormonal signals match consummatory responses to the general energy status of the organism, thus they are considered to be long-term feeding regulatory factors, the GI tract exerts endocrine control that aligns with transient eating behavioral and nutritional changes that reflect a more contemporaneous meal regimen.

Cholecystokinin

Cholecystokinin (CCK), the first identified GI hormone synthesized primarily in the jejunum and duodenum, is released into the blood within minutes after food enters the small intestine (Gibbs et al. 1973; Buffa et al. 1976; Liddle et al. 1985). This endogenous ligand of CCK1 and CCK2 receptors (Dufresne et al. 2006) decreases food intake in a dose-dependent manner when injected peripherally and centrally, by affecting meal size (West et al. 1984; Kissileff et al. 1981)as well as by inhibiting gastric emptying (though the latter effect has not been shown to be accompanied by body weight changes in human trials) (Cummings and Overduin 2007; Castillo et al. 2004; Jordan et al. 2008). Circulating CCK binds the CCK1 receptor on brain stem-targeting vagal afferents (Noetzel et al. 2009). The brain stem, including its nucleus of the solitary tract (NTS) and dorsal motor nucleus of the vagus (DMNV), relays this information into other feeding-related forebrain areas, including the hypothalamic PVN and DMH as evidenced by studies utilizing immediate early gene mapping (Noetzel et al. 2009; Kobelt et al. 2006; Monnikes et al. 1997). Otsuka Long-Evans Tokushima Fatty (OLETF) rats lacking the CCK1 receptor show excessive food intake and obesity (Covasa and Ritter 2001; Moran et al. 1998). Already at the pre-obese stage, young OLETF rats express much higher levels of NPY mRNA in the DMH (Moran and Bi 2006). CCK injections augment appetite and body weight-lowering effects of adipokines in diet-induced obesity animal models, most likely by strengthening adipokine action at the caudal brain stem (Trevaskis et al. 2010; Matson et al. 2002).

Preproglucagon-Cleaved Hormone Molecules

Posttranslational processing of the preproglucagon gene product gives rise to glucagon-like peptides 1/2 (GLP-1/2) and oxyntomodulin. Enteroendocrine L cells of the intestine release these hormones into the general circulation in a biphasic manner between 10 and 60 min after meal ingestion (Holst 2007). Interestingly, the amount secreted into the blood parallels the amount of ingested food (Dakin et al.

2004; Cohen et al. 2003). Both peripheral and central administration of these hormones reduces food intake and has a beneficial effect on body weight (Cohen et al. 2003; Dakin et al. 2001). Unlike GLP, oxyntomodulin's receptor target is unclear as, although it does bind the GLP-1 receptor with a relatively low affinity, it is very likely to act simultaneously via the glucagon/GLP receptors and an alternative cellular mechanism (Dakin et al. 2001). Consequently, not all actions of GLP and oxyntomodulin are identical. For example, oxyntomodulin does not affect stomach emptying, whereas GLP-1 modifies gastric motility (Maida et al. 2008).

Hypothalamic changes occur in response to oxyntomodulin and GLP-1/2. Electrophysiological studies indicate that GLP-1 and oxyntomodulin act via GLP-1 receptor expressing ARC neurons, at least to some extent, by interfering with orexigenic ghrelin-induced cellular actions (Riediger et al. 2010). Both peptides administered peripherally induce c-Fos immunoreactivity in the PVN, NTS, and AP. GLP-1 in the PVN elicits activation of the anorexigenic corticotropin-releasing hormone (CRH), nesfatin-1, and oxytocin (OT) systems (Bojanowska and Stempniak 2000; Katsurada et al. 2014).

Peptide YY

Intestinal L cells also synthesize peptide YY (PYY), which belongs to the PP-fold family of proteins and binds the Y2 receptor (Halatchev and Cone 2005). Most PYY in the circulation is the truncated 34-amino acid PYY(3-36) molecule, and the increase in this peptide concentration in the plasma can be observed – in a manner proportional to the amount of ingested energy (Batterham et al. 2002) – within 60 min postprandially (Grandt et al. 1994). Peripheral injections of this hormone decrease food intake in normal-weight and obese subjects (Batterham et al. 2003). Peripheral PYY(3-36) increases c-Fos levels in the ARC and decreases hypothalamic NPY mRNA expression (Halatchev et al. 2004). As PYY(3-36) inhibits electrical activity of NPY nerve terminals, an initial hypothesis was coined that this might underlie activation of adjacent POMC neurons (Batterham et al. 2002). Unlike peripheral PYY, however, centrally administered peptide was found to not activate but rather inhibit POMC neurons, which may explain disparate effects seen after peripheral versus central administration in pharmacological in vivo studies (Ghamari-Langroudi et al. 2005; Boggiano et al. 2005). The fact that PYY decreases food intake in animals with genetically inactivated melanocortin receptor further supports the notion that POMC neurons are not involved in the mediation of the anorexigenic effect of this hormone and other hypothalamic neurons should be viewed as mediators of PYY-induced hypophagia (Halatchev et al. 2004).

Peptide PP

Peripheral injections of peptide PP, the ligand of the Y receptors (particularly, the Y4 subtype) synthesized (other than in the pancreas) in the colon and rectum, have been shown to decrease food intake and body weight in laboratory animals and humans (Balasubramaniam et al. 2006; Malaisse-Lagae et al. 1977). This reduction occurs in both obese and normal-weight individuals (Asakawa et al. 2003). Anorexigenic effects of circulating PP are likely mediated by the brain stem-hypothalamic circuit (Asakawa et al. 2003). In line with this notion, vagotomy abolishes PP-induced hypophagia, whereas continuous 24-h peripheral administration of PP lowers hypothalamic NPY mRNA levels (Asakawa et al. 2003). Sainsbury and colleagues reported that peripheral PP elevates c-Fos immunoreactivity in the ARC and VMH (Sainsbury et al. 2010), and Lin et al. showed increased c-Fos levels in the AP, NTS, PVN, and ARC, including in those ARC neurons that synthesize alpha-melanocyte-stimulating hormone (alpha-MSH) (Lin et al. 2009). In a subsequent study, more detailed c-Fos mapping in Y4 KO animals revealed enhanced activity of the lateral aspect of the ARC, VMH, DMH, and LH in response to PP treatment despite the lack of the functional Y4 receptor (Shi et al. 2013). PP has been also found to downregulate expression of an LH appetite stimulant, orexin, while simultaneously increasing mRNA levels of VMH brain-derived neurotrophic factor (BDNF) and of ARC POMC (Sainsbury et al. 2010).

Oleoylethanolamide

A quite recent discovery of oleoylethanolamide (OEA), a hormone synthesized in the upper part of the small intestine, has generated a lot of interest in the research community as OEA was determined to be released upon the absorption of a specific group of dietary macronutrients, lipids (Fu et al. 2005, 2007). Nutrient availability controls OEA mobilization in the intestinal mucosa through a coordinated regulation of OEA biosynthesis and degradation (Fu et al. 2007). This compound acts via the nuclear receptor peroxisome proliferator-activated receptor alpha (PPAR-alpha), and activation of this receptor affects the expression of genes involved in fat absorption and fatty acid metabolism (Yang et al. 2007; Guzman et al. 2004; Fu et al. 2003). While lipolysis is one likely route through which OEA affects energy balance (Guzman et al. 2004), another key parameter is satiety. In their elegant study, Fu and collaborators constructed adenoviral vector (Ad-NPLD) that causes overexpression of N-acylphosphatidylethanolamine (NAPE)-phospholipase D (PLD) enzyme, which catalyzes the hydrolysis of NAPE to OEA. Intraduodenal injection of the vector leading to an increase in NAPE-PLD and OEA synthesis promoted a reduction in energy intake through increased feeding latency and postmeal interval (Fu et al. 2008). Peripheral injections and oral administration of OEA decrease food intake and body weight in rodents (Rodriguez de Fonseca et al. 2001; Gaetani et al. 2003). OEA treatment activates the NTS, AP, and PVN, and the effect on the brain is most likely mediated via the vagal afferents (Umehara et al. 2016). Consequently, ICV OEA is ineffective, and peripheral OEA shows diminished effectiveness in animals with disrupted vagal input as well as AP lesions (Rodriguez de Fonseca et al. 2001; Romano et al. 2017). It has been reported that histaminergic and OT signaling in the brain is required for the hypophagic action of OEA to occur (Provensi et al. 2014).

Ghrelin

An endogenous ligand of the growth hormone secratagogue (GHS) receptor, ghrelin, is produced mainly by the stomach (although populations of ghrelin-synthesizing CNS neurons also play a role in feeding control) (Kojima et al. 1999). Ghrelin administered peripherally and directly into the brain induces an acute and robust feeding response, similar in magnitude than the effect of NPY. Chronic peripheral injections of this peptide stimulate feeding and increase body weight (Bailey et al. 1999; Wren et al. 2001). The GHS receptor is expressed predominantly in the hypothalamus, particularly in the ARC, PVN, and VMH, as well as in several brain stem areas, including the NTS, AP, and DMNV (Zigman et al. 2006). In line with that, peripheral ghrelin has been reported to induce c-Fos immunoreactivity in the ARC, PVN, NTS, DMNV, and AP (Takayama et al. 2007; Ruter et al. 2003; Hewson and Dickson 2000). Ghrelin promotes NPY release from hypothalamic explants in vitro (Wren et al. 2002). At the central level, ghrelin increases ARC NPY and AgRP gene expression (Kamegai et al. 2001), and postnatal ablation of AgRP/NPY neurons abolishes orexigenic effects of peripheral ghrelin (Luquet et al. 2007). Furthermore, peripheral ghrelin blockade by specific antibodies reduces the orexigenic effect of 2-DG, producing a decrease in the number of c-Fos-immunoreactive orexin neurons in the LH (Solomon et al. 2007).

2.2 *Hypothalamic Circuit*

2.2.1 Key Molecular and Neuroanatomical Players in Hypothalamic Control of Appetite

Arcuate Nucleus as a Component of the Feeding-Related Hypothalamic Circuit

In the hypothalamic circuit, the ARC integrates peripheral input related to appetite. Though our knowledge of mechanisms underpinning ARC's ability to efficiently integrate dynamic blood-borne signaling remains incomplete, the location of the ARC, as well as cytoarchitecture and molecular processes in cells forming the BBB, is of importance. On the one hand, ARC's proximity to the median eminence, a circumventricular organ composed of fenestrated capillaries that project toward the ventromedial part of the nucleus, might facilitate the passive entry of molecules (as shown for, e.g., ghrelin) (Schaeffer et al. 2013). On the other hand, transcytosis and transendothelial transporters selectively control the passage of circulating molecules into the ARC via vesicle-mediated transport across the endothelium (Tuma and Hubbard 2003; Byun et al. 2014; Dietrich et al. 2008). Finally, pathological and metabolic states during various phases of ontogeny affect molecular "sealing off" of the CNS, which further complicates the ability of the ARC to sense peripheral changes related to hunger and satiety (Siegenthaler et al. 2013; Kim et al. 2016).

ARC neuronal populations critical for the regulation of food intake communicate via rich reciprocal innervation with other hypothalamic feeding-related sites, including the VMH, DMH, and PVN, as well as with extrahypothalamic areas related to feeding reward, such as the ventral tegmental area (VTA) (Wang et al. 2015; Gruber et al. 1987). Although – somewhat counterintuitively – only limited input from the caudal brain stem has been detected in the ARC, DVC-relayed signaling reaches this nucleus via multisynaptic pathways that include, e.g., the PVN (Wang et al. 2015).

The ARC encompasses two functionally distinct neuronal populations involved in appetite regulation: one that co-expresses NPY/AgRP (as well as GABA) and promotes food intake and the other that supports termination of consummatory behavior, in which CART and POMC are colocalized. From the standpoint of epidemiological contribution to the obesity "epidemic," neurons synthesizing POMC, whose main anorexigenic posttranslational processing product is alpha-MSH, have gained most attention. Alpha-MSH acts as an agonist of the melanocortin-4 (MC-4) receptor. In humans, single-nucleotide polymorphisms and larger-scale mutations within and near the MC4 encoding gene are associated with excessive appetite and body weight, being the most common monogenic cause of obesity (Scherag et al. 2010; Vaisse et al. 2000; Wang et al. 2017). Laboratory animal studies have shown that injections of alpha-MSH as well as synthetic MC4 receptor agonists (e.g., melanotan II) reduce food intake and body weight, activation of alpha-MSH ARC neurons coincides with termination of food intake, and POMC and MC-4 gene expression levels are modified by energy status (and by physiologically relevant energy needs, e.g., during lactation) (Wirth et al. 2001; Olszewski et al. 2001; Suzuki et al. 2014; Huszar et al. 1997). POMC-deficient mice exhibit hyperphagia and obesity, and administration of MC-4 agonists reverses this abnormal phenotype (Yaswen et al. 1999). Also, postnatal ablation of ARC POMC leads to overeating and increased body weight (Gropp et al. 2005). It should be noted that alpha-MSH binds also the MC-3 receptor. However, this receptor's involvement in appetite control remains disputed as MC-3 agonists are not particularly effective in decreasing appetite (Butler and Cone 2003).

ARC POMC neurons synthesize also CART, and – since initial studies involving lateral and third ventricular administration of this peptide showed a decrease in food intake, including that induced by NPY – CART was viewed exclusively as an anorexigen, possibly having a synergistic effect with alpha-MSH (Kristensen et al. 1998). However, more recent data indicate that anorexigenic effects of CART are mediated mainly by the hindbrain, whereas CART in the hypothalamus promotes consumption (by, among others, engaging circuits that release NPY and AgRP) (Hou et al. 2010). Intraparenchymal administration of CART in the PVN, VMH, or ARC of food-deprived rats elevates food intake at refeeding (Abbott et al. 2001). Thus, ARC-derived POMC/CART input into intrahypothalamic circuit seems to generate the balance between hypo- and hyperphagic action rather than a single directional anorexigenic response.

The feeding-related melanocortin system is unique in that it relies on two antagonistically acting ligands, both synthesized in the ARC (albeit, as mentioned above, in distinct neuronal subpopulations). Aside from MSH peptides (with alpha-MSH being most commonly studied), the MC-4 receptor is targeted by an endogenous antagonist, AgRP. In vivo pharmacological studies show that MC-4 receptor blockade leads to overeating (Hagan et al. 2000). Even though inactivation of the AgRP-encoding gene does not affect food intake or body weight in mice (of note, the lack of effect is similar to a KO of another key orexigen, NPY), it is most likely caused by developmental redundancy, because postnatal ablation of AgRP neurons produces the expected hypophagic and underweight phenotype (Gropp et al. 2005; Qian et al. 2002). AgRP and alpha-MSH neurons execute their opposite functions by projecting to common targets, particularly the PVN, and the distribution of the MC-4 receptor is well aligned in those areas (Wang et al. 2015; Garfield et al. 2015). Intraparenchymal PVN administration of MC-4 receptor agonists decreases food intake, whereas antagonists (such as AgRP) increase consumption (Olszewski et al. 2001, 2003; Shrestha et al. 2004). Alpha-MSH injected in the PVN increases the percentage of c-Fos-positive OT PVN neurons, and the release of OT occurs mainly at central targets rather than into the general circulation (Olszewski et al. 2001; Sabatier et al. 2003). Central administration of AgRP blocks activation of PVN OT neurons in response to anorexigenic LiCl treatment (Wirth et al. 2002). Overall, MC-4 blockade is thought to produce not only a delay in satiety but also overconsumption of rewarding tastants (Pandit et al. 2015; Navarro et al. 2011). In line with that notion, AgRP appears to stimulate intake of high-calorie food as well as tastants that are energy-dilute yet palatable (Wirth and Giraudo 2001). Moreover, AgRP neuronal activity is a necessary factor in the process of alcohol-induced excessive food intake (Cains et al. 2017). AgRP promotes operant responding for a fat reinforcer and to fat-paired stimuli in the absence of ingestive stimulation, which indicates a potential role of this peptide in overconsumption of fat (Tracy et al. 2008). In this context, one should note that AgRP has been proposed to act as one of the key common denominator molecules determining behavioral intent and intensity. Through its interactions in the circuitry encompassing, among others, GABA, glutamate, and serotonin, AgRP appears to direct complex motivated behaviors toward specific goals by, for example, balancing the need to replenish lacking calories after starvation and to protect internal milieu from potentially tainted foods (Sternson 2013; Sternson et al. 2013).

Interestingly, ARC AgRP neurons co-express other orexigenic factors, most notably, NPY. NPY is one of the most potent orexigens identified to date, and its feeding stimulatory action is mediated via Y1 and Y5 receptors in the hypothalamus as well as by inhibitory action at ARC alpha-MSH/CART neurons (Stanley et al. 1986). Acute and chronic injections of NPY produce overeating, and the PVN is regarded as one of the primary areas mediating NPY-induced hyperphagia (Billington et al. 1994). It has been proposed that in the context of a meal, NPY is likely responsible for rapid initiation of feeding whereas AgRP, for meal prolongation, thereby collectively contributing to a vigorous and sustained consummatory activity (Krashes et al. 2013). Furthermore, colocalization of NPY with AgRP in the

ARC might be the basis for the redundancy of the system, ensuring resistance of orexigenic responsiveness to genetic/molecular pathophysiological changes.

Other Hypothalamic Systems Controlling Appetite

As emphasized earlier, ARC does not act alone in the process of food intake regulation, but it should rather be regarded as an essential part of a broader hypothalamic network of sites. In line with that, ARC NPY/AgRP and POMC/CART neurons project to the PVN, VMH, DMH, and LH, where they form synaptic connections with neurons synthesizing feeding neuroregulators. For example, in the PVN, CRH and thyrotropin-releasing hormone cells are components of this circuit, and the appetite-related functional relevance of pathways that these cells form with ARC-derived axons has been documented (Baker and Herkenham 1995; Kawano and Masuko 2000; Diano et al. 1998; Swanson and Simmons 1989; Fekete et al. 2000). Recent discoveries pointed to the importance of PVN OT as a broad-spectrum satiety system and the fact that OT neurons receive strong input from the ARC, and they express, among others, MC-4 and NPY receptors (Garza et al. 2008). OT neurons are activated during a meal and contribute to termination of food intake regardless of a reason underlying the anorexigenic effect. This activation and concurrent OT release occur upon excessive stomach distension, increase in plasma osmolality (commonly associated with consumption of osmotically challenging ingestants), toxicity, as well as the intake of solid and liquid diets (Olszewski et al. 2016). In animal models, PVN OT has been found to be particularly effective in reducing consumption of sweet and carbohydrate foods (Herisson et al. 2014; Olszewski et al. 2010). Not surprisingly, therefore, the ability of ARC-derived input to affect responsiveness of PVN OT cells underscores the critical role of this particular pathway in the regulation of feeding. Interestingly, in the case of OT, which also regulates a variety of social behaviors, the social context of a meal can influence effectiveness of oxytocin receptor ligands on food intake. For example, while blocking the OT receptor in laboratory mice consistently increases consumption of sugar in these animals, the same treatment is ineffective in subordinate animals pair housed with a dominant animal (Olszewski et al. 2015). In addition, in male rats placed in a subdivided social feeding apparatus, OT administration directly into the nucleus accumbens core failed to reduce the intake of a palatable meal (Herisson et al. 2016).

ARC NPY/AgRP and α-MSH densely innervate also LH neurons which express hyperphagia-promoting orexins and melanin-concentrating hormone (MCH). MCH and orexin neurons are targets for stimulatory and inhibitory stimulation from other brain regions (including the ARC) and from the general circulation (Harthoorn et al. 2005). Orexin-induced feeding and body weight gain are associated with an increase in behavioral activity in the context of the sleep-wake cycle and with the decrease in energy expenditure (Mavanji et al. 2015; Sweet et al. 1999). Mice with a genetic deletion of the MCH receptor have increased energy expenditure, and they are resistant to diet-induced obesity, whereas MCH injections elevate consumption.

(Chen et al. 2002). Interestingly, hypothalamic MCH concentrations are affected by body weight in laboratory animals, adding to the evidence that the activity of the MCH system is further propelled by dietary obesity (Elliott et al. 2004).

Finally, in order to add to the cytoarchitectural and molecular complexity of the hypothalamic circuit relevant to neuroendocrine regulation of eating behavior, it should be noted that the VMH (with its brain-derived neurotrophic factor, BDNF) and DMH (with a pool of NPY neurons) also receive innervation from the ARC and communicate reciprocally with other hypothalamic sites, including the PVN (Broberger et al. 1998; Jacobowitz and O'Donohue 1978; Minokoshi et al. 2004; Xu et al. 2003).

3 Failure to Control Appetite in an Obesogenic Environment: Is Neuronal Plasticity the Culprit?

Based on the complexity of parallel (and seemingly redundant) effects of the aforementioned neural and endocrine processes that govern appetite, one could easily presume that energy intake and, consequentially, energy balance should fall within a range that is conducive to maintaining homeostasis. However, this might be true in an environment in which food is a scarce commodity rather than an ever-present resource. An obesogenic environment, unprecedented in the evolutionary history, reveals multiple vulnerabilities in mechanisms that promote excessive food intake. While maladaptive peripheral processes (in the context of overabundance of food) integrated by the hypothalamus have been quite well described, exciting new evidence points to changes in synaptic plasticity in the hypothalamus itself as the neurophysiological foundation of abnormally high food intake.

3.1 Peripheral Signals: A Brief Overview

At the level of adiposity-related signaling, leptin resistance is one of the most recognized endocrine pathologies in obesity: individuals with excessive body weight display hyperphagia despite high leptin levels, and they fail to respond to the exogenous hormone. This resistance has been hypothesized to be driven by, among others, maladaptive sterile inflammation in the hypothalamus, inhibited LepRb signaling owing to abnormal overexpression of suppressor of cytokine signaling 3 and protein tyrosine phosphatases (PTPases), hypothalamic endoplasmic reticulum stress, and impaired BBB passage of the molecule (Cui et al. 2017). In light of the findings showing that obese mice overeating high-energy food display normal sensitivity to endogenous leptin despite having impaired sensitivity to the exogenous hormone (Ottaway et al. 2015), the recently proposed PTPase-driven mechanism has generated a lot of attention. In this process, for leptin signaling to occur, ARC neurons require a system to temporarily inactivate PTPases. PTPases dephosphorylate tyrosyl residues in the leptin signaling cascade, affecting the action of this hormone (Zhang et al. 2015). CNS PTPase deletion promotes leptin signaling via the ARC, leading to reduced food intake, increased energy expenditure, and lower adiposity (Dodd et al. 2015), whereas an increase in hypothalamic PTPase levels promotes diet-induced obesity and central leptin resistance (Bence et al. 2006; Loh et al. 2011).

Dysregulation of energy balance results not only in malfunctioning of the leptin system alone, but it has a simultaneous effect on insulin. Increased adiposity lowers insulin sensitivity and – eventually – promotes insulin resistance (Adam et al. 2009). As the effect of another adipokine, adiponectin, on food intake depends on blood glucose levels, therefore, whether adiponectin decreases or increases appetite (and whether it aids leptin in conveying the effect on ARC neurons) might be greatly affected by metabolic disturbances, including those associated with insulin resistance.

Numerous variables have been also found to interfere with effectiveness of the gut-derived anorexigenic endocrine system. For example, anxiety reduced anorexigenic effects of PYY as well as of oxyntomodulin (Abbott et al. 2006). The timing of PYY treatment in relation to the behaviorally active phase has also been shown as a critical parameter: while PYY produces acutely hypophagia, Parkinson et al. (Parkinson et al. 2008) found that it causes a delayed orexigenic effect in ad libitum-fed mice treated in the early light phase. Early light-phase administration of PYY resulted in a trend toward increased NPY/AgRP and a decrease in POMC mRNA in the hypothalamus at the onset of the dark phase (thus, at the time when a natural feeding cycle begins). Circulating ghrelin levels were also higher in those PYY-treated mice (Parkinson et al. 2008).

Furthermore, composition of a diet affects responsiveness of the feeding-related brain circuit and – consequently – the ability of OEA to induce hypophagia. Romano et al. reported a weaker NTS activation and stronger activation of the PVN in rats fed a high-fat diet than in chow-fed controls. High-fat diet-fed rats were also more sensitive to OEA's immediate anorexigenic action (Romano et al. 2014). Since the ability of OT to reduce food intake is dependent on the social environment (Olszewski et al. 2015; Herisson et al. 2016; Olszewski and Levine 2016), it is quite possible that hypophagic properties of OEA might also be hampered by social stressors. This is quite possible considering that in human subjects, serum OEA concentrations were significantly lower in response to the Trier social stress test (Hill et al. 2009).

Importantly, factors that diminish the effectiveness of peripheral anorexigens oftentimes further potentiate hyperphagia induced by an orexigenic gut hormone, ghrelin. For example, stressors (including psychosocial stress) rely on ghrelin signaling to induce overeating (Chuang et al. 2011). Chronic stress has been proposed to elevate food intake and body weight by enhancing ghrelin release (Labarthe et al. 2014).

It seems that aside from having a generalized orexigenic effect, peripheral ghrelin is capable of increasing eating for pleasure. Liu et al. found that injection of GHRP-2, a GHS receptor agonist, in fat-preferring Osborne-Mendel rats and in S5B/P1 rats that favor low-fat foods, primarily stimulated intake of the preferred macronutrient (Liu et al. 2004). Conversely, wild-type mice injected with ghrelin receptor antagonist and GHS receptor KO mice both failed to show conditioned place preference to a high-fed diet typically seen under energy restriction (Perello et al. 2010). These findings are well aligned with the fact that the GHS receptor is expressed also in reward sites and that central neuroanatomical and molecular mediators of eating for pleasure appear to convey ghrelin-induced overeating (Zigman et al. 2006; Romero-Pico et al. 2013; Skibicka et al. 2013). Finally, studies in human subjects indicate that – unlike insulin and leptin – ghrelin retains its orexigenic properties even in obese individuals, which indicates that the current status of energy stores is not an effective mechanism preventing ghrelin from inducing food intake regardless of the actual long-term energy needs (Druce et al. 2005). The attenuated postprandial decrease in circulating ghrelin levels in high-BMI patients suggests that abnormal GI tract endocrine signaling is an additional contributing factor in the development of excessive food intake in obesity (English et al. 2002; le Roux et al. 2005).

3.2 Neuronal Plasticity in the Hypothalamus: A Path Forward in Deciphering Causative Factors Underlying Hyperphagia?

Exposure to an environment in which energy-dense and tasty food is readily available leads to excessive consumption. Concurrently, dysregulation of peripheral endocrine mechanisms occurs, creating a self-perpetuating sequence of orexigenic physiological and behavioral events. The question arises as to whether the hypothalamic circuit involved in the regulation of feeding is a "passive" recipient of appetite-related peripheral signals and it therefore generates only transient responses to momentary hormonal cues or whether it undergoes more significant functional reorganization within relevant pathways. The latter is particularly conceivable as, in order to adapt to changing energy needs, the CNS should presumably be able to adjust the efficiency of interneuronal communication. Synaptic plasticity, a change in synaptic strength, would then reflect this dynamic and ongoing adaptive process.

The need to determine whether and how synaptic plasticity affects feeding became apparent when conceptualizing functioning of the ARC-derived pathways that encompass the endogenous MC-4 agonist-antagonist system. As mentioned earlier, the ARC AgRP/NPY neurons promote feeding, whereas the POMC population decreases it via overlapping target cells. Hence, maintaining a balance between NPY/AgRP and POMC firing (and, consequently, the strength as well as spatial and temporal organization of the synaptic input) might be of crucial value to sustain adequate energy intake. That the hypothesis was worth pursuing was further supported by the findings that cellular/molecular responses to nutrient and hormonal signals reaching ARC differ in the fasted versus fed state and that the responses are opposite in AgRP/NPY versus POMC populations (Cone et al. 2001). For example, leptin was found to inhibit fasting-induced NPY/AgRP neuronal activity, whereas it stimulated POMC cells (Cowley et al. 2001). In addition, NPY/AgRP neurons were shown to inhibit POMC neurons directly through axons that release NPY and GABA into synapses formed on POMC perikarya (Cowley et al. 2001; Vong et al. 2011). This suggested that the orexigenic action of ARC NPY/AgRP neurons is potentiated even further, as they affect MC-4 target cells both directly and indirectly by inhibiting POMC excitatory input on the same MC-4 target cells.

Data (albeit, still at a relatively early stage) indicate that synaptic plasticity is indeed continuous in order to adapt to internal and environmental challenges and to (seemingly) protect the organism. However, similarly to peripheral endocrine processes, many synaptic plasticity changes in circuits regulating food intake appear to promote a drive to eat regardless of whether the encountered challenges that initiated select plasticity changes involved scarcity or overabundance of food. Pinto et al. (2004) were first to report electrophysiological recordings showing leptin's effects on synaptic plasticity on NPY/AgRP and POMC neurons in a murine model of leptin deficiency. In this elegant study, transgenic mice with bacterial artificial chromosome genes expressing green fluorescent protein (GFP) under the transcriptional control of NPY (tau-sapphire GFP) and POMC (tau-topaz GFP) were generated. Immunofluorescent analysis revealed that Lep-null mice had more excitatory and less inhibitory synaptic input onto NPY perikarya as well as fewer excitatory synaptic inputs onto POMC perikarya compared to wild-type controls. Patch-clamp-detected postsynaptic currents in acute ARC slices also differed between leptin-deficient and wild-type mice. Systemic administration of leptin in Lep-null mice normalized synaptic density on NPY and POMC neurons reaching values seen in wild types, and inhibitory postsynaptic currents in POMC neurons decreased to control levels just after 2 days of the leptin treatment. Thus, leptin was found to be able to regulate synaptic plasticity of POMC and NPY/AgRP neurons (Pinto et al. 2004). Notably, leptin is not the only hormone with this ability, and, in the following years, other hormones have been found to change synaptic plasticity of ARC POMC and NPY/AgRP neurons in a similar manner. For example, corticosterone replacement in adrenalectomized mice restored the reduced number of inhibitory synapses on POMC and excitatory synapses on NPY/AgRP to normal levels (Gyengesi et al. 2010). Estradiol increased the number of excitatory synapses on POMC neurons (Horvath 2006), and ghrelin stimulated the electrophysiological activity of NPY/AgRP neuron as well as reduced the activity of POMC neurons by increasing the release of GABA onto POMC perikarya and by hyperpolarizing these neurons (Cowley et al. 2003).

In a recent study defining the role of excitatory glutamatergic input, Liu et al. generated mice lacking the NMDA receptor on NPY/AgRP or on POMC neurons. Deletion of NMDA receptors on NPY/AgRP neurons reduced food intake, body weight, and body fat in mice, but there was no effect stemming from NMDA receptor deletion on POMC neurons. Remarkably, fasting-induced increase in glutamatergic input onto NPY/AgRP neurons correlated with an increase in dendritic spines, suggesting that negative energy balance induces synaptogenesis and spinogenesis of ARC NPY/AgRP cells. POMC neurons, on the other hand, have no dendritic spines and thus cannot be excited by glutamatergic input (Liu et al. 2012). In addition, Yang et al. used patch-clamp recordings and found that ghrelin increases synaptic activity in NPY/AgRP neurons during fasting. This effect is mediated via an AMP-activated protein kinase (AMPK)-dependent positive feedback loop, allowing ghrelin's influence to persist for hours even after removal of this hormone (Yang et al. 2011). Leptin pretreatment causes a release of opioids that prevents this feedback loop from occurring (Fig. 2).

Importantly, some synaptic plasticity changes in the melanocortin system might be irreversible in vulnerable subjects, eventually leading to diet-induced obesity (DIO). Animals susceptible to DIO exposed to a high-fat diet lose inhibitory synapses on POMC cells as well as excitatory synapses on NPY/AgRP neurons. This loss of synapses upon high-fat diet intake also correlates with increased glial ensheathment of POMC perikarya in these animals, which makes POMC cell bodies and dendrites less accessible to circulating satiety signals. Interestingly, this process does not occur in animals resistant to DIO (Horvath et al. 2010) (Fig. 2).

It should be noted that even if these synaptic plasticity changes were irreversible in DIO-susceptible individuals, neurogenesis might still provide a reversal-conducive scenario. Pierce et al. noted that while acute ablation of ARC NPY/AgRP neurons causes severe hypophagia and weight loss, progressive degeneration of these neurons appears to have little effect, suggesting that the brain can adapt through compensatory mechanisms. These authors studied this phenomenon in transgenic mice experiencing progressive neurodegeneration due to deletion of the

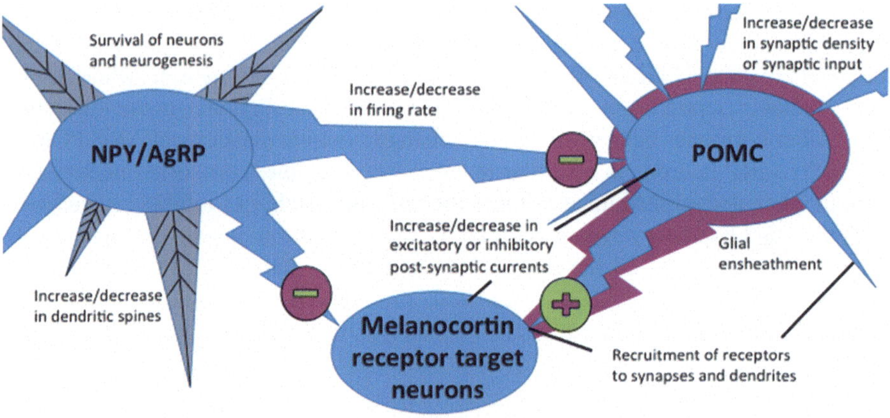

Fig. 2 Mechanisms of synaptic plasticity in the melanocortin system. Synaptic efficiency can be facilitated through a variety of mechanisms. They include a shift in synaptic density and input on dendrites and perikarya of target neurons, occurrence of dendritic spines on target neurons, changes in neuronal firing rate, as well as postsynaptic inhibitory and excitatory currents. Other mechanisms include endocytosis and exocytosis of receptors, glial ensheathment of neuronal perikarya or axons, and neurogenesis

mitochondrial transcription factor A in NPY/AgRP neurons. They found increased cell proliferation with a subset of new cells differentiating into NPY/AgRP neurons capable of responding to leptin in these animals. This suggests that populations of feeding-related ARC neurons lost due to neurodegenerative mechanisms can be regenerated (Pierce and Xu 2010). McNay and colleagues confirmed that although there is a substantial turnover of hypothalamic neurons and that remodeling of the energy-balance circuit occurs continuously, under the DIO conditions, neurogenesis in the ARC is significantly slowed down. The authors note that DIO mice do not lack hypothalamic stem cells but rather have a decreased number of actively proliferating ones and show increased apoptosis in newly differentiated cells while retaining more of the old cells. Interestingly, they also reported that a period of calorie restriction can restore neurogenesis in DIO animals (22201680).

Synaptic plasticity changes in the feeding circuit are not restricted to the ARC-localized cells. OT neurons in the PVN and SON, for example, are also regulated by inhibitory inputs mediated via postsynaptic GABA receptors. This becomes apparent during gestation and lactation, when the responsiveness of the OT system is somewhat weakened. To shed more light on this phenomenon, Theodosis et al. have published numerous studies delineating synaptic plasticity in hypothalamic cells, especially those synthesizing OT, in pregnant and lactating rats. As early as in 1984, they described that, for example, 2 weeks into gestation in rats, the percentage of neurosecretory soma and dendritic profiles was low. However, a day before parturition and throughout lactation, synaptic density on OT neurons in the SON increased dramatically (Theodosis and Poulain 1984). Interestingly, low excitability of OT neurons during pregnancy might be one of the underlying factors of gestational hyperphagia (Douglas et al. 2007).

There is some evidence that some nutrients can alter synaptic plasticity. Wan et al. showed in the chicken embryo that insulin modifies synaptic plasticity by rapidly recruiting intracellular $GABA_A$ receptors to the membranes and dendrites of post-synaptic neurons and thereby increases miniature inhibitory postsynaptic currents in these cells. Insulin and insulin-like growth factor have also been shown to support the survival and growth of neurons (Gu et al. 2014). On the other hand, DMH GABAergic neurons are regulated by glucose and leptin (low glucose levels depolarize, and leptin hyperpolarizes these cells) and innervate the PVN, where they inhibit neurons and promote food intake (Wan et al. 1997). MC-4 in dopamine receptors D1 and D2 expressing neurons appears to change synaptic strength by causing either endocytosis or increased surface expression of AMPA receptors (Caruso et al. 2014).

Intake of a high-fat diet in pregnant dams appears to also affect plasticity of the offspring. This occurs when the dam is chronically fed a high-fat diet and suffers from DIO at the time of conception, as well as when high-fat diet exposure during pregnancy is brief, indicating that maternal obesity per se might not be the driving factor behind synaptic changes in the offspring (Rivera et al. 2015; Desai et al. 2014; Rajia et al. 2010; Khalyfa et al. 2013). It has been suggested that there is a link between maternal obesity and synaptic plasticity defects in children, leading to adverse metabolic outcomes, including obesity, adiposity, hyperinsulinemia, and hyperleptinemia (Rajia et al. 2010; Camacho et al. 2017; Levin and Dunn-Meynell 2002; Reynolds et al. 2010). However, the causality of this link is still unclear, as dams in maternal obesity rodent studies are typically fed a high-fat hypercaloric diet before and during gestation and lactation. In human studies, gestational weight gain in obese mothers exceeds that in normal-weight subjects, indicating excessive energy intake (Reynolds et al. 2010). However, in human observations, macronutrient composition of food is not strictly controlled throughout such a prolonged time period. Thus, it is difficult to pinpoint whether maternal obesity or rather maternal high-fat diet causes synaptic plasticity alterations in the offspring. Levin et al. provided some additional insight into this issue by selectively breeding rats for their propensity for high (DIO susceptible) and low (DIO resistant) weight gain on a high-fat diet (Levin and Dunn-Meynell 2002). The two resulting animal cohorts thus inherited their energy balance/appetite characteristics as a polygenetic trait. The authors manipulated the diet of the dams to make them either lean or obese during gestation and until weaning. They found that regardless of the mothers' state of adiposity, offspring of DIO-susceptible dams became more obese than offspring of resistant dams, while maternal obesity in the resistant dams had no effect on body weight in their offspring. Furthermore, when comparing the offspring of both obese DIO-susceptible and DIO-resistant dams, only the offspring of obese DIO-susceptible dams developed high adiposity, hyperleptinemia, and hyperinsulinemia at an adult age (Levin and Dunn-Meynell 2002). Furthermore, maternal high-fat diet intake increases proliferation of orexigenic neurons in the developing PVN and LH (Chang et al. 2008). Chronic consumption of this diet during pregnancy decreased the density of AgRP fibers innervating the PVN in the offspring, enhancing the strength of orexigenic synaptic input (Kirk et al. 2009; Grayson et al. 2010).

While many studies focus on the effect of maternal high-fat diet intake on synaptic plasticity in the offspring, little is known about the effects of maternal chronic sugar intake or maternal food restriction. Epigenetic studies indicate that food intake and body weight in the offspring can be elevated due to epigenetic changes in DNA methylation if the mother has been subjected to food restriction or exposed to high-carbohydrate diets during pregnancy. For example, carbohydrate intake at the beginning of the pregnancy is associated with higher methylation of the retinoid X receptor alpha gene, which in turn is associated with a higher BMI and increased fat mass in the offspring (Godfrey et al. 2011). Food restriction during pregnancy, however, is linked to obesity in the offspring (Zambrano et al. 2006). Thus, it is reasonable to presume that both prenatal chronic sugar intake and food restriction also affect neuromodulation in the offspring as critical plasticity processes occur in the fetus during this developmental period.

Finally, it should be emphasized that non-neuronal CNS cells – particularly, astrocytes – contribute to synaptic plasticity and energy homeostasis, as well. Astrocytes, whose excitability relies on intracellular Ca2+ concentrations instead of membrane potential, modulate synaptic transmission and synaptic strength. Neurotransmitters released from a neuron into the synaptic cleft can "spill over" to nearby astrocytes, increasing their internal Ca2+ concentration, triggering the release of transmitters (e.g., glutamate) from pedicels. This chemical change feeds back to the presynaptic terminal and, in turn, affects neuronal activity and promotes synaptic remodeling. Astrocytes have been shown to sense nutrients (glucose and some amino acids), hormones (OT, vasopressin, leptin, insulin, and glucagon), neurotransmitters, and ions and to regulate activity of secretory neurons. Recently, Yang et al. have reported that astrocytes inhibit ARC AgRP neurons by releasing adenosine and that activation of astrocytes decreases ghrelin-induced food intake (Yang et al. 2015). Conditional deletion of the astrocyte-specific leptin receptor in the hypothalamus leads to an increase in synaptic inputs on hypothalamic neurons, to diminished leptin-induced suppression of feeding, and to elevated food intake in response to ghrelin or fasting (Kim et al. 2014). In nonfeeding contexts, hypothalamic astrocytes have been linked to PVN/SON functioning. In time of high neurohypophyseal release demand (parturition, lactation, or chronic dehydration), astrocytic processes in the PVN or SON retract from magnocellular neurons which send projections to the pituitary, thereby exposing neuronal membranes to targeting molecules (Gordon et al. 2009) (Fig. 3).

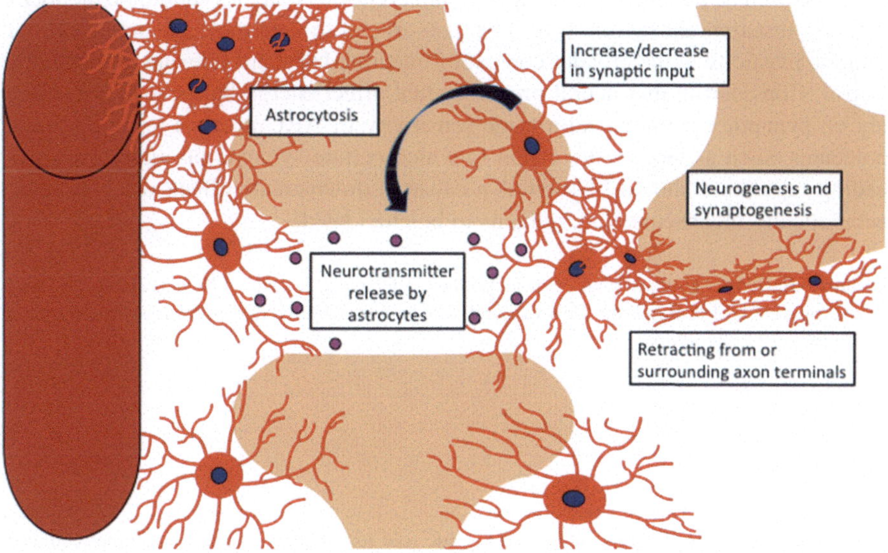

Fig. 3 Astrocyte-mediated synaptic plasticity. Astrocytes can change synaptic plasticity by communicating with pre- and postsynaptic neurons by secreting neurotransmitters or ions. Astrocytes can also increase or decrease synaptic input on neuronal perikarya and dendrites as well as neurogenesis and synaptogenesis. In addition, diet-induced astrocytosis changes synaptic plasticity through impaired communication between neurons and peripheral signals, and astrocytes can even change synaptic plasticity by covering or retracting from axons

4 Concluding Remarks

Historically, the hypothalamus has been viewed as a brain region that – through integrating peripheral endocrine signals related to adiposity and GI status – serves as an "on-off" switch for energy consumption. While this notion may still hold true as a simplistic theory, it is also clear that multiple orexigenic and anorexigenic neural and endocrine processes dynamically shape a drive to eat. What seems particularly important in understanding our vulnerability to overconsume energy (even to the point of jeopardizing our homeostasis) is that peripheral and central mechanisms that regulate appetite are biased toward increasing food intake. This bias might have been evolutionarily favored as it ensured sufficient energy intake in the environment where food was scarce. However, in the obesogenic "environment of plenty," it propels a vicious cycle of neuroendocrine and behavioral changes that promote extreme hyperphagia and dysregulation of appetitive and metabolic pathways. Based on emerging evidence, it appears that altered synaptic plasticity in the hypothalamus is one of the key contributors to neuroendocrine pathologies underlying excessive energy intake. The exact nature of changes, their functional significance, as well as potential for therapeutic intervention strategies should be the focus of future research endeavors. It will be also crucial to determine the actual role of synaptic plasticity in the broader context of integrated systems relevant to metabolic control. Considering that neural processes that affect energy balance do not solely rely on synaptic communication (vide sensitivity of ARC neurons to blood-borne molecules, such as leptin), the extent to which certain synaptic plasticity events in feeding regulation should be seen as a cause, a downstream effect, or even being metabolically inconsequential, remains to be elucidated.

Acknowledgments This work has been supported by the Royal Society of New Zealand and Dairy Goat Coop. (NZ), Ltd., Hamilton, New Zealand. The authors would like to thank Erin Wood for her help in preparing the figures.

References

Abbott CR et al (2001) Evidence of an orexigenic role for cocaine- and amphetamine-regulated transcript after administration into discrete hypothalamic nuclei. Endocrinology 142 (8):3457–3463

Abbott CR et al (2006) The importance of acclimatisation and habituation to experimental conditions when investigating the anorectic effects of gastrointestinal hormones in the rat. Int J Obes 30(2):288–292

Adam TC et al (2009) Insulin sensitivity as an independent predictor of fat mass gain in Hispanic adolescents. Diabetes Care 32(11):2114–2115

Ahima RS et al (1996) Role of leptin in the neuroendocrine response to fasting. Nature 382 (6588):250–252

Alsio J et al (2012) Feed-forward mechanisms: addiction-like behavioral and molecular adaptations in overeating. Front Neuroendocrinol 33(2):127–139

Andiran N, Celik N, Andiran F (2011) Homozygosity for two missense mutations in the leptin receptor gene (P316:W646C) in a Turkmenian girl with severe early-onset obesity. J Pediatr Endocrinol Metab 24(11–12):1043–1045
Asakawa A et al (2003) Characterization of the effects of pancreatic polypeptide in the regulation of energy balance. Gastroenterology 124(5):1325–1336
Bagnasco M et al (2002) Evidence for the existence of distinct central appetite, energy expenditure, and ghrelin stimulation pathways as revealed by hypothalamic site-specific leptin gene therapy. Endocrinology 143(11):4409–4421
Bailey AR et al (1999) Chronic central infusion of growth hormone secretagogues: effects on fos expression and peptide gene expression in the rat arcuate nucleus. Neuroendocrinology 70 (2):83–92
Baker RA, Herkenham M (1995) Arcuate nucleus neurons that project to the hypothalamic paraventricular nucleus: neuropeptidergic identity and consequences of adrenalectomy on mRNA levels in the rat. J Comp Neurol 358(4):518–530
Balasubramaniam A et al (2006) Neuropeptide Y (NPY) Y4 receptor selective agonists based on NPY(32-36): development of an anorectic Y4 receptor selective agonist with picomolar affinity. J Med Chem 49(8):2661–2665
Batterham RL et al (2002) Gut hormone PYY(3-36) physiologically inhibits food intake. Nature 418(6898):650–654
Batterham RL et al (2003) Inhibition of food intake in obese subjects by peptide YY3-36. N Engl J Med 349(10):941–948
Bence KK et al (2006) Neuronal PTP1B regulates body weight, adiposity and leptin action. Nat Med 12(8):917–924
Benoit SC et al (2002) The catabolic action of insulin in the brain is mediated by melanocortins. J Neurosci 22(20):9048–9052
Billington CJ et al (1994) Neuropeptide Y in hypothalamic paraventricular nucleus: a center coordinating energy metabolism. Am J Phys 266(6 Pt 2):R1765–R1770
Blake CB, Smith BN (2012) Insulin reduces excitation in gastric-related neurons of the dorsal motor nucleus of the vagus. Am J Physiol Regul Integr Comp Physiol 303(8):R807–R814
Boggiano MM et al (2005) PYY3-36 as an anti-obesity drug target. Obes Rev 6(4):307–322
Bojanowska E, Stempniak B (2000) Effects of centrally or systemically injected glucagon-like peptide-1 (7-36) amide on release of neurohypophysial hormones and blood pressure in the rat. Regul Pept 91(1–3):75–81
Broberger C et al (1998) The neuropeptide Y/agouti gene-related protein (AGRP) brain circuitry in normal, anorectic, and monosodium glutamate-treated mice. Proc Natl Acad Sci U S A 95 (25):15043–15048
Bruning JC et al (2000) Role of brain insulin receptor in control of body weight and reproduction. Science 289(5487):2122–2125
Buffa R, Solcia E, Go VL (1976) Immunohistochemical identification of the cholecystokinin cell in the intestinal mucosa. Gastroenterology 70(4):528–532
Burdyga G et al (2002) Expression of the leptin receptor in rat and human nodose ganglion neurons. Neuroscience 109(2):339–347
Butler AA, Cone RD (2003) Knockout studies defining different roles for melanocortin receptors in energy homeostasis. Ann N Y Acad Sci 994:240–245
Byun K et al (2014) Clusterin/ApoJ enhances central leptin signaling through Lrp2-mediated endocytosis. EMBO Rep 15(7):801–808
Cains S et al (2017) Agrp neuron activity is required for alcohol-induced overeating. Nat Commun 8:14014
Camacho A et al (2017) Obesogenic diet intake during pregnancy programs aberrant synaptic plasticity and addiction-like behavior to a palatable food in offspring. Behav Brain Res 330:46–55
Caruso V et al (2014) Synaptic changes induced by melanocortin signalling. Nat Rev Neurosci 15 (2):98–110

Castillo EJ et al (2004) Effect of oral CCK-1 agonist GI181771X on fasting and postprandial gastric functions in healthy volunteers. Am J Physiol Gastrointest Liver Physiol 287(2):G363–G369
Challier J et al (2003) Placental leptin receptor isoforms in normal and pathological pregnancies. Placenta 24(1):92–99
Chang GQ et al (2008) Maternal high-fat diet and fetal programming: increased proliferation of hypothalamic peptide-producing neurons that increase risk for overeating and obesity. J Neurosci 28(46):12107–12119
Chen Y et al (2002) Targeted disruption of the melanin-concentrating hormone receptor-1 results in hyperphagia and resistance to diet-induced obesity. Endocrinology 143(7):2469–2477
Chuang JC et al (2011) Ghrelin mediates stress-induced food-reward behavior in mice. J Clin Invest 121(7):2684–2692
Cohen MA et al (2003) Oxyntomodulin suppresses appetite and reduces food intake in humans. J Clin Endocrinol Metab 88(10):4696–4701
Cone RD et al (2001) The arcuate nucleus as a conduit for diverse signals relevant to energy homeostasis. Int J Obes Relat Metab Disord 25(Suppl 5):S63–S67
Covasa M, Ritter RC (2001) Attenuated satiation response to intestinal nutrients in rats that do not express CCK-A receptors. Peptides 22(8):1339–1348
Cowley MA et al (2001) Leptin activates anorexigenic POMC neurons through a neural network in the arcuate nucleus. Nature 411(6836):480–484
Cowley MA et al (2003) The distribution and mechanism of action of ghrelin in the CNS demonstrates a novel hypothalamic circuit regulating energy homeostasis. Neuron 37 (4):649–661
Cui H, Lopez M, Rahmouni K (2017) The cellular and molecular bases of leptin and ghrelin resistance in obesity. Nat Rev Endocrinol 13(6):338–351
Cummings DE, Overduin J (2007) Gastrointestinal regulation of food intake. J Clin Invest 117 (1):13–23
Dakin CL et al (2001) Oxyntomodulin inhibits food intake in the rat. Endocrinology 142 (10):4244–4250
Dakin CL et al (2004) Peripheral oxyntomodulin reduces food intake and body weight gain in rats. Endocrinology 145(6):2687–2695
Date Y (2012) Ghrelin and the vagus nerve. Methods Enzymol 514:261–269
Desai M et al (2014) Maternal obesity and high-fat diet program offspring metabolic syndrome. Am J Obstet Gynecol 211(3):237.e1–237.e13
Diano S et al (1998) Monosynaptic pathway between the arcuate nucleus expressing glial type II iodothyronine 5′-deiodinase mRNA and the median eminence-projective TRH cells of the rat paraventricular nucleus. J Neuroendocrinol 10(10):731–742
Dietrich MO et al (2008) Megalin mediates the transport of leptin across the blood-CSF barrier. Neurobiol Aging 29(6):902–912
Dodd GT et al (2015) Leptin and insulin act on POMC neurons to promote the browning of white fat. Cell 160(1–2):88–104
Douglas AJ, Johnstone LE, Leng G (2007) Neuroendocrine mechanisms of change in food intake during pregnancy: a potential role for brain oxytocin. Physiol Behav 91(4):352–365
Druce MR et al (2005) Ghrelin increases food intake in obese as well as lean subjects. Int J Obes 29 (9):1130–1136
Dufresne M, Seva C, Fourmy D (2006) Cholecystokinin and gastrin receptors. Physiol Rev 86 (3):805–847
Ellacott KL, Halatchev IG, Cone RD (2006) Characterization of leptin-responsive neurons in the caudal brainstem. Endocrinology 147(7):3190–3195
Elliott JC et al (2004) Increases in melanin-concentrating hormone and MCH receptor levels in the hypothalamus of dietary-obese rats. Brain Res Mol Brain Res 128(2):150–159
Elmquist JK et al (1997) Leptin activates neurons in ventrobasal hypothalamus and brainstem. Endocrinology 138(2):839–842

English PJ et al (2002) Food fails to suppress ghrelin levels in obese humans. J Clin Endocrinol Metab 87(6):2984
Farooqi IS et al (2007) Clinical and molecular genetic spectrum of congenital deficiency of the leptin receptor. N Engl J Med 356(3):237–247
Fekete C et al (2000) Alpha-melanocyte-stimulating hormone is contained in nerve terminals innervating thyrotropin-releasing hormone-synthesizing neurons in the hypothalamic paraventricular nucleus and prevents fasting-induced suppression of prothyrotropin-releasing hormone gene expression. J Neurosci 20(4):1550–1558
Frederich RC et al (1995) Expression of ob mRNA and its encoded protein in rodents. Impact of nutrition and obesity. J Clin Invest 96(3):1658–1663
Friedman JM, Halaas JL (1998) Leptin and the regulation of body weight in mammals. Nature 395 (6704):763–770
Fu J et al (2003) Oleoylethanolamide regulates feeding and body weight through activation of the nuclear receptor PPAR-alpha. Nature 425(6953):90–93
Fu J et al (2005) Oleoylethanolamide, an endogenous PPAR-alpha agonist, lowers body weight and hyperlipidemia in obese rats. Neuropharmacology 48(8):1147–1153
Fu J et al (2007) Food intake regulates oleoylethanolamide formation and degradation in the proximal small intestine. J Biol Chem 282(2):1518–1528
Fu J et al (2008) Targeted enhancement of oleoylethanolamide production in proximal small intestine induces across-meal satiety in rats. Am J Physiol Regul Integr Comp Physiol 295(1): R45–R50
Gaetani S, Oveisi F, Piomelli D (2003) Modulation of meal pattern in the rat by the anorexic lipid mediator oleoylethanolamide. Neuropsychopharmacology 28(7):1311–1316
Garfield AS et al (2015) A neural basis for melanocortin-4 receptor-regulated appetite. Nat Neurosci 18(6):863–871
Garza JC et al (2008) Adeno-associated virus-mediated knockdown of melanocortin-4 receptor in the paraventricular nucleus of the hypothalamus promotes high-fat diet-induced hyperphagia and obesity. J Endocrinol 197(3):471–482
Ghamari-Langroudi M, Colmers WF, Cone RD (2005) PYY3-36 inhibits the action potential firing activity of POMC neurons of arcuate nucleus through postsynaptic Y2 receptors. Cell Metab 2 (3):191–199
Gibbs J, Young RC, Smith GP (1973) Cholecystokinin elicits satiety in rats with open gastric fistulas. Nature 245(5424):323–325
Godfrey KM et al (2011) Epigenetic gene promoter methylation at birth is associated with child's later adiposity. Diabetes 60(5):1528–1534
Gordon GR et al (2009) Astrocyte-mediated distributed plasticity at hypothalamic glutamate synapses. Neuron 64(3):391–403
Grandt D et al (1994) Two molecular forms of peptide YY (PYY) are abundant in human blood: characterization of a radioimmunoassay recognizing PYY 1-36 and PYY 3-36. Regul Pept 51 (2):151–159
Grayson BE et al (2010) Changes in melanocortin expression and inflammatory pathways in fetal offspring of nonhuman primates fed a high-fat diet. Endocrinology 151(4):1622–1632
Gropp E et al (2005) Agouti-related peptide-expressing neurons are mandatory for feeding. Nat Neurosci 8(10):1289–1291
Gruber K et al (1987) Forebrain and brainstem afferents to the arcuate nucleus in the rat: potential pathways for the modulation of hypophyseal secretions. Neurosci Lett 75(1):1–5
Gu T, Zhao T, Hewes RS (2014) Insulin signaling regulates neurite growth during metamorphic neuronal remodeling. Biol Open 3(1):81–93
Guzman M et al (2004) Oleoylethanolamide stimulates lipolysis by activating the nuclear receptor peroxisome proliferator-activated receptor alpha (PPAR-alpha). J Biol Chem 279 (27):27849–27854
Gyengesi E et al (2010) Corticosterone regulates synaptic input organization of POMC and NPY/AgRP neurons in adult mice. Endocrinology 151(11):5395–5402

Hagan MM et al (2000) Long-term orexigenic effects of AgRP-(83---132) involve mechanisms other than melanocortin receptor blockade. Am J Physiol Regul Integr Comp Physiol 279(1): R47–R52

Halatchev IG, Cone RD (2005) Peripheral administration of PYY(3-36) produces conditioned taste aversion in mice. Cell Metab 1(3):159–168

Halatchev IG et al (2004) Peptide YY3-36 inhibits food intake in mice through a melanocortin-4 receptor-independent mechanism. Endocrinology 145(6):2585–2590

Harthoorn LF et al (2005) Multi-transcriptional profiling of melanin-concentrating hormone and orexin-containing neurons. Cell Mol Neurobiol 25(8):1209–1223

Herisson FM et al (2014) Functional relationship between oxytocin and appetite for carbohydrates versus saccharin. Neuroreport 25(12):909–914

Herisson FM et al (2016) Oxytocin acting in the nucleus accumbens core decreases food intake. J Neuroendocrinol 28(4)

Hewson AK, Dickson SL (2000) Systemic administration of ghrelin induces Fos and Egr-1 proteins in the hypothalamic arcuate nucleus of fasted and fed rats. J Neuroendocrinol 12(11):1047–1049

Hill MN et al (2009) Circulating endocannabinoids and N-acyl ethanolamines are differentially regulated in major depression and following exposure to social stress. Psychoneuroendocrinology 34(8):1257–1262

Holst JJ (2007) The physiology of glucagon-like peptide 1. Physiol Rev 87(4):1409–1439

Horvath TL (2006) Synaptic plasticity in energy balance regulation. Obesity (Silver Spring) 14 (Suppl 5):228S–233S

Horvath TL et al (2010) Synaptic input organization of the melanocortin system predicts diet-induced hypothalamic reactive gliosis and obesity. Proc Natl Acad Sci U S A 107 (33):14875–14880

Hou J et al (2010) Orexigenic effect of cocaine- and amphetamine-regulated transcript (CART) after injection into hypothalamic nuclei in streptozotocin-diabetic rats. Clin Exp Pharmacol Physiol 37(10):989–995

Huszar D et al (1997) Targeted disruption of the melanocortin-4 receptor results in obesity in mice. Cell 88(1):131–141

Jacobowitz DM, O'Donohue TL (1978) Alpha-melanocyte stimulating hormone: immunohistochemical identification and mapping in neurons of rat brain. Proc Natl Acad Sci U S A 75 (12):6300–6304

Jordan J et al (2008) Stimulation of cholecystokinin-A receptors with GI181771X does not cause weight loss in overweight or obese patients. Clin Pharmacol Ther 83(2):281–287

Kamegai J et al (2001) Chronic central infusion of ghrelin increases hypothalamic neuropeptide Y and Agouti-related protein mRNA levels and body weight in rats. Diabetes 50(11):2438–2443

Katsurada K et al (2014) Endogenous GLP-1 acts on paraventricular nucleus to suppress feeding: projection from nucleus tractus solitarius and activation of corticotropin-releasing hormone, nesfatin-1 and oxytocin neurons. Biochem Biophys Res Commun 451(2):276–281

Kawano H, Masuko S (2000) Beta-endorphin-, adrenocorticotrophic hormone- and neuropeptide y-containing projection fibers from the arcuate hypothalamic nucleus make synaptic contacts on to nucleus preopticus medianus neurons projecting to the paraventricular hypothalamic nucleus in the rat. Neuroscience 98(3):555–565

Khalyfa A et al (2013) Effects of late gestational high-fat diet on body weight, metabolic regulation and adipokine expression in offspring. Int J Obes 37(11):1481–1489

Kim JG et al (2014) Leptin signaling in astrocytes regulates hypothalamic neuronal circuits and feeding. Nat Neurosci 17(7):908–910

Kim DW et al (2016) Maternal obesity in the mouse compromises the blood-brain barrier in the arcuate nucleus of offspring. Endocrinology 157(6):2229–2242

Kirchgessner AL, Sclafani A (1988) PVN-hindbrain pathway involved in the hypothalamic hyperphagia-obesity syndrome. Physiol Behav 42(6):517–528

Kirk SL et al (2009) Maternal obesity induced by diet in rats permanently influences central processes regulating food intake in offspring. PLoS One 4(6):e5870

Kissileff HR et al (1981) C-terminal octapeptide of cholecystokinin decreases food intake in man. Am J Clin Nutr 34(2):154–160
Kobelt P et al (2006) Peripheral injection of CCK-8S induces Fos expression in the dorsomedial hypothalamic nucleus in rats. Brain Res 1117(1):109–117
Koda S et al (2005) The role of the vagal nerve in peripheral PYY3-36-induced feeding reduction in rats. Endocrinology 146(5):2369–2375
Kojima M et al (1999) Ghrelin is a growth-hormone-releasing acylated peptide from stomach. Nature 402(6762):656–660
Krashes MJ et al (2013) Rapid versus delayed stimulation of feeding by the endogenously released AgRP neuron mediators GABA, NPY, and AgRP. Cell Metab 18(4):588–595
Kristensen P et al (1998) Hypothalamic CART is a new anorectic peptide regulated by leptin. Nature 393(6680):72–76
Kubota N et al (2007) Adiponectin stimulates AMP-activated protein kinase in the hypothalamus and increases food intake. Cell Metab 6(1):55–68
Labarthe A et al (2014) Ghrelin-derived peptides: a link between appetite/reward, GH axis, and psychiatric disorders? Front Endocrinol (Lausanne) 5:163
le Roux CW et al (2005) Postprandial plasma ghrelin is suppressed proportional to meal calorie content in normal-weight but not obese subjects. J Clin Endocrinol Metab 90(2):1068–1071
Leibowitz SF, Hammer NJ, Chang K (1981) Hypothalamic paraventricular nucleus lesions produce overeating and obesity in the rat. Physiol Behav 27(6):1031–1040
Levin BE, Dunn-Meynell AA (2002) Maternal obesity alters adiposity and monoamine function in genetically predisposed offspring. Am J Physiol Regul Integr Comp Physiol 283(5):R1087–R1093
Liddle RA et al (1985) Cholecystokinin bioactivity in human plasma. Molecular forms, responses to feeding, and relationship to gallbladder contraction. J Clin Invest 75(4):1144–1152
Lin S et al (2009) Critical role of arcuate Y4 receptors and the melanocortin system in pancreatic polypeptide-induced reduction in food intake in mice. PLoS One 4(12):e8488
Liu X, York DA, Bray GA (2004) Regulation of ghrelin gene expression in stomach and feeding response to a ghrelin analogue in two strains of rats. Peptides 25(12):2171–2177
Liu T et al (2012) Fasting activation of AgRP neurons requires NMDA receptors and involves spinogenesis and increased excitatory tone. Neuron 73(3):511–522
Loh K et al (2011) Elevated hypothalamic TCPTP in obesity contributes to cellular leptin resistance. Cell Metab 14(5):684–699
Luquet S, Phillips CT, Palmiter RD (2007) NPY/AgRP neurons are not essential for feeding responses to glucoprivation. Peptides 28(2):214–225
Maffei M et al (1995) Increased expression in adipocytes of ob RNA in mice with lesions of the hypothalamus and with mutations at the db locus. Proc Natl Acad Sci U S A 92(15):6957–6960
Maida A et al (2008) The glucagon-like peptide-1 receptor agonist oxyntomodulin enhances beta-cell function but does not inhibit gastric emptying in mice. Endocrinology 149(11):5670–5678
Malaisse-Lagae F et al (1977) Pancreatic polypeptide: a possible role in the regulation of food intake in the mouse. Hypothesis. Experientia 33(7):915–917
Marks JL et al (1990) Localization of insulin receptor mRNA in rat brain by in situ hybridization. Endocrinology 127(6):3234–3236
Matson CA, Reid DF, Ritter RC (2002) Daily CCK injection enhances reduction of body weight by chronic intracerebroventricular leptin infusion. Am J Physiol Regul Integr Comp Physiol 282 (5):R1368–R1373
Mavanji V et al (2015) Promotion of wakefulness and energy expenditure by orexin-a in the ventrolateral preoptic area. Sleep 38(9):1361–1370
Mayer J, Thomas DW (1967) Regulation of food intake and obesity. Science 156(3773):328–337
Mazen I et al (2011) Homozygosity for a novel missense mutation in the leptin receptor gene (P316T) in two Egyptian cousins with severe early onset obesity. Mol Genet Metab 102 (4):461–464

McGowan MK, Andrews KM, Grossman SP (1992a) Chronic intrahypothalamic infusions of insulin or insulin antibodies alter body weight and food intake in the rat. Physiol Behav 51 (4):753–766

McGowan MK, Andrews KM, Grossman SP (1992b) Role of intrahypothalamic insulin in circadian patterns of food intake, activity, and body temperature. Behav Neurosci 106(2):380–385

Minokoshi Y et al (2004) AMP-kinase regulates food intake by responding to hormonal and nutrient signals in the hypothalamus. Nature 428(6982):569–574

Miyata S (2015) New aspects in fenestrated capillary and tissue dynamics in the sensory circumventricular organs of adult brains. Front Neurosci 9:390

Monnikes H, Lauer G, Arnold R (1997) Peripheral administration of cholecystokinin activates c-fos expression in the locus coeruleus/subcoeruleus nucleus, dorsal vagal complex and paraventricular nucleus via capsaicin-sensitive vagal afferents and CCK-A receptors in the rat. Brain Res 770(1–2):277–288

Moran TH, Bi S (2006) Hyperphagia and obesity in OLETF rats lacking CCK-1 receptors. Philos Trans R Soc Lond Ser B Biol Sci 361(1471):1211–1218

Moran TH et al (1998) Disordered food intake and obesity in rats lacking cholecystokinin A receptors. Am J Phys 274(3 Pt 2):R618–R625

Mori H et al (2004) Socs3 deficiency in the brain elevates leptin sensitivity and confers resistance to diet-induced obesity. Nat Med 10(7):739–743

Moriarty P et al (1997) Characterization of cholecystokininA and cholecystokininB receptors expressed by vagal afferent neurons. Neuroscience 79(3):905–913

Morita S et al (2016) Heterogeneous vascular permeability and alternative diffusion barrier in sensory circumventricular organs of adult mouse brain. Cell Tissue Res 363(2):497–511

Murphy KG, Bloom SR (2004) Gut hormones in the control of appetite. Exp Physiol 89(5):507–516

Nakagawa A et al (2004) Receptor gene expression of glucagon-like peptide-1, but not glucose-dependent insulinotropic polypeptide, in rat nodose ganglion cells. Auton Neurosci 110 (1):36–43

Navarro M et al (2011) Assessment of voluntary ethanol consumption and the effects of a melanocortin (MC) receptor agonist on ethanol intake in mutant C57BL/6J mice lacking the MC-4 receptor. Alcohol Clin Exp Res 35(6):1058–1066

Nelson DW et al (2007) Localization and activation of glucagon-like peptide-2 receptors on vagal afferents in the rat. Endocrinology 148(5):1954–1962

Noetzel S et al (2009) CCK-8S activates c-Fos in a dose-dependent manner in nesfatin-1 immunoreactive neurons in the paraventricular nucleus of the hypothalamus and in the nucleus of the solitary tract of the brainstem. Regul Pept 157(1–3):84–91

Obici S et al (2002) Decreasing hypothalamic insulin receptors causes hyperphagia and insulin resistance in rats. Nat Neurosci 5(6):566–572

Olney JW (1969) Brain lesions, obesity, and other disturbances in mice treated with monosodium glutamate. Science 164(3880):719–721

Olszewski PK, Levine AS (2007) Central opioids and consumption of sweet tastants: when reward outweighs homeostasis. Physiol Behav 91(5):506–512

Olszewski PK, Levine AS (2016) Basic research on appetite regulation: social context of a meal is missing. Pharmacol Biochem Behav 148:106–107

Olszewski PK et al (2001) Role of alpha-MSH in the regulation of consummatory behavior: immunohistochemical evidence. Am J Physiol Regul Integr Comp Physiol 281(2):R673–R680

Olszewski P et al (2003) Agouti-related protein: appetite or reward? Ann N Y Acad Sci 994:187–191

Olszewski PK et al (2010) Molecular, immunohistochemical, and pharmacological evidence of oxytocin's role as inhibitor of carbohydrate but not fat intake. Endocrinology 151 (10):4736–4744

Olszewski PK, Allen K, Levine AS (2015) Effect of oxytocin receptor blockade on appetite for sugar is modified by social context. Appetite 86:81–87

Olszewski PK, Klockars A, Levine AS (2016) Oxytocin: a conditional anorexigen whose effects on appetite depend on the physiological, behavioural and social contexts. J Neuroendocrinol 28(4)
Ottaway N et al (2015) Diet-induced obese mice retain endogenous leptin action. Cell Metab 21 (6):877–882
Pandit R et al (2015) Central melanocortins regulate the motivation for sucrose reward. PLoS One 10(3):e0121768
Parkinson JR et al (2008) PYY3-36 injection in mice produces an acute anorexigenic effect followed by a delayed orexigenic effect not observed with other anorexigenic gut hormones. Am J Physiol Endocrinol Metab 294(4):E698–E708
Perello M et al (2010) Ghrelin increases the rewarding value of high-fat diet in an orexin-dependent manner. Biol Psychiatry 67(9):880–886
Pierce AA, Xu AW (2010) De novo neurogenesis in adult hypothalamus as a compensatory mechanism to regulate energy balance. J Neurosci 30(2):723–730
Pinto S et al (2004) Rapid rewiring of arcuate nucleus feeding circuits by leptin. Science 304 (5667):110–115
Provensi G et al (2014) Satiety factor oleoylethanolamide recruits the brain histaminergic system to inhibit food intake. Proc Natl Acad Sci U S A 111(31):11527–11532
Qian S et al (2002) Neither agouti-related protein nor neuropeptide Y is critically required for the regulation of energy homeostasis in mice. Mol Cell Biol 22(14):5027–5035
Rajia S, Chen H, Morris MJ (2010) Maternal overnutrition impacts offspring adiposity and brain appetite markers-modulation by postweaning diet. J Neuroendocrinol 22(8):905–914
Reynolds RM et al (2010) Maternal BMI, parity, and pregnancy weight gain: influences on offspring adiposity in young adulthood. J Clin Endocrinol Metab 95(12):5365–5369
Riediger T et al (2010) Effects of glucagon-like peptide 1 and oxyntomodulin on neuronal activity of ghrelin-sensitive neurons in the hypothalamic arcuate nucleus. Am J Physiol Regul Integr Comp Physiol 298(4):R1061–R1067
Rivera HM et al (2015) Maternal high-fat diet and obesity impact palatable food intake and dopamine signaling in nonhuman primate offspring. Obesity (Silver Spring) 23(11):2157–2164
Rocha PM et al (2011) Visceral abdominal and subfascial femoral adipose tissue have opposite associations with liver fat in overweight and obese premenopausal caucasian women. J Lipids 2011:154672
Rodriguez de Fonseca F et al (2001) An anorexic lipid mediator regulated by feeding. Nature 414 (6860):209–212
Rodriguez-Pacheco F et al (2007) Regulation of pituitary cell function by adiponectin. Endocrinology 148(1):401–410
Romano A et al (2014) High dietary fat intake influences the activation of specific hindbrain and hypothalamic nuclei by the satiety factor oleoylethanolamide. Physiol Behav 136:55–62
Romano A et al (2017) Role of the area postrema in the hypophagic effects of oleoylethanolamide. Pharmacol Res 122:20–34
Romero-Pico A et al (2013) Hypothalamic kappa-opioid receptor modulates the orexigenic effect of ghrelin. Neuropsychopharmacology 38(7):1296–1307
Ruter J et al (2003) Intraperitoneal injection of ghrelin induces Fos expression in the paraventricular nucleus of the hypothalamus in rats. Brain Res 991(1–2):26–33
Sabatier N et al (2003) Alpha-melanocyte-stimulating hormone stimulates oxytocin release from the dendrites of hypothalamic neurons while inhibiting oxytocin release from their terminals in the neurohypophysis. J Neurosci 23(32):10351–10358
Sainsbury A et al (2010) Y4 receptors and pancreatic polypeptide regulate food intake via hypothalamic orexin and brain-derived neurotropic factor dependent pathways. Neuropeptides 44(3):261–268
Schaeffer M et al (2013) Rapid sensing of circulating ghrelin by hypothalamic appetite-modifying neurons. Proc Natl Acad Sci U S A 110(4):1512–1517

Scherag A et al (2010) Two new loci for body-weight regulation identified in a joint analysis of genome-wide association studies for early-onset extreme obesity in French and German study groups. PLoS Genet 6(4):e1000916
Schioth HB et al (2005) Evolutionary conservation of the structural, pharmacological, and genomic characteristics of the melanocortin receptor subtypes. Peptides 26(10):1886–1900
Schwartz GJ (2000) The role of gastrointestinal vagal afferents in the control of food intake: current prospects. Nutrition 16(10):866–873
Schwartz GJ (2010) Brainstem integrative function in the central nervous system control of food intake. Forum Nutr 63:141–151
Schwartz GJ et al (1999) Gut vagal afferent lesions increase meal size but do not block gastric preload-induced feeding suppression. Am J Phys 276(6 Pt 2):R1623–R1629
Shi YC et al (2013) PYY3-36 and pancreatic polypeptide reduce food intake in an additive manner via distinct hypothalamic dependent pathways in mice. Obesity (Silver Spring) 21(12):E669–E678
Shor-Posner G et al (1985) Deficits in the control of food intake after hypothalamic paraventricular nucleus lesions. Physiol Behav 35(6):883–890
Shrestha YB, Wickwire K, Giraudo SQ (2004) Action of MT-II on ghrelin-induced feeding in the paraventricular nucleus of the hypothalamus. Neuroreport 15(8):1365–1367
Siegenthaler JA, Sohet F, Daneman R (2013) 'Sealing off the CNS': cellular and molecular regulation of blood-brain barriergenesis. Curr Opin Neurobiol 23(6):1057–1064
Sims JS, Lorden JF (1986) Effect of paraventricular nucleus lesions on body weight, food intake and insulin levels. Behav Brain Res 22(3):265–281
Skibicka KP et al (2013) Divergent circuitry underlying food reward and intake effects of ghrelin: dopaminergic VTA-accumbens projection mediates ghrelin's effect on food reward but not food intake. Neuropharmacology 73:274–283
Sobhani I et al (2000) Leptin secretion and leptin receptor in the human stomach. Gut 47 (2):178–183
Solomon A, De Fanti BA, Martinez JA (2007) Peripheral ghrelin interacts with orexin neurons in glucostatic signalling. Regul Pept 144(1–3):17–24
Stanley BG et al (1986) Neuropeptide Y chronically injected into the hypothalamus: a powerful neurochemical inducer of hyperphagia and obesity. Peptides 7(6):1189–1192
Sternson SM (2013) Hypothalamic survival circuits: blueprints for purposive behaviors. Neuron 77 (5):810–824
Sternson SM, Nicholas Betley J, Cao ZF (2013) Neural circuits and motivational processes for hunger. Curr Opin Neurobiol 23(3):353–360
Sun J et al (2016) Adiponectin potentiates the acute effects of leptin in arcuate Pomc neurons. Mol Metab 5(10):882–891
Suyama S et al (2016) Glucose level determines excitatory or inhibitory effects of adiponectin on arcuate POMC neuron activity and feeding. Sci Rep 6:30796
Suyama S et al (2017) Adiponectin at physiological level glucose-independently enhances inhibitory postsynaptic current onto NPY neurons in the hypothalamic arcuate nucleus. Neuropeptides 65:1–9
Suzuki Y et al (2014) Changes in mRNA expression of arcuate nucleus appetite-regulating peptides during lactation in rats. J Mol Endocrinol 52(2):97–109
Swanson LW, Simmons DM (1989) Differential steroid hormone and neural influences on peptide mRNA levels in CRH cells of the paraventricular nucleus: a hybridization histochemical study in the rat. J Comp Neurol 285(4):413–435
Sweet DC et al (1999) Feeding response to central orexins. Brain Res 821(2):535–538
Takayama K et al (2007) Expression of c-Fos protein in the brain after intravenous injection of ghrelin in rats. Neurosci Lett 417(3):292–296
Theodosis DT, Poulain DA (1984) Evidence that oxytocin-secreting neurons are involved in the ultrastructural reorganisation of the rat supraoptic nucleus apparent at lactation. Cell Tissue Res 235(1):217–219

Tracy AL et al (2008) The melanocortin antagonist AgRP (83-132) increases appetitive responding for a fat, but not a carbohydrate, reinforcer. Pharmacol Biochem Behav 89(3):263–271
Trevaskis JL et al (2010) Multi-hormonal weight loss combinations in diet-induced obese rats: therapeutic potential of cholecystokinin? Physiol Behav 100(2):187–195
Tuma P, Hubbard AL (2003) Transcytosis: crossing cellular barriers. Physiol Rev 83(3):871–932
Umehara H et al (2016) The hypophagic factor oleoylethanolamide differentially increases c-fos expression in appetite regulating centres in the brain of wild type and histamine deficient mice. Pharmacol Res 113(Pt A):100–107
Vaisse C et al (2000) Melanocortin-4 receptor mutations are a frequent and heterogeneous cause of morbid obesity. J Clin Invest 106(2):253–262
Vong L et al (2011) Leptin action on GABAergic neurons prevents obesity and reduces inhibitory tone to POMC neurons. Neuron 71(1):142–154
Wan Q et al (1997) Recruitment of functional GABA(A) receptors to postsynaptic domains by insulin. Nature 388(6643):686–690
Wang D et al (2015) Whole-brain mapping of the direct inputs and axonal projections of POMC and AgRP neurons. Front Neuroanat 9:40
Wang S et al (2017) Rs12970134 near MC4R is associated with appetite and beverage intake in overweight and obese children: a family-based association study in Chinese population. PLoS One 12(5):e0177983
West DB, Fey D, Woods SC (1984) Cholecystokinin persistently suppresses meal size but not food intake in free-feeding rats. Am J Phys 246(5 Pt 2):R776–R787
Wilkinson M et al (2007) Adipokine gene expression in brain and pituitary gland. Neuroendocrinology 86(3):191–209
Wirth MM, Giraudo SQ (2001) Effect of Agouti-related protein delivered to the dorsomedial nucleus of the hypothalamus on intake of a preferred versus a non-preferred diet. Brain Res 897(1–2):169–174
Wirth MM et al (2001) Paraventricular hypothalamic alpha-melanocyte-stimulating hormone and MTII reduce feeding without causing aversive effects. Peptides 22(1):129–134
Wirth MM et al (2002) Effect of Agouti-related protein on development of conditioned taste aversion and oxytocin neuronal activation. Neuroreport 13(10):1355–1358
Wren AM et al (2001) Ghrelin causes hyperphagia and obesity in rats. Diabetes 50(11):2540–2547
Wren AM et al (2002) The hypothalamic mechanisms of the hypophysiotropic action of ghrelin. Neuroendocrinology 76(5):316–324
Xu B et al (2003) Brain-derived neurotrophic factor regulates energy balance downstream of melanocortin-4 receptor. Nat Neurosci 6(7):736–742
Yang Y et al (2007) Mechanism of oleoylethanolamide on fatty acid uptake in small intestine after food intake and body weight reduction. Am J Physiol Regul Integr Comp Physiol 292(1):R235–R241
Yang Y et al (2011) Hunger states switch a flip-flop memory circuit via a synaptic AMPK-dependent positive feedback loop. Cell 146(6):992–1003
Yang L, Qi Y, Yang Y (2015) Astrocytes control food intake by inhibiting AGRP neuron activity via adenosine A1 receptors. Cell Rep 11(5):798–807
Yaswen L et al (1999) Obesity in the mouse model of pro-opiomelanocortin deficiency responds to peripheral melanocortin. Nat Med 5(9):1066–1070
Zambrano E et al (2006) A low maternal protein diet during pregnancy and lactation has sex- and window of exposure-specific effects on offspring growth and food intake, glucose metabolism and serum leptin in the rat. J Physiol 571(Pt 1):221–230
Zhang Y et al (1994) Positional cloning of the mouse obese gene and its human homologue. Nature 372(6505):425–432
Zhang ZY, Dodd GT, Tiganis T (2015) Protein tyrosine phosphatases in hypothalamic insulin and leptin signaling. Trends Pharmacol Sci 36(10):661–674
Zigman JM et al (2006) Expression of ghrelin receptor mRNA in the rat and the mouse brain. J Comp Neurol 494(3):528–548

The Impact of Stressor Exposure and Glucocorticoids on Anxiety and Fear

J. E. Hassell Jr., K. T. Nguyen, C. A. Gates, and C. A. Lowry

Contents

1 Definitions of Stress, Anxiety, and Fear 274
2 Neurocircuits Controlling HPA Axis, Anxiety, and Fear Responses 278
2.1 Neurocircuitry of HPA Axis Responses to Aversive Stimuli 278
2.2 Neurocircuitry of Anxiety 279
2.3 Neurocircuitry of Fear 283
3 Stress-Induced Exaggeration of Anxiety- and Fear-Related Responses: Introduction to the Hypothalamic-Pituitary-Adrenal Axis 287
4 Models of Stress-Induced Vulnerability to Exaggerated Anxiety- and Fear-Related Behavioral Responses 289
4.1 Maternal Separation: A Model of Adverse early Life Experience Leading to Lifelong Vulnerability to Exaggerated Anxiety- and Fear-Related Behavioral Responses 289

J. E. Hassell Jr.
Department of Integrative Physiology, University of Colorado Boulder, Boulder, CO, USA

Center for Neuroscience, University of Colorado Boulder, Boulder, CO, USA
e-mail: james.hassell@colorado.edu

K. T. Nguyen and C. A. Gates
Department of Integrative Physiology, University of Colorado Boulder, Boulder, CO, USA
e-mail: kadi.nguyen@colorado.edu; chloe.gates@colorado.edu

C. A. Lowry (✉)
Department of Integrative Physiology, University of Colorado Boulder, Boulder, CO, USA

Center for Neuroscience, University of Colorado Boulder, Boulder, CO, USA

Department of Physical Medicine and Rehabilitation, University of Colorado Anschutz Medical Campus, Aurora, CO, USA

Center for Neuroscience, University of Colorado Anschutz Medical Campus, Aurora, CO, USA

Veterans Health Administration, Rocky Mountain Mental Illness Research Education and Clinical Center, Denver Veterans Affairs Medical Center (VAMC), Denver, CO, USA

Military and Veteran Microbiome Consortium for Research and Education (MVM-CoRE), Denver, CO, USA
e-mail: christopher.lowry@colorado.edu

Curr Topics Behav Neurosci (2019) 43: 271–322
DOI 10.1007/7854_2018_63
Published Online: 25 October 2018

4.2 Adolescent Social Isolation: A Model of Stress-Induced Vulnerability to Exaggerated Anxiety- and Fear-Related Behavioral Responses 290
4.3 Learned Helplessness: A Model of Stress-Induced Vulnerability to Exaggerated Anxiety- and Fear-Related Behavioral Responses 291
4.4 Stress-Enhanced Fear Learning (SEFL): A Model of Stress-Induced Vulnerability to Exaggerated Anxiety- and Fear-Related Behavioral Responses 294
4.5 Chronic Subordinate Colony Housing (CSC) Model: A Model of Stress-Induced Vulnerability to Exaggerated Anxiety- and Fear-Related Behavioral Responses . . . 295
5 Focus on Glucocorticoid-Dependent Mechanisms of Stress-Induced Vulnerability to Exaggerated Anxiety-Like and Fear-Related Behavioral Responses 296
5.1 Glucocorticoids and Memory 296
5.2 Glucocorticoid Effects on CRH mRNA Expression in the Amygdala and BNST . . 298
5.3 Stress, Glucocorticoids, and Organic Cation Transporter 3 (OCT3) 299
5.4 Chronic Glucocorticoid Administration Increases Tryptophan Hydroxylase 2 (Tph2) mRNA and Protein Expression and Tph2 Activity in Association with Exaggerated Anxiety-Like Behavioral Responses 300
6 Conclusions 305
References 306

Abstract Anxiety disorders and trauma- and stressor-related disorders, such as posttraumatic stress disorder (PTSD), are common and are associated with significant economic and social burdens. Although trauma and stressor exposure are recognized as a risk factors for development of anxiety disorders and trauma or stressor exposure is recognized as essential for diagnosis of PTSD, the mechanisms through which trauma and stressor exposure lead to these disorders are not well characterized. An improved understanding of the mechanisms through which trauma or stressor exposure leads to development and persistence of anxiety disorders or PTSD may result in novel therapeutic approaches for the treatment of these disorders. Here, we review the current state-of-the-art theories, with respect to mechanisms through which stressor exposure leads to acute or chronic exaggeration of avoidance or anxiety-like defensive behavioral responses and fear, endophenotypes in both anxiety disorders and trauma- and stressor-related psychiatric disorders. In this chapter, we will explore physiological responses and neural circuits involved in the development of acute and chronic exaggeration of anxiety-like defensive behavioral responses and fear states, focusing on the role of the hypothalamic-pituitary-adrenal (HPA) axis and glucocorticoid hormones.

Keywords Amygdala · Anxiety · Corticosterone · Corticotropin-releasing factor · Fear · Hypothalamic-pituitary-adrenal axis · Memory consolidation · Microbiome · Organic cation transporter 3

Abbreviations

5-HT	5-Hydroxytryptamine (serotonin)
5-HT_{1A}	Serotonin 1A receptor
5-HTT	Serotonin transporter

aBNST	Anterior bed nucleus of the stria terminalis
ACTH	Adrenocorticotropic hormone
alBNST	Anterolateral bed nucleus of the stria terminalis
amBNST	Anteromedial bed nucleus of the stria terminalis
AMPA receptors	α-Amino-3-hydroxy-5-methyl-4-isoxazolepropionic acid receptor
BLA	Basolateral amygdala
BMI	Bicuculline methiodide
BNST	Bed nucleus of the stria terminalis
CeA	Central nucleus of the amygdala
CeAL	Central nucleus of the amygdala, lateral division
CeAM	Central nucleus of the amygdala, medial division
CNS	Central nervous system
CPFE	Context preexposure facilitation effect
CRF	Corticotropin-releasing factor
CRH	Corticotropin-releasing hormone
CRHR	Corticotropin-releasing hormone receptor
CS	Conditioned stimulus
CSC	Chronic subordinate colony housing
ES	Escapable shock
GABA	Gamma-aminobutyric acid
GF	Germ-free
HPA	Hypothalamic-pituitary-adrenal
IL	Interleukin
IS	Inescapable shock
LA	Lateral amygdala
L-AG	L-allylglycine
mPFCv	Ventral medial prefrontal cortex
NMDA	*N*-methyl-D-aspartate
OCT	Organic cation transporter
PVN	Paraventricular nucleus of the hypothalamus
RVLM	Rostral ventrolateral medulla
SEFL	Stress-enhanced fear learning
SHC	Single-housed control
SPF	Specific-pathogen-free
tph2	Tryptophan hydroxylase 2 gene
UCN	Urocortin
US	Unconditioned stimulus
vlPAG	Ventrolateral periaqueductal gray

1 Definitions of Stress, Anxiety, and Fear

Concepts such as homeostasis (Cannon 1929), briefly defined as coordinated physiological processes that maintain life, and stress (Selye 1950), a non-specific multi-system physiological reaction to an overwhelming real or perceived threat (Day 2005), are critical concepts that have been at the center of biomedicine. Traditionally, the term "stressor" refers to a stimulus that elicits a stress response, with a stress response being a physiological reaction to stressors. Animals have developed behavioral and physiological strategies to survive and meet the demands of their environments to maintain homeostasis (Medzhitov 2008; LeDoux 2012). The next frontier in stress research may be understanding physiological and behavioral consequences of chronic adverse experience in the context of either allostasis, the process by which animals maintain stability, or homeostasis, through change [i.e., the collective adaptive behavioral, immune (e.g. inflammatory), autonomic, and neuroendocrine mediator responses (i.e., those involved in growth, reproduction, and metabolism)] (McEwen and Wingfield 2003; McEwen 1998b; Miller and Raison 2016). For definitions of key terms related to stress physiology and allostasis, see Table 1.

Animals have developed behavioral and physiological strategies to survive and meet the demands of their environments to maintain homeostasis (Medzhitov 2008; LeDoux 2012). The allostasis model explores how fluctuating energy demands due to external stressors facilitate the use of physiological mediators to reestablish homeostasis through adaptive responses. Allostatic load refers to the energy expended to accommodate predictable and unpredictable conditions in order to maintain homeostasis. An overwhelming allostatic load leads to Type 1 allostatic overload, defined as a negative energy balance, in which energy demands exceed energy income or energy that can be mobilized from stores (McEwen and Wingfield 2003); Type 2 allostatic overload is the chronic dysregulation of physiological mediators of allostasis, such as glucocorticoids (McEwen 1998b; McEwen and Wingfield 2003, 2010). The allostasis model attempts to explain physiological/behavioral reactions to, or anticipation of, harmful stimuli. The resulting physiological and/or behavioral changes, specifically increases in physiological mediators (e.g., glucocorticoids), aim to alleviate the stress of allostatic load; the aim is not necessarily to restore homeostatic parameters directly.

One of the key ideas of the allostasis model is that allostatic mediators, including glucocorticoids, act to facilitate adaptation to an adverse situation. However, repeated engagement or dysregulation of these mediators may result in "wear and tear" to various physiological systems, which contribute to allostatic overload (McEwen 1998b). Prolonged predictable/unpredictable allostatic load can progress into Type 1 allostatic overload. Type 2 allostatic overload, characterized by a dysregulation of these chronically present mediators, which could induce a pathological state, is a chronic maladaptive response. In this case, the allostatic mediators themselves become a source of harm, potentially resulting in a state of anxiety/fear (McEwen and Wingfield 2003, 2010). For our discussion purposes, we will address stress in the context of the allostatic model.

Table 1 Definitions of key terms related to stress physiology and allostasis

Stress	A physiological state associated with disruption of homeostasis due to environmental, physical, or psychological stimuli (i.e., stressors) that elicit adaptive physiological and behavioral responses to restore homeostasis (i.e., the stress response) (Glaser and Kiecolt-Glaser 2005)
Stressors	Stressors can be described as selective pressures from the physical and social environment that threaten or challenge an organism and elicit compensatory response patterns (Weiner 1991)
Stress response	A stress response is the body's multi-system response to any challenge or stressor that overwhelms, or is judged likely to overwhelm, selective homeostatic response mechanisms (Day 2005)
Homeostasis	Physiological mechanisms by which the body maintains stability (body temperature, pH, glucose levels, etc.) despite fluctuating environmental conditions and stressful stimuli (McEwen and Wingfield 2007)
Allostasis	The process by which animals maintain stability, or homeostasis, through change [i.e., the collective adaptive behavioral, immune (e.g. inflammatory), autonomic, and neuroendocrine mediator responses (e.g., those involved in growth, reproduction, and metabolism)] (McEwen and Wingfield 2007)
Allostasis model	This model explores how fluctuating energy demands due to external stressors facilitate the use of mediators to reestablish homeostasis through adaptive responses (McEwen and Wingfield 2003)
Allostatic load	The wear and tear that the body experiences due to repeated cycles of allostasis (McEwen 1998a) as well as the inefficient turning on or shutting off of these responses (McEwen and Stellar 1993)
Type 1 allostatic overload	Type 1 allostatic overload occurs when energy demand exceeds supply, resulting in activation of the emergency life history stage. This serves to direct the animal away from normal life history stages into a survival mode that decreases allostatic load and regains positive energy balance. The normal life cycle can be resumed when the perturbation passes (Sterling 1988; McEwen and Wingfield 2003)
Type 2 allostatic overload	Type 2 allostatic overload occurs when there is sufficient or even excess energy consumption accompanied by social conflict and other types of social dysfunction. The latter is the case in human society and certain situations affecting animals in captivity. In all cases, secretion of glucocorticoids and activity of other mediators of allostasis such as the autonomic nervous system, central nervous system neurotransmitters, and inflammatory cytokines wax and wane with allostatic load. If allostatic load is chronically high, then pathologies develop. Type 2 allostatic overload does not trigger an escape response and can only be counteracted through learning and changes in the social structure (Sterling 1988; McEwen and Wingfield 2003)

These terms provide a comprehensive framework in which researchers can separate the protective effects of physiological mediators during acute stress exposure from the detrimental effects of these same mediators during chronic or repeated stress exposure. It is believed that allostatic systems, specifically the nervous, endocrine, and immune systems, mediate these adaptive responses during acute stressors (Danese and McEwen 2012). For example, normal stimulation of the hypothalamic-pituitary-adrenal (HPA) axis under acute stress induces energy mobilization to address increased energy demands (Dallman et al. 1995). However, during chronic or repeated stress, prolonged activation of the nervous, endocrine, and immune systems can produce excess mediators that prove harmful to the body. For example, in the nervous system, chronic exposure to stressors can lead to functional impairment of the prefrontal cortex, the amygdala, and the hippocampus (McEwen and Gianaros 2010). In the endocrine system, chronic stress results in increased corticotropin-releasing hormone (CRH) levels due to a dysregulation of the HPA axis (Miller et al. 2007). Furthermore, hyperactive HPA axis functionality can lead to "wear and tear" on the organism, due to excessive exposure to catabolic properties of glucocorticoids (Stephens and Wand 2012).

The "wear and tear" effect of allostatic load induced by elevated CRH and systemic glucocorticoids can result in chronic disorders that compromise immune function, metabolic activity, and general physical and mental health (Schulkin et al. 1998). One stressor that results in vulnerability to psychopathology in adult life is early life adversity (Aisa et al. 2007). Studies suggest that stressful, traumatic, or anxiety-inducing early childhood experiences may have enduring consequences on the development and function of allostatic systems (Danese and McEwen 2012). The repeated external threat of childhood stress signals to the developing brain, producing an allostatic load that could increase risk for anxiety disorders, such as generalized anxiety disorder, trauma- and stressor-related disorders, such as PTSD, and stress-related affective disorders, such as major depressive disorder. These stress- and anxiety-related disorders stemming from early life adversity are associated with dysregulated HPA axis function and elevated/depressed basal cortisol levels (among other allostatic mediators) (Danese and McEwen 2012).

Anxiety-related behavioral responses are defensive behavioral responses characterized by increased autonomic and behavioral arousal (Lowry et al. 2005) associated with a diffuse, unspecified threat in time and space (Walker and Davis 2008). In McNaughton and Corr's (2004) work, anxiety and fear are categorized into two separate defensive behaviors that recruit different neural networks, with some overlap, to either accomplish defensive approach (i.e., anxiety) or defensive avoidance (i.e., fear; McNaughton and Corr 2004) (see Fig. 1). This description of anxiety is in line with those of Gray and McNaughton, in which anxiety-related defensive behaviors are associated with a state of conflict between approaching a potential rewarding outcome and avoiding a potential aversive outcome. Inherent in this conflict between approach and avoidance is a high level of unpredictability or uncertainty of outcomes (Gray and McNaughton 2003). Defensive approach/avoidance is guided by a subjective defensive distance, which is a dimension that animals use to assess the imminence of a real or perceived threat (Blanchard and Blanchard 1989). A critical distinction between anxiety and fear is the "defensive direction." In anxiety, due to a conflict between potential rewarding outcomes and potential aversive outcomes, an animal is driven to move toward potential threat, thereby decreasing defensive distance using risk assessment. With fear, the defensive direction of the animal is to move away from the threat, thereby increasing defensive distance with characteristic behaviors such as fight, flight, or freezing (Blanchard

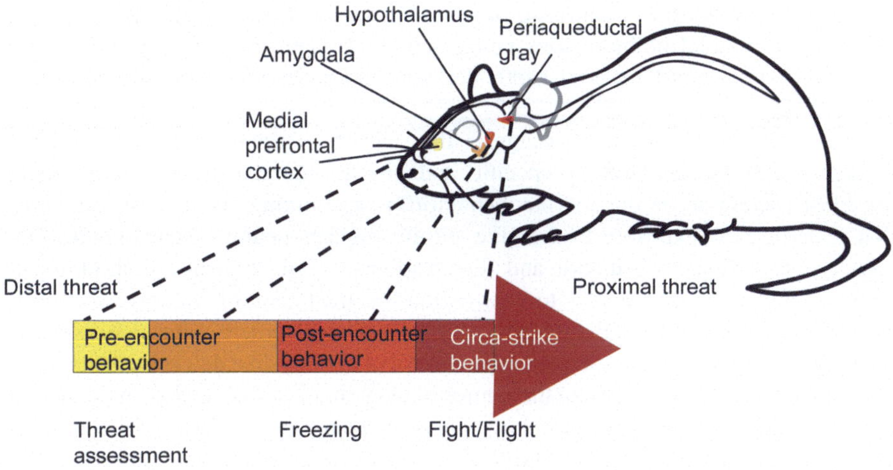

Fig. 1 Theoretical model of defensive behaviors, based on McNaughton and Corr (2004). Distal perceived threats are detected, and pre-encounter behaviors, such as threat assessment, are mobilized by rostral regions of the brain, such as the prefrontal cortex. As perceived threats become closer, post-encounter behaviors, like freezing, become mobilized, and more caudal regions of the brain, such as the amygdala, organize behavioral and physiological processes. When a perceived threat is most proximal, caudal regions of the brain, such as the periaqueductal gray, mobilize circa-strike behaviors, such as fight-or-flight behaviors

et al. 2011; McNaughton and Corr 2004). Evidence suggests that a neural hierarchy is responsible for controlling defensive approach and defensive avoidance, with rostral parts of the brain, such as the prefrontal cortex, involved in threat assessment and more caudal parts of the brain, such as the periaqueductal gray, executing behavioral responses to more proximate threats, including approach/escape behaviors. In a similar vein of defensive distance, Fanselow and Lester proposed a threat imminence continuum that animals use in order to engage defensive behaviors (Fanselow and Lester 1988). This continuum is similar to McNaughton and Corr's defensive distance, whereby it incorporates a physical distance, a temporal distance, and a subjective psychological distance in order to assess the predator threat imminence. Three main categories of behavior are associated with the predator threat imminence: pre-encounter defensive behavior(s), post-encounter defensive behavior(s), and circa-strike defensive behavior(s) (Fanselow and Lester 1988). Similar to the above model, which proposes that rostral portions of the brain control threat assessment and caudal portions of the brain control the proximate behaviors mentioned above, overlaying these categories of behaviors are brain networks that would coordinate the appropriate behavior for pre-encounter defensive behavior(s). Networks coordinating pre-encounter defensive behavior(s) would primarily involve the prefrontal cortex, while networks coordinating circa-strike behavior(s) would primarily involve hindbrain nuclei, such as the periaqueductal gray (Perusini and Fanselow 2015) (Fig. 1). Human research has supported these notions of anxiety and fear with functional magnetic resonance imaging (fMRI) studies showing rostral regions of the brain being activated in association with a perceived large defensive distance and caudal brain regions being activated when the threat became more proximal (Bolles and Fanselow 1980; Walker and Davis 2008; Mobbs et al. 2007, 2009, 2010).

This theoretical framework suggests that anxiety-, stress-, and trauma-related disorders are characterized by specific and unique neural networks involved in pre-/post-encounter or circa-strike behavior(s). Examining the neural structures that underlie how exposure to aversive stimuli impacts neural systems controlling anxiety-like defensive behavior and fear responses could inform our thinking of novel therapeutic approaches to prevention or treatment of anxiety disorders, trauma- and stressor-related disorders, and affective disorders. First, we examine the neurocircuitry that is associated with control of anxiety-like defensive behavioral responses. Second, we examine the neurocircuitry that is associated with control of fear responses. We then evaluate how stress exposure can lead to sensitization of anxiety-like defensive behavior and fear responses and mechanisms underlying the persistence or chronicity of these altered states, including glucocorticoid-dependent and glucocorticoid-independent mechanisms.

2 Neurocircuits Controlling HPA Axis, Anxiety, and Fear Responses

Brain circuits involved in physiological and behavioral responses to stressful stimuli are different, depending on the type of stressor (Dayas et al. 2001; Dayas and Day 2002), acute versus chronic stress (Ostrander et al. 2009; Flak et al. 2012; McKlveen et al. 2016), and developmental and genetic history of the organism (Caspi et al. 2003; Liu et al. 1997, 2000). Brain circuits controlling the HPA axis, as well as those controlling anxiety-related defensive behavioral responses and fear-related behavioral states, have many overlapping regions in the brain (Poulos et al. 2010; Herman and Cullinan 1997), which can be influenced by allostatic mediators, such as glucocorticoids. In some, but not all, cases, glucocorticoids are involved in the stress-induced exaggeration of anxiety- and fear-related behavioral responses. Before discussing specific examples, first we explore neural circuitry mediating HPA axis, anxiety, and fear responses.

2.1 Neurocircuitry of HPA Axis Responses to Aversive Stimuli

Traditionally, studies of neuroendocrine and autonomic responses to aversive stimuli have focused on the paraventricular nucleus of the hypothalamus (PVN) and rostral ventrolateral medulla, respectively, as the proximate effector brain regions controlling these responses. In addition, a hierarchical system of brain nuclei is implicated in glucocorticoid-mediated negative feedback inhibition of the HPA axis, including the PVN itself (Herman et al. 2016), the dorsomedial hypothalamus (DMH) (Stamper et al. 2015), the prefrontal cortex, and the ventral subiculum (reviewed by Herman et al. 2016). This section will describe the diverse neural regions that have been shown to coordinate the neurocircuitry of HPA axis responses to aversive stimuli.

The primary neuroendocrine outputs that activate the HPA axis are CRH neurons within the PVN. These CRH neurons release CRH from axon terminals into the zona externa of the median eminence where CRH enters the hypothalamic-hypophyseal portal circulation and ultimately stimulates corticotrophs in the anterior pituitary to release ACTH.

Other areas of the brain involved in HPA axis responses to aversive stimuli include the dorsal hippocampus and the BNST, which exert regulatory control of hypothalamic nuclei mediating HPA axis responses (Myers et al. 2014; Petrov et al. 1994). Specifically, the aBNST and the pBNST, both parts of the extended amygdala, activate and inhibit the HPA axis, respectively, during an acute stressor (Choi et al. 2007). In addition, neurons within the BNST are also thought to play a role in stress recovery and regulating the extent of the HPA axis response to acute stressors (Henckens et al. 2017). Hypothalamic nuclei may also regulate the HPA axis by stimulating or inhibiting the PVN, such as the medial preoptic area and DMH, both of which are activated following stressor exposure (Fontes et al. 2011).

These data show that the regulation of HPA axis responses to aversive stimuli depends on a balance between activation and inhibition of CRH-synthesizing neurons in the PVN by specific brain nuclei (e.g., BNST or DMH) that are part of hierarchical systems that control activation of the HPA axis. This relationship between the activity of individual brain regions and circuits as a whole allows the brain to react to stressors of different modalities and different intensities in an appropriate manner. However, loss of balance between activation and inhibition could result in dysregulation and possibly stress-related disorders.

2.2 *Neurocircuitry of Anxiety*

Anxiety-related defensive behaviors are evident when a threat, real or perceived, is considered diffuse in time and space (Perusini and Fanselow 2015; Walker and Davis 2008). The prefrontal cortex, ventral hippocampus, and bed nucleus of the stria terminalis (BNST) are brain regions that integrate threatening information for initiating anxiety-like behaviors (Calhoon and Tye 2015). The basolateral amygdala complex (BL), which consists of a lateral nucleus (LA) and basal nucleus, receives sensory information from the cortex and thalamus (LeDoux et al. 1990a, b). The basolateral nucleus of the amygdala (BLA) sends projections to multiple areas of the brain, including the lateral division (CeL) of the central nucleus of the amygdala (CeA) (Tye et al. 2011), as well as other brain areas, such as the ventral hippocampus (vHPC) (Felix-Ortiz et al. 2013), medial prefrontal cortex (mPFC) (Felix-Ortiz et al. 2016), and BNST (Kim et al. 2013). These projections and their effects on anxiety-like behaviors are shown in Fig. 2.

Using optogenetics, Tye et al. demonstrated that the BLA sends excitatory inputs to the CeL, which then sends feedforward inhibitory inputs to CeM neurons, resulting in an anxiolytic (i.e., anxiety inhibiting), phenotype (Tye et al. 2011) (Fig. 2). Interestingly, direct optogenetic inhibition of BLA-CeL projection neurons results in inhibition of the feedforward inhibitory inputs to CeM neurons, and thus disinhibition of CeM output and anxiety-like behavioral responses. Disinhibition of CeM neurons corresponded with an anxiogenic, anxiety inducing, phenotype, as shown in the elevated plus-maze and open-field tests (Tye et al. 2011). These results suggest a bidirectional balance within the BLA-CeA circuits that is maintained in order to express anxiety-related behavior.

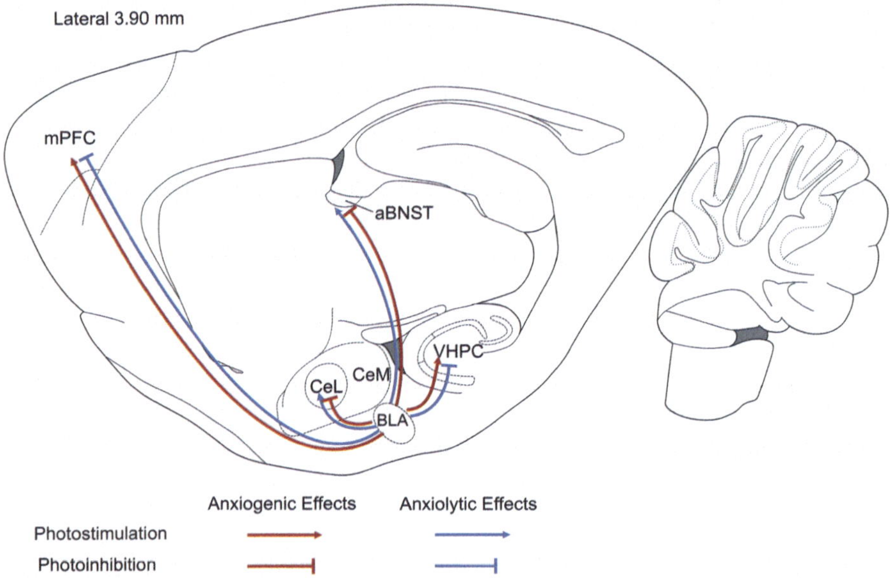

Fig. 2 The projections of the basolateral amygdala (BLA) in the neurocircuitry of anxiety-like defensive behavioral responses. Neurons from the BLA project to various regions to regulate anxiety-like behavioral responses, including the medial prefrontal cortex (mPFC) (Felix-Ortiz et al. 2016), the central nucleus of the amygdala, lateral region (CeL), the ventral hippocampus (vHPC) (Felix-Ortiz et al. 2013), and the bed nucleus of the stria terminalis (BNST) (Kim et al. 2013). Optogenetic stimulation of the BLA projections to the mPFC or the vHPC results in anxiogenic effects. In contrast, optogenetic inhibition of the same pathways results in anxiolytic effects. Optogenetic stimulation of the BLA projections to the anterior BNST (aBNST) (Kim et al. 2013) or the CeL (Tye et al. 2011; Haubensak et al. 2010) [the latter of which projects on to the central nucleus of the amygdala, medial region (CeM)] results in anxiolytic-like effects. In contrast, optogenetic inhibition of the aBNST results in anxiogenic effects. Sagittal brain diagram adapted from Paxinos and Watson (2007). Abbreviations: *aBNST* anterior subnucleus of the bed nucleus of the stria terminalis, *BLA* basolateral amygdala, *CeL* central nucleus of the amygdala, lateral part, *CeM* central nucleus of the amygdala, medial part, *mPFC* medial prefrontal cortex, *VHPC* ventral hippocampus

A BLA to vHPC circuit also contributes to control of anxiety-like behavior (Felix-Ortiz et al. 2013). Optogenetic activation of the BLA-vHPC circuit induces an anxiogenic phenotype, and optogenetic inhibition induces anxiolytic behavior on the elevated plus-maze and open-field tests (Felix-Ortiz et al. 2013) (Fig. 2).

A BLA to mPFC circuit also contributes to control of anxiety-like behavior. Optogenetic stimulation of BLA terminals within the mPFC produces anxiogenic behavior, while optogenetic inhibition of BLA terminals within the mPFC induces anxiolytic behavior on the elevated plus-maze and open-field tests (Felix-Ortiz et al. 2016) (Fig. 2).

In support of the hypothesis that the BLA is a nodal structure in the neural circuitry controlling anxiety-related behavioral responses, pharmacological approaches to (1) acute activation of the BLA or (2) priming of the BLA lead to acute anxiety-like responses or chronic vulnerability to exaggerated anxiety-like behavioral responses, respectively. Physiological aspects of anxiety include increased heart rate, increased blood pressure, and increased respiratory rate (Sajdyk and Shekhar 2000). Acute infusion of bicuculline methiodide (BMI), a gamma-aminobutyric acid $(GABA)_A$ receptor antagonist, into the BLA was found to significantly increase heart rate and blood pressure (Sanders and Shekhar 1991), block the anxiolytic effects of the antianxiety drug chlordiazepoxide (Sanders and Shekhar 1995a), and increase anxiety-like behavior (Sanders and Shekhar 1995b). Sanders and Shekhar noticed that, with subsequent bouts of disinhibition of the BLA, heart rate and blood pressure continued to increase above and beyond the previous BMI microinfusion. This led to the idea that the BLA could be "primed" with consecutive subthreshold infusions of BMI, and indeed repeated subthreshold microinfusions of BMI lead to long-term changes that last for at least 6 weeks after "priming," including increases in heart rate and blood pressure in response to a subsequent challenge with a subthreshold infusion of BMI (Sanders et al. 1995; Sajdyk and Shekhar 2000). Priming the BLA once daily for at least 3–5 days led to anxiogenic behavior such as decreased social interaction and increased conflict anxiety-like behavior (Sanders et al. 1995; Sajdyk and Shekhar 2000). Similar to GABA disinhibition of the BLA with BMI, repeated daily microinfusions of subthreshold CRF or urocortin 1 (Ucn1), a neuropeptide with affinity for both CRHR1 and CRHR2 receptors (Vaughan et al. 1995), into the BLA also increase heart rate, blood pressure, and anxiety-like behavior (Sajdyk et al. 1999). In fact, "priming" of the BLA with Ucn1 was found to decrease GABAergic inhibitory inputs within the BLA, and priming could be prevented by blocking either *N*-methyl-D-aspartate (NMDA) receptors or calcium-calmodulin kinase II (CaMKII), one signaling cascade of NMDA receptors (Rainnie et al. 2004). Corticotropin-releasing hormone primarily binds to two G-protein-coupled receptors (CRHR1 and CRHR2) with CRHR1 thought to be responsible for anxiogenic effects of Ucn1 and CRH (Bale and Vale 2004). Supporting this notion, the anxiogenic effect of repeated stress and intra-BLA Ucn1 was blocked with a prior systemic CRHR1 antagonist (Gehlert et al. 2005). One consequence of BLA priming is the activation of serotonergic neurons within the dorsal raphe nucleus and median raphe nucleus, suggesting that the BLA stimulates serotonin release from these regions (Spiga et al. 2006). Additionally, intra-BLA Ucn1 priming increases *tph2* mRNA expression within the DRVL, a region of the dorsal raphe nucleus associated with inhibition of panic-like escape behaviors (Spiacci et al. 2016; Donner et al. 2012a; Johnson et al. 2004). As noted earlier, the BLA is in a central position to influence areas of the brain involved with anxiety-related physiological and behavioral responses such as the CeA (Tye et al. 2011), vHPC (Felix-Ortiz et al. 2013), mPFC (Felix-Ortiz et al. 2016), and BNST, and that a "primed" and hyperactive BLA could be a mechanism for the development of stress-induced anxiety states.

Finally, a BLA to BNST circuit contributes to control of anxiety-like behavior. The BNST has several subnuclei, such as the anterior nucleus (aBNST) and posterior nucleus (pBNST). The aBNST can be further subdivided into anterolateral (alBNST), anteromedial (amBNST), and oval nuclei (ovBNST) (Gungor and Paré 2016). The alBNST is involved in control of behavioral responses in a sustained fear-potentiated startle paradigm, mimicking a diffuse threat, suggestive of an anxiety-like state (Daldrup et al. 2015, 2016). The BLA sends projections to the aBNST, and optogenetic stimulation of BLA projections to the aBNST induces anxiolytic behavioral responses; optogenetic inhibition of the BLA-aBNST corresponds with anxiogenic behavior on the elevated plus-maze and open-field tests (Kim et al. 2013) (Fig. 2). However, the ovBNST region has an opposing, anxiogenic nature in which optogenetic stimulation of its projections to the aBNST increases anxiety-like behavior and optogenetic inhibition decreases anxiety-like behavior. Taken together, the BLA is positioned to modulate anxiety-like behavior through projections to several nodes within a distributed network of brain nuclei that control anxiety-related behavior in a bidirectional manner.

The BNST is a neural hub for control of anxiety-like behaviors, as shown with light-enhanced startle (Walker and Davis 1997), CRH-enhanced startle (Lee and Davis 1997), and the elevated plus-maze (Sahuque et al. 2006). Similar to the BLA, decreased inhibitory tone of the BNST using the GABA synthesis inhibitor, L-AG, results in decreased social interaction measured 2 weeks later and increased anxiety-like behavior on the elevated plus-maze nearly a month after initiation of suppression of GABA synthesis (Sajdyk et al. 2008). In contrast to the BLA, however, is the apparent lack of autonomic responses, such as increases in heart rate and blood pressure in response to intravenous sodium lactate (Sajdyk et al. 2008), suggesting that the BNST may be selectively involved in control of conflict anxiety-like behavioral responses. The same pattern emerges with a once-a-day administration, for a period of 5 days, of intra-BNST Ucn1, whereby rats display a decreased social interaction but a lack of autonomic responses to intravenous sodium lactate (Lee et al. 2008).

Data suggest that the DMH may play a role in the autonomic and fight-or-flight responses to anxiogenic or panicogenic agents. The DMH is a hub for coordinating defensive reactions in response to stressors (Kataoka et al. 2014; Fontes et al. 2011; Zaretskaia et al. 2008). The DMH is also under tonic inhibitory control (Johnson et al. 2008b). Acute disinhibition of the DMH with BMI elicits a characteristic defensive "escape" behavior with increased locomotor activity and rearing in an open-field (Shekhar and DiMicco 1987; Zaretskaia et al. 2008). To disambiguate the increased locomotor behavior in response to intra-DMH BMI from appetitive and aversive behaviors, rats were tested on a modified conflict anxiety paradigm to test whether inhibiting or disinhibiting the DMH increased or decreased anxiety-like behavior (Shekhar et al. 1990). Inhibiting the DMH with muscimol, a $GABA_A$ receptor agonist, decreases conflict anxiety-like behavior, while disinhibiting the DMH with $GABA_A$ receptor antagonists, BMI or picrotoxin, increases conflict anxiety-like behavior (Shekhar et al. 1990). Additionally, within the DMH, muscimol increases the amount of time spent in the open arms of an elevated plus-maze and increases social interaction time on the social interaction test, both indicative of anxiolytic-like behavior (Shekhar 1993; Shekhar and Katner 1995). Conversely, disinhibiting the DMH with intra-DMH BMI decreases the amount of time spent in the open arms of the elevated plus-maze and decreases social interaction time in the social interaction test (Shekhar 1993; Shekhar and Katner 1995), which is indicative of anxiogenic behavioral responses. These data led to the development of an animal model of panic disorder using continuous release of $_L$-allylglycine ($_L$-AG; an inhibitor of glutamic acid decarboxylase after transformation from $_L$-AG to 2-keto-4-pentenoic acid) into the DMH (Shekhar et al. 1996; Johnson and Shekhar 2012). Loss of local GABA synthesis within the DMH over the course of days increases anxiety-like behavior and sympathetic output, which, in turn, makes rats vulnerable to panic-like physiological responses to intravenous infusions of sodium lactate (Shekhar et al. 1996; Johnson and Shekhar 2006), a substance demonstrated to elicit panic symptoms in patients with panic disorder in a laboratory setting (Goetz et al. 1996). Evidence for a functional change in neural systems that normally inhibit panic-like behaviors, such as serotonergic neurons in the dorsal raphe nucleus ventrolateral part (DRVL) (Johnson et al. 2004), was provided by an attenuation of the activation of serotonergic neurons within the DRVL (Johnson et al. 2008a) in response to intravenous sodium lactate after intra-DMH $_L$-AG (Johnson et al. 2008a). Additionally, areas such as the CeL, CeM, and BNSTL were shown to have increases in c-Fos expression, a neuronal marker of activation (Sagar et al. 1988), after intra-DMH $_L$-AG and subsequent intravenous sodium lactate challenge, which confirms a role for the DMH in control of neural systems regulating anxiety- and fear-related responses (Johnson et al. 2008b). Stressors may influence neural circuits, leading to disinhibition of the DMH, which may predispose or increase vulnerability to increased anxiety-related behaviors and physiology.

2.3 Neurocircuitry of Fear

In Pavlovian fear conditioning, auditory sensory information from the medial geniculate nucleus of the thalamus and cortex conveys conditioned stimulus (CS) information, and the spinal cord conveys unconditioned stimulus (US) somatosensory information, in the case of footshock, to the LA (Romanski et al. 1993; LeDoux et al. 1990a; Quirk et al. 1997). In the LA, synaptic plasticity involving NMDA and α-amino-3-hydroxy-5-methyl-4-isoxazolepropionic acid

(AMPA) receptors are thought to be one of the mechanisms whereby the CS-US association occurs during Pavlovian fear conditioning (Rogan et al. 1997; Fanselow and LeDoux 1999; Bauer et al. 2002; Rumpel et al. 2005). Neurons in the LA will increase their firing rates when the CS is present, after fear conditioning (Quirk et al. 1997), and elicit a fear response, i.e., freezing behavior (Rogan et al. 1997; Goosens et al. 2003) and increased blood pressure (LeDoux et al. 1990a). Projections from the CeA to the ventrolateral periaqueductal gray (vlPAG) in particular were found to be integral to post-encounter defensive behavior such as freezing (Isosaka et al. 2015). Optogenetic stimulation of glutamatergic neurons in the vlPAG induces unconditioned freezing behavior in a novel environment (Tovote et al. 2016). These findings demonstrate that the amygdala is a nodal structure in the circuitry of fear-related behaviors.

2.3.1 Fear-Potentiated Startle

The startle reflex, a circa-strike defensive behavior, has been investigated since at least the experiments of Sherrington in the 1910s (Sherrington 1910). The acoustic startle reflex was elucidated in the rat through a series of lesion and electrical stimulation experiments. These experiments identified the ventral cochlear nucleus, dorsal lateral lemniscus, and ventral reticularis pontis caudalis, which projects through the medial longitudinal fasciculus to the lumbar spinal cord and quadriceps femoris muscles, as essential for the acoustic startle reflex to occur (Davis et al. 1982). In 1951, Brown et al. described the "energizing" effect of rats being in a fear state by using the startle reflex as a behavioral measure, which was thought to be relevant to the observation that clinically anxious patients startled more frequently (Brown et al. 1951). Based on Hull's work on emotional states (Hull 1943), Brown et al. suggested that fear states would enhance the startle reflex in rats that were conditioned to associate a buzzer/light to an electric footshock (Brown et al. 1951). Brown et al. subsequently found that rats that underwent fear conditioning did in fact startle more in the presence of the buzzer/light and suggested that the reason for an increased startle reflex was because the rat was in an emotional fear state that was enhancing startle responses. The Davis group found that the CeA plays a role in fear-potentiated startle, based on findings that lesions of the CeA abolished the expression of fear-potentiated startle (Hitchcock and Davis 1986), electrical stimulation of the CeA enhanced acoustic startle (Rosen and Davis 1988), and anterograde/retrograde tracing revealed a direct connection between the CeA and the reticularis pontis caudalis (Rosen et al. 1991), an essential nucleus for acoustic startle to occur (Fig. 3).

NMDA receptors within the amygdala have been found to be involved in control of fear acquisition (Miserendino et al. 1990) and fear extinction (Falls et al. 1992) using fear-potentiated startle as a model of learning and memory. Fear-potentiated startle thus follows similar learning principles involving Pavlovian learning, i.e., acquisition, consolidation, extinction, and spontaneous recovery (Koch 1999). The startle reflex, as alluded to earlier, is present throughout the animal kingdom, ranging from *Aplysia* to humans, giving the fear-potentiated startle paradigm high validity as

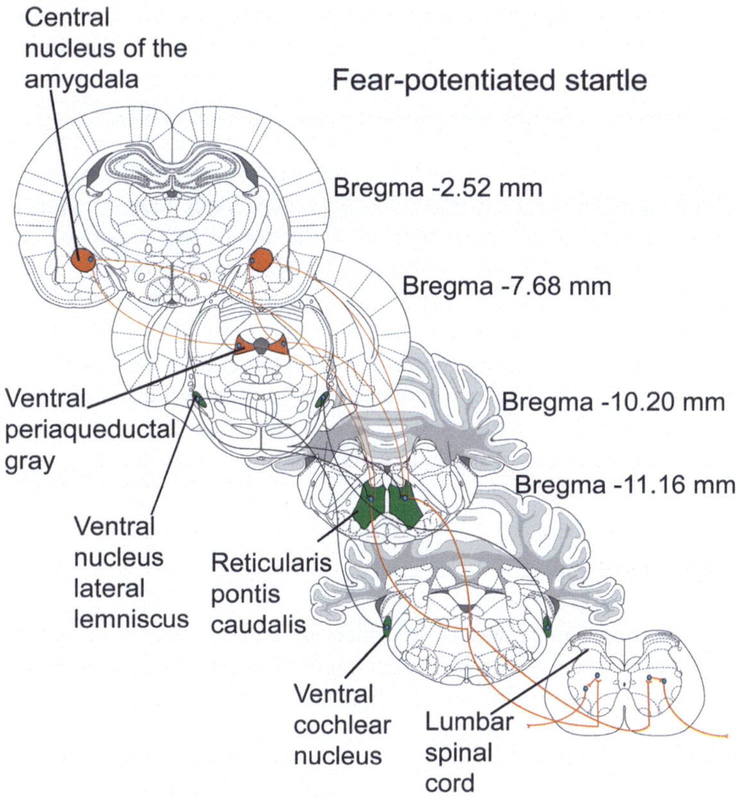

Fig. 3 Hypothetical model illustrating fear potentiation of the acoustic startle circuit (Davis et al. 1982). The central nucleus of the amygdala and ventral nucleus of the periaqueductal gray (orange) send direct projections to the reticularis pontis caudalis (green) to increase motor output at the level of the lumbar spinal cord (Hitchcock and Davis 1991; Rosen et al. 1991; Lingenhohl and Friauf 1994). Coronal brain diagrams adapted from Paxinos and Watson (2007)

a translational model (Fendt and Koch 2013). In humans, the fear-potentiated startle paradigm has been adopted with electromyograms measuring eyeblink contractions from the orbicularis oculi muscle in response to sudden acoustic startle. Individuals that encounter a traumatic experience will undergo stereotypical responses but afterwards begin to extinguish the traumatic experience. This does not happen with everyone. For those individuals that do not extinguish the traumatic memory readily, they continue to remember and be affected by the memory; in other words, they fail to extinguish traumatic memories (Jovanovic and Norrholm 2011). Indeed, individuals that report having experienced a traumatic episode are resistant to extinguishing the conditioned stimuli that signaled an aversive stimulus, such as shock (Jovanovic et al. 2010). Many correlations have been found between symptom severity of PTSD, a trauma- and stressor-related disorder, and failure to extinguish fear using the fear-potentiated startle paradigm (Jovanovic et al. 2009). Yang et al. (2006) explored the effects of glucocorticoids on extinguishing fear memories using the fear-potentiated startle paradigm in rats (Yang et al. 2006). Systemic injections of the synthetic glucocorticoid, dexamethasone, increased extinction in a dose-dependent manner, with higher doses increasing extinction of fear memory most efficiently (Yang et al. 2006). Specifically, intra-BLA synthetic glucocorticoids also facilitated more efficient fear extinction in a dose-dependent manner, while blocking intra-BLA glucocorticoid receptors blocked extinction of fear memory (Yang et al. 2006). The observation that glucocorticoids can influence extinction points to an interaction of stress and fear processes. Fear-potentiated startle has proven to be a useful model for elucidation of fear circuitry and allows researchers to explore treatment efficacy by examining the effects of interventions designed to enhance fear extinction. This is the strategy behind exposure therapy for trauma victims and possible strategies to help treat PTSD. Interventions that enhance fear extinction have potential for treatment of individuals with a diagnosis of PTSD.

2.3.2 Contextual Fear

Classical fear conditioning models, in which an aversive unconditioned stimulus, such as footshock, is presented in association with a neutral conditioned stimulus (CS), often an auditory tone or flash of light, have elucidated many aspects of the emotional state of fear and principles in associative learning (Pavlov 1927; LeDoux 1998; Goosens et al. 2003). If rats experience an aversive stimulus, such as electric footshock, in an apparatus that has a particular context, but not a discrete CS presented in association with the footshock, and are tested for freezing behavior shortly thereafter in a different apparatus with different contextual cues, then freezing behavior is attenuated, relative to rats that are exposed to fear conditioning and testing in the same apparatus (Blanchard and Blanchard 1969). In other words, animals condition to contextual cues and exhibit conditioned fear responses to the context in which the aversive footshock occurred (Blanchard and Blanchard 1969; Fanselow 1980). This paradigm is known as contextual fear conditioning. Optimal learning in the contextual fear conditioning paradigm requires at least 2 min to transpire between placement in the apparatus and presentation of the US. Otherwise an associative learning deficit occurs, as evidenced through significantly reduced freezing behavior, an effect referred to as the immediate shock deficit (Fanselow 1986, 1990). One explanation proposed by Fanselow (1990) for the immediate shock deficit was inspired by Pavlov (Pavlov 1962), as it takes time to systematize and equilibrate all of the different stimuli until a "dynamic stereotype" is formed, or it takes time to form a representation of the context (Fanselow 1990). The immediate shock deficit can be overcome if the rat is pre-exposed to the context 24 h prior to contextual fear conditioning, thereby allowing a memory of the context to be formed. This effect is referred to as the context preexposure facilitation effect (CPFE) (Fanselow 1986, 1990; Rudy et al. 2004).

The dorsal hippocampus has been identified as a brain region that plays a role in some forms of contextual fear conditioning (Phillips and LeDoux 1992; Kim and Fanselow 1992; Fanselow and Dong 2010). The hippocampus has an inhibitory role on the activation of the HPA axis, and hippocampal glucocorticoid receptors are involved (Sapolsky et al. 1984; McEwen et al. 1968). Additionally, glucocorticoids have a facilitating or permissive effect on the ability to form a contextual fear memory. Systemic blockade of glucocorticoid receptors (Pugh et al. 1997a), adrenalectomy, effective removal of systemic glucocorticoids (Pugh et al. 1997b), and systemic dehydroepiandrosterone sulfate (DHEA-S), an effective glucocorticoid receptor antagonist, all attenuate contextual fear conditioning. These results suggest that stress mediators, such as glucocorticoids, facilitate formation of contextual fear memories. In fact, Cordero et al. (2003) demonstrated that acute restraint stress in rats, administered 24 h before fear conditioning, enhanced contextual fear conditioning (Cordero et al. 2003). This stress-enhanced fear process, as shown through contextual fear conditioning, can help us understand how stress circuitry, i.e., dorsal hippocampus, can influence fear circuitry. Contextual fear conditioning has been useful for elucidating neural mechanisms of PTSD.

3 Stress-Induced Exaggeration of Anxiety- and Fear-Related Responses: Introduction to the Hypothalamic-Pituitary-Adrenal Axis

Activation of the HPA axis is a hallmark endocrine response to stressful stimuli. Selye (1950) identified the HPA axis as a critical neuroendocrine system that organizes and enhances bodily defenses against stressors with glucocorticoids being the principal molecules responsible for the coordination (Selye 1950). The characterization of corticotropin-releasing factor (CRF; also referred to as CRH) by Vale et al. (1981) as the factor responsible for Guillemin and Rosenberg's hypothalamic factor controlling the release of ACTH from the anterior pituitary (Guillemin and Rosenberg 1955) set the stage for detailed, mechanistic studies of the HPA axis. Soon afterward, it was discovered that, in addition to their role in controlling the HPA axis, CRH neurons are widely distributed throughout the central nervous system and play a role in controlling stress-related, anxiety-related, and fear-related behavioral responses (Dunn and Berridge 1990). Parvocellular neurons of the PVN secrete CRH and arginine vasopressin into the zona externa of the median eminence (Bruhn et al. 1984) where CRH enters the hypothalamic-hypophyseal portal circulation. Corticotropin-releasing hormone then stimulates corticotrophs (Aguilera et al. 1983) by binding to corticotropin-releasing hormone receptor Type 1 (CRHR1) (Chang et al. 1993; Vita et al. 1993), resulting in release of ACTH (De Souza et al. 1984; Potter et al. 1994; Vale et al. 1981). Adrenocorticotropic hormone then travels through systemic circulation, ultimately, through activation of melanocortin 2 (MC2) receptors, stimulating cells in the adrenal cortex to synthesize and release glucocorticoid hormones (Haynes et al. 1959). The adrenal cortex is subdivided into different layers, with the zona fasciculata and zona reticularis having receptors for ACTH (Vinson 2016; Arnold 1866). Upon stimulation of MC2 receptors by ACTH, cells within the zona fasciculata synthesize and release glucocorticoids, primarily corticosterone in rodents and cortisol in humans (Chida et al. 2007). These glucocorticoids travel in the blood, with the majority bound to corticosteroid binding globulin and albumin (Lewis et al. 2005; Moisan et al. 2014). After glucocorticoids dissociate from plasma binding proteins, they readily cross plasma membranes and bind to intracellular receptors, i.e., mineralocorticoid receptors and glucocorticoid receptors. Glucocorticoids preferentially bind to the mineralocorticoid receptor and, at higher concentrations, will bind to the glucocorticoid receptor (Reul and de Kloet 1985, 1986). These activated receptors will dimerize, translocate into the nucleus, and affect transcription of target genes, such as nuclear factor kappa-light-chain-enhancer of activated B cells (NF-κB) (Presman et al. 2016; Presman and Hager 2017; Adcock et al. 1999; Barnes and Karin 1997), a mechanism underlying the immunosuppressive effects of glucocorticoids.

Under basal (i.e., unstressed) conditions, the dynamics of the HPA axis are characterized by both a circadian rhythm (i.e., a rhythm with a period of approximately 24 h) and an ultradian rhythm (i.e., a recurrent rhythm with a period of less than 24 h) of hormone secretion (Spiga et al. 2014). The PVN receives circadian input originating from the suprachiasmatic nucleus (SCN), which is thought to involve both direct projections from the SCN to the PVN and indirect projections relayed by the medial preoptic area, subparaventricular nucleus of the hypothalamus, and DMH (Watts and Swanson 1987). Circadian rhythms of HPA axis activity allow the organism to anticipate and prepare for predictable environmental changes (i.e., the light/dark cycle, food availability, and risk of predation). This results in an optimization of energy expenditure by coordination of physiological processes. Of potential interest, recent studies suggest that neural systems that encode fear are part of the neural circuitry that constitutes the circadian system (Kim et al. 2014; Pellman et al. 2015). Consistent with these findings, the prefrontal cortex has rhythmic clock gene expression (Chun et al. 2015) that is regulated by the circadian pattern of corticosterone (Woodruff et al. 2016). Furthermore, performance in a prefrontal cortex-dependent emotional learning task varies with time of day, and the time of day variation is absent in rats that lack circulating corticosterone (Woodruff et al. 2015). Although the ultradian rhythm of corticosterone secretion has been shown to be a determinant of stress-induced activation of the HPA axis (Windle et al. 1998), it is as yet unclear if the ultradian rhythm of corticosterone secretion has an impact on anxiety- and fear-related behavioral responses.

Ultradian and circadian patterns of ACTH and corticosterone secretion, and stress-induced activation of the HPA axis, show sex differences in rodents and humans (Windle et al. 1998; Seale et al. 2004). Although beyond the scope of this review, these sex differences are likely to be relevant to understanding sex differences in stress-induced anxiety- and fear-related responses in rodents and humans and anxiety disorders, trauma- and stressor-induced psychiatric disorders, and affective disorders, which are known to show major sex differences (Donner and Lowry 2013).

Although stress-induced activation of the HPA axis undoubtedly has many adaptive functions, Munck and colleagues proposed that the function of the HPA axis was to prevent the bodily defenses, including immune and inflammatory responses, from overshooting and to reduce bystander damage to the body (Munck et al. 1984). Recently, Frank et al. (2013) further suggested that a function of the HPA axis is to act as an endocrine alarm signal for the innate immune system (Frank et al. 2013). Although glucocorticoids during an aversive event are initially anti-inflammatory, they can also prime the immune system to respond to future potential infection or injury (Frank et al. 2013).

While glucocorticoid action in the brain has a major role in feedback inhibition of the HPA axis, i.e., acting directly on CRH-synthesizing neurons in the PVN and ACTH-synthesizing neurons in the pituitary (collectively referred to as long-loop negative feedback), and indirectly on a hierarchical neural system controlling the HPA axis, glucocorticoids also have impacts on many parts of the circuitry mediating fear and anxiety. This will be specifically discussed in the sections below.

4 Models of Stress-Induced Vulnerability to Exaggerated Anxiety- and Fear-Related Behavioral Responses

4.1 Maternal Separation: A Model of Adverse early Life Experience Leading to Lifelong Vulnerability to Exaggerated Anxiety- and Fear-Related Behavioral Responses

Brain development during childhood is thought to be especially sensitive to stress, particularly the development of the hippocampus, amygdala, and frontal lobes and the neurocircuitry that connects them (Lupien et al. 2009). The effects of early life adversity have also been shown to persist later on in life, causing increased reactivity to stress and increased vulnerability to stress- and anxiety-related disorders. While the production of glucocorticoids is essential for maintaining energy balance and coordinating neuronal responses to external/internal stimuli, both insufficient and excessive glucocorticoid production as a result of early life stress can impact brain development and function (Lupien et al. 2009). One interesting paradigm of early life adversity in rodents is maternal separation during critical postnatal development periods. These separation periods are typically for 3 h or more per day for the first 2–3 weeks of life. Notably, maternal separation has been shown to activate the HPA axis in the pup, causing increases in CRH and corticosterone (Vazquez et al. 2000), as well as increased plasma levels of ACTH (Vazquez et al. 2000) and glucocorticoids (Levine and Wiener 1988; Schulkin et al. 1998; Lupien et al. 2009). Maternal separation results in a dysregulation of the circadian rhythm of glucocorticoid hormone secretion and a sensitization of the HPA axis responses to stressors during adulthood (Francis et al. 2002). This excessive stress hormone exposure can induce the "wear and tear" associated with allostatic load and can ultimately result in chronic maladaptive states during adulthood.

One maladaptive consequence of maternal separation is increased expression of CRH within the amygdala and hypothalamus, which is associated with an increase in anxiety-like behaviors (Schulkin et al. 1998). Daniels et al. (2004) found that rats that underwent maternal separation showed more anxiety-like behaviors during an elevated plus-maze test (Daniels et al. 2004). Ladd et al. (2004) suggest that maternal separation results in functional impairment of mineralocorticoid receptors or glucocorticoid receptors, ultimately leading to impaired inhibition of the HPA axis and hyperreactivity of the endocrine stress response (Ladd et al. 2004). The consequent overproduction of stress hormones from a hyperactive HPA axis can result in behavioral abnormalities and anxiogenic phenotypes; thus, maternal separation paradigms are associated with increased vulnerability to anxiety-like behavior during adulthood.

Although, to the best of our knowledge, the effects of maternal separation on subsequent panic-like responses have not been studied in rodent models, recent studies have shown that female rats previously exposed to neonatal maternal separation show an increased hypercapnic ventilator response to the panicogenic agent CO_2 as adults (Genest et al. 2007; Dumont et al. 2011). In addition, recent studies in mice have demonstrated that an unstable maternal environment [cross-fostering during postnatal day (PD)2-PD5] results in increased CO_2 sensitivity during development and adulthood, relative to normally reared mice (D'Amato et al. 2011). Finally, in humans, events involving childhood separation from caregivers or an unstable parental environment are associated with heightened CO_2 sensitivity and increased risk for panic disorder in adulthood (Battaglia et al. 1995; Klein 1995).

4.2 Adolescent Social Isolation: A Model of Stress-Induced Vulnerability to Exaggerated Anxiety- and Fear-Related Behavioral Responses

Similar to maternal separation during the first few weeks of life, stressor exposure during adolescence, including adolescent social isolation, social defeat, and social instability stress, has been shown to induce vulnerability to exaggerated anxiety- and fear-related behavioral responses later in life. For a comprehensive review, see Burke et al. (2017).

4.3 Learned Helplessness: A Model of Stress-Induced Vulnerability to Exaggerated Anxiety- and Fear-Related Behavioral Responses

Stressor controllability, in addition to stressor exposure, has become an important concept in stress physiology (Koolhaas et al. 2011; Day and Walker 2007). The learned helplessness paradigm compares two conditions in which animals experience an aversive event (typically tail shock) and a third condition in which animals do not experience the aversive event. Within the two groups of animals that experience the tail shock, one of the groups of animals has control over the termination of the shocks [escapable shock (ES)]. In comparison, the other group does not have this control, but is yoked to the escapable group [inescapable shock (IS)]. For these animals, both groups are exposed to the same frequency and duration of the shock(s), with the only difference between the two being controllability over termination of the stressor. A third group of animals are maintained in the home cage condition or are simply restrained for a similar duration of time as the ES and IS animals (Seligman and Maier 1967). Behavioral consequences of IS include increased conflict anxiety in a juvenile social exploration paradigm (Christianson et al. 2010), potentiation of contextual and auditory cued fear conditioning (Amat et al. 2005, 2008; Maier 1990; Baratta et al. 2007), and an impaired ability to learn escape in a shuttle box escape task (Amat et al. 2005, 2008; Seligman and Maier 1967; Maier et al. 1973; Maier and Testa 1975). One aspect of the learned helplessness paradigm is trans-situationality whereby the uncontrollable stressor occurs in one context and the behavior deficits occur in a completely different context. A consequence of maintaining separate context from the uncontrollable stressor and behavioral testing is that shuttle box escape deficits are only present within 24–48 h after exposure to IS and are no longer present after 48 h (Maier et al. 1979; Maier 2001).

Rats that are exposed to IS have greater activation of serotonergic neurons in the caudal part of the dorsal raphe nucleus (DR) (Grahn et al. 1999; Amat et al. 2005) and increased extracellular concentrations of serotonin within the DR (Maswood et al. 1998) compared to rats exposed to ES or home cage controls. The overactivation of DR serotonergic neurons and the downstream consequences are essential for the behavioral sequelae of IS (Maier and Watkins 2005). Consistent with these findings, serotonin has been implicated in many physiological and behavioral functions, including anxiety- and fear-related behavioral responses (Abrams et al. 2005; Hale et al. 2012). In contrast, prevention of the overactivation of DR serotonergic neurons during ES is responsible for protection from negative sequelae. The ventral medial prefrontal cortex (mPFCv) projects to the DR (Baratta et al. 2009), and chemical inactivation of this projection in rats during ES results in a significant increase in extracellular concentrations of serotonin and increased activation of serotonin neurons in the DR, such that rats exposed to ES now resemble those exposed to IS (Amat et al. 2005). In contrast, activation of the mPFCv in IS animals leads to decreased concentrations of extracellular serotonin in the DR and a behavioral phenotype similar to ES animals (Amat et al. 2008). Both of these results suggest that the mPFCv mediates the protective effects of stressor controllability by inhibiting activation of the DR, as shown in Fig. 4.

A number of brain regions have been implicated in activation of DR serotonergic neurons during IS. For example, the BNST projects to the DR (Marcinkiewcz et al. 2016) and activation of the BNST is necessary and sufficient for the behavioral consequences of IS (Hammack et al. 2003b, 2004). One hypothesis is that CRH neurons within the BNST that project to the caudal DR are responsible for the overactivation of DR serotonergic neurons during IS. Lowry et al. (2000) found that 5d of restraint stress leads to sensitization of serotonergic neurons in the caudal portion of the DR (DRC) to CRH (Lowry et al. 2000). Hammack et al. (2002) demonstrated that intra-DR administration of the CRH receptor antagonist, $_{D}$-Phe $CRH_{(12–41)}$, blocked the IS-induced behavioral deficits. Intra-DR administration of high concentrations of CRH was sufficient in mimicking IS-induced behavioral deficits (Hammack et al. 2002), specifically by binding to CRH receptor subtype 2 (CRHR2) within the DR (Hammack et al. 2003b). Subsequently, Hammack et al. (2003a, b) found that intra-DR administration of low concentrations of CRH prevented IS-induced behavioral deficits. The authors proposed that relatively low concentrations of CRH within the DR preferentially bind to CRHR1 found on GABA interneurons, presumably inhibiting serotonin neurons, resulting in an overall inhibitory function in the presence of low concentrations of CRH (Hammack et al. 2003a). Increasing concentrations of CRH within the DR would shift away from an overall inhibitory signal to a stimulatory signal by binding to CRHR2 on serotonin neurons (Lukkes et al. 2011), thereby activating serotonin neurons (Hammack et al. 2003a).

The overactivation of DR serotonergic neurons following IS is thought to lead to desensitization of autoinhibitory serotonin 1A ($5\text{-}HT_{1A}$) receptors on DR serotonergic neurons and hyperreactivity of anxiety-related DR serotonergic neurons following subsequent exposure to anxiogenic stimuli 24–48 h later. Twenty-four hours after exposure to IS, rats indeed show desensitization of $5\text{-}HT_{1A}$ receptor function in the dorsomedial DR relative to ES or home cage control rats (Rozeske et al. 2011). Taken together, the lack of glutamate input from the mPFCv onto inhibitory GABA neurons, the shift of CRH from being inhibitory to stimulatory within the dorsomedial DR, and desensitized serotonin autoinhibition result in overproduction of serotonin in the DR and projection regions such as the BLA, effects that typically persist for 24–48 h, corresponding with the short-term behavioral deficits found with uncontrollable stress (Amat et al. 1998, 2004; Maswood et al. 1998). In conclusion, the IS paradigm is a model of stress-induced exaggeration of anxiety- and fear-related behavioral responses (an example of Type 1 allostatic overload), with relevance to stress-related psychiatric disorders, including anxiety disorders, trauma- and stressor-related psychiatric disorders, and affective disorders (Maier and Seligman 2016; Maier and Watkins 2005).

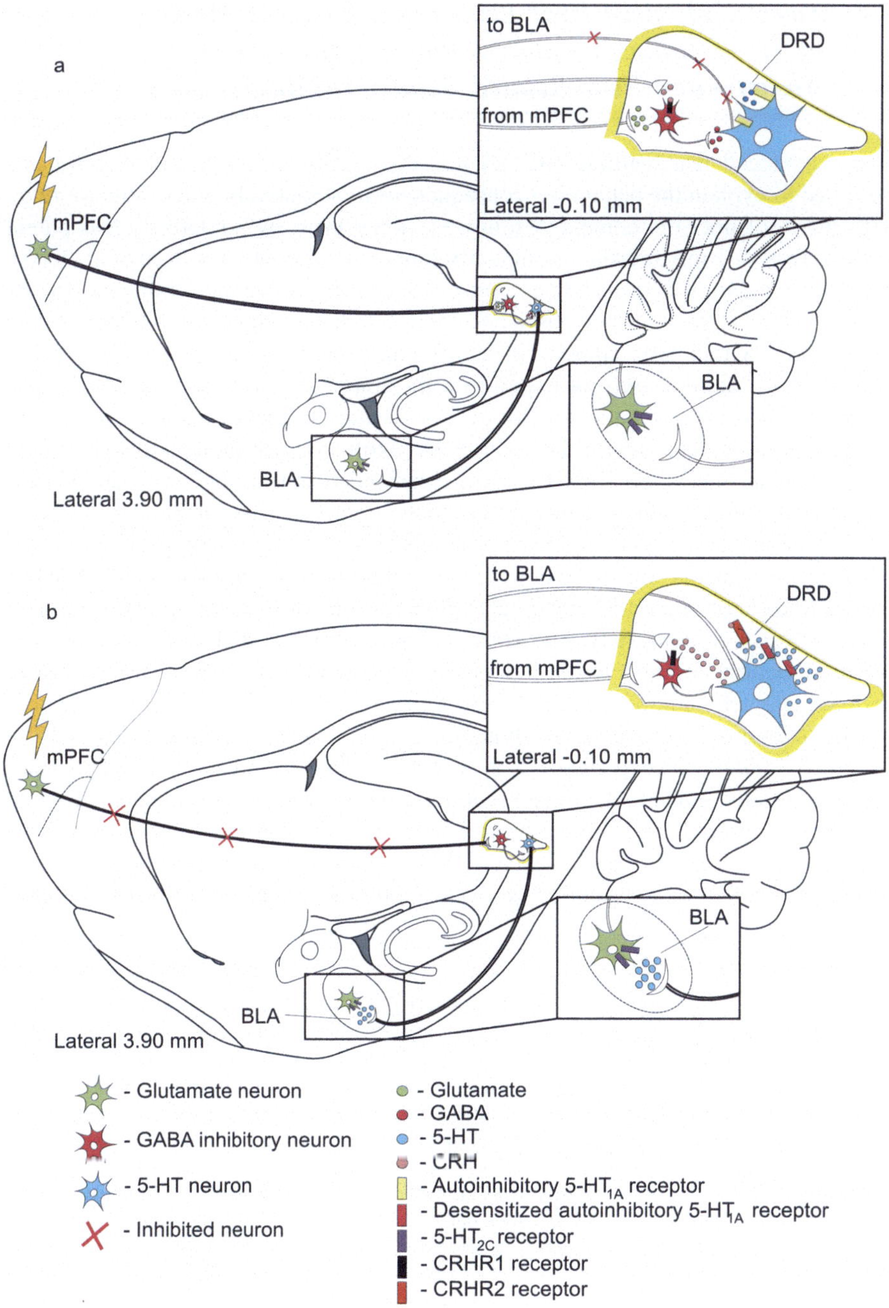

Fig. 4 Neural mechanisms underlying sensitization of anxiety-like behavioral responses following inescapable shock (IS) in a model of learned helplessness. (**a**) During escapable shock (ES), in which the animal has control over the termination of the stressor, glutamatergic neurons located in

4.4 Stress-Enhanced Fear Learning (SEFL): A Model of Stress-Induced Vulnerability to Exaggerated Anxiety- and Fear-Related Behavioral Responses

Stress-enhanced fear learning (SEFL) was discovered in a series of experiments designed to explain the behavioral consequences of a relatively potent stressor (e.g., a 15 mA footshock) in context A followed, a day later, by a relatively less potent stressor (e.g., a 1 mA footshock), in a different context, context B (Rau et al. 2005). Freezing behavior was measured as a proxy for learning about the contingency of the stressor to the context. It was discovered that rats freeze significantly more if they experienced a traumatic stressor in a different context (context A) a day prior, compared to rats that did not experience the traumatic stressor a day prior. The order in which the traumatic stressor occurs is vital; it must occur first, in order to elicit the stress-enhanced effect. Traditionally, extinction was thought to be able to at least attenuate this effect, but the stress-enhanced fear learning of a new context remained, despite rats fully extinguishing to the traumatic stressor (Rau et al. 2005). Even more intriguing was that memory acquisition of the traumatic stressor through NMDA mechanisms was not required, since pharmacological blockade of NMDA receptors did not block the stress-induced enhancement of learning about the less potent stressor in context B. The stress enhancement has such potency that even pairing a neutral stimulus to the less potent stressor in context B will significantly increase fear behavior in a neutral context C, when rats are presented with the paired neutral stimulus. This SEFL is thought to model one characteristic of PTSD, in which individuals experience a very potent stressor and are primed to learn about a usually innocuous stressor within a novel context. Stress-enhanced fear learning has long-lasting effects, previously measured up to 90 days after the traumatic exposure (Rau and Fanselow 2009). Glucocorticoids do not seem to be involved in SEFL as a systemic injection of the glucocorticoid synthesis inhibitor,

---

Fig. 4 (continued) layer V of the prelimbic part of the medial prefrontal cortex (mPFC) (Varela et al. 2012), projecting to the dorsal raphe nucleus (DR), are activated (Baratta et al. 2009), which stimulate local γ-amino-butyric acid (GABA) inhibitory interneurons. These GABA interneurons inhibit anxiety-promoting serotonergic neurons located in the dorsomedial part of the DR (Jankowski and Sesack 2004) projecting to the basolateral amygdala (Christianson et al. 2010). (**b**) During inescapable shock (IS), mPFC projections to the DR are not stimulated (Amat et al. 2005, 2008). Exposure to IS activates a number of brain regions that give rise to excitatory input to serotonergic neurons located in the dorsomedial part of the DR, including the bed nucleus of the stria terminalis (Hammack et al. 2004), lateral habenula (Amat et al. 2001; Dolzani et al. 2016), and locus coeruleus (McDevitt et al. 2009). Hyperactivation of serotonergic neurons located in the dorsomedial part of the DR (Grahn et al. 1999) results in sensitization of these neurons through a functional desensitization of autoinhibitory 5-HT_{1A} receptors (Rozeske et al. 2011); consequently, 24–48 h after IS (Maier et al. 1979), rats respond with increased anxiety-like behavioral responses, which are dependent on serotonergic projections to the basolateral amygdala and activation of 5-HT_{2C} receptors (Christianson et al. 2010). Sagittal brain diagrams adapted from Paxinos and Watson (2007)

metyrapone, administered prior to stress exposure, did not influence stress-enhanced fear learning (Ryoke et al. 2014). Exposure therapy for PTSD patients relies on extinction principles, but this model implies that processes that are glucocorticoid independent, resistant to extinction learning, and NMDA independent may be involved. These data indicate that the SEFL phenomenon requires the development of novel therapeutic strategies for the treatment of PTSD.

4.5 *Chronic Subordinate Colony Housing (CSC) Model: A Model of Stress-Induced Vulnerability to Exaggerated Anxiety- and Fear-Related Behavioral Responses*

Another proposed animal model for PTSD is the chronic subordinate colony housing (CSC) model in mice (Reber et al. 2016a). One characteristic of PTSD is that, in many situations, the psychological stressor or reminders of the stressor are still present after the initial trauma. In the CSC paradigm, mice are exposed to a traumatic experience by physically being subordinated by a larger aggressive mouse (establishment of the social hierarchy) and are chronically reminded of the trauma with constant unpredictable attacks from the larger aggressive mouse (maintenance of the social hierarchy). The CSC paradigm takes advantage of ethologically relevant stressors, similar to other resident-intruder paradigms that have robust stress responses. Briefly, in the CSC model, four younger male mice are placed into the cage of a prescreened, aggressive male mouse's cage to undergo chronic and unpredictable stress from the aggressor mouse. A social hierarchy will typically be established with the aggressive male physically and socially dominating the younger intruder mice. On days 8 and 15, the younger intruder mice are moved to the home cage of a different prescreened aggressive mouse to then undergo the same chronic, unpredictable stress but in another new aggressor's cage. Following 19 days of CSC exposure, subordinate mice have reduced body weight gain [when compared to single-housed control (SHC) mice], decreased thymus weight, increased adrenal gland weight, significant histological damage to the colon (spontaneous colitis), exaggerated release of proinflammatory cytokines from freshly isolated mesenteric lymph node cells stimulated with anti-CD3 antibody ex vivo, exaggerated chemically induced colitis in a model of inflammatory bowel disease, and increased anxiety-like behavior (Reber et al. 2007, 2008; Singewald et al. 2009; Uschold-Schmidt et al. 2012) without affecting depression-like behavior (Slattery et al. 2012). Mice exposed to the CSC paradigm have significantly elevated plasma corticosterone and elevated CRH mRNA in the PVN on the second day of CSC exposure, although both parameters return to pre-stressor basal levels 24 h later and remain at these levels at 7 days, 14 days, and 19 days of the CSC protocol, with the exception of an evening hypocorticosteronemia (Reber et al. 2007). Additionally, 19 days of CSC, relative to SHC control conditions, reduces the adrenal gland's ability to secrete corticosterone in response to ACTH stimulation ex vivo (Reber et al. 2007;

Uschold-Schmidt et al. 2012). Interestingly, mice that had unresponsive adrenal glands upon stimulation with ACTH ex vivo had sensitized responses to a heterotypic stressor (an elevated platform exposure), with significantly increased plasma corticosterone, despite similar concentrations of plasma ACTH (Uschold-Schmidt et al. 2012). Areas of the brain that are activated in response to 19 days of CSC followed by an acute heterotypic stressor, (open-arm exposure on an elevated plus-maze), include the dorsal periaqueductal gray [a brain region associated with flight/escape behavior (Yamashita et al. 2011; Miguel et al. 2010)], and lower activation in the lateral septum and PVN, compared to SHC mice (Singewald et al. 2009). Clinically, individuals with a diagnosis of PTSD have hypocortisolism (Yehuda et al. 1990), an enhanced negative feedback of the HPA axis in response to the dexamethasone suppression tests (Yehuda et al. 1993), and higher concentrations of CRH found in cerebrospinal fluid (Kataoka et al. 2014; Fontes et al. 2011; Zaretskaia et al. 2008). The CSC model recapitulates the evening hypocorticosteronemia and significant increase in CRH mRNA in the PVN after undergoing chronic subordination, similar to physiological parameters as found in patients diagnosed with PTSD (Bremner et al. 1997; Yehuda et al. 1990; Reber et al. 2007).

5 Focus on Glucocorticoid-Dependent Mechanisms of Stress-Induced Vulnerability to Exaggerated Anxiety-Like and Fear-Related Behavioral Responses

5.1 Glucocorticoids and Memory

Evaluating the significance of an event after it has occurred and determining whether to forget the event or store the event as a salient experience in long-term memory has been suggested to be evolutionarily advantageous (Kety 1972). Stress-related hormones play a role in biasing toward long-term memory of events. Experiments exploring memory enhancement have found that systemic injections of epinephrine, norepinephrine, and glucocorticoids after an inhibitory avoidance task are capable of modulating memory storage in a dose- and time-dependent manner (Roozendaal et al. 1997; Gold and Van Buskirk 1975, 1976a, b). Experimentally, memory enhancement is revealed when treatments are immediately administered after the passive avoidance training has occurred. Evidence for central actions was shown when intracerebroventricular epinephrine was found to also enhance memory after an inhibitory avoidance task (Haycock et al. 1977). Later, the role of the amygdala in memory enhancement was revealed through lesions of the amygdala, immediately after footshock (Liang et al. 1982). Stress-induced enhancement of memory was impaired only in highly arousing situations, such as a footshock (Cahill and McGaugh 1990). Additionally, intra-amygdala infusions of BMI, a $GABA_A$ receptor antagonist, facilitate stress-induced enhancement of memory consolidation, while intra-amygdala infusions of muscimol, a $GABA_A$ receptor agonist, and baclofen, a

$GABA_B$ receptor agonist, impair stress-induced enhancement of memory consolidation in an inhibitory avoidance task (Castellano et al. 1989; Brioni et al. 1989). More specifically, the BLA, and not the CeA, was revealed to be critical for the stress-induced memory enhancement (Parent and McGaugh 1994). The amygdala also enhances memory consolidation through actions of the adrenal hormones, i.e., corticosterone and epinephrine (McGaugh et al. 1990). The importance of the role of glucocorticoids in stress-induced consolidation, which was found to be required for the memory enhancement effects of systemic injection of epinephrine, was supported by studies showing that pharmacologically suppressing glucocorticoid synthesis blocks epinephrine-induced memory enhancement (Roozendaal et al. 1996). Furthermore, an intact BLA and glucocorticoid receptor binding are both required for systemic glucocorticoid-induced memory enhancement (Roozendaal and McGaugh 1996, 1997). The nucleus of the solitary tract (NTS), a source for central norepinephrine, is crucial for stress-induced enhancement of memory consolidation (Williams and McGaugh 1992). The NTS projects to the BLA, whereby extracellular concentrations of norepinephrine increase in response to systemic injections of epinephrine (Williams et al. 1998), previously shown to enhance memory storage (Gold and Van Buskirk 1975). Establishing an interaction between glucocorticoids and norepinephrine acting within the BLA to facilitate stress-induced memory enhancement was accomplished using intra-BLA infusions of various norepinephrine antagonists (Quirarte et al. 1997). For example, propranolol, a nonselective β-adrenergic receptor antagonist; atenolol, a specific β_1-adrenergic receptor antagonist; and zinterol, a specific β_2 receptor antagonist, all blocked systemic dexamethasone-induced enhancement of memory in adrenalectomized rats (Quirarte et al. 1997). Enhanced norepinephrine release in the BLA, activation of postsynaptic β adrenoceptors, and (through facilitation of local norepinephrine release) blockade of autoinhibitory α_2 adrenoceptors are proposed mechanisms of action for memory consolidation of inhibitory avoidance training (Quirarte et al. 1998; Galvez et al. 1996; Hatfield et al. 1999; Ferry et al. 2015).

At a neural systems level, the amygdala is proposed to modulate other brain regions involved in memory consolidation, such as the hippocampus and caudate putamen (Packard et al. 1994). Support for this hypothesis includes intact expression of freezing after Pavlovian fear conditioning following BLA lesions, indicating that the BLA is not necessary for the expression of freezing behavior (Cahill et al. 2000). Cahill et al. (2000) suggested that the memory trace or necessary information is distributed to other areas of the brain in addition to the BLA. Much like the proposed mechanism described earlier, systemic administration of corticosterone enhances memory consolidation of inhibitory avoidance through increasing local extracellular norepinephrine within the BLA, but also enhances activation of dorsal hippocampal neurons (McReynolds et al. 2010). The mPFC is functionally connected to the BLA and influences memory consolidation (Arruda-Carvalho and Clem 2014). Recently, Kitamura et al. (2017) examined neuronal circuits during a contextual fear conditioning task and found that hippocampal dentate gyrus, mPFC, and BLA cells (specifically neurons that are activated by a learning event) have enduring cellular changes and are reactivated by the original stimuli used to induce the learning. These cells, i.e., engram cells, were activated as a consequence of contextual fear conditioning and reactivated when re-exposed to the context in which conditioning occurred (Kitamura et al. 2017). Medial prefrontal cortex engram cells were found to be critical for contextual memory formation at 13 days (remote memory), but not 2 days (recent memory) (Kitamura et al. 2017). In contrast, dentate gyrus engram cells were critical for expression of recent memory of the contextual fear conditioning. Medial prefrontal cortex engram cell formation was blocked by optogenetically inhibiting BLA inputs to the mPFC during contextual fear conditioning, supporting a role for the BLA in establishing the formation of contextual fear memory (Kitamura et al. 2017). Additionally, BLA inactivation significantly lowers the expression of unconditioned freezing (Vazdarjanova et al. 2001).

In contrast to the memory storage enhancement effects of glucocorticoids immediately after a stressor, glucocorticoids can impair memory retrieval (de Quervain et al. 1998). Evidence for this phenomenon was found in a spatial memory task in which glucocorticoids, given 30 min prior to testing, impaired spatial memory performance (de Quervain et al. 1998). In humans, oral administration of cortisone impaired declarative memory when administered just prior to testing (de Quervain et al. 2000). The evidence described above has led to the idea that glucocorticoids, in concert with norepinephrine, in the BLA activate synaptic and cellular processes to exclusively attend to the current stressor at hand and form memories about the stressor (Roozendaal 2002; Schwabe et al. 2012). At the same time, memory retrieval mechanisms are inhibited in order to not interfere with the formation of the new memory.

5.2 *Glucocorticoid Effects on CRH mRNA Expression in the Amygdala and BNST*

Corticotropin-releasing hormone mRNA levels in the PVN are reduced over the course of days in adrenal intact male rats as a consequence of daily injections of plasma glucocorticoids (Makino et al. 1994a). This corresponds with the classical long-loop negative feedback of glucocorticoids and inhibition of the HPA axis. In contrast, steroid hormones, including glucocorticoids, can induce synthesis of specific neuropeptides and/or their receptors, resulting in feed-forward regulation of behavior (Schulkin 2003). For example, daily systemic injections of glucocorticoids increase CRH mRNA expression in the BNSTd and CeA (Makino et al. 1994a, b). Behaviorally, a potentiation of fear behavior after Pavlovian fear conditioning was demonstrated through increased levels of freezing behavior with similar doses of chronic systemic administration of corticosterone in association with increased CRH mRNA within the CeA (Corodimas et al. 1994). Consistent with these findings, chronic systemic corticosterone also increased CRH mRNA within the CeA in association with increased fear-related behavioral responses in a contextual fear conditioning paradigm (Thompson et al. 2000; Schulkin 2003). Chronic intra-CeA corticosterone increases CRH mRNA within the CeA in response to exposure to the elevated plus-maze (Shepard et al. 2000). Chronic intra-CeA corticosterone also potentiated increases in CRH mRNA in the BNSTd in response to exposure to the elevated plus-maze (Shepard et al. 2006). Therefore, glucocorticoids increase CRH mRNA expression in the CeA and BNSTd, leading to exaggeration of anxiety- and fear-related behavioral responses.

5.3 Stress, Glucocorticoids, and Organic Cation Transporter 3 (OCT3)

Studies of regulation of monoamine clearance in the central nervous system have focused on low-capacity, high-affinity uptake$_1$ mechanisms, such as the norepinephrine transporter, dopamine transporter, and serotonin transporter (5-HTT) (Torres et al. 2003). In contrast, high-capacity, low-affinity uptake$_2$ mechanisms such as organic cation transporter (OCT) 1, OCT2, and OCT3 have garnered less attention (Iversen and Salt 1970; Zhu et al. 2012; Schildkraut and Mooney 2004; Wu et al. 1998; Gasser and Lowry 2018). Organic cation transporter 3 is notable, as its transport activity is sensitive to blockade by glucocorticoids (Gasser et al. 2006; Hill et al. 2011). Schmitt et al. (2003) reported an upregulation of OCT3 in the hippocampus of 5-HTT-deficient mice as a possible explanation for lack of efficacy of antidepressants (Schmitt et al. 2003). Baganz et al. (2008) further proposed that OCT3 was a compensatory uptake mechanism for 5-HT when the 5-HTT was genetically knocked out in mice, suggesting that OCT3 maintains 5-HT homeostasis if 5-HTTs were to become less available (Baganz et al. 2008). Interestingly, Baganz et al. (2010) found that OCT3 expression and function are downregulated through corticosterone-dependent mechanisms within the hippocampus in animals that underwent repeated swim stress (Baganz et al. 2010).

Blockade of OCT3 alone with decynium-22, an OCT3 blocker, or blockade of OCT3 during acute restraint stress, also leads to increases in extracellular serotonin in the DMH, despite acute restraint stress alone having no effect (Feng et al. 2009). Glucocorticoid/OCT3 interactions in the DMH have been proposed as a mechanism contributing to negative feedback control of the HPA axis (Stamper et al. 2015). Blockade of OCT3 by glucocorticoids increases serotonergic neurotransmission, which may lead to activation of inhibitory 5-HT$_{1A}$ receptors within the DMH. This mechanism may contribute, at the intermediate time scale, to glucocorticoid-mediated negative feedback control of HPA axis activity (Stamper et al. 2015). Indeed, when a 5-HT$_{1A}$ receptor agonist was infused into the DMH, there was a suppression of acute restraint-induced plasma glucocorticoids, similar to what was found previously with intra-DMH corticosterone infusions (Stamper et al. 2017). Acutely, glucocorticoids inhibit reuptake of 5-HT, and other monoamines, which would extend synaptic actions of monoamines and enhance stress-induced anxiety. Chronic stress-induced glucocorticoids downregulate OCT3 (Baganz et al. 2010), which would have a long-term effect on monoamine synaptic actions and possibly extend the memory-enhancing effects of stress. This mechanism is depicted in Fig. 5.

5.4 Chronic Glucocorticoid Administration Increases Tryptophan Hydroxylase 2 (Tph2) mRNA and Protein Expression and Tph2 Activity in Association with Exaggerated Anxiety-Like Behavioral Responses

Administration of dissolved crystalline corticosterone via drinking water to adrenal-intact rats over the course of 3 weeks increases anxiety-like behavior, as measured in the open-field test, elevated plus-maze test, and social interaction test (Donner et al. 2012b). Typically, the HPA axis is under circadian control, with ACTH and corticosterone having a higher plasma concentration during the active phase, relative to the inactive phase of the diurnal activity rhythm. Chronic administration of corticosterone abolishes the diurnal rhythms of plasma concentrations of ACTH and corticosterone (Donner et al. 2012b, 2016). Additionally, continuous corticosterone induces thymus involution and adrenal gland atrophy, with the adrenal medulla remaining relatively unaffected but with reduced volumes of the zona glomerulosa, zona fasciculata, and zona reticularis of the adrenal cortex (Donner et al. 2012b, 2016). Chronic corticosterone also renders the adrenal glands unresponsive to ex vivo ACTH stimulation (Donner et al. 2016). Accompanying these changes in physiology and behavior are increases in tryptophan hydroxylase 2 (*tph2*) mRNA expression within the DR. Expression of *tph2*, the gene encoding the rate-limiting enzyme for the synthesis of serotonin, is sensitive to both external stressors and glucocorticoid hormones (Donner et al. 2012b). Data suggest that chronic exposure to glucocorticoids sensitizes stress- and anxiety-related serotonergic systems through the enhancement of Tph2 activity (Donner et al. 2016). As mentioned above, chronic glucocorticoid administration increases *tph2* mRNA expression in the DR during the inactive phase, resulting in a loss of the circadian rhythm of *tph2* mRNA expression (Donner et al. 2012b). Increased *tph2* mRNA expression was evident in the dorsomedial subregion of the DR (Donner et al. 2012b), thought to be responsible for stress-induced exaggeration of anxiety- and fear-related behavioral responses (Lowry et al. 2008). These effects were associated with increases in anxiety-like behavioral responses, as measured in the elevated plus-maze and social interaction test. Donner et al. (2016) found that chronic corticosterone administration increases *tph2* mRNA expression, as well as Tph2 protein expression in the anxiety-related dorsomedial subregion of the DR. These effects were associated with increases in anxiety-like behavioral responses, as measured in the elevated plus-maze. Furthermore, chronic corticosterone administration sensitized serotonergic systems, as measured by tryptophan hydroxylase activity, to acoustic startle, in neural systems associated with anxiety- and fear-related behavioral responses, including the lateral orbitofrontal cortex, infralimbic

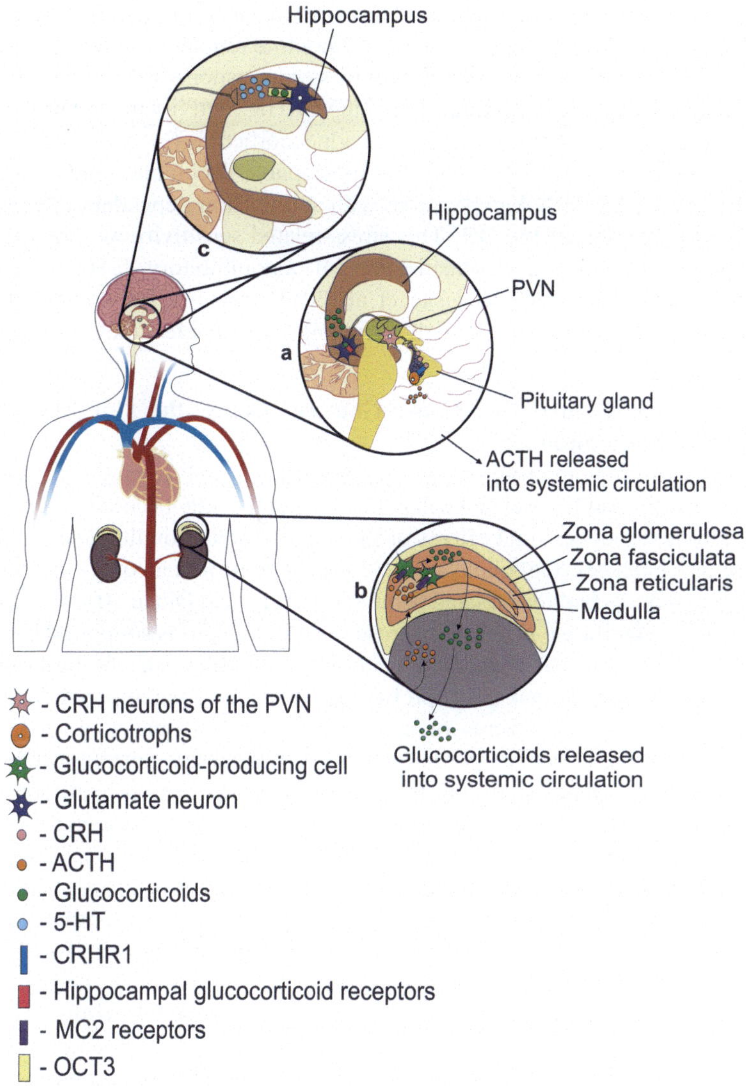

Fig. 5 Hypothetical model illustrating glucocorticoid-dependent mechanisms underlying stress-induced exaggeration of anxiety- and fear-related behavioral responses. (**a**) Stressful stimuli promote the synthesis of corticotropin-releasing hormone (CRH) from the paraventricular nucleus (PVN) of the hypothalamus (Rivier et al. 1984). Axons arising from CRH neurons in the PVN project to the median eminence, where CRH is released from axon terminals, enters the hypothalamic-hypophyseal portal system, and is delivered to the anterior pituitary via the portal circulation. CRH then binds to corticotropin-releasing hormone receptor Type 1 receptors (CRHR1) on corticotrophs in the anterior pituitary (Aguilera et al. 1983), which stimulates the corticotrophs to release adrenocorticotropic hormone (ACTH). (**b**) ACTH travels through systemic circulation and ultimately binds to melanocortin 2 (MC2) receptors on glucocorticoid-synthesizing cells (Chida et al. 2007), predominantly in the zona fasciculata of the adrenal cortex. ACTH induces the synthesis and release of glucocorticoid hormones from adrenocortical cells. These glucocorticoids

cortex, dorsal hippocampus, and caudal pontine reticular formation. Interestingly, tryptophan hydroxylase activity in the caudal pontine reticular formation, known to be a critical substrate for fear potentiation of startle responses (Davis et al. 1982), was correlated with startle amplitude. Blocking CRHR2 activity in the DR decreased sensitivity to stress after chronic exposure to glucocorticoids (Donner et al. 2016). In addition, Donner et al. (2018) found that inescapable shock and cold swim stress elevated overall *tph2* mRNA expression, with subdivision-dependent effects in the dorsomedial subregion of the DR. This stress-related sensitivity may be due to an inescapable shock-induced desensitization of autoinhibitory 5-HT_{1A} receptors within the dorsomedial DR (Rozeske et al. 2011), making the dorsomedial DR more vulnerable to anxiogenic stimuli following inescapable shock (Donner et al. 2018).

Perspectives and Future Directions: Glucocorticoids, the Gut Microbiome, and Psychoneuroimmunology
The influence of the microbiota on the physiology and behavior of the host, including anxiety- and fear-related behaviors, is only recently becoming appreciated. Studies in rodents suggest that effects of the gut microbiota on the host range from control of neurogenesis (Ogbonnaya et al. 2015) to control of anxiety-like defensive behavioral responses and fear-related behaviors (Cryan and Dinan 2012). A potential role of the gut microbiome in emotional responses in humans is supported by a study of a cohort of women that consumed a probiotic, in which specific brain regions, including the periaqueductal gray, prefrontal cortex, and parahippocampal gyrus were less responsive to pictures of emotional faces (Tillisch et al. 2013). One strategy to study the effects of the microbiota is to generate germ-free (GF) mice. One of the earliest studies using GF mice found that they exhibit increased plasma ACTH and corticosterone concentrations, compared to specific-pathogen-free (SPF) mice following exposure to 1 h of restraint stress (Sudo et al. 2004). Although no morphological differences were found in the adrenal or pituitary glands between GF and SPF mice, CRH mRNA expression and CRH peptide in the hypothalamus were significantly increased in GF mice, compared to SPF mice. Gut reconstitution with *Bifidobacterium infantis*, a bacterium found early in development in the gut, into GF mice reduced both plasma levels of ACTH and corticosterone following exposure to 1 h of restraint stress. In contrast, reconstitution with the enteropathogen, *Escherichia coli*, in GF mice increased ACTH and corticosterone responses, following exposure to 1 h of restraint stress compared to GF mice (Sudo et al. 2004). In a separate study, mice that were found to have a resilient phenotype versus a

---

Fig. 5 (continued) (primarily corticosterone in rodents and cortisol in humans) travel in the systemic circulation to organize and enhance various bodily defenses against stressors. (**c**) In the hippocampus, cortisol blocks 5-hydroxytryptamine (5-HT) reuptake by organic cation transporters 3 (OCT3) in hippocampal synapses, therefore delaying 5-HT clearance and producing behavioral effects (Baganz et al. 2010). Glucocorticoids also act on an interconnected network, including the basolateral amygdala, hippocampus, and medial prefrontal cortex to enhance memory consolidation

susceptible phenotype had a higher relative abundance of *Bifidobacterium* in the fecal microbiome after chronic social defeat stress (Yang et al. 2017), while, in humans, higher maternal stress and higher maternal cortisol have been found to be associated with reductions in *Bifidobacterium* (Zijlmans et al. 2015). Future studies are required to delineate the specific mechanisms through which the gut microbiome can affect the HPA axis and emotional behavior.

The immune system and the central nervous system (CNS) communicate bidirectionally, and one of the consequences of an acute stress response is the redistribution of immune cells within the body (Dhabhar et al. 2012; Dhabhar 2009). Norepinephrine, epinephrine, and corticosterone all influence which immune cell populations are trafficked from certain compartments into the blood (Dhabhar et al. 2012). The mechanisms through which stress exposure leads to peripheral inflammation, and particularly neuroinflammation, are not thoroughly understood, but recent evidence suggests that chronic psychosocial stress induces production of inflammatory monocytes in the bone marrow that can traffic to the brain, where they release interleukin-1β (IL-1β) and promote neuroinflammation (Niraula et al. 2018), as well as increased anxiety (Wohleb et al. 2014a, b; McKim et al. 2017). Additionally, blood monocyte cell numbers increase in response to repeated social defeat and are correlated with increased microglia activation in the prefrontal cortex, amygdala, and hippocampus, leading to neuroinflammation in limbic forebrain areas that are involved in HPA axis-, anxiety-, and fear-related responses (Wohleb et al. 2014a). In another study, peripheral monocytes were found in the prefrontal cortex, amygdala, and BNST after repeated social defeat stress, highlighting how the immune system can influence parts of the brain involved in HPA axis-, anxiety-, and fear-related responses in a direct manner (Wohleb et al. 2013). Recently, McKim et al. (2017) demonstrated that after repeated social defeat stress, bone marrow-derived monocytes expressing IL-1β were recruited by endothelial cells to areas such as the prefrontal cortex, hippocampus, and amygdala (McKim et al. 2017). This specific recruitment activated local microglia in the brain to possibly drive the anxiety-like behavior (McKim et al. 2017; Nie et al. 2018).

There are clear examples of interactions among glucocorticoid hormones, the gut microbiota, and anxiety- and fear-related behavioral responses that deserve further study. For example, exposure of mice to the CSC paradigm and other stressors results in expansion of pathobionts, such as *Helicobacter* spp. (Reber et al. 2016b; Guo et al. 2009), which induce intestinal inflammation in mice with impaired immunoregulation (Bassett et al. 2015). Furthermore, expansion of *Helicobacter* spp. following exposure to psychological stress is dependent on glucocorticoid receptors (Guo et al. 2009), possibly due to the immunosuppressive effects of glucocorticoids. Chronic subordinate colony housing-induced spontaneous colitis and CSC-induced exaggeration of anxiety-like behavioral responses both require the presence of *Helicobacter* spp. (Langgartner et al. 2017; Guo et al. 2009; Reber et al. 2016b), supporting the interactions among glucocorticoid hormones, the gut microbiota, and anxiety- and fear-related behavioral responses.

Recent studies provide new insight into how chronic psychosocial stress leads to increases in monocyte mobilization from the bone marrow, induces glucocorticoid insensitivity of these cells, and increases neuroinflammation. Mice subjected to six daily 2-h social defeat exposures in a murine model of repeated social defeat (RSD) respond with increased plasma corticosterone and IL-6 concentrations immediately after stress exposure and, when assessed 14 h after the final stress exposure, respond with increases in $Ly6C^{hi}$ monocytes in the blood and increased $CD11b^{+}/CD45^{hi}$ monocytes in the brain. These responses were attenuated by blockade of glucocorticoid signaling, either by adrenalectomy or by treatment with metyrapone, a corticosterone synthesis inhibitor. Treatment with metyrapone also prevented (1) stress-induced glucocorticoid resistance, (2) stress-induced exaggeration of inflammatory responses to lipopolysaccharide in splenocytes, (3) stress-induced morphological changes in microglia, and (4) increases in *IL-1β* mRNA expression in the brain. Together, these findings point toward a role of stress-induced activation of the HPA axis and glucocorticoids in neuroimmune signaling from the periphery to the brain and subsequent neuroinflammatory responses that are relevant to stress-related psychiatric disorders.

A number of mechanisms associated with microbiome-gut-brain axis signaling are important areas for further study. These include afferent neural pathways, including afferent vagal neurons in the nodose and jugular ganglia that signal to the nucleus of the solitary tract and other brain stem structures (reviewed in Pavlov and Tracey 2012), a pathway that is essential for the communication of proinflammatory signals to the CNS (Watkins et al. 1995) and mediates gut-brain axis signaling (Breit et al. 2018). Other neural afferents are likely to be involved, including sympathetic afferents and afferent fibers that travel within the sympathetic nerve bundles, with cell bodies in the dorsal root ganglia. These afferents have the potential to relay signals from viscera to the CNS via spinothalamic, spinoreticular, spinomesencephalic, spinoparabrachial, spinohypothalamic, spinocervical, spinovestibular, spino-olivary, and other spinal afferent pathways to the CNS (Hendry and Hsiao 2013; Kayalioglu 2009; Mayer 2011; Mayer et al. 2015). Alternatively, bacterially derived metabolites, such as tryptophan metabolites, may alter mucosal immunity and/or cross the blood-brain barrier to directly affect brain function (Jin et al. 2014). Finally, peripheral immune cells, such as bone marrow- or spleen-derived monocytes (discussed above), macrophages, dendritic cells, or T cells, may traffic into the CNS or signal through the meningeal compartment (Derecki et al. 2010; McKim et al. 2016; Wohleb et al. 2014a, b). Although T cells do not typically access the CNS under healthy conditions, there is evidence that aged animals exhibit a gradual enhancement of T cell trafficking to the brain (Gemechu and Bentivoglio 2012). T cells can modulate innate immune cells in the CNS (Walsh et al. 2014; Xie et al. 2015), and redirecting the population of T cells that access the CNS promotes a reparative environment in some contexts. Future studies could explore how microbiome-based interventions alter T cell density and phenotype in the CNS, neuroinflammatory responses within the CNS, and the consequences for trauma- and stressor-related outcomes (Lowry et al. 2016).

Additional work is also needed to translate the findings discussed above in rodent model systems to clinical populations. In combat-exposed veterans with a diagnosis of PTSD, there is a significant overall proinflammatory bias, with increases in plasma concentrations of IL-1β, IL-6, C-reactive protein, tumor necrosis factor, IL-10, and interferon γ (Lindqvist et al. 2014, 2017). In a recent study, Hodes et al. (2014) tested the hypothesis that inflammation is a predisposing factor for stress-induced etiologies, such as depression and PTSD. Patients with major depressive disorder were found to have elevated concentrations of IL-6 in plasma, compared to control levels (Hodes et al. 2014). In the same study, animals that were repeatedly socially defeated and later developed a susceptible phenotype had higher concentrations of the proinflammatory cytokine IL-6 in plasma, which correlated with lower social interaction behavior indicative of increased anxiety-like behavior (Hodes et al. 2014). Blood drawn prior to social defeat stress revealed that monocytes, and not T cell or B cell lineages, were significantly increased in animals and were predictive of development of a susceptible phenotype (Hodes et al. 2014).

6 Conclusions

The role that glucocorticoids play in stress-related physiological and behavioral responses, including anxiety and fear responses, cannot be overstated. Glucocorticoids have a multitude of effects on many different physiological and behavioral systems. Glucocorticoids can enhance memory consolidation (Roozendaal et al. 1997) and impair memory retrieval (de Quervain et al. 1998). Although the mechanisms through which glucocorticoids alter anxiety- and fear-related behavioral responses remain the subject of intense study, glucocorticoid-mediated blockade of OCT3-mediated transport of monoamines (Baganz et al. 2010), facilitation of trafficking of peripheral immune cells to the CNS (Dhabhar et al. 2012), and altered serotonergic signaling may be involved (Donner et al. 2012b, 2016, 2018). All of these mechanisms could be working in concert to influence allostatic processes. By exploring allostatic mediators, such as glucocorticoids, and their ability to modulate different brain processes, we may be able to better understand the etiology, pathophysiology, and symptomatology of anxiety disorders, trauma- and stressor-related disorders, and affective disorders, leading to novel therapeutic approaches to prevention and treatment.

Acknowledgments We gratefully acknowledge Akaysha D. Joiner for critical evaluation of the manuscript.

References

Abrams JK, Johnson PL, Hay-Schmidt A, Mikkelsen JD, Shekhar A, Lowry CA (2005) Serotonergic systems associated with arousal and vigilance behaviors following administration of anxiogenic drugs. Neuroscience 133(4):983–997

Adcock IM, Nasuhara Y, Stevens DA, Barnes PJ (1999) Ligand-induced differentiation of glucocorticoid receptor (GR) trans-repression and transactivation: preferential targetting of NF-κB and lack of I-κB involvement. Br J Pharmacol 127(4):1003–1011

Aguilera G, Harwood JP, Wilson JX, Morell J, Brown JH, Catt KJ (1983) Mechanisms of action of corticotropin-releasing factor and other regulators of corticotropin release in rat pituitary cells. J Biol Chem 258(13):8039–8045

Aisa B, Tordera R, Lasheras B, Del Rio J, Ramirez MJ (2007) Cognitive impairment associated to HPA axis hyperactivity after maternal separation in rats. Psychoneuroendocrinology 32(3):256–266

Amat J, Matus-Amat P, Watkins LR, Maier SF (1998) Escapable and inescapable stress differentially alter extracellular levels of 5-HT in the basolateral amygdala of the rat. Brain Res 812(1–2):113–120

Amat J, Sparks PD, Matus-Amat P, Griggs J, Watkins LR, Maier SF (2001) The role of the habenular complex in the elevation of dorsal raphe nucleus serotonin and the changes in the behavioral responses produced by uncontrollable stress. Brain Res 917(1):118–126

Amat J, Tamblyn JP, Paul ED, Bland ST, Amat P, Foster AC, Watkins LR, Maier SF (2004) Microinjection of urocortin 2 into the dorsal raphe nucleus activates serotonergic neurons and increases extracellular serotonin in the basolateral amygdala. Neuroscience 129(3):509–519

Amat J, Baratta MV, Paul E, Bland ST, Watkins LR, Maier SF (2005) Medial prefrontal cortex determines how stressor controllability affects behavior and dorsal raphe nucleus. Nat Neurosci 8(3):365–371

Amat J, Paul E, Watkins LR, Maier SF (2008) Activation of the ventral medial prefrontal cortex during an uncontrollable stressor reproduces both the immediate and long-term protective effects of behavioral control. Neuroscience 154(4):1178–1186

Arnold J (1866) Ein Beitrag zu der feineren Structur und dem Chemismus der Nebennieren. Virchows Arch 35(1):64–107

Arruda-Carvalho M, Clem RL (2014) Pathway-selective adjustment of prefrontal-amygdala transmission during fear encoding. J Neurosci 34(47):15601–15609

Baganz NL, Horton RE, Calderon AS, Owens WA, Munn JL, Watts LT, Koldzic-Zivanovic N, Jeske NA, Koek W, Toney GM, Daws LC (2008) Organic cation transporter 3: keeping the brake on extracellular serotonin in serotonin-transporter-deficient mice. Proc Natl Acad Sci U S A 105(48):18976–18981

Baganz NL, Horton R, Martin K, Holmes A, Daws LC (2010) Repeated swim impairs serotonin clearance via a corticosterone-sensitive mechanism: organic cation transporter 3, the smoking gun. J Neurosci 30(45):15185–15195

Bale TL, Vale WW (2004) CRF and CRF receptors: role in stress responsivity and other behaviors. Annu Rev Pharmacol Toxicol 44:525–557

Baratta MV, Christianson JP, Gomez DM, Zarza CM, Amat J, Masini CV, Watkins LR, Maier SF (2007) Controllable versus uncontrollable stressors bi-directionally modulate conditioned but not innate fear. Neuroscience 146(4):1495–1503

Baratta MV, Zarza CM, Gomez DM, Campeau S, Watkins LR, Maier SF (2009) Selective activation of dorsal raphe nucleus-projecting neurons in the ventral medial prefrontal cortex by controllable stress. Eur J Neurosci 30(6):1111–1116

Barnes PJ, Karin M (1997) Nuclear factor-κB: a pivotal transcription factor in chronic inflammatory diseases. N Engl J Med 336(15):1066–1071

Bassett SA, Young W, Barnett MP, Cookson AL, McNabb WC, Roy NC (2015) Changes in composition of caecal microbiota associated with increased colon inflammation in interleukin-10 gene-deficient mice inoculated with *Enterococcus* species. Nutrients 7(3):1798–1816

Battaglia M, Bertella S, Politi E, Bernardeschi L, Perna G, Gabriele A, Bellodi L (1995) Age at onset of panic disorder: influence of familial liability to the disease and of childhood separation anxiety disorder. Am J Psychiatry 152(9):1362–1364
Bauer EP, Schafe GE, LeDoux JE (2002) NMDA receptors and L-type voltage-gated calcium channels contribute to long-term potentiation and different components of fear memory formation in the lateral amygdala. J Neurosci 22(12):5239–5249
Blanchard RJ, Blanchard DC (1969) Crouching as an index of fear. J Comp Physiol Psychol 67(3):370–375
Blanchard RJ, Blanchard DC (1989) Attack and defense in rodents as ethoexperimental models for the study of emotion. Prog Neuro-Psychopharmacol Biol Psychiatry 13(Suppl):S3–S14
Blanchard DC, Griebel G, Pobbe R, Blanchard RJ (2011) Risk assessment as an evolved threat detection and analysis process. Neurosci Biobehav Rev 35(4):991–998
Bolles RC, Fanselow MS (1980) A perceptual-defensive-recuperative model of fear and pain. Behav Brain Sci 3(2):291–301
Breit S, Kupferberg A, Rogler G, Hasler G (2018) Vagus nerve as modulator of the brain-gut axis in psychiatric and inflammatory disorders. Front Psych 9:44
Bremner JD, Licinio J, Darnell A, Krystal JH, Owens MJ, Southwick SM, Nemeroff CB, Charney DS (1997) Elevated CSF corticotropin-releasing factor concentrations in posttraumatic stress disorder. Am J Psychiatry 154(5):624–629
Brioni JD, Nagahara AH, McGaugh JL (1989) Involvement of the amygdala GABAergic system in the modulation of memory storage. Brain Res 487(1):105–112
Brown JS, Kalish HI, Farber IE (1951) Conditioned fear as revealed by magnitude of startle response to an auditory stimulus. J Exp Psychol 41(5):317–328
Bruhn TO, Plotsky PM, Vale WW (1984) Effect of paraventricular lesions on corticotropin-releasing factor (CRF)-like immunoreactivity in the stalk-median eminence: studies on the adrenocorticotropin response to ether stress and exogenous CRF. Endocrinology 114(1):57–62
Burke AR, McCormick CM, Pellis SM, Lukkes JL (2017) Impact of adolescent social experiences on behavior and neural circuits implicated in mental illnesses. Neurosci Biobehav Rev 76(Pt B):280–300
Cahill L, McGaugh JL (1990) Amygdaloid complex lesions differentially affect retention of tasks using appetitive and aversive reinforcement. Behav Neurosci 104(4):532–543
Cahill L, Vazdarjanova A, Setlow B (2000) The basolateral amygdala complex is involved with, but is not necessary for, rapid acquisition of Pavlovian 'fear conditioning'. Eur J Neurosci 12(8):3044–3050
Calhoon GG, Tye KM (2015) Resolving the neural circuits of anxiety. Nat Neurosci 18(10):1394–1404
Cannon WB (1929) Organization for physiological homeostasis. Physiol Rev 9(3):399–431
Caspi A, Sugden K, Moffitt TE, Taylor A, Craig IW, Harrington H, McClay J, Mill J, Martin J, Braithwaite A, Poulton R (2003) Influence of life stress on depression: moderation by a polymorphism in the 5-HTT gene. Science 301(5631):386–389
Castellano C, Brioni JD, Nagahara AH, McGaugh JL (1989) Post-training systemic and intra-amygdala administration of the GABA-B agonist baclofen impairs retention. Behav Neural Biol 52(2):170–179
Chang CP, Pearse RV 2nd, O'Connell S, Rosenfeld MG (1993) Identification of a seven transmembrane helix receptor for corticotropin-releasing factor and sauvagine in mammalian brain. Neuron 11(6):1187–1195
Chida D, Nakagawa S, Nagai S, Sagara H, Katsumata H, Imaki T, Suzuki H, Mitani F, Ogishima T, Shimizu C, Kotaki H, Kakuta S, Sudo K, Koike T, Kubo M, Iwakura Y (2007) Melanocortin 2 receptor is required for adrenal gland development, steroidogenesis, and neonatal gluconeogenesis. Proc Natl Acad Sci U S A 104(46):18205–18210
Choi DC, Furay AR, Evanson NK, Ostrander MM, Ulrich-Lai YM, Herman JP (2007) Bed nucleus of the stria terminalis subregions differentially regulate hypothalamic-pituitary-adrenal axis activity: implications for the integration of limbic inputs. J Neurosci 27(8):2025–2034

Christianson JP, Ragole T, Amat J, Greenwood BN, Strong PV, Paul ED, Fleshner M, Watkins LR, Maier SF (2010) 5-Hydroxytryptamine 2C receptors in the basolateral amygdala are involved in the expression of anxiety after uncontrollable traumatic stress. Biol Psychiatry 67(4):339–345
Chun LE, Woodruff ER, Morton S, Hinds LR, Spencer RL (2015) Variations in phase and amplitude of rhythmic clock gene expression across prefrontal cortex, hippocampus, amygdala, and hypothalamic paraventricular and suprachiasmatic nuclei of male and female rats. J Biol Rhythm 30(5):417–436
Cordero MI, Venero C, Kruyt ND, Sandi C (2003) Prior exposure to a single stress session facilitates subsequent contextual fear conditioning in rats. Evidence for a role of corticosterone. Horm Behav 44:338–345
Corodimas KP, LeDoux JE, Gold PW, Schulkin J (1994) Corticosterone potentiation of conditioned fear in rats. Ann N Y Acad Sci 746:392–393
Cryan JF, Dinan TG (2012) Mind-altering microorganisms: the impact of the gut microbiota on brain and behaviour. Nat Rev Neurosci 13(10):701–712
D'Amato FR, Zanettini C, Lampis V, Coccurello R, Pascucci T, Ventura R, Puglisi-Allegra S, Spatola CA, Pesenti-Gritti P, Oddi D, Moles A, Battaglia M (2011) Unstable maternal environment, separation anxiety, and heightened CO_2 sensitivity induced by gene-by-environment interplay. PLoS One 6(4):e18637
Daldrup T, Remmes J, Lesting J, Gaburro S, Fendt M, Meuth P, Kloke V, Pape HC, Seidenbecher T (2015) Expression of freezing and fear-potentiated startle during sustained fear in mice. Genes Brain Behav 14(3):281–291
Daldrup T, Lesting J, Meuth P, Seidenbecher T, Pape HC (2016) Neuronal correlates of sustained fear in the anterolateral part of the bed nucleus of stria terminalis. Neurobiol Learn Mem 131:137–146
Dallman MF, Akana SF, Strack AM, Hanson ES, Sebastian RJ (1995) The neural network that regulates energy balance is responsive to glucocorticoids and insulin and also regulates HPA axis responsivity at a site proximal to CRF neurons. Ann N Y Acad Sci 771:730–742
Danese A, McEwen BS (2012) Adverse childhood experiences, allostasis, allostatic load, and age-related disease. Physiol Behav 106(1):29–39
Daniels WM, Pietersen CY, Carstens ME, Stein DJ (2004) Maternal separation in rats leads to anxiety-like behavior and a blunted ACTH response and altered neurotransmitter levels in response to a subsequent stressor. Metab Brain Dis 19(1–2):3–14
Davis M, Gendelman DS, Tischler MD, Gendelman PM (1982) A primary acoustic startle circuit: lesion and stimulation studies. J Neurosci 2(6):791–805
Day TA (2005) Defining stress as a prelude to mapping its neurocircuitry: no help from allostasis. Prog Neuro-Psychopharmacol Biol Psychiatry 29(8):1195–1200
Day TA, Walker FR (2007) More appraisal please: a commentary on Pfaff et al. (2007) "Relations between mechanisms of CNS arousal and mechanisms of stress". Stress 10(4):311–313
Dayas CV, Day TA (2002) Opposing roles for medial and central amygdala in the initiation of noradrenergic cell responses to a psychological stressor. Eur J Neurosci 15(10):1712–1718
Dayas CV, Buller KM, Crane JW, Xu Y, Day TA (2001) Stressor categorization: acute physical and psychological stressors elicit distinctive recruitment patterns in the amygdala and in medullary noradrenergic cell groups. Eur J Neurosci 14(7):1143–1152
de Quervain DJ, Roozendaal B, McGaugh JL (1998) Stress and glucocorticoids impair retrieval of long-term spatial memory. Nature 394(6695):787–790
de Quervain DJ, Roozendaal B, Nitsch RM, McGaugh JL, Hock C (2000) Acute cortisone administration impairs retrieval of long-term declarative memory in humans. Nat Neurosci 3(4):313–314
De Souza EB, Perrin MH, Rivier J, Vale WW, Kuhar MJ (1984) Corticotropin-releasing factor receptors in rat pituitary gland: autoradiographic localization. Brain Res 296(1):202–207
Derecki NC, Cardani AN, Yang CH, Quinnies KM, Crihfield A, Lynch KR, Kipnis J (2010) Regulation of learning and memory by meningeal immunity: a key role for IL-4. J Exp Med 207(5):1067–1080

Dhabhar FS (2009) A hassle a day may keep the pathogens away: the fight-or-flight stress response and the augmentation of immune function. Integr Comp Biol 49(3):215–236
Dhabhar FS, Malarkey WB, Neri E, McEwen BS (2012) Stress-induced redistribution of immune cells – from barracks to boulevards to battlefields: a tale of three hormones – Curt Richter Award winner. Psychoneuroendocrinology 37(9):1345–1368
Dolzani SD, Baratta MV, Amat J, Agster KL, Saddoris MP, Watkins LR, Maier SF (2016) Activation of a Habenulo-Raphe circuit is critical for the behavioral and neurochemical consequences of uncontrollable stress in the male rat. eNeuro 3(5)
Donner NC, Lowry CA (2013) Sex differences in anxiety and emotional behavior. Pflügers Arch 465(5):601–626
Donner NC, Johnson PL, Fitz SD, Kellen KE, Shekhar A, Lowry CA (2012a) Elevated *tph2* mRNA expression in a rat model of chronic anxiety. Depress Anxiety 29(4):307–319
Donner NC, Montoya CD, Lukkes JL, Lowry CA (2012b) Chronic non-invasive corticosterone administration abolishes the diurnal pattern of *tph2* expression. Psychoneuroendocrinology 37(5):645–661
Donner NC, Siebler PH, Johnson DT, Villarreal MD, Mani S, Matti AJ, Lowry CA (2016) Serotonergic systems in the balance: CRHR1 and CRHR2 differentially control stress-induced serotonin synthesis. Psychoneuroendocrinology 63:178–190
Donner NC, Kubala KH, Hassell JE Jr, Lieb MW, Nguyen KT, Heinze JD, Drugan RC, Maier SF, Lowry CA (2018) Two models of inescapable stress increase *tph2* mRNA expression in the anxiety-related dorsomedial part of the dorsal raphe nucleus. Neurobiol Stress 8:68–81
Dumont FS, Biancardi V, Kinkead R (2011) Hypercapnic ventilatory response of anesthetized female rats subjected to neonatal maternal separation: insight into the origins of panic attacks? Respir Physiol Neurobiol 175(2):288–295
Dunn AJ, Berridge CW (1990) Is corticotropin-releasing factor a mediator of stress responses? Ann N Y Acad Sci 579:183–191
Falls WA, Miserendino MJ, Davis M (1992) Extinction of fear-potentiated startle: blockade by infusion of an NMDA antagonist into the amygdala. J Neurosci 12(3):854–863
Fanselow MS (1980) Conditional and unconditional components of post-shock freezing. Pavlov J Biol Sci 15(4):177–182
Fanselow MS (1986) Associative vs topographical accounts of the immediate shock freezing deficit in rats – implications for the response selection-rules governing species-specific defensive reactions. Learn Motiv 17(1):16–39
Fanselow MS (1990) Factors governing one-trial contextual conditioning. Anim Learn Behav 18(3):264–270
Fanselow MS, Dong HW (2010) Are the dorsal and ventral hippocampus functionally distinct structures? Neuron 65(1):7–19
Fanselow MS, LeDoux JE (1999) Why we think plasticity underlying Pavlovian fear conditioning occurs in the basolateral amygdala. Neuron 23(2):229–232
Fanselow MS, Lester LS (1988) A functional behavioristic approach to aversively motivated behavior: predatory imminence as a determinant of the topography of defensive behavior. In: Bolles RC, Beecher MD (eds) Evolution and learning. Lawrence Erlbaum Associates, Hillsdale
Felix-Ortiz AC, Beyeler A, Seo C, Leppla CA, Wildes CP, Tye KM (2013) BLA to vHPC inputs modulate anxiety-related behaviors. Neuron 79(4):658–664
Felix-Ortiz AC, Burgos-Robles A, Bhagat ND, Leppla CA, Tye KM (2016) Bidirectional modulation of anxiety-related and social behaviors by amygdala projections to the medial prefrontal cortex. Neuroscience 321:197–209
Fendt M, Koch M (2013) Translational value of startle modulations. Cell Tissue Res 354(1):287–295
Feng N, Telefont M, Kelly KJ, Orchinik M, Forster GL, Renner KJ, Lowry CA (2009) Local perfusion of corticosterone in the rat medial hypothalamus potentiates D-fenfluramine-induced elevations of extracellular 5-HT concentrations. Horm Behav 56(1):149–157

Ferry B, Parrot S, Marien M, Lazarus C, Cassel JC, McGaugh JL (2015) Noradrenergic influences in the basolateral amygdala on inhibitory avoidance memory are mediated by an action on alpha2-adrenoceptors. Psychoneuroendocrinology 51:68–79

Flak JN, Solomon MB, Jankord R, Krause EG, Herman JP (2012) Identification of chronic stress-activated regions reveals a potential recruited circuit in rat brain. Eur J Neurosci 36(4):2547–2555

Fontes MA, Xavier CH, de Menezes RC, Dimicco JA (2011) The dorsomedial hypothalamus and the central pathways involved in the cardiovascular response to emotional stress. Neuroscience 184:64–74

Francis DD, Diorio J, Plotsky PM, Meaney MJ (2002) Environmental enrichment reverses the effects of maternal separation on stress reactivity. J Neurosci 22(18):7840–7843

Frank MG, Watkins LR, Maier SF (2013) Stress-induced glucocorticoids as a neuroendocrine alarm signal of danger. Brain, Behavior, and Immunity 33:1–6

Galvez R, Mesches MH, McGaugh JL (1996) Norepinephrine release in the amygdala in response to footshock stimulation. Neurobiol Learn Mem 66(3):253–257

Gasser PJ, Lowry CA (2018) Organic cation transporter 3: a cellular mechanism underlying rapid, non-genomic glucocorticoid regulation of monoaminergic neurotransmission, physiology, and behavior. Horm Behav

Gasser PJ, Lowry CA, Orchinik M (2006) Corticosterone-sensitive monoamine transport in the rat dorsomedial hypothalamus: potential role for organic cation transporter 3 in stress-induced modulation of monoaminergic neurotransmission. J Neurosci 26(34):8758–8766

Gehlert DR, Shekhar A, Morin SM, Hipskind PA, Zink C, Gackenheimer SL, Shaw J, Fitz SD, Sajdyk TJ (2005) Stress and central Urocortin increase anxiety-like behavior in the social interaction test via the CRF_1 receptor. Eur J Pharmacol 509(2–3):145–153

Gemechu JM, Bentivoglio M (2012) T cell recruitment in the brain during normal aging. Front Cell Neurosci 6:38

Genest SE, Balon N, Laforest S, Drolet G, Kinkead R (2007) Neonatal maternal separation and enhancement of the hypoxic ventilatory response in rat: the role of GABAergic modulation within the paraventricular nucleus of the hypothalamus. J Physiol 583(Pt 1):299–314

Glaser R, Kiecolt-Glaser JK (2005) Stress-induced immune dysfunction: implications for health. Nat Rev Immunol 5(3):243–251

Goetz RR, Klein DF, Gorman JM (1996) Symptoms essential to the experience of sodium lactate-induced panic. Neuropsychopharmacology 14(5):355–366

Gold PE, Van Buskirk RB (1975) Facilitation of time-dependent memory processes with posttrial epinephrine injections. Behav Biol 13(2):145–153

Gold PE, van Buskirk R (1976a) Effects of posttrial hormone injections on memory processes. Horm Behav 7(4):509–517

Gold PE, Van Buskirk R (1976b) Enhancement and impairment of memory processes with post-trial injections of adrenocorticotrophic hormone. Behav Biol 16(4):387–400

Goosens KA, Hobin JA, Maren S (2003) Auditory-evoked spike firing in the lateral amygdala and Pavlovian fear conditioning: mnemonic code or fear bias? Neuron 40(5):1013–1022

Grahn RE, Will MJ, Hammack SE, Maswood S, McQueen MB, Watkins LR, Maier SF (1999) Activation of serotonin-immunoreactive cells in the dorsal raphe nucleus in rats exposed to an uncontrollable stressor. Brain Res 826(1):35–43

Gray JA, McNaughton N (2003) The neuropsychology of anxiety: an enquiry into the function of the septo-hippocampal system. Oxford University Press, New York

Guillemin R, Rosenberg B (1955) Humoral hypothalamic control of anterior pituitary: a study with combined tissue cultures. Endocrinology 57(5):599–607

Gungor NZ, Paré D (2016) Functional heterogeneity in the bed nucleus of the stria terminalis. J Neurosci 36(31):8038–8049

Guo G, Jia KR, Shi Y, Liu XF, Liu KY, Qi W, Guo Y, Zhang WJ, Wang T, Xiao B, Zou QM (2009) Psychological stress enhances the colonization of the stomach by *Helicobacter pylori* in the BALB/c mouse. Stress 12(6):478–485

Hale MW, Shekhar A, Lowry CA (2012) Stress-related serotonergic systems: implications for symptomatology of anxiety and affective disorders. Cell Mol Neurobiol 32(5):695–708
Hammack SE, Richey KJ, Schmid MJ, LoPresti ML, Watkins LR, Maier SF (2002) The role of corticotropin-releasing hormone in the dorsal raphe nucleus in mediating the behavioral consequences of uncontrollable stress. J Neurosci 22(3):1020–1026
Hammack SE, Pepin JL, DesMarteau JS, Watkins LR, Maier SF (2003a) Low doses of corticotropin-releasing hormone injected into the dorsal raphe nucleus block the behavioral consequences of uncontrollable stress. Behav Brain Res 147(1–2):55–64
Hammack SE, Schmid MJ, LoPresti ML, Der-Avakian A, Pellymounter MA, Foster AC, Watkins LR, Maier SF (2003b) Corticotropin releasing hormone type 2 receptors in the dorsal raphe nucleus mediate the behavioral consequences of uncontrollable stress. J Neurosci 23(3):1019–1025
Hammack SE, Richey KJ, Watkins LR, Maier SF (2004) Chemical lesion of the bed nucleus of the stria terminalis blocks the behavioral consequences of uncontrollable stress. Behav Neurosci 118(2):443–448
Hatfield T, Spanis C, McGaugh JL (1999) Response of amygdalar norepinephrine to footshock and GABAergic drugs using in vivo microdialysis and HPLC. Brain Res 835(2):340–345
Haubensak W, Kunwar PS, Cai H, Ciocchi S, Wall NR, Ponnusamy R, Biag J, Dong HW, Deisseroth K, Callaway EM, Fanselow MS, Lüthi A, Anderson DJ (2010) Genetic dissection of an amygdala microcircuit that gates conditioned fear. Nature 468(7321):270–276
Haycock JW, Van Buskirk R, Ryan JR, McGaugh JL (1977) Enhancement of retention with centrally administered catecholamines. Exp Neurol 54(2):199–208
Haynes RC Jr, Koritz SB, Péron FG (1959) Influence of adenosine 3′,5′-monophosphate on corticoid production by rat adrenal glands. J Biol Chem 234(6):1421–1423
Henckens MJ, Printz Y, Shamgar U, Dine J, Lebow M, Drori Y, Kuehne C, Kolarz A, Eder M, Deussing JM, Justice NJ, Yizhar O, Chen A (2017) CRF receptor type 2 neurons in the posterior bed nucleus of the stria terminalis critically contribute to stress recovery. Mol Psychiatry 22:1691–1700
Hendry S, Hsiao S (2013) The somatosensory system. In: Squire LR, Berg D, Bloom FE, du Lac S, Ghosh A, Spitzer NC (eds) Fundamental neuroscience (4th edn). Academic, San Diego
Herman JP, Cullinan WE (1997) Neurocircuitry of stress: central control of the hypothalamo-pituitary-adrenocortical axis. Trends Neurosci 20(2):78–84
Herman JP, McKlveen JM, Ghosal S, Kopp B, Wulsin A, Makinson R, Scheimann J, Myers B (2016) Regulation of the hypothalamic-pituitary-adrenocortical stress response. Compr Physiol 6(2):603–621
Hill JE, Makky K, Shrestha L, Hillard CJ, Gasser PJ (2011) Natural and synthetic corticosteroids inhibit uptake$_2$-mediated transport in CNS neurons. Physiol Behav 104(2):306–311
Hitchcock J, Davis M (1986) Lesions of the amygdala, but not of the cerebellum or red nucleus, block conditioned fear as measured with the potentiated startle paradigm. Behav Neurosci 100(1):11–22
Hitchcock JM, Davis M (1991) Efferent pathway of the amygdala involved in conditioned fear as measured with the fear-potentiated startle paradigm. Behav Neurosci 105(6):826–842
Hodes GE, Pfau ML, Leboeuf M, Golden SA, Christoffel DJ, Bregman D, Rebusi N, Heshmati M, Aleyasin H, Warren BL, Lebonté B, Horn S, Lapidus KA, Stelzhammer V, Wong EH, Bahn S, Krishnan V, Bolaños-Guzman CA, Murrough JW, Merad M, Russo SJ (2014) Individual differences in the peripheral immune system promote resilience versus susceptibility to social stress. Proc Natl Acad Sci U S A 111(45):16136–16141
Hull C (1943) Principles of behavior. Appleton-Century, Oxford
Isosaka T, Matsuo T, Yamaguchi T, Funabiki K, Nakanishi S, Kobayakawa R, Kobayakawa K (2015) Htr2a-expressing cells in the central amygdala control the hierarchy between innate and learned fear. Cell 163(5):1153–1164
Iversen LL, Salt PJ (1970) Inhibition of catecholamine Uptake-2 by steroids in the isolated rat heart. Br J Pharmacol 40(3):528–530

Jankowski MP, Sesack SR (2004) Prefrontal cortical projections to the rat dorsal raphe nucleus: ultrastructural features and associations with serotonin and gamma-aminobutyric acid neurons. J Comp Neurol 468(4):518–529

Jin UH, Lee SO, Sridharan G, Lee K, Davidson LA, Jayaraman A, Chapkin RS, Alaniz R, Safe S (2014) Microbiome-derived tryptophan metabolites and their aryl hydrocarbon receptor-dependent agonist and antagonist activities. Mol Pharmacol 85(5):777–788

Johnson PL, Shekhar A (2006) Panic-prone state induced in rats with GABA dysfunction in the dorsomedial hypothalamus is mediated by NMDA receptors. J Neurosci 26(26):7093–7104

Johnson PL, Shekhar A (2012) An animal model of panic vulnerability with chronic disinhibition of the dorsomedial/perifornical hypothalamus. Physiol Behav 107(5):686–698

Johnson PL, Lightman SL, Lowry CA (2004) A functional subset of serotonergic neurons in the rat ventrolateral periaqueductal gray implicated in the inhibition of sympathoexcitation and panic. Ann N Y Acad Sci 1018:58–64

Johnson PL, Lowry CA, Truitt W, Shekhar A (2008a) Disruption of GABAergic tone in the dorsomedial hypothalamus attenuates responses in a subset of serotonergic neurons in the dorsal raphe nucleus following lactate-induced panic. J Psychopharmacol 22(6):642–652

Johnson PL, Truitt WA, Fitz SD, Lowry CA, Shekhar A (2008b) Neural pathways underlying lactate-induced panic. Neuropsychopharmacology 33(9):2093–2107

Jovanovic T, Norrholm SD (2011) Neural mechanisms of impaired fear inhibition in posttraumatic stress disorder. Front Behav Neurosci 5:44

Jovanovic T, Norrholm SD, Fennell JE, Keyes M, Fiallos AM, Myers KM, Davis M, Duncan EJ (2009) Posttraumatic stress disorder may be associated with impaired fear inhibition: relation to symptom severity. Psychiatry Res 167(1–2):151–160

Jovanovic T, Norrholm SD, Blanding NQ, Davis M, Duncan E, Bradley B, Ressler KJ (2010) Impaired fear inhibition is a biomarker of PTSD but not depression. Depress Anxiety 27(3):244–251

Kataoka N, Hioki H, Kaneko T, Nakamura K (2014) Psychological stress activates a dorsomedial hypothalamus-medullary raphe circuit driving brown adipose tissue thermogenesis and hyperthermia. Cell Metab 20(2):346–358

Kayalioglu G (2009) Projections from the spinal cord to the brain. In: Spinal cord: a Christopher and Dana Reeve Foundation Text and Atlas. Academic, Cambridge, pp 148–167

Kety S (1972) Brain catecholamines, affective states and memory. In: McGaugh JL (ed) The chemistry of mood, motivation, and memory: the proceedings of an interdisciplinary conference on the Chemistry of Mood, Motivation, and Memory held at the University of California, San Francisco, in October 1971. Springer, Boston

Kim JJ, Fanselow MS (1992) Modality-specific retrograde amnesia of fear. Science 256(5057):675–677

Kim SY, Adhikari A, Lee SY, Marshel JH, Kim CK, Mallory CS, Lo M, Pak S, Mattis J, Lim BK, Malenka RC, Warden MR, Neve R, Tye KM, Deisseroth K (2013) Diverging neural pathways assemble a behavioural state from separable features in anxiety. Nature 496(7444):219–223

Kim E, Kim EJ, Yeh R, Shin M, Bobman J, Krasne FB, Kim JJ (2014) Amygdaloid and non-amygdaloid fear both influence avoidance of risky foraging in hungry rats. Proc Biol Sci 281(1790):20133357

Kitamura T, Ogawa SK, Roy DS, Okuyama T, Morrissey MD, Smith LM, Redondo RL, Tonegawa S (2017) Engrams and circuits crucial for systems consolidation of a memory. Science 356(6333):73–78

Klein RG (1995) Is panic disorder associated with childhood separation anxiety disorder? Clin Neuropharmacol 18:S7–S14

Koch M (1999) The neurobiology of startle. Prog Neurobiol 59(2):107–128

Koolhaas JM, Bartolomucci A, Buwalda B, de Boer SF, Flugge G, Korte SM, Meerlo P, Murison R, Olivier B, Palanza P, Richter-Levin G, Sgoifo A, Steimer T, Stiedl O, van Dijk G, Wöhr M, Fuchs E (2011) Stress revisited: a critical evaluation of the stress concept. Neurosci Biobehav Rev 35(5):1291–1301

Ladd CO, Huot RL, Thrivikraman KV, Nemeroff CB, Plotsky PM (2004) Long-term adaptations in glucocorticoid receptor and mineralocorticoid receptor mRNA and negative feedback on the

hypothalamo-pituitary-adrenal axis following neonatal maternal separation. Biol Psychiatry 55(4):367–375
Langgartner D, Peterlik D, Foertsch S, Füchsl AM, Brokmann P, Flor PJ, Shen Z, Fox JG, Uschold-Schmidt N, Lowry CA, Reber SO (2017) Individual differences in stress vulnerability: the role of gut pathobionts in stress-induced colitis. Brain Behav Immun 64:23–32
LeDoux J (1998) Fear and the brain: where have we been, and where are we going? Biol Psychiatry 44(12):1229–1238
LeDoux JE (2012) Rethinking the emotional brain. Neuron 73(4):653–676
LeDoux JE, Cicchetti P, Xagoraris A, Romanski LM (1990a) The lateral amygdaloid nucleus: sensory interface of the amygdala in fear conditioning. J Neurosci 10(4):1062–1069
LeDoux JE, Farb C, Ruggiero DA (1990b) Topographic organization of neurons in the acoustic thalamus that project to the amygdala. J Neurosci 10(4):1043–1054
Lee Y, Davis M (1997) Role of the hippocampus, the bed nucleus of the stria terminalis, and the amygdala in the excitatory effect of corticotropin-releasing hormone on the acoustic startle reflex. J Neurosci 17(16):6434–6446
Lee Y, Fitz S, Johnson PL, Shekhar A (2008) Repeated stimulation of CRF receptors in the BNST of rats selectively induces social but not panic-like anxiety. Neuropsychopharmacology 33(11):2586–2594
Levine S, Wiener SG (1988) Psychoendocrine aspects of mother-infant relationships in nonhuman primates. Psychoneuroendocrinology 13(1–2):143–154
Lewis JG, Bagley CJ, Elder PA, Bachmann AW, Torpy DJ (2005) Plasma free cortisol fraction reflects levels of functioning corticosteroid-binding globulin. Clin Chim Acta 359(1–2):189–194
Liang KC, McGaugh JL, Martinez JL Jr, Jensen RA, Vasquez BJ, Messing RB (1982) Post-training amygdaloid lesions impair retention of an inhibitory avoidance response. Behav Brain Res 4(3):237–249
Lindqvist D, Wolkowitz OM, Mellon S, Yehuda R, Flory JD, Henn-Haase C, Bierer LM, Abu-Amara D, Coy M, Neylan TC, Makotkine I, Reus VI, Yan X, Taylor NM, Marmar CR, Dhabhar FS (2014) Proinflammatory milieu in combat-related PTSD is independent of depression and early life stress. Brain Behav Immun 42:81–88
Lindqvist D, Dhabhar FS, Mellon SH, Yehuda R, Grenon SM, Flory JD, Bierer LM, Abu-Amara D, Coy M, Makotkine I, Reus VI, Bersani FS, Marmar CR, Wolkowitz OM (2017) Increased pro-inflammatory milieu in combat related PTSD – a new cohort replication study. Brain Behav Immun 59:260–264
Lingenhohl K, Friauf E (1994) Giant neurons in the rat reticular formation: a sensorimotor interface in the elementary acoustic startle circuit? J Neurosci 14(3 Pt 1):1176–1194
Liu D, Diorio J, Tannenbaum B, Caldji C, Francis D, Freedman A, Sharma S, Pearson D, Plotsky PM, Meaney MJ (1997) Maternal care, hippocampal glucocorticoid receptors, and hypothalamic-pituitary-adrenal responses to stress. Science 277(5332):1659–1662
Liu D, Diorio J, Day JC, Francis DD, Meaney MJ (2000) Maternal care, hippocampal synaptogenesis and cognitive development in rats. Nat Neurosci 3(8):799–806
Lowry CA, Rodda JE, Lightman SL, Ingram CD (2000) Corticotropin-releasing factor increases in vitro firing rates of serotonergic neurons in the rat dorsal raphe nucleus: evidence for activation of a topographically organized mesolimbocortical serotonergic system. J Neurosci 20(20):7728–7736
Lowry CA, Johnson PL, Hay-Schmidt A, Mikkelsen J, Shekhar A (2005) Modulation of anxiety circuits by serotonergic systems. Stress 8(4):233–246
Lowry CA, Hale MW, Evans AK, Heerkens J, Staub DR, Gasser PJ, Shekhar A (2008) Serotonergic systems, anxiety, and affective disorder: focus on the dorsomedial part of the dorsal raphe nucleus. Ann N Y Acad Sci 1148:86–94
Lowry CA, Smith DG, Siebler PH, Schmidt D, Stamper CE, Hassell JE Jr, Yamashita PS, Fox JH, Reber SO, Brenner LA, Hoisington AJ, Postolache TT, Kinney KA, Marciani D, Hernandez M, Hemmings SM, Malan-Muller S, Wright KP, Knight R, Raison CL, Rook GA (2016) The

microbiota, immunoregulation, and mental health: implications for public health. Curr Environ Health Rep 3(3):270–286

Lukkes JL, Staub DR, Dietrich A, Truitt W, Neufeld-Cohen A, Chen A, Johnson PL, Shekhar A, Lowry CA (2011) Topographical distribution of corticotropin-releasing factor type 2 receptor-like immunoreactivity in the rat dorsal raphe nucleus: co-localization with tryptophan hydroxylase. Neuroscience 183:47–63

Lupien SJ, McEwen BS, Gunnar MR, Heim C (2009) Effects of stress throughout the lifespan on the brain, behaviour and cognition. Nat Rev Neurosci 10(6):434–445

Maier SF (1990) Role of fear in mediating shuttle escape learning deficit produced by inescapable shock. J Exp Psychol Anim Behav Process 16(2):137–149

Maier SF (2001) Exposure to the stressor environment prevents the temporal dissipation of behavioral depression/learned helplessness. Biol Psychiatry 49(9):763–773

Maier SF, Seligman ME (2016) Learned helplessness at fifty: insights from neuroscience. Psychol Rev 123(4):349–367

Maier SF, Testa TJ (1975) Failure to learn to escape by rats previously exposed to inescapable shock is partly produced by associative interference. J Comp Physiol Psychol 88(2):554–564

Maier SF, Watkins LR (2005) Stressor controllability and learned helplessness: the roles of the dorsal raphe nucleus, serotonin, and corticotropin-releasing factor. Neurosci Biobehav Rev 29(4–5):829–841

Maier SF, Albin RW, Testa TJ (1973) Failure to learn to escape in rats previously exposed to inescapable shock depends on nature of escape response. J Comp Physiol Psychol 85(3):581–592

Maier SF, Coon DJ, Mcdaniel MA, Jackson RL, Grau J (1979) The time course of learned helplessness, inactivity, and nociceptive deficits in rats. Learn Motiv 10(4):467–487

Makino S, Gold PW, Schulkin J (1994a) Corticosterone effects on corticotropin-releasing hormone mRNA in the central nucleus of the amygdala and the parvocellular region of the paraventricular nucleus of the hypothalamus. Brain Res 640(1–2):105–112

Makino S, Gold PW, Schulkin J (1994b) Effects of corticosterone on CRH mRNA and content in the bed nucleus of the stria terminalis; comparison with the effects in the central nucleus of the amygdala and the paraventricular nucleus of the hypothalamus. Brain Res 657(1–2):141–149

Marcinkiewcz CA, Mazzone CM, D'Agostino G, Halladay LR, Hardaway JA, DiBerto JF, Navarro M, Burnham N, Cristiano C, Dorrier CE, Tipton GJ, Ramakrishnan C, Kozicz T, Deisseroth K, Thiele TE, McElligott ZA, Holmes A, Heisler LK, Kash TL (2016) Serotonin engages an anxiety and fear-promoting circuit in the extended amygdala. Nature 537(7618):97–101

Maswood S, Barter JE, Watkins LR, Maier SF (1998) Exposure to inescapable but not escapable shock increases extracellular levels of 5-HT in the dorsal raphe nucleus of the rat. Brain Res 783(1):115–120

Mayer EA (2011) Gut feelings: the emerging biology of gut-brain communication. Nat Rev Neurosci 12(8):453–466

Mayer EA, Tillisch K, Gupta A (2015) Gut/brain axis and the microbiota. J Clin Invest 125(3):926–938

McDevitt RA, Szot P, Baratta MV, Bland ST, White SS, Maier SF, Neumaier JF (2009) Stress-induced activity in the locus coeruleus is not sensitive to stressor controllability. Brain Res 1285:109–118

McEwen BS (1998a) Protective and damaging effects of stress mediators. N Engl J Med 338(3):171–179

McEwen BS (1998b) Stress, adaptation, and disease. Allostasis and allostatic load. Ann N Y Acad Sci 840:33–44

McEwen BS, Gianaros PJ (2010) Central role of the brain in stress and adaptation: links to socioeconomic status, health, and disease. Ann N Y Acad Sci 1186:190–222

McEwen BS, Stellar E (1993) Stress and the individual. Mechanisms leading to disease. Arch Intern Med 153(18):2093–2101

McEwen BS, Wingfield JC (2003) The concept of allostasis in biology and biomedicine. Horm Behav 43(1):2–15
McEwen BS, Wingfield J (2007) Allostasis and allostatic load. In: Fink G (ed) Encyclopedia of stress. Academic, New York, pp 135–141
McEwen BS, Wingfield JC (2010) What is in a name? Integrating homeostasis, allostasis and stress. Horm Behav 57(2):105–111
McEwen BS, Weiss JM, Schwartz LS (1968) Selective retention of corticosterone by limbic structures in rat brain. Nature 220(5170):911–912
McGaugh JL, Introini-Collison IB, Nagahara AH, Cahill L, Brioni JD, Castellano C (1990) Involvement of the amygdaloid complex in neuromodulatory influences on memory storage. Neurosci Biobehav Rev 14(4):425–431
McKim DB, Patterson JM, Wohleb ES, Jarrett BL, Reader BF, Godbout JP, Sheridan JF (2016) Sympathetic release of splenic monocytes promotes recurring anxiety following repeated social defeat. Biol Psychiatry 79(10):803–813
McKim DB, Weber MD, Niraula A, Sawicki CM, Liu X, Jarrett BL, Ramirez-Chan K, Wang Y, Roeth RM, Sucaldito AD, Sobol CG, Quan N, Sheridan JF, Godbout JP (2017) Microglial recruitment of IL-1β-producing monocytes to brain endothelium causes stress-induced anxiety. Mol Psychiatry 23:1421–1431
McKlveen JM, Morano RL, Fitzgerald M, Zoubovsky S, Cassella SN, Scheimann JR, Ghosal S, Mahbod P, Packard BA, Myers B, Baccei ML, Herman JP (2016) Chronic stress increases prefrontal inhibition: a mechanism for stress-induced prefrontal dysfunction. Biol Psychiatry 80(10):754–764
McNaughton N, Corr PJ (2004) A two-dimensional neuropsychology of defense: fear/anxiety and defensive distance. Neurosci Biobehav Rev 28(3):285–305
McReynolds JR, Donowho K, Abdi A, McGaugh JL, Roozendaal B, McIntyre CK (2010) Memory-enhancing corticosterone treatment increases amygdala norepinephrine and Arc protein expression in hippocampal synaptic fractions. Neurobiol Learn Mem 93(3):312–321
Medzhitov R (2008) Origin and physiological roles of inflammation. Nature 454(7203):428–435
Miguel TL, Pobbe RL, Spiacci Junior A, Zangrossi Junior H (2010) Dorsal raphe nucleus regulation of a panic-like defensive behavior evoked by chemical stimulation of the rat dorsal periaqueductal gray matter. Behav Brain Res 213(2):195–200
Miller AH, Raison CL (2016) The role of inflammation in depression: from evolutionary imperative to modern treatment target. Nat Rev Immunol 16(1):22–34
Miller GE, Chen E, Zhou ES (2007) If it goes up, must it come down? Chronic stress and the hypothalamic-pituitary-adrenocortical axis in humans. Psychol Bull 133(1):25–45
Miserendino MJ, Sananes CB, Melia KR, Davis M (1990) Blocking of acquisition but not expression of conditioned fear-potentiated startle by NMDA antagonists in the amygdala. Nature 345(6277):716–718
Mobbs D, Petrovic P, Marchant JL, Hassabis D, Weiskopf N, Seymour B, Dolan RJ, Frith CD (2007) When fear is near: threat imminence elicits prefrontal-periaqueductal gray shifts in humans. Science 317(5841):1079–1083
Mobbs D, Marchant JL, Hassabis D, Seymour B, Tan G, Gray M, Petrovic P, Dolan RJ, Frith CD (2009) From threat to fear: the neural organization of defensive fear systems in humans. J Neurosci 29(39):12236–12243
Mobbs D, Yu R, Rowe JB, Eich H, FeldmanHall O, Dalgleish T (2010) Neural activity associated with monitoring the oscillating threat value of a tarantula. Proc Natl Acad Sci U S A 107 (47):20582–20586
Moisan MP, Minni AM, Dominguez G, Helbling JC, Foury A, Henkous N, Dorey R, Béracochéa D (2014) Role of corticosteroid binding globulin in the fast actions of glucocorticoids on the brain. Steroids 81:109–115
Munck A, Guyre PM, Holbrook NJ (1984) Physiological functions of glucocorticoids in stress and their relation to pharmacological actions. Endocrine Reviews 5:25–44

Myers B, Mark Dolgas C, Kasckow J, Cullinan WE, Herman JP (2014) Central stress-integrative circuits: forebrain glutamatergic and GABAergic projections to the dorsomedial hypothalamus, medial preoptic area, and bed nucleus of the stria terminalis. Brain Struct Funct 219:1287–1303

Nie X, Kitaoka S, Tanaka K, Segi-Nishida E, Imoto Y, Ogawa A, Nakano F, Tomohiro A, Nakayama K, Taniguchi M, Mimori-Kiyosue Y, Kakizuka A, Narumiya S, Furuyashiki T (2018) The innate immune receptors TLR2/4 mediate repeated social defeat stress-induced social avoidance through prefrontal microglial activation. Neuron 99:1–16

Niraula A, Wang Y, Godbout JP, Sheridan JF (2018) Corticosterone production during repeated social defeat causes monocyte mobilization from the bone marrow, glucocorticoid resistance, and neurovascular adhesion molecule expression. J Neurosci 38(9):2328–2340

Ogbonnaya ES, Clarke G, Shanahan F, Dinan TG, Cryan JF, O'Leary OF (2015) Adult hippocampal neurogenesis is regulated by the microbiome. Biol Psychiatry 78(4):e7–e9

Ostrander MM, Ulrich-Lai YM, Choi DC, Flak JN, Richtand NM, Herman JP (2009) Chronic stress produces enduring decreases in novel stress-evoked c-*fos* mRNA expression in discrete brain regions of the rat. Stress 12(6):469–477

Packard MG, Cahill L, McGaugh JL (1994) Amygdala modulation of hippocampal-dependent and caudate nucleus-dependent memory processes. Proc Natl Acad Sci U S A 91(18):8477–8481

Parent MB, McGaugh JL (1994) Posttraining infusion of lidocaine into the amygdala basolateral complex impairs retention of inhibitory avoidance training. Brain Res 661(1–2):97–103

Pavlov IP (1927) Conditioned reflexes: an investigation of the physiological activity of the cerebral cortex. Oxford University Press, Oxford

Pavlov IP (1962) Essays in psychology and psychiatry, including a section on sleep and hypnosis. Citadel Press, New York

Pavlov VA, Tracey KJ (2012) The vagus nerve and the inflammatory reflex – linking immunity and metabolism. Nat Rev Endocrinol 8(12):743–754

Paxinos G, Watson C (2007) The rat brain in stereotaxic coordinates. Academic, San Diego

Pellman BA, Kim E, Reilly M, Kashima J, Motch O, de la Iglesia HO, Kim JJ (2015) Time-specific fear acts as a non-photic entraining stimulus of circadian rhythms in rats. Sci Rep 5:14916

Perusini JN, Fanselow MS (2015) Neurobehavioral perspectives on the distinction between fear and anxiety. Learn Mem 22(9):417–425

Petrov T, Jhamandas JH, Krukoff TL (1994) Electrical stimulation of the central nucleus of the amygdala induces fos-like immunoreactivity in the hypothalamus of the rat: a quantitative study. Brain Res Mol Brain Res 22(1–4):333–340

Phillips RG, LeDoux JE (1992) Differential contribution of amygdala and hippocampus to cued and contextual fear conditioning. Behav Neurosci 106(2):274–285

Potter E, Sutton S, Donaldson C, Chen R, Perrin M, Lewis K, Sawchenko PE, Vale W (1994) Distribution of corticotropin-releasing factor receptor mRNA expression in the rat brain and pituitary. Proc Natl Acad Sci U S A 91(19):8777–8781

Poulos AM, Ponnusamy R, Dong HW, Fanselow MS (2010) Compensation in the neural circuitry of fear conditioning awakens learning circuits in the bed nuclei of the stria terminalis. Proc Natl Acad Sci U S A 107(33):14881–14886

Presman DM, Hager GL (2017) More than meets the dimer: what is the quaternary structure of the glucocorticoid receptor? Transcription 8(1):32–39

Presman DM, Ganguly S, Schiltz RL, Johnson TA, Karpova TS, Hager GL (2016) DNA binding triggers tetramerization of the glucocorticoid receptor in live cells. Proc Natl Acad Sci U S A 113(29):8236–8241

Pugh CR, Fleshner M, Rudy JW (1997a) Type II glucocorticoid receptor antagonists impair contextual but not auditory-cue fear conditioning in juvenile rats. Neurobiol Learn Mem 67(1):75–79

Pugh CR, Tremblay D, Fleshner M, Rudy JW (1997b) A selective role for corticosterone in contextual-fear conditioning. Behav Neurosci 111(3):503–511

Quirarte GL, Roozendaal B, McGaugh JL (1997) Glucocorticoid enhancement of memory storage involves noradrenergic activation in the basolateral amygdala. Proc Natl Acad Sci U S A 94(25):14048–14053
Quirarte GL, Galvez R, Roozendaal B, McGaugh JL (1998) Norepinephrine release in the amygdala in response to footshock and opioid peptidergic drugs. Brain Res 808(2):134–140
Quirk GJ, Armony JL, LeDoux JE (1997) Fear conditioning enhances different temporal components of tone-evoked spike trains in auditory cortex and lateral amygdala. Neuron 19(3):613–624
Rainnie DG, Bergeron R, Sajdyk TJ, Patil M, Gehlert DR, Shekhar A (2004) Corticotrophin releasing factor-induced synaptic plasticity in the amygdala translates stress into emotional disorders. J Neurosci 24(14):3471–3479
Rau V, DeCola JP, Fanselow MS (2005) Stress-induced enhancement of fear learning: an animal model of posttraumatic stress disorder. Neurosci Biobehav Rev 29(8):1207–1223
Rau V, Fanselow MS (2009) Exposure to a stressor produces a long lasting enhancement of fear learning in rats. Stress 12(2):125–133
Reber SO, Birkeneder L, Veenema AH, Obermeier F, Falk W, Straub RH, Neumann ID (2007) Adrenal insufficiency and colonic inflammation after a novel chronic psycho-social stress paradigm in mice: implications and mechanisms. Endocrinology 148(2):670–682
Reber SO, Langgartner D, Foertsch S, Postolache TT, Brenner LA, Guendel H, Lowry CA (2016a) Chronic subordinate colony housing paradigm: a mouse model for mechanisms of PTSD vulnerability, targeted prevention, and treatment-2016 Curt Richter Award Paper. Psychoneuroendocrinology 74:221–230
Reber SO, Obermeier F, Straub RH, Veenema AH, Neumann ID (2008) Aggravation of DSS-induced colitis after chronic subordinate colony (CSC) housing is partially mediated by adrenal mechanisms. Stress 11(3):225–234
Reber SO, Siebler PH, Donner NC, Morton JT, Smith DG, Kopelman JM, Lowe KR, Wheeler KJ, Fox JH, Hassell JE Jr, Greenwood BN, Jansch C, Lechner A, Schmidt D, Uschold-Schmidt N, Fuchsl AM, Langgartner D, Walker FR, Hale MW, Lopez Perez G, Van Treuren W, Gonzalez A, Halweg-Edwards AL, Fleshner M, Raison CL, Rook GA, Peddada SD, Knight R, Lowry CA (2016b) Immunization with a heat-killed preparation of the environmental bacterium *Mycobacterium vaccae* promotes stress resilience in mice. Proc Natl Acad Sci U S A 113(22):E3130–E3139
Reul JM, de Kloet ER (1985) Two receptor systems for corticosterone in rat brain: microdistribution and differential occupation. Endocrinology 117(6):2505–2511
Reul JM, de Kloet ER (1986) Anatomical resolution of two types of corticosterone receptor sites in rat brain with in vitro autoradiography and computerized image analysis. J Steroid Biochem 24(1):269–272
Rivier C, Bruhn T, Vale W (1984) Effect of ethanol on the hypothalamic-pituitary-adrenal axis in the rat: role of corticotropin-releasing factor (CRF). J Pharmacol Exp Ther 229(1):127–131
Rogan MT, Staubli UV, LeDoux JE (1997) Fear conditioning induces associative long-term potentiation in the amygdala. Nature 390(6660):604–607
Romanski LM, Clugnet MC, Bordi F, LeDoux JE (1993) Somatosensory and auditory convergence in the lateral nucleus of the amygdala. Behav Neurosci 107(3):444–450
Roozendaal B (2002) Stress and memory: opposing effects of glucocorticoids on memory consolidation and memory retrieval. Neurobiol Learn Mem 78(3):578–595
Roozendaal B, Carmi O, McGaugh JL (1996) Adrenocortical suppression blocks the memory-enhancing effects of amphetamine and epinephrine. Proc Natl Acad Sci U S A 93(4):1429–1433
Roozendaal B, McGaugh JL (1996) Amygdaloid nuclei lesions differentially affect glucocorticoid-induced memory enhancement in an inhibitory avoidance task. Neurobiol Learn Mem 65(1):1–8
Roozendaal B, McGaugh JL (1997) Glucocorticoid receptor agonist and antagonist administration into the basolateral but not central amygdala modulates memory storage. Neurobiol Learn Mem 67(2):176–179

Roozendaal B, Quirarte GL, McGaugh JL (1997) Stress-activated hormonal systems and the regulation of memory storage. Ann N Y Acad Sci 821:247–258

Rosen JB, Davis M (1988) Temporal characteristics of enhancement of startle by stimulation of the amygdala. Physiol Behav 44(1):117–123

Rosen JB, Hitchcock JM, Sananes CB, Miserendino MJ, Davis M (1991) A direct projection from the central nucleus of the amygdala to the acoustic startle pathway: anterograde and retrograde tracing studies. Behav Neurosci 105(6):817–825

Rozeske RR, Evans AK, Frank MG, Watkins LR, Lowry CA, Maier SF (2011) Uncontrollable, but not controllable, stress desensitizes 5-HT_{1A} receptors in the dorsal raphe nucleus. J Neurosci 31(40):14107–14115

Rudy JW, Huff NC, Matus-Amat P (2004) Understanding contextual fear conditioning: insights from a two-process model. Neurosci Biobehav Rev 28(7):675–685

Rumpel S, LeDoux J, Zador A, Malinow R (2005) Postsynaptic receptor trafficking underlying a form of associative learning. Science 308(5718):83–88

Ryoke R, Yamada K, Ichitani Y (2014) Long-term effects of traumatic stress on subsequent contextual fear conditioning in rats. Physiol Behav 129:30–35

Sagar SM, Sharp FR, Curran T (1988) Expression of c-fos protein in brain: metabolic mapping at the cellular level. Science 240(4857):1328–1331

Sahuque LL, Kullberg EF, McGeehan AJ, Kinder JR, Hicks MP, Blanton MG, Janak PH, Olive MF (2006) Anxiogenic and aversive effects of corticotropin-releasing factor (CRF) in the bed nucleus of the stria terminalis in the rat: role of CRF receptor subtypes. Psychopharmacology 186(1):122–132

Sajdyk T, Johnson P, Fitz S, Shekhar A (2008) Chronic inhibition of GABA synthesis in the bed nucleus of the stria terminalis elicits anxiety-like behavior. J Psychopharmacol 22(6):633–641

Sajdyk TJ, Schober DA, Gehlert DR, Shekhar A (1999) Role of corticotropin-releasing factor and urocortin within the basolateral amygdala of rats in anxiety and panic responses. Behav Brain Res 100(1–2):207–215

Sajdyk TJ, Shekhar A (2000) Sodium lactate elicits anxiety in rats after repeated GABA receptor blockade in the basolateral amygdala. Eur J Pharmacol 394(2–3):265–273

Sanders SK, Morzorati SL, Shekhar A (1995) Priming of experimental anxiety by repeated subthreshold GABA blockade in the rat amygdala. Brain Res 699(2):250–259

Sanders SK, Shekhar A (1991) Blockade of GABAA receptors in the region of the anterior basolateral amygdala of rats elicits increases in heart rate and blood pressure. Brain Res 567(1):101–110

Sanders SK, Shekhar A (1995a) Anxiolytic effects of chlordiazepoxide blocked by injection of GABAA and benzodiazepine receptor antagonists in the region of the anterior basolateral amygdala of rats. Biol Psychiatry 37(7):473–476

Sanders SK, Shekhar A (1995b) Regulation of anxiety by GABAA receptors in the rat amygdala. Pharmacol Biochem Behav 52(4):701–706

Sapolsky RM, Krey LC, McEwen BS (1984) Glucocorticoid-sensitive hippocampal neurons are involved in terminating the adrenocortical stress response. Proc Natl Acad Sci U S A 81(19):6174–6177

Schildkraut JJ, Mooney JJ (2004) Toward a rapidly acting antidepressant: the normetanephrine and extraneuronal monoamine transporter (Uptake 2) hypothesis. Am J Psychiatry 161(5):909–911

Schmitt A, Mossner R, Gossmann A, Fischer IG, Gorboulev V, Murphy DL, Koepsell H, Lesch KP (2003) Organic cation transporter capable of transporting serotonin is up-regulated in serotonin transporter-deficient mice. J Neurosci Res 71(5):701–709

Schulkin J (2003) Rethinking homeostasis: allostatic regulation in physiology and pathophysiology. MIT Press, Cambridge

Schulkin J, Gold PW, McEwen BS (1998) Induction of corticotropin-releasing hormone gene expression by glucocorticoids: implication for understanding the states of fear and anxiety and allostatic load. Psychoneuroendocrinology 23(3):219–243

Schwabe L, Joels M, Roozendaal B, Wolf OT, Oitzl MS (2012) Stress effects on memory: an update and integration. Neurosci Biobehav Rev 36(7):1740–1749

Seale JV, Wood SA, Atkinson HC, Bate E, Lightman SL, Ingram CD, Jessop DS, Harbuz MS (2004) Gonadectomy reverses the sexually diergic patterns of circadian and stress-induced hypothalamic-pituitary-adrenal axis activity in male and female rats. J Neuroendocrinol 16(6):516–524

Seligman ME, Maier SF (1967) Failure to escape traumatic shock. J Exp Psychol 74(1):1–9

Selye H (1950) Stress and the general adaptation syndrome. Br Med J 1(4667):1383–1392

Shekhar A (1993) GABA receptors in the region of the dorsomedial hypothalamus of rats regulate anxiety in the elevated plus-maze test. I. Behavioral measures. Brain Res 627(1):9–16

Shekhar A, DiMicco JA (1987) Defense reaction elicited by injection of GABA antagonists and synthesis inhibitors into the posterior hypothalamus in rats. Neuropharmacology 26(5):407–417

Shekhar A, Hingtgen JN, DiMicco JA (1990) GABA receptors in the posterior hypothalamus regulate experimental anxiety in rats. Brain Res 512(1):81–88

Shekhar A, Katner JS (1995) Dorsomedial hypothalamic GABA regulates anxiety in the social interaction test. Pharmacol Biochem Behav 50(2):253–258

Shekhar A, Keim SR, Simon JR, McBride WJ (1996) Dorsomedial hypothalamic GABA dysfunction produces physiological arousal following sodium lactate infusions. Pharmacol Biochem Behav 55(2):249–256

Shepard JD, Barron KW, Myers DA (2000) Corticosterone delivery to the amygdala increases corticotropin-releasing factor mRNA in the central amygdaloid nucleus and anxiety-like behavior. Brain Res 861(2):288–295

Shepard JD, Schulkin J, Myers DA (2006) Chronically elevated corticosterone in the amygdala increases corticotropin releasing factor mRNA in the dorsolateral bed nucleus of stria terminalis following duress. Behav Brain Res 174(1):193–196

Sherrington CS (1910) Flexion-reflex of the limb, crossed extension-reflex, and reflex stepping and standing. J Physiol 40(1–2):28–121

Singewald GM, Nguyen NK, Neumann ID, Singewald N, Reber SO (2009) Effect of chronic psychosocial stress-induced by subordinate colony (CSC) housing on brain neuronal activity patterns in mice. Stress 12(1):58–69

Slattery DA, Uschold N, Magoni M, Bar J, Popoli M, Neumann ID, Reber SO (2012) Behavioural consequences of two chronic psychosocial stress paradigms: anxiety without depression. Psychoneuroendocrinology 37(5):702–714

Spiacci A Jr, Pobbe RL, Matthiesen M, Zangrossi H Jr (2016) 5-HT1A receptors of the rat dorsal raphe lateral wings and dorsomedial subnuclei differentially control anxiety- and panic-related defensive responses. Neuropharmacology 107:471–479

Spiga F, Lightman SL, Shekhar A, Lowry CA (2006) Injections of urocortin 1 into the basolateral amygdala induce anxiety-like behavior and c-Fos expression in brainstem serotonergic neurons. Neuroscience 138(4):1265–1276

Spiga F, Walker JJ, Terry JR, Lightman SL (2014) HPA axis-rhythms. Compr Physiol 4(3):1273–1298

Stamper CE, Hassell JE Jr, Kapitz AJ, Renner KJ, Orchinik M, Lowry CA (2017) Activation of 5-HT1A receptors in the rat dorsomedial hypothalamus inhibits stress-induced activation of the hypothalamic-pituitary-adrenal axis. Stress 20(2):223–230

Stamper CE, Hennessey PA, Hale MW, Lukkes JL, Donner NC, Lowe KR, Paul ED, Spencer RL, Renner KJ, Orchinik M, Lowry CA (2015) Role of the dorsomedial hypothalamus in glucocorticoid-mediated feedback inhibition of the hypothalamic-pituitary-adrenal axis. Stress 18:76–87

Stephens MA, Wand G (2012) Stress and the HPA axis: role of glucocorticoids in alcohol dependence. Alcohol Res 34(4):468–483

Sterling P (1988) Allostasis: a new paradigm to explain arousal pathology. In: Fisher S, Reason J (eds) Handbook of life stress, cognition and health. Wiley, New York

Sudo N, Chida Y, Aiba Y, Sonoda J, Oyama N, Yu XN, Kubo C, Koga Y (2004) Postnatal microbial colonization programs the hypothalamic-pituitary-adrenal system for stress response in mice. J Physiol 558(Pt 1):263–275

Thompson B, Schulkin J, Rosen J (2000) Chronic corticosterone enhances contextual fear conditioning and CRH mRNA expression in the amygdala. Society for Neuroscience, New Orleans

Tillisch K, Labus J, Kilpatrick L, Jiang Z, Stains J, Ebrat B, Guyonnet D, Legrain-Raspaud S, Trotin B, Naliboff B, Mayer EA (2013) Consumption of fermented milk product with probiotic modulates brain activity. Gastroenterology 144(7):1394–1401, 1401.e1–4

Torres GE, Gainetdinov RR, Caron MG (2003) Plasma membrane monoamine transporters: structure, regulation and function. Nat Rev Neurosci 4(1):13–25

Tovote P, Esposito MS, Botta P, Chaudun F, Fadok JP, Markovic M, Wolff SB, Ramakrishnan C, Fenno L, Deisseroth K, Herry C, Arber S, Luthi A (2016) Midbrain circuits for defensive behaviour. Nature 534(7606):206–212

Tye KM, Prakash R, Kim SY, Fenno LE, Grosenick L, Zarabi H, Thompson KR, Gradinaru V, Ramakrishnan C, Deisseroth K (2011) Amygdala circuitry mediating reversible and bidirectional control of anxiety. Nature 471(7338):358–362

Uschold-Schmidt N, Nyuyki KD, Fuchsl AM, Neumann ID, Reber SO (2012) Chronic psychosocial stress results in sensitization of the HPA axis to acute heterotypic stressors despite a reduction of adrenal in vitro ACTH responsiveness. Psychoneuroendocrinology 37(10):1676–1687

Vale W, Spiess J, Rivier C, Rivier J (1981) Characterization of a 41-residue ovine hypothalamic peptide that stimulates secretion of corticotropin and β-endorphin. Science 213 (4514):1394–1397

Varela JA, Wang J, Christianson JP, Maier SF, Cooper DC (2012) Control over stress, but not stress per se increases prefrontal cortical pyramidal neuron excitability. J Neurosci 32(37):12848–12853

Vaughan J, Donaldson C, Bittencourt J, Perrin MH, Lewis K, Sutton S, Chan R, Turnbull AV, Lovejoy D, Rivier C et al (1995) Urocortin, a mammalian neuropeptide related to fish urotensin I and to corticotropin-releasing factor. Nature 378(6554):287–292

Vazdarjanova A, Cahill L, McGaugh JL (2001) Disrupting basolateral amygdala function impairs unconditioned freezing and avoidance in rats. Eur J Neurosci 14(4):709–718

Vazquez DM, Lopez JF, Van Hoers H, Watson SJ, Levine S (2000) Maternal deprivation regulates serotonin 1A and 2A receptors in the infant rat. Brain Res 855(1):76–82

Vinson GP (2016) Functional zonation of the adult mammalian adrenal cortex. Front Neurosci 10:238

Vita N, Laurent P, Lefort S, Chalon P, Lelias JM, Kaghad M, Le Fur G, Caput D, Ferrara P (1993) Primary structure and functional expression of mouse pituitary and human brain corticotrophin releasing factor receptors. FEBS Lett 335(1):1–5

Walker DL, Davis M (1997) Double dissociation between the involvement of the bed nucleus of the stria terminalis and the central nucleus of the amygdala in startle increases produced by conditioned versus unconditioned fear. J Neurosci 17(23):9375–9383

Walker DL, Davis M (2008) Role of the extended amygdala in short-duration versus sustained fear: a tribute to Dr. Lennart Heimer. Brain Struct Funct 213(1–2):29–42

Walsh JT, Zheng J, Smirnov I, Lorenz U, Tung K, Kipnis J (2014) Regulatory T cells in central nervous system injury: a double-edged sword. J Immunol 193(10):5013–5022

Watkins LR, Goehler LE, Relton JK, Tartaglia N, Silbert L, Martin D, Maier SF (1995) Blockade of interleukin-1 induced hyperthermia by subdiaphragmatic vagotomy: evidence for vagal mediation of immune-brain communication. Neurosci Lett 183(1–2):27–31

Watts AG, Swanson LW (1987) Efferent projections of the suprachiasmatic nucleus: II. Studies using retrograde transport of fluorescent dyes and simultaneous peptide immunohistochemistry in the rat. J Comp Neurol 258(2):230–252

Weiner H (1991) Behavioral biology of stress and psychosomatic medicine. In: Brown MR, Koob GF, Rivier C (eds) Stress: neurobiology and neuroendocrinology. Dekker, New York, pp 23–51

Williams CL, McGaugh JL (1992) Reversible inactivation of the nucleus of the solitary tract impairs retention performance in an inhibitory avoidance task. Behav Neural Biol 58(3):204–210

Williams CL, Men D, Clayton EC, Gold PE (1998) Norepinephrine release in the amygdala after systemic injection of epinephrine or escapable footshock: contribution of the nucleus of the solitary tract. Behav Neurosci 112(6):1414–1422

Windle RJ, Wood SA, Shanks N, Lightman SL, Ingram CD (1998) Ultradian rhythm of basal corticosterone release in the female rat: dynamic interaction with the response to acute stress. Endocrinology 139(2):443–450

Wohleb ES, McKim DB, Shea DT, Powell ND, Tarr AJ, Sheridan JF, Godbout JP (2014a) Re-establishment of anxiety in stress-sensitized mice is caused by monocyte trafficking from the spleen to the brain. Biol Psychiatry 75(12):970–981

Wohleb ES, McKim DB, Sheridan JF, Godbout JP (2014b) Monocyte trafficking to the brain with stress and inflammation: a novel axis of immune-to-brain communication that influences mood and behavior. Front Neurosci 8:447

Wohleb ES, Powell ND, Godbout JP, Sheridan JF (2013) Stress-induced recruitment of bone marrow-derived monocytes to the brain promotes anxiety-like behavior. J Neurosci 33(34):13820–13833

Woodruff ER, Chun LE, Hinds LR, Spencer RL (2016) Diurnal corticosterone presence and phase modulate clock gene expression in the male rat prefrontal cortex. Endocrinology 157(4):1522–1534

Woodruff ER, Greenwood BN, Chun LE, Fardi S, Hinds LR, Spencer RL (2015) Adrenal-dependent diurnal modulation of conditioned fear extinction learning. Behav Brain Res 286:249–255

Wu X, Kekuda R, Huang W, Fei YJ, Leibach FH, Chen J, Conway SJ, Ganapathy V (1998) Identity of the organic cation transporter OCT3 as the extraneuronal monoamine transporter (uptake$_2$) and evidence for the expression of the transporter in the brain. J Biol Chem 273(49):32776–32786

Xie L, Choudhury GR, Winters A, Yang SH, Jin K (2015) Cerebral regulatory T cells restrain microglia/macrophage-mediated inflammatory responses via IL-10. Eur J Immunol 45(1):180–191

Yamashita PSM, de Bortoli VC, Zangrossi H Jr (2011) 5-HT2C receptor regulation of defensive responses in the rat dorsal periaqueductal gray. Neuropharmacology 60(2–3):216–222

Yang C, Fujita Y, Ren Q, Ma M, Dong C, Hashimoto K (2017) Bifidobacterium in the gut microbiota confer resilience to chronic social defeat stress in mice. Sci Rep 7:45942

Yang YL, Chao PK, Lu KT (2006) Systemic and intra-amygdala administration of glucocorticoid agonist and antagonist modulate extinction of conditioned fear. Neuropsychopharmacology 31(5):912–924

Yehuda R, Southwick SM, Krystal JH, Bremner D, Charney DS, Mason JW (1993) Enhanced suppression of cortisol following dexamethasone administration in posttraumatic stress disorder. Am J Psychiatry 150(1):83–86

Yehuda R, Southwick SM, Nussbaum G, Wahby V, Giller EL Jr, Mason JW (1990) Low urinary cortisol excretion in patients with posttraumatic stress disorder. J Nerv Ment Dis 178(6):366–369

Zaretskaia MV, Zaretsky DV, Sarkar S, Shekhar A, DiMicco JA (2008) Induction of Fos-immunoreactivity in the rat brain following disinhibition of the dorsomedial hypothalamus. Brain Res 1200:39–50

Zhu HJ, Appel DI, Grundemann D, Richelson E, Markowitz JS (2012) Evaluation of organic cation transporter 3 (SLC22A3) inhibition as a potential mechanism of antidepressant action. Pharmacol Res 65(4):491–496

Zijlmans MA, Korpela K, Riksen-Walraven JM, de Vos WM, de Weerth C (2015) Maternal prenatal stress is associated with the infant intestinal microbiota. Psychoneuroendocrinology 53:233–245

Circadian Regulation of the Brain and Behavior: A Neuroendocrine Perspective

Ilia N. Karatsoreos

Contents

1 General Introduction ... 324
 1.1 Circadian Rhythms: A Definition ... 324
 1.2 The Suprachiasmatic Nucleus and the Molecular Clock ... 325
 1.3 A Network of Central and Peripheral Oscillators ... 328
2 Circadian Control of Reproduction ... 331
 2.1 Rhythms of the HPG Axis ... 331
 2.2 From SCN to HPG: Neural Control of HPG Function by the Circadian Clock ... 332
 2.3 From HPG to SCN: Reciprocal Feedback to the Clock ... 335
3 Hypothalamic-Pituitary-Adrenal (HPA) Axis ... 336
 3.1 Rhythms of the HPA Axis ... 336
 3.2 HPA Rhythms: Role of the Suprachiasmatic Nucleus ... 337
 3.3 Effects of Adrenal Rhythms on the Brain and Periphery ... 338
4 Interlocking Loops: Physiological Significance ... 339
5 Summary ... 341
References ... 342

Abstract Neuroendocrine systems are key regulators of brain and body functions, providing an important nexus between internal states and the external world, which then modulates appropriate behavioral outputs. Circadian (daily) rhythms are endogenously generated rhythms of approximately 24 h that help to synchronize internal physiological processes and behavioral states to the external environmental light-dark cycle. Given the importance of timing (hours, days, annual) in many different neuroendocrine axes, understanding how the circadian timing system regulates neuroendocrine function is particularly critical. Similarly, neuroendocrine signals can significantly affect circadian timing, and understanding these mechanisms can provide insights into general concepts of neuroendocrine regulation of brain circuits and behavior. This chapter will review the circadian timing system and its control of two key neuroendocrine systems: the hypothalamic-pituitary-gonadal (HPG) axis

I. N. Karatsoreos (✉)
Department of Integrative Physiology and Neuroscience, Washington State University, Pullman, WA, USA
e-mail: ilia.karatsoreos@wsu.edu

Curr Topics Behav Neurosci (2019) 43: 323–352
DOI 10.1007/7854_2019_115
Published Online: 5 October 2019

and the hypothalamic-pituitary-adrenal (HPA) axis. It will also discuss how outputs from these axes feedback to affect the circadian clock. Given that disruption of circadian timing is a central component of many mental and physical health conditions and that neuroendocrine function is similarly implicated in many of the same conditions, understanding these links will help illuminate potentially shared causality and perhaps lead to a better understanding of how to manipulate these systems when they begin to malfunction.

Keywords Androgen · Corticosterone · Cortisol · Estrogen · Reproduction · Stress · Suprachiasmatic

1 General Introduction

Neuroendocrine systems are key regulators of brain and body functions, providing an important nexus between internal states and the external world, which then modulates appropriate behavioral outputs. For instance, as nicely demonstrated in this volume, reproductive hormones impact several brain areas to prime them to respond to signs and signals in the environment that indicate it is the appropriate time to engage in reproductive behaviors. This is important because the tightly orchestrated set of responses ensures that energy is not wasted at times when reproduction would at best be unfruitful or at worst negatively impact survival of mother and/or offspring. Thus, a common theme in neuroendocrine regulation of behavior is the tight coordination of physiological/body function with environmental context. In this regard, the circadian (daily) timing system is a key regulator not only of neuroendocrine function but of general neurobehavioral and physiological systems. This chapter will explore a few key aspects of what is known about the generation, regulation, and implementation of circadian rhythms in the brain and body. It will also spend some time investigating direct regulation of the hypothalamic-pituitary-gonadal (HPG) and hypothalamic-pituitary-adrenal (HPA) axes by the circadian clock and their bidirectional impacts on circadian rhythms.

1.1 Circadian Rhythms: A Definition

Circadian rhythms have been found in nearly all organisms that have life spans longer than a day. These rhythms are endogenously generated (i.e., they persist even in the absence of external cues) and allow organisms to anticipate the daily change in the environment that is a result of the approximately 24 h period defined by the rotation of the Earth about its axis. It is important to note that circadian rhythms are but one of a set of "biological rhythms" that have varying periods and are also important for neuroendocrine function. For instance, the circannual timing system is influenced by photoperiod and helps seasonal animals predict the onset of changes in the climate that is important for breeding seasons. However, the circadian system is perhaps one of the most well characterized of these biological timing systems, with its molecular underpinnings, neurobiological basis, and system-wide organization particularly well-understood.

1.2 The Suprachiasmatic Nucleus and the Molecular Clock

The focus of this chapter will be on the vertebrate circadian system and more specifically on the mammalian circadian timing system. From many experiments undertaken in the early part of the twentieth century, it was clear that mammals had a circadian clock and that the seat of this clock must be somewhere in the brain. However, it was unclear where in the brain the clock was located or, indeed, if the circadian clock was actually localized to a single brain nucleus. Early work attempted to find the circadian clock using a series of tract-tracing studies from the retina. The hypothesis was that since circadian rhythms are influenced by the light-dark cycle, then there should be a direct pathway from the retina to the locus of the clock. Work by Moore and Klein (Klein and Moore 1979; Moore and Klein 1974) finally determined that circadian rhythms are entrained (or synchronized) to external environmental time by photic (light) information conveyed by the retinohypothalamic tract (RHT) to a small collection of cells above the optic chiasm in the anterior hypothalamus, known as the suprachiasmatic nucleus (SCN) (Klein and Moore 1979; Moore and Klein 1974). Importantly, this pathway is nonimage forming but is necessary and sufficient for entrainment to light. If the primary visual pathway is transected at the level of the optic tract beyond the optic chiasm (i.e., thalamic relays caudal to the SCN), then animals are visually blind, but the circadian system is able to respond to photic cues, and the entrainment is still possible (Klein and Moore 1979; Johnson et al. 1988). This pathway was further elaborated with the discovery of dedicated pathway of nonimage forming retinal cells that convey photic information to the SCN. While both rods and cones contribute to entrainment, a novel subset of intrinsically photosensitive retinal ganglion cells (ipRGCs) containing melanopsin was also discovered to be central to entrainment (Panda et al. 2002b, 2003; Ruby et al. 2002; Hattar et al. 2003; Lucás et al. 2003). All three photoreceptors must be knocked out to abolish entrainment (Hattar et al. 2003); however, specific ablation of these cells prevents entrainment even while preserving vision (Guler et al. 2008). That the SCN receives photic input is but one of the myriad pieces of evidence that have been collected demonstrating its position as the master clock in mammals. The SCN shows rhythms of electrical activity and gene expression both in vivo and in vitro (Inouye and Kawamura 1979; Green and Gillette 1982; Groos and Hendriks 1982; Shibata et al. 1982; Yamazaki et al. 2000; Ono et al. 2015), and SCN rhythms persist for many cycles in vitro, demonstrating that input from other brain areas is not necessary. In addition, lesioning the SCN abolishes circadian rhythms (Moore and Lenn 1972; Stephan and Zucker 1972), and transplants of SCN tissue restore rhythmicity in hosts whose own rhythms have been eliminated by SCN lesions (Lehman et al. 1987; Silver et al. 1990). While restoring host rhythms indicates SCN tissue may sustain rhythms, it does not fully demonstrate that the SCN is the clock. Follow-up experiments using SCNs from hamsters with a naturally occurring mutation that affected period length ("tau" mutant, 20 h) demonstrated that transplanted SCNs restore the period of the activity rhythm to that of the donor, rather than of the host – a remarkable demonstration of the ability of a brain area to determine behavioral outputs (Ralph et al. 1990; Silver et al. 1996).

The SCN is a compact structure of only about 10,000 cells, but this seemingly small collection of cells belies an incredibly complex and elegant neuroanatomical and functional organization. SCN cells are heterogeneous in neurotransmitter and peptidergic content, as well as in their cellular morphologies, and show distinct differences in their afferent and efferent connections. This heterogenous organization of the SCN has been shown to be central to its special abilities to act as a clock. At the tissue level, in vivo the SCN shows a coordinated pattern of activity, with a period of about 24 h. Puzzlingly, dissociated individual SCN neurons exhibit different free-running periods (Welsh et al. 1995), which has led to the hypothesis that SCN cells must communicate among themselves in an organized circuit that is functionally important. Generally speaking, the SCN has a ventromedial subregion that has come to be known as the "core" and dorsolateral region known as the "shell." This organizational scheme is based on anatomical, neurochemical, and functional studies (reviewed in Antle and Silver 2005). Functionally, the shell shows high-amplitude oscillations of many core "clock genes" including the *Period1* and *Period2* genes (*Per1*, *Per2*; discussed further below in the context of molecular clocks), while in the core, these rhythms are very low amplitude or in some studies undetectable. Conversely, when the animal is exposed to a light pulse, there is a rapid and transient induction of *Per1* and *Per2*, while in the shell compartment, *Per1* and *Per2* are not induced by light (Hamada et al. 2001). Intense study of the SCN over the past decade, using both advanced analytical tools and cutting-edge imaging techniques, has made this brain structure a powerful model to understand how small networks of neurons interact in order to have a cohesive output. A full discussion of how this phenomenon is manifest in the SCN is beyond the scope of this chapter, but current models suggest that the spatiotemporal pattern of SCN activity is a dynamic interaction between the various SCN components but can also be influenced by the environmental light-dark cycle (Hamada et al. 2001; Herzog 2007; Yan et al. 2007; Pauls et al. 2014; Evans et al. 2013). However, the specific mechanism that underlies this unique ability of the SCN remains unknown, though multiple mechanisms have been proposed for coupling among SCN neurons (Antle et al. 2003, 2007).

While there has been tremendous work undertaken to unravel the complex organization of the SCN at a circuit level, even more work has been undertaken at the level of the molecular clock. The incredible findings at the molecular level is based largely upon work conducted in *Drosophila* (fruit fly), a fact that was recently acknowledged with the awarding of a Nobel Prize in Physiology or Medicine to Rosbash, Young, and Hall in 2017.

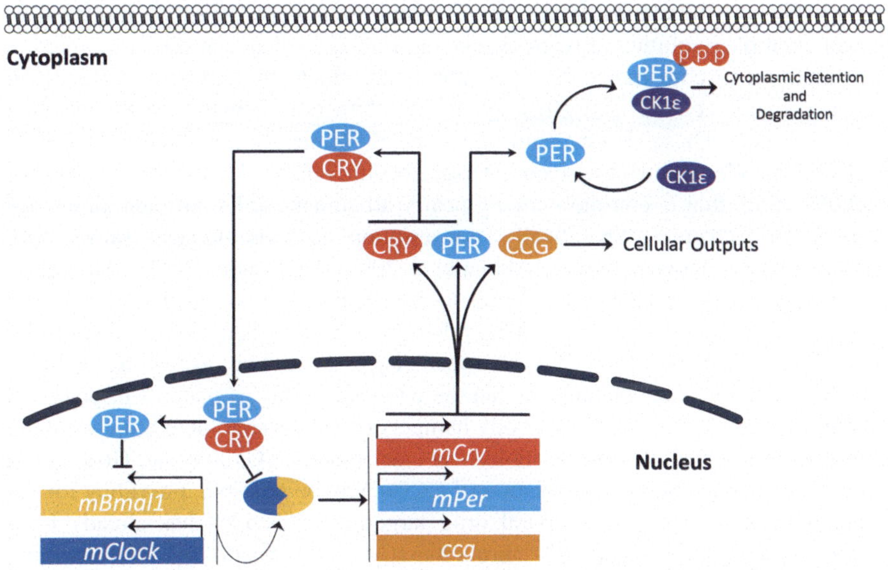

Fig. 1 Diagram of the molecular circadian clock. A simplified depiction of the molecular circadian clock mechanism showing the transcriptional-translational feedback loop (TTFL) in mammals. In the nucleus, BMAL1-CLOCK protein complex binds to the promoter region of genes of the *Per* and *Cry* family, as well as to clock-controlled genes (*ccg*). This increased transcriptional activity results in production of their protein products, which then shuttle to the cytoplasm, where PER/CRY protein dimers translocate back to the nucleus. Once nuclear transport has occurred, they act to inhibit BMAL1/CLOCK, thus reducing their own transcription. In addition to the TTFL, kinases pathways, including casein kinase 1 epsilon (CK1e), phosphorylate PER in the cytosol, which targets the protein for degradation. This provides a time delay that enables the molecular clock to cycle with a period of approximately 24 h

In the SCN, rhythms are generated by interactions between cells. But at the level of individual cells, circadian rhythms are generated by a clocklike molecular timing mechanism (Fig. 1). This is what underlies the "cell-autonomous" nature of circadian rhythms: while networks of neurons are needed to time behavior and physiology, individual cells can show clear circadian rhythms. Circadian clocks in animals have a highly conserved molecular mechanism based on self-regulating transcriptional-translational feedback loops that underlie rhythmic expression of core "clock" genes/proteins. In mammals, this consists of a set of transcriptional activators and transcriptional repressors. The transcriptional activator proteins BMAL1 and CLOCK bind enhancer elements (E-boxes) on DNA and promote the transcription of *period* and *cryptochrome* genes. These genes encode for repressor proteins, which feed back to inhibit BMAL1/CLOCK function. Inhibiting the enhancers thereby reduces their own expression: a classic "negative feedback" loop. In addition to these core clock genes and proteins, several kinases are important to help set the "pace" of the reactions, including casein kinase 1/ε (CK1/ε). These factors provide fine control of variables such as period length and amplitude of the core clock machinery.

Remarkably, this biochemical cascade takes approximately 24 h to complete – an elegant example evolution at work (Koike et al. 2012; Takahashi 2017).

In addition to core molecular elements of the clock, the circadian system can influence many cellular pathways by direct interaction which "clock-controlled genes" (CCGs). These downstream genes are rhythmically regulated by the core clock and encode proteins involved in tissue-specific effects. This schema allows for a cellular clock that is common among nearly all somatic cells but then allows for tissue-specific changes. Specifically, transcription analysis suggests only ~10% overlap of CCGs between tissues (Akhtar et al. 2002; Panda et al. 2002a). Moreover, even when the same CCGs are found to be regulated rhythmically in different tissues, in many cases they have different distributions of peak phase and amplitudes in different tissues. This level of tissue specificity provides both a challenge but also a potential benefit. The challenge is that the wide variability of patterns of gene and protein expression throughout the body hampers our abilities to determine what the basic operating principles of the molecular clocks and CCGs may be. However, a benefit is that tissue-specific regulation of CCGs may be possible, providing for the potential of tissue-specific temporal drug targeting. This idea has already been implemented in the concept of "chronotherapy" for cancers. By understanding how tumors may present different circadian activity than surrounding healthy tissue, it is possible to time chemotherapy or radiotherapy to have maximal effect on the cancer with less effects on healthy tissues (Levi 2006).

1.3 A Network of Central and Peripheral Oscillators

Though it is undeniable that the SCN is the master circadian clock, as in nearly all other neuroendocrine pathways, the circadian system is comprised of a "pseudo-hierarchical" network of control nodes throughout the brain and body. Originally, the system was conceived as purely hierarchical, with the SCN sitting atop the hierarchy. However, it has become clear that while the SCN is certainly the master circadian clock, it controls a network of circadian oscillators throughout the body, which also interact with each other. But how does the SCN control these myriad oscillators? The current models involve two non-mutually exclusive methods: direct neural control and humoral regulation, with indirect regulation (which may be a combination of the previous two) also playing a role.

1.3.1 Direct Neural Control

Tract tracing has been used to carefully delineate neural outputs from the SCN to other brain areas (Morin 2013). As discussed earlier, the SCN has several component parts, each showing different peptidergic contents, and anatomical projections. However, neural efferents originate in both the SCN core and shell. What has been discovered is that many SCN outputs are monosynaptic projections targeting key neuroendocrine cell populations that are involved in the production of essential releasing hormones. Direct projections from the SCN have been traced from the SCN to the medial preoptic area (MPOA), supraoptic nucleus (SON), anteroventral periventricular nucleus (AVPV), the paraventricular nucleus (PVN), dorsomedial nucleus of the hypothalamus (DMH), lateral septum, and the arcuate nucleus (Arc) (van der Beek et al. 1993, 1997; Gerhold et al. 2001; Horvath et al. 1998; Buijs et al. 1998, 2003; Kalsbeek et al. 1996, 2000; Vrang et al. 1995; Kriegsfeld et al. 2002; Egli et al. 2004; Kalsbeek and Buijs 2002). Given these brain nuclei regulate secretion of factors into the cerebrospinal fluid, pituitary portal system, and general circulation, SCN-derived signals can control widespread systems in the brain and body (Skinner and Malpaux 1999; Skinner and Caraty 2002; Reiter and Tan 2002; Tricoire et al. 2003).

An elegant example of how the SCN can control downstream targets via neural projections comes from work done in the regulation of reproductive hormones. Initial indications that endocrine function may be under direct neural control of the SCN is that hormonal rhythms are eliminated after the severing SCN efferents (Hakim et al. 1991; Nunez and Stephan 1977). In addition, while behavioral rhythms are restored in SCN-lesioned animals following SCN transplants, hormonal rhythms are not (Nunez and Stephan 1977; Meyer-Bernstein et al. 1999; Silver et al. 1996). Making use of the phenomenon known as "splitting" which occurs in hamsters housed in constant light (LL), several groups have specifically probed how the SCN regulates rhythms of reproductive hormones. In "splitting," hamsters begin to show two separate bouts of activity within each 24 h period. These split rhythms seem like in each 24 h day, the split hamster experiences two 12 h days. Supporting this theory, it was demonstrated that split female hamsters show two daily preovulatory surges of luteinizing hormone (LH) and that the plasma concentration of LH in each of these surges was only about ½ of the plasma concentration of a control female (Swann and Turek 1985). The discovery of rhythmic clock gene expression in the SCN allowed further probing of this phenomenon, and it was found that in split hamsters, the left and right SCNs showed activity that was 180° out of phase with each other (de la Iglesia et al. 2000). The LH surge is regulated by activity of gonadotropin-releasing hormone (GnRH) neurons, and split hamsters were found to have activity in GnRH neurons only in one side of the brain at a time – and in the ipsilateral side to the active SCN (De la Iglesia et al. 2003). These findings support the conclusion that the timing of the LH surge must be from a neural signal in the SCN, which is communicated to ipsilateral GnRH neurons. This had to be a neural signal since a diffusible signal from the SCN would impact both hemispheres of the brain indiscriminately.

1.3.2 Humoral Regulation

While the neural regulation of physiology and behavior by the SCN is clear, a remarkable set of studies demonstrate that the SCN can also exert its influence via humoral or diffusible signals. Lesions of the SCN definitively show that the SCN is necessary for normal circadian rhythms to be expressed within an animal; however, SCN transplants into SCN-lesioned hamsters showed more mixed results.

It is well known that plasma melatonin rises at the start of the dark phase in humans and nonhuman animals and can be considered a key vertebrate timing hormone (reviewed in Simonneaux and Ribelayga 2003; Johnston and Skene 2015). The pineal gland is the main driver of this timed melatonin release, by a projection from the SCN via the superior cervical ganglion. Thus, the dusk rise in plasma melatonin has been shown to be driven by an interaction between the light-dark cycle and the endogenous clock and serves as an excellent phase marker (Lewy and Sack 1989; Lewy 2007) and importantly can serve as a critical driver of photoperiodic changes (Simonneaux and Ribelayga 2003; Hazlerigg 2012; Hazlerigg and Wagner 2006; Nishiwaki-Ohkawa and Yoshimura 2016). Many studies have demonstrated that melatonin can serve as a phase-resetting agent acting through melatonin receptors (MT1 and MT2) found throughout the brain and the periphery of mammals (Dubocovich and Markowska 2005). As such, there is significant interest in the use of melatonin as pharmacological tool to reset the circadian clock after an acute circadian disruption (e.g., transmeridian flight, aka "jetlag"). However, the precise mechanism of action (if there are any effects) remains elusive. Melatonin can alter core body temperature (Johnston and Skene 2015), and it is possible that these changes in core body temperature can also serve as synchronizing cues to peripheral clocks (Schibler et al. 2015), providing yet another pathway by which melatonin can act as a timing hormone. In humans the links between melatonin, body temperature, and sleepiness are less clear (Lok et al. 2019). A few caveats are certainly required in this brief discussion of melatonin. It is likely that nearly 99% of melatonin in the body is not of pineal origin and never released, instead being contained within mitochondria where the molecule can act as a free-radical scavenger independent of its activity at MT1 or MT2 (Zhao et al. 2019). Remarkably, and perhaps unknown to many researchers, many strains of laboratory mice (e.g., C57BL/6) have negligible pineal melatonin production due to a point mutation in arylalkylamine *N*-acetyltransferase (AANAT), a critical enzyme in the melatonin synthesis (Ebihara et al. 1986; Roseboom et al. 1998; von Gall et al. 2000). These animals have stable and precise circadian rhythms, but their rhythms can indeed be shifted by exogenous melatonin (Dubocovich et al. 2005), demonstrating that melatonin is not necessary for normal circadian rhythms to be expressed but can still influence circadian timing, largely through the MT1 and/or MT2 receptors.

As discussed above, though SCN transplants restored many behavioral rhythms (e.g., locomotion, feeding, drinking), they did not restore hormonal rhythms or rhythms in neuroendocrine axis activity. Specifically, Meyer-Bernstein et al. (1999) showed that SCN-lesioned hamsters that had behavioral rhythms restored did not have their estrous cycle rhythms rescued nor could estradiol induce LH surges at the appropriate circadian time. Rhythms in adrenal hormones were absent in SCN-transplanted hamsters and plasma melatonin did not rise at the expected time in the late subjective night. This led to the conclusion that is while the neural integration of the graft was enough to restore locomotor rhythms, more extensive (or complex) connections were needed to restore endocrine function. However, in a heroic set of experiments, the necessity of neural integration for restoration of behavioral rhythms was tested by transplanting encapsulating donor SCN tissue in a membrane that prevented neural outgrowth while allowing signals to diffuse between graft and host (Silver et al. 1996). These experiments demonstrated that even with an encapsulated graft, behavioral rhythms could be restored, showing that neural connections from graft to host were not necessary. To further probe the role of diffusible signals in regulation of SCN function, an in vitro co-culture technique was developed by Maywood and colleagues. In these experiments, a "control" wild-type SCN was placed upon a membrane that enabled communication via shared culture media with a target host SCN that was rendered arrhythmic via genetic deletion of different neuropeptides or molecular components of the core circadian clockwork (Maywood et al. 2011). These paracrine signaling experiments demonstrated that a wild-type SCN could rescue circadian pacemaking ability in cultured SCN slices which were deficient in vasoactive intestinal peptide (VIP), which is key for communication between cellular oscillators in the SCN. Moreover, they also demonstrated that arrhythmic SCN tissue carrying a *Cry*-null mutation (thus missing a core clock element) could be induced to become rhythmic when co-cultured with the wild-type SCN.

Thus, not only are diffusible humoral signals sufficient to regulate rhythms in behavior, they are also able to act at the level of the SCN to synchronize SCN oscillators. Conversely, neural outputs seem necessary to generate rhythms in neuroendocrine signals and endocrine axes more generally.

2 Circadian Control of Reproduction

2.1 *Rhythms of the HPG Axis*

The HPG is a primary regulator of reproductive function. It is also evident that circadian rhythms are present at many levels of the HPG, albeit with significant species differences (Khan and Kauffman 2012). In many species, including rodents, these rhythms are under control of the SCN clock, as both SCN lesions and genetic manipulations of core clock disrupt HPG function (Chu et al. 2013; Gray et al. 1978; Miller et al. 2004). In mice, mutations of the core clock genes *Clock* or *Bmal1* lead to altered LH surge rhythmicity and abnormal estrous cyclicity (Chu et al. 2013; Gray et al. 1978; van der Horst et al. 1999). Remarkably, human females who have a single-nucleotide polymorphisms in ARNTL (*Bmal1*) have more miscarriages (Kovanen et al. 2010), and though this effect is likely mediated by a non-CNS site of action, it further substantiates a role for the clock in regulation of the basic function of the HPG. The circadian contribution to HPG regulation has been well characterized in rodent models, with the Syrian hamster being a gold-standard model system. In entrained (i.e., 24 h light-dark cycle), the estrous cycle is nearly exactly 4 days long. Early studies clearly demonstrated the link between the endogenous circadian period and estrous cycle length, showing that in free-running (i.e., in constant conditions) hamsters, the estrous cycle is a multiple of four of the free-running period (Fitzgerald and Zucker 1976). If hamsters are entrained to a non-24 h light-dark cycle (e.g., 22 h) or if the circadian period is lengthened (e.g., using heavy water), the duration of the estrous cycle is still, reliably, a quadruple multiple of the circadian day.

2.2 From SCN to HPG: Neural Control of HPG Function by the Circadian Clock

As described above, the SCN is a compact yet particularly heterogenous structure in terms of neuropeptidergic content. Of the many types of cells within the SCN, two seem particularly important in the regulation of HPG function: vasoactive intestinal peptide (VIP) and arginine vasopressin (AVP) containing neurons. In both cases, these cell types seem to exercise their regulation of the HPG via actions, directly or indirectly, on gonadotrophin-releasing hormone (GnRH) cells (Fig. 2).

VIP and AVP neurons are spatially distinct within the SCN, with the former being located more ventrally and the latter being located more dorsally. Efferents of SCN VIP neurons contact GnRH neurons (Van der Beek et al. 1997), with female rats showing higher GnRH innervation by VIP than do males (Horvath et al. 1998). Developmentally, it has been demonstrated that in hamster, VIP contacts increase following puberty (Kriegsfeld et al. 2002). Ex vivo, VIP can affect firing rates of GnRH neurons (Piet et al. 2016), and importantly, VIP has a time-of-day sensitive effect on GnRH neuron firing, driving an increase in firing rate only at around the time of the LH surge but much less of an effect at other times (Christian and Moenter 2008). Similarly, in female mice, pharmacological blockade of the VPAC2 receptor attenuates neuronal firing of GnRH cells only during the afternoon proestrus surge.

AVP in the plasma peaks in a coincident manner with the preovulatory LH surge (Schwartz et al. 1983). In rat, AVP injected into the medial preoptic area (MPOA) induces an LH surge in SCN-lesioned animals. In *Clock* mutant mice, LH surges are not detectable, but can be induced by central injections of AVP (Miller et al. 2006). Even in in vitro models, AVP has clear effects on GnRH rhythms. Preoptic area and SCN co-cultures show that GnRH release is in phase with AVP release (Funabashi et al. 2000). However, what is interesting about AVP cells in their regulation of HPG function is that, unlike VIP, there is as yet no direct evidence of AVP innervation of GnRH neurons, and, perhaps more importantly, GnRH neurons only express the AVP receptor (V1a) sparsely, if at all (Kalamatianos et al. 2004). How does AVP have such a pronounced impact on the LH surge without communicating directly to GnRH cells? An intermediate player had to be at work in this circuitry.

One potential way this mystery could be solved is that the GnRH cells themselves are capable of generating part of this rhythmic sensitivity. There is evidence of time-of-day-dependent sensitivity of GnRH cells to VIP and AVP (Williams et al. 2011).

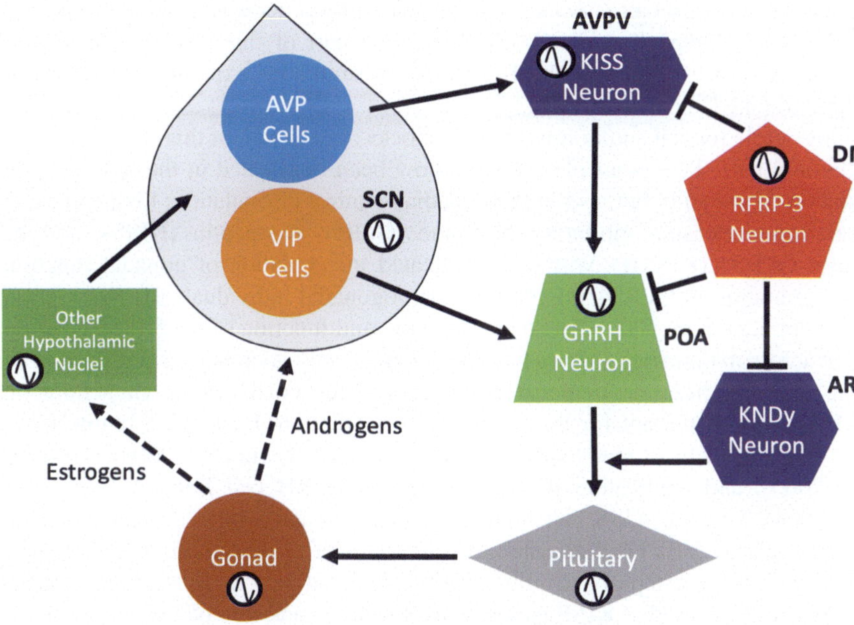

Fig. 2 Circadian control of the hypothalamic-pituitary-gonadal (HPG) axis. A depiction of the main nodes of the circadian control of the HPG axis. While vasoactive intestinal polypeptide (VIP) neurons of the suprachiasmatic nucleus (SCN) project directly to gonadotropin-releasing hormone (GnRH) cells in the preoptic area (POA), arginine vasopressin (AVP) cells project to kisspeptin (KISS) neurons in the anteroventral periventricular area (AVPV), which then synapse onto GnRH neurons. Gonadotropin inhibitory hormone/RFRP-3 (RFRP-3) neurons in the dorsomedial hypothalamus (DMH) play an important role in the circadian gating of responses of both GnRH and KISS neurons. Another population of gonadal hormone-sensitive KISS neurons (aka *K*isspeptin, *N*eurokinin B, *Dy*norphin; KNDy neurons) in the arcuate nucleus (ARC) are also inhibited by GnIH and project onto terminals of GnRH neurons in the median eminence and act as important regulators of GnRH secretion. Outputs of the gonads, particularly androgens and estrogens, can impact circadian timing. While androgen receptors (AR) have been found within the core region of the SCN, suggesting a direct effect, estrogen type 1 receptor (ESR1) is only sparsely expressed in the SCN, with most effects likely being mediated by extra-SCN hypothalamic relays. Beyond the scope of this review, gonadal hormones also directly act upon many of the neuron types described in this pathway. Oscillations in clock genes (depicted by the small oscillator) have been found in many levels of the HPG axis, suggesting local timing control is a key mechanism by which the SCN can regulate the overall activity of the axis

However, there is not strong evidence for functional consequences of a cell-autonomous GnRH circadian clock. While GnRH neurons in vivo do not have endogenous rhythms in clock gene expression, GnRH mRNA content does show a diurnal pattern (Kriegsfeld 2006; Schirman-Hildesheim et al. 2006). The data from in vitro preparations are clearer. GnRH neurons show circadian rhythms in firing rates in hypothalamic slice cultures, even without an SCN present (Christian et al. 2005). In GnRH secreting GT1-7 cell lines, rhythms can be detected in both GnRH mRNA and many clock genes (Gillespie et al. 2003), and dominant negative overexpression of mutant *Clock-Δ19* in these cell lines decreases the frequency of GnRH pulses (Chappell et al. 2003). Thus, one part of this missing connection between SCN AVP and GnRH/HPG function could be explained by intrinsic rhythms of GnRH cells.

While a role for cell autonomous GnRH clocks may be part of this story, it cannot be the only part. Two other players have now been implicated in the both general regulation of the HPG but also in the circadian control of ovulation: kisspeptin and gonadotropin-releasing inhibitory hormone (GnIH). Kisspeptin (KISS), and its receptor GPR54 (KISS1r), was first implicated in regulation of gonadal function when a mutation in GPR54 was found in hypogonadal individuals (de Roux et al. 2003; Seminara et al. 2003). The neuroanatomy and function of the KISS signaling system across mammalian species have been extensively documented (Lehman et al. 2013). Many studies have demonstrated that the effects of KISS are via actions on GnRH and are important for the regulation of ovulation. First, KISS1r mRNA is found in GnRH cells of many species (Han et al. 2005; Irwig et al. 2004; Messager et al. 2005), and application of KISS increases GnRH cell firing in vitro (Liu et al. 2008). Second, KISS cell bodies are concentrated within the anteroventral periventricular (AVPV) and arcuate (ARC) regions of the rodent brain and express many sex steroid receptors (Smith 2008). Third, studies with transgenic KISS knockout mice show that KISS is necessary for the expression of LH surges (Dror et al. 2013). In the species explored, most ARC KISS cells co-express neurokinin B (NKB) and dynorphin, positive and negative regulators of the reproductive axis, respectively (Goodman et al. 2007). These neurons have been coined "KNDy" (K = KISS, N = NKB, and Dy = dynorphin) (Lehman et al. 2010a, b). Interestingly, these neurons also co-express receptors for NKB and dynorphin and are reciprocally interconnected, which underpins their unique ability to propagate stimulatory or inhibitory signaling (Lehman et al. 2010a, b), and also express progesterone and estrogen receptors, which enables steroid negative feedback (Foradori et al. 2002; Smith et al. 2007). As for the circadian control of ovulation and these neurons, KISS mRNA expression is highest at the time of the LH surge in the AVPV and KISS cells expresses the highest levels of FOS expression on the afternoon of proestrus (Williams et al. 2011; Robertson et al. 2009). The SCN projects to the AVPV and regulate the function of resident KISS neurons (Kriegsfeld et al. 2004; Leak and Moore 2001), further supporting the notion that the timing of the LH surge may occur via projections to AVPV KISS cells. More recently, Ca^{2+} imaging approaches have elegantly demonstrated both AVP and VIP can directly modulate ARC GnRH neurons in a sex-dependent manner, which shows a rostro-caudal anatomical organization in the ARC (Schafer et al. 2018).

In addition to KISS and KNDy neurons, GnIH plays an important part of the circadian regulation of ovulation. GnIH was first described in avian species (Tsutsui et al. 2000, 2009), and more recently the mammalian ortholog (Arg)(Phe)-related peptide-3 (RFRP-3) has been uncovered (Tsutsui and Osugi 2009; Tsutsui et al. 2018). RFRP-3 cells are largely concentrated in the dorsomedial hypothalamus (DMH) of rodents, with extensive hypothalamic and limbic projections, and project to GnRH cells and to the median eminence (Kriegsfeld 2006). Functionally, a role for RFRP-3 has been defined by demonstrations that RFRP-3 injections cause a rapid LH suppression in several rodent species. Similar to KNDy neurons, RFRP-3-expressing cells also possess estrogen receptors and are an important target for steroid-negative feedback (Kriegsfeld 2006). With regard to circadian control, it has been well established that the SCN has significant efferent projections to the DMH region that contains the RFRP-3 cells (Kriegsfeld et al. 2004; Leak and Moore 2001). Moreover, a large proportion of RFRP-3 cells (in excess of 60%) are contacted by terminals of SCN origin (Gibson et al. 2008; Russo et al. 2015). This system has been characterized in a functional context as well, with the SCN acting to suppress activity of RFRP-3 neurons during the LH surge, which allows for the time-coordinated release of the GnRH system from negative feedback (Gibson et al. 2008; Russo et al. 2015). It is important to note that the role for RFRP-3 in mammalian reproduction is not without some controversy or caveats (Reviewed in Angelopoulou et al. 2019).

2.3 *From HPG to SCN: Reciprocal Feedback to the Clock*

It has been observed that gonadal hormones can modulate circadian rhythms, suggesting that gonadal hormones may feed back to the circadian clock. For instance, cycling hamsters show a clear phase advance on the day of estrus when levels of E2 peak. Conversely, clamping E2 by pellet implant shortens free-running rhythms (Morin et al. 1977). Since these aspects of circadian rhythms (phase and period) are known to be driven by the SCN, it suggests that there must be E2 effects on the SCN directly. However, estrogen receptors (ESR) are only sparsely expressed in the SCN (Shughrue et al. 1997; Hileman et al. 1999; Gundlah et al. 2000; Vida et al. 2008). Thus, this suggests an indirect circuit may be responsible. Several populations of ESR1-positive cells provide input to the SCN, including the ARC, amygdala, bed nucleus of the stria terminalis (BNST), and the POA (De La Iglesia et al. 1999), providing the basis for such a circuit. Indirect E2 effects may still occur in the SCN, perhaps without ESR activity through gap junction effects on glial cells and neurons. Cultured SCN cells show the presence of gap junctions, both by dye and electrical coupling (Colwell 2000; Jiang et al. 1997; Long et al. 2005), and blockade of gap junctions in vitro affects SCN electrical activity (Shinohara et al. 2000; Prosser et al. 1994). In vivo, male connexin-36-knockout mice show reduced amplitude of locomotor activity rhythms (Long et al. 2005). In female rats, E2 leads to increased expression of the interneuronal gap junction subunit connexin-36 in the SCN (Shinohara et al. 2001; Rash et al. 2007). While this hypothesis has yet to be completely tested, it remains a plausible way by which E2 can affect circadian function.

Gonadal androgens also show strong influences on circadian locomotor activity. Castration of male mice results in a longer free-running period, decreased precision, and significantly reduced consolidation of daily activity (Karatsoreos et al. 2007; Daan et al. 1975). Replacement of either testosterone (T) or the non-aromatizable dihydrotestosterone (DHT) rescues this phenotype, which suggests that the androgen receptor (AR) mediates the effects of androgens on circadian rhythms, rather than conversion to E2, and effects through the ESR. In male mice, androgens also drive an increased sensitivity of the SCN to light (Butler et al. 2012), which seems to be related to the effect on period. Remarkably, in mice, the SCN contains a highly level of AR, which are concentrated in the core region (Karatsoreos et al. 2007), and local administration of T to the SCN restores intact-typical circadian periods, strongly indicating a direct effect of androgens in the SCN (Karatsoreos et al. 2007; Model et al. 2015). In other species, while AR can be found in the SCN, it seems to be more diffuse than observed in mouse (see Karatsoreos and Silver 2007 for review). It is interesting to note that there seems to be a clear sex difference in the expression of SCN AR in both humans (Fernandez-Guasti et al. 2000) and in rodents (Iwahana et al. 2008). In mice, ARs are more highly expressed in male than in females. This sex difference in SCN AR is paralleled by functional sex differences, as ovariectomy does not have the same effects on period or the onset activity bout as castration does in males. Thus, effects of androgens appear to be mediated largely by direct effects at the level of the SCN, while the effects of estrogens may occur primarily outside the SCN.

3 Hypothalamic-Pituitary-Adrenal (HPA) Axis

3.1 Rhythms of the HPA Axis

The primary output of the HPA axis are glucocorticoids, such as cortisol (corticosterone in most rodents; CORT), and circadian rhythms of CORT in blood have been well described in many species (Dickmeis 2009), including humans, nonhuman primates, and rodents (Weitzman et al. 1971; Moore and Eichler 1972; Sachar et al. 1973; Dubey et al. 1983; Czeisler and Klerman 1999; Van Cauter and Refetoff 1985). An important aspect of CORT in the plasma is that its phasing does not seem to be linked to the light-dark cycle per se but instead to the activity phase of the animal. For instance, between nocturnal and diurnal species, CORT levels rise before waking in both, which results in a peak during the day in diurnal animals and a peak during the night in nocturnal animals (Wong et al. 1983; Albers et al. 1985; Ottenweller et al. 1987). Rhythms of CORT in humans have been a major target of study because they are associated with several neuropsychiatric disorders. As in other diurnal species, cortisol levels in humans peak around the morning wake-up time.

A discussion about rhythms in CORT and the HPA axis would not be complete without addressing the findings that circulating CORT rhythms are actually generated by changes in the pulsatile secretion of the hormone that occur in an ultradian fashion, with about a 1 h period (Walker et al. 2012; Spiga et al. 2014; Russell et al. 2015). Remarkably, and differently than the HPG, data show that this ultradian pattern of CORT secretion is driven by feed-forward mechanisms and timed delays in biological processes at the level of both the pituitary and adrenal glands; it does not require a hypothalamic pulse generator (Walker et al. 2010, 2012). The functional significance of these ultradian pulses is an area of significant research, with findings demonstrating that the adrenal responds optimally to a pulsatile pituitary adrenocorticotropic hormone (ACTH) profile, given that constant ACTH infusion results in significantly reduced CORT levels (Spiga et al. 2011).

3.2 HPA Rhythms: Role of the Suprachiasmatic Nucleus

As noted above, there has been significant work aimed at understanding how rhythms in the HPA are generated, with the pulsatile profile seemingly regulated extra-hypothalamically. For instance, the adrenal gland shows high amplitude rhythms in core clock genes, which form the basis for the rhythmic responsiveness to ACTH and to both physiological and physical stressors (Ungar and Halberg 1962; Kalsbeek et al. 2003; Bittman et al. 2003; Oster et al. 2006). Thus, at the level of the target glad (in this case the adrenal), circadian clocks gate sensitivity to other signals on a daily basis, much like the elegant control of ovulatory rhythms as discussed above.

While peripheral clocks are clearly involved in regulation of HPA rhythms, the SCN is necessary for overt circadian rhythms in CORT. As noted above, lesions of the SCN completely eliminate CORT rhythms, and these rhythms are not restored by SCN transplants (Meyer-Bernstein et al. 1999; Moore and Eichler 1972). Moreover, a period of circadian "forced desynchrony" to disrupt SCN function has been shown to drive changes in corticosterone rhythms in rats (Wotus et al. 2013). Though the adrenal gland contains the molecular machinery to express circadian rhythms, adrenal rhythms of clock gene expression are dependent upon the SCN (Guo et al. 2006). Together, these lines of evidence strongly implicate a neural projection from the SCN to the adrenal that can regulate rhythms in this tissue (Fig. 3). There are two efferent pathways from the SCN that have been implicated in its regulation of HPA rhythms. The first involves monosynaptic projections from the SCN to the a neighboring hypothalamic region, specifically onto corticotropin-releasing hormone (CRH) neurons in the PVN (Vrang et al. 1995; Kalsbeek et al. 1996; Buijs et al. 1998). Output from these neurons is clearly important both in the generation of circulating CORT rhythms and for the regulation of response to acute stressors. Complimenting this monosynaptic pathway, a multisynaptic pathway from the SCN to the adrenal cortex itself has also been discovered (Buijs et al. 1999), the specific role of which has not been fully characterized.

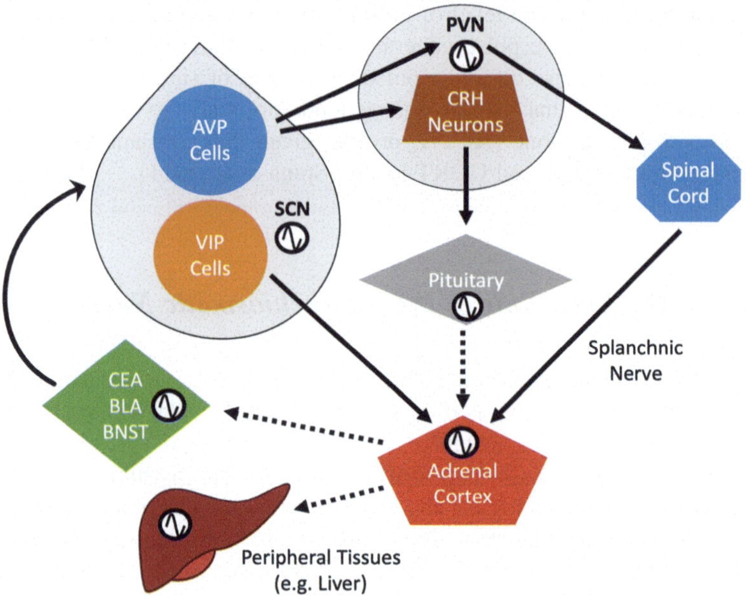

Fig. 3 Circadian control of the hypothalamic-pituitary-adrenal axis. The SCN projects to corticotropin-releasing hormone (CRH) neurons of the paraventricular nucleus (PVN) which then stimulate release of adrenocorticotropic hormone (ACTH) from the pituitary that then causes release of corticosterone/cortisol (CORT) from the adrenal cortex. CORT can then affect peripheral organs (such as the liver) and other brain regions by causing transcriptional activation of many hundreds of genes, including clock genes. While there is no apparent effect of glucocorticoid receptor (GR) activation in the adult SCN, this indirect action of CORT on peripheral tissues and brain regions can subsequently affect the SCN. There is also neural innervation of the adrenal cortex from the SCN through a multisynaptic projection via the PVN, to the spinal cord and via the splanchnic nerve to the adrenal cortex. Remarkably, a direct monosynaptic projection from the SCN to the adrenal cortex has also been discovered. It is important to recognize that oscillators (depicted by the small oscillator symbol) exist at nearly every level of this neuroendocrine axis, providing an important local timing cue that can affect how upstream signals are processed and downstream signals are communicated

3.3 *Effects of Adrenal Rhythms on the Brain and Periphery*

While the SCN is essential in driving rhythms in the HPA, the SCN itself does not express glucocorticoid receptors (GR) (Balsalobre et al. 2000; Rosenfeld et al. 1988). Much as E2, adrenal hormones have indirect effects on the SCN. For instance, treatment with glucocorticoids affects sleep in humans and also changes AVP expression within the SCN (Liu et al. 2006). Similarly, rodent studies indicate glial fibrillary acidic protein is upregulated in the SCN following treatment with glucocorticoids (Maurel et al. 2000). However, the mechanisms by which these effects occur, and their overall significance for circadian function, have yet to be fully elucidated.

In describing how the SCN can regulate peripheral clocks, perhaps by way of diffusible signals, it has become increasingly clear the glucocorticoids, and rhythms in glucocorticoids, are an important route. Several studies have shown that PER2 is rhythmic in extra-SCN brain areas, including the oval nucleus of the BNST (BNST-OV) and the central (CEA) and basolateral nuclei of the amygdala (BLA) (Amir et al. 2004; Lamont et al. 2005). Remarkably, these rhythms are not monolithic in nature, in that while rhythms in the BNST-OV and CEA depend on an intact adrenal, PER2 rhythms in the BLA are not affected by adrenalectomy. That CORT is a key player in this rhythm is evidenced by experiments that showed PER2 rhythms could be restored with CORT in the drinking water (i.e., a rhythmic route of administration), but not by pellet implantation (i.e., tonic CORT) (Segall et al. 2006). More recently, a role for diurnal changes in CORT has been uncovered in cortical regions, with diurnal CORT rhythms seemingly important for spine turnover in motor cortex, which seems to drive performance in motor learning tasks (Liston and Gan 2011).

Outside of the brain, adrenal corticosteroids are important in the communication of circadian time to the periphery (Balsalobre et al. 2000). Intraperitoneal injection of dexamethasone (DEX), a potent GR agonist, can cause phase shifts in hepatic clock gene expression, demonstrating that GR signaling can synchronize liver clocks. More broad assays of hepatic gene expression demonstrate that rhythms of 100 genes in the liver lose their rhythmicity following ADX (Oishi et al. 2005). It is interesting to observe that nearly 60% of hepatic transcriptome rhythms that are eliminated by SCN lesions can be restored by DEX treatment (Reddy et al. 2007). The mechanisms by which these effects of GR are translated to changes in transcription are myriad. Since only 2/3 of the genes that had their rhythms restored by DEX contain glucocorticoid response elements (GRE) (Reddy et al. 2007), other indirect pathways must be important and likely rely on interactions between other clock genes and clock-controlled genes.

4 Interlocking Loops: Physiological Significance

In closing this chapter, it is important to consider the physiological significance of the interactions between biological rhythms and the HPG/HPA axes. What are the potential advantages that this tight integration would provide for an organism? Biological rhythms that are endogenously generated, such as circadian rhythms, provide a mechanism through which recurring events in physiology and the environment can be predicted. Thus, organisms need not simply reactively respond to these regular occurrences but can instead anticipate them. It is thought that this ability provides more efficient regulation of physiological and behavioral processes, thereby ensuring that resource usage is maximized. This hypothesis is logically appealing, and both the HPG and HPA axes provide excellent models to test its validity.

With regard to the HPG and the ovulatory cycle, biological rhythms are critical both on the seasonal and circadian timescales, since ovulation only happens when several different cyclical factors (e.g., estradiol and luteinizing hormone levels) occur in a particular sequence and interval. If these events occur in a different temporal order, ovulation fails, and fertility is thus reduced (Angelopoulou et al. 2019; Beymer et al. 2016; Simonneaux et al. 2017). In species that have limited windows of fertility (e.g., spontaneous ovulators), it is critical to ensure that sexual motivation is timed to occur at maximal fertility, and the sensitivity to estradiol levels is important to ensure that the follicle is mature before ovulation. Similarly, in the context of HPA function, it is hypothesized that the daily rise in corticosteroids at the time of waking permits a greater mobilization of glucose when the animal is beginning the active phase. Many circadian "clock genes" have GREs in their promoter regions (Reddy et al. 2007, 2012), permitting interaction between plasma glucocorticoids and local molecular circadian clocks in tissues, thereby adjusting local timing at the tissue level. For example, the daily rise in corticosteroids may help synchronize peripheral clocks in the gut so they are prepared to receive nutrients. Conversely, the lack of GR in the SCN seems to provide the master circadian clock some protection against acute stressor-induced phase shifts.

Finally, there is cross talk between the HPA and HPG, with the aspects of the HPG axis being sensitive to stress mediators related to HPA activity and metabolism. RFRP-3 has been shown to affect, and be affected by, both stress and metabolic signals (Takayanagi and Onaka 2010; Schneider et al. 2017), and both KISS and RFRP-3 can modulate body mass in several species (Cazarez-Marquez et al. 2019; Talbi et al. 2016), and KISS1 signaling is important in linking fertility and nutritional state (Padilla et al. 2017, 2019). GnRH neurons are also sensitive to glucose (Roland and Moenter 2011a, b), further linking HPG, HPA, circadian, seasonal, and metabolic systems in interlocking loops of regulation. Many aspects of this model for the adaptive significance of rhythms in the HPG and HPA remain experimentally untested and may well be nearly impossible to dissect. While laboratory experiments that disrupt the temporal organization of these systems do show negative outcomes, as discussed above, naturalistic studies are far more difficult to undertake. Seasonality (with regard to reproduction) is an excellent example of a natural phenomenon that can be modeled in the laboratory. While a detailed discussion of seasonality is beyond the scope of this chapter, the changing duration and intensity of light over the year, coupled with changes in temperature and food availability, provide an important cue to animals about changing environmental circumstances that could affect reproductive success (Revel et al. 2007; Simonneaux et al. 2012; Dardente et al. 2019). For instance, it may not be best to give birth at a time when food will be scarce (e.g., the winter), so fertility is highest at times of year such that parturition occurs at an optimal time to make use of environmental resources. A similar set of processes can happen in more acute situations. For example, it may not be wise to spend energy on reproductive function if the organism is coping with a stressor in the environment that immediately threatens survival (e.g., forest fire, major storm), and hence HPA outputs can inhibit HPG function. More mechanistic work still needs to be done to understand the functional significance of the circadian clocks at the different levels of both the HPG and HPA axes and how they contribute directly to some of these adaptive processes, but the preponderance of the evidence strongly supports an important physiologic role for this tight temporal regulation.

5 Summary

Seasonal reproductive rhythms in animals have been observed for millennia. It is only more recently that circadian rhythms in neuroendocrine function have emerged as an area of focused research. It has become clear that the circadian clock regulates rhythms in the HPG and HPA axes through a variety of control systems (Fig. 4), from a central timekeeping mechanism in the SCN, to circadian clocks at the level of important neuroendocrine brain regions, and finally to circadian clocks in the endocrine tissues themselves. These rhythms in endocrine function have significant impacts on the health and well-being of animals, impacting everything from

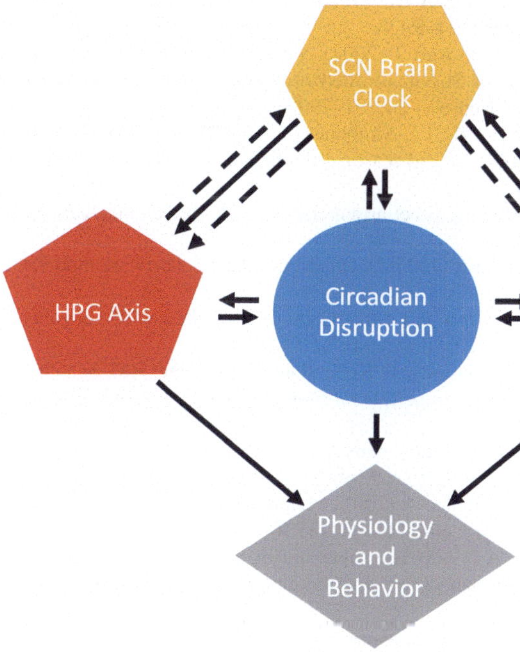

Fig. 4 The circadian system and neuroendocrine regulation. This schematic depicts that the SCN brain clock can regulate both the HPG and HPA axes through neural (solid arrows) and diffusible (dotted arrows) signals, which then impact physiology and behavior. Similarly, the outputs of the HPG and HPA can have effects back upon the SCN clock. These interrelated loops ensure that physiology, behavior, and recurring environmental events are synchronized to promote optimal function. Circadian disruption is an insidious factor that can drive changes in these carefully balanced systems, leading to disrupted rhythms in HPG and HPA function that could lead to a variety of mental and physical health issues, from depression to infertility

reproduction, to metabolism, to stress and immune function. They also feed back to the circadian system, both directly and indirectly. Thus, they form an interconnected loop that intimately links circadian rhythms and neuroendocrine function. In the past decades, it has become apparent that this loop can be disrupted, with consequences for both mental and physical health (Karatsoreos 2012, 2014; Karatsoreos and McEwen 2011; Oliver et al. 2012). It is critical to continue investigating the specific mechanisms that regulate these interacting loops so that we may be able to develop countermeasures or treatments when they become desynchronized with each other and with the solar day.

References

Akhtar RA, Reddy AB, Maywood ES, Clayton JD, King VM, Smith AG, Gant TW, Hastings MH, Kyriacou CP (2002) Circadian cycling of the mouse liver transcriptome, as revealed by cDNA microarray, is driven by the suprachiasmatic nucleus. Curr Biol 12:540–550

Albers HE, Yogev L, Todd RB, Goldman BD (1985) Adrenal corticoids in hamsters: role in circadian timing. Am J Phys 248:R434–R438

Amir S, Lamont EW, Robinson B, Stewart J (2004) A circadian rhythm in the expression of PERIOD2 protein reveals a novel SCN-controlled oscillator in the oval nucleus of the bed nucleus of the stria terminalis. J Neurosci 24:781–790

Angelopoulou E, Quignon C, Kriegsfeld LJ, Simonneaux V (2019) Functional implications of RFRP-3 in the central control of daily and seasonal rhythms in reproduction. Front Endocrinol (Lausanne) 10:183

Antle MC, Silver R (2005) Orchestrating time: arrangements of the brain circadian clock. Trends Neurosci 28:145–151

Antle MC, Foley DK, Foley NC, Silver R (2003) Gates and oscillators: a network model of the brain clock. J Biol Rhythm 18:339–350

Antle MC, Foley NC, Foley DK, Silver R (2007) Gates and oscillators II: zeitgebers and the network model of the brain clock. J Biol Rhythm 22:14–25

Balsalobre A, Brown SA, Marcacci L, Tronche F, Kellendonk C, Reichardt HM, Schutz G, Schibler U (2000) Resetting of circadian time in peripheral tissues by glucocorticoid signaling. Science 289:2344–2347

Beymer M, Henningsen J, Bahougne T, Simonneaux V (2016) The role of kisspeptin and RFRP in the circadian control of female reproduction. Mol Cell Endocrinol 438:89–99

Bittman EL, Doherty L, Huang L, Paroskie A (2003) Period gene expression in mouse endocrine tissues. Am J Physiol Regul Integr Comp Physiol 285:R561–R569

Buijs RM, Hermes MH, Kalsbeek A (1998) The suprachiasmatic nucleus-paraventricular nucleus interactions: a bridge to the neuroendocrine and autonomic nervous system. Prog Brain Res 119:365–382

Buijs RM, Wortel J, Van Heerikhuize JJ, Feenstra MG, Ter Horst GJ, Romijn HJ, Kalsbeek A (1999) Anatomical and functional demonstration of a multisynaptic suprachiasmatic nucleus adrenal (cortex) pathway. Eur J Neurosci 11:1535–1544

Buijs RM, van Eden CG, Goncharuk VD, Kalsbeek A (2003) The biological clock tunes the organs of the body: timing by hormones and the autonomic nervous system. J Endocrinol 177:17–26

Butler MP, Karatsoreos IN, LeSauter J, Silver R (2012) Dose-dependent effects of androgens on the circadian timing system and its response to light. Endocrinology 153:2344–2352

Cazarez-Marquez F, Milesi S, Laran-Chich MP, Klosen P, Kalsbeek A, Simonneaux V (2019) Kisspeptin and RFRP3 modulate body mass in Phodopus sungorus via two different neuroendocrine pathways. J Neuroendocrinol 31:e12710

Chappell PE, White RS, Mellon PL (2003) Circadian gene expression regulates pulsatile gonadotropin-releasing hormone (GnRH) secretory patterns in the hypothalamic GnRH-secreting GT1-7 cell line. J Neurosci 23:11202–11213
Christian CA, Moenter SM (2008) Vasoactive intestinal polypeptide can excite gonadotropin-releasing hormone neurons in a manner dependent on estradiol and gated by time of day. Endocrinology 149:3130–3136
Christian CA, Mobley JL, Moenter SM (2005) Diurnal and estradiol-dependent changes in gonadotropin-releasing hormone neuron firing activity. Proc Natl Acad Sci U S A 102:15682–15687
Chu A, Zhu L, Blum ID, Mai O, Leliavski A, Fahrenkrug J, Oster H, Boehm U, Storch KF (2013) Global but not gonadotrope-specific disruption of Bmal1 abolishes the luteinizing hormone surge without affecting ovulation. Endocrinology 154:2924–2935
Colwell CS (2000) Rhythmic coupling among cells in the suprachiasmatic nucleus. J Neurobiol 43:379–388
Czeisler CA, Klerman EB (1999) Circadian and sleep-dependent regulation of hormone release in humans. Recent Prog Horm Res 54:97–130. Discussion 30–32
Daan S, Damassa D, Pittendrigh CS, Smith ER (1975) An effect of castration and testosterone replacement on a circadian pacemaker in mice (Mus musculus). Proc Natl Acad Sci U S A 72:3744–3747
Dardente H, Wood S, Ebling F, Saenz de Miera C (2019) An integrative view of mammalian seasonal neuroendocrinology. J Neuroendocrinol 31:e12729
De La Iglesia HO, Blaustein JD, Bittman EL (1999) Oestrogen receptor-alpha-immunoreactive neurones project to the suprachiasmatic nucleus of the female Syrian hamster. J Neuroendocrinol 11:481–490
de la Iglesia HO, Meyer J, Carpino A Jr, Schwartz WJ (2000) Antiphase oscillation of the left and right suprachiasmatic nuclei. Science 290:799–801
De la Iglesia HO, Meyer J, Schwartz WJ (2003) Lateralization of circadian pacemaker output: activation of left- and right-sided luteinizing hormone-releasing hormone neurons involves a neural rather than a humoral pathway. J Neurosci 23:7412–7414
de Roux N, Genin E, Carel JC, Matsuda F, Chaussain JL, Milgrom E (2003) Hypogonadotropic hypogonadism due to loss of function of the KiSS1-derived peptide receptor GPR54. Proc Natl Acad Sci U S A 100:10972–10976
Dickmeis T (2009) Glucocorticoids and the circadian clock. J Endocrinol 200:3–22
Dror T, Franks J, Kauffman AS (2013) Analysis of multiple positive feedback paradigms demonstrates a complete absence of LH surges and GnRH activation in mice lacking kisspeptin signaling. Biol Reprod 88:146
Dubey AK, Puri CP, Puri V, Anand Kumar TC (1983) Day and night levels of hormones in male rhesus monkeys kept under controlled or constant environmental light. Experientia 39:207–209
Dubocovich ML, Markowska M (2005) Functional MT1 and MT2 melatonin receptors in mammals. Endocrine 27:101–110
Dubocovich ML, Hudson RL, Sumaya IC, Masana MI, Manna E (2005) Effect of MT1 melatonin receptor deletion on melatonin-mediated phase shift of circadian rhythms in the C57BL/6 mouse. J Pineal Res 39:113–120
Ebihara S, Marks T, Hudson DJ, Menaker M (1986) Genetic control of melatonin synthesis in the pineal gland of the mouse. Science 231:491–493
Egli M, Bertram R, Sellix MT, Freeman ME (2004) Rhythmic secretion of prolactin in rats: action of oxytocin coordinated by vasoactive intestinal polypeptide of suprachiasmatic nucleus origin. Endocrinology 145:3386–3394
Evans JA, Leise TL, Castanon-Cervantes O, Davidson AJ (2013) Dynamic interactions mediated by nonredundant signaling mechanisms couple circadian clock neurons. Neuron 80:973–983
Fernandez-Guasti A, Kruijver FP, Fodor M, Swaab DF (2000) Sex differences in the distribution of androgen receptors in the human hypothalamus. J Comp Neurol 425:422–435

Fitzgerald K, Zucker I (1976) Circadian organization of the estrous cycle of the golden hamster. Proc Natl Acad Sci U S A 73:2923–2927

Foradori CD, Coolen LM, Fitzgerald ME, Skinner DC, Goodman RL, Lehman MN (2002) Colocalization of progesterone receptors in parvicellular dynorphin neurons of the ovine preoptic area and hypothalamus. Endocrinology 143:4366–4374

Funabashi T, Shinohara K, Mitsushima D, Kimura F (2000) Gonadotropin-releasing hormone exhibits circadian rhythm in phase with arginine-vasopressin in co-cultures of the female rat preoptic area and suprachiasmatic nucleus. J Neuroendocrinol 12:521–528

Gerhold LM, Horvath TL, Freeman ME (2001) Vasoactive intestinal peptide fibers innervate neuroendocrine dopaminergic neurons. Brain Res 919:48–56

Gibson EM, Humber SA, Jain S, Williams WP 3rd, Zhao S, Bentley GE, Tsutsui K, Kriegsfeld LJ (2008) Alterations in RFamide-related peptide expression are coordinated with the preovulatory luteinizing hormone surge. Endocrinology 149:4958–4969

Gillespie JM, Chan BP, Roy D, Cai F, Belsham DD (2003) Expression of circadian rhythm genes in gonadotropin-releasing hormone-secreting GT1-7 neurons. Endocrinology 144:5285–5292

Goodman RL, Lehman MN, Smith JT, Coolen LM, de Oliveira CV, Jafarzadehshirazi MR, Pereira A, Iqbal J, Caraty A, Ciofi P, Clarke IJ (2007) Kisspeptin neurons in the arcuate nucleus of the ewe express both dynorphin A and neurokinin B. Endocrinology 148:5752–5760

Gray GD, Soderstein P, Tallentire D, Davidson JM (1978) Effects of lesions in various structures of the suprachiasmatic-preoptic region on LH regulation and sexual behavior in female rats. Neuroendocrinology 25:174–191

Green DJ, Gillette R (1982) Circadian rhythm of firing rate recorded from single cells in the rat suprachiasmatic brain slice. Brain Res 245:198–200

Groos G, Hendriks J (1982) Circadian rhythms in electrical discharge of rat suprachiasmatic neurones recorded *in vitro*. Neurosci Lett 34:283–288

Guler AD, Ecker JL, Lall GS, Haq S, Altimus CM, Liao HW, Barnard AR, Cahill H, Badea TC, Zhao H, Hankins MW, Berson DM, Lucas RJ, Yau KW, Hattar S (2008) Melanopsin cells are the principal conduits for rod-cone input to non-image-forming vision. Nature 453:102–105

Gundlah C, Kohama SG, Mirkes SJ, Garyfallou VT, Urbanski HF, Bethea CL (2000) Distribution of estrogen receptor beta (ERbeta) mRNA in hypothalamus, midbrain and temporal lobe of spayed macaque: continued expression with hormone replacement. Brain Res Mol Brain Res 76:191–204

Guo H, Brewer JM, Lehman MN, Bittman EL (2006) Suprachiasmatic regulation of circadian rhythms of gene expression in hamster peripheral organs: effects of transplanting the pacemaker. J Neurosci 26:6406–6412

Hakim H, DeBernardo AP, Silver R (1991) Circadian locomotor rhythms, but not photoperiodic responses, survive surgical isolation of the SCN in hamsters. J Biol Rhythm 6:97–113

Hamada T, LeSauter J, Venuti JM, Silver R (2001) Expression of period genes: rhythmic and nonrhythmic compartments of the suprachiasmatic nucleus pacemaker. J Neurosci 21:7742–7750

Han SK, Gottsch ML, Lee KJ, Popa SM, Smith JT, Jakawich SK, Clifton DK, Steiner RA, Herbison AE (2005) Activation of gonadotropin-releasing hormone neurons by kisspeptin as a neuroendocrine switch for the onset of puberty. J Neurosci 25:11349–11356

Hattar S, Lucas RJ, Mrosovsky N, Thompson S, Douglas RH, Hankins MW, Lem J, Biel M, Hofmann F, Foster RG, Yau KW (2003) Melanopsin and rod-cone photoreceptive systems account for all major accessory visual functions in mice. Nature 424:76–81

Hazlerigg D (2012) The evolutionary physiology of photoperiodism in vertebrates. Prog Brain Res 199:413–422

Hazlerigg DG, Wagner GC (2006) Seasonal photoperiodism in vertebrates: from coincidence to amplitude. Trends Endocrinol Metab 17:83–91

Herzog ED (2007) Neurons and networks in daily rhythms. Nat Rev Neurosci 8:790–802

Hileman SM, Handa RJ, Jackson GL (1999) Distribution of estrogen receptor-beta messenger ribonucleic acid in the male sheep hypothalamus. Biol Reprod 60:1279–1284

Horvath TL, Cela V, van der Beek EM (1998) Gender-specific apposition between vasoactive intestinal peptide-containing axons and gonadotrophin-releasing hormone-producing neurons in the rat. Brain Res 795:277–281
Inouye ST, Kawamura H (1979) Persistence of circadian rhythmicity in a mammalian hypothalamic "island" containing the suprachiasmatic nucleus. Proc Natl Acad Sci U S A 76:5962–5966
Irwig MS, Fraley GS, Smith JT, Acohido BV, Popa SM, Cunningham MJ, Gottsch ML, Clifton DK, Steiner RA (2004) Kisspeptin activation of gonadotropin releasing hormone neurons and regulation of KiSS-1 mRNA in the male rat. Neuroendocrinology 80:264–272
Iwahana E, Karatsoreos I, Shibata S, Silver R (2008) Gonadectomy reveals sex differences in circadian rhythms and suprachiasmatic nucleus androgen receptors in mice. Horm Behav 53:422–430
Jiang ZG, Yang YQ, Allen CN (1997) Tracer and electrical coupling of rat suprachiasmatic nucleus neurons. Neuroscience 77:1059–1066
Johnson RF, Moore RY, Morin LP (1988) Loss of entrainment and anatomical plasticity after lesions of the hamster retinohypothalamic tract. Brain Res 460:297–313
Johnston JD, Skene DJ (2015) 60 YEARS OF NEUROENDOCRINOLOGY: regulation of mammalian neuroendocrine physiology and rhythms by melatonin. J Endocrinol 226:T187–T198
Kalamatianos T, Kallo I, Goubillon ML, Coen CW (2004) Cellular expression of V1a vasopressin receptor mRNA in the female rat preoptic area: effects of oestrogen. J Neuroendocrinol 16:525–533
Kalsbeek A, Buijs RM (2002) Output pathways of the mammalian suprachiasmatic nucleus: coding circadian time by transmitter selection and specific targeting. Cell Tissue Res 309:109–118
Kalsbeek A, van Heerikhuize JJ, Wortel J, Buijs RM (1996) A diurnal rhythm of stimulatory input to the hypothalamo-pituitary-adrenal system as revealed by timed intrahypothalamic administration of the vasopressin V1 antagonist. J Neurosci 16:5555–5565
Kalsbeek A, Fliers E, Franke AN, Wortel J, Buijs RM (2000) Functional connections between the suprachiasmatic nucleus and the thyroid gland as revealed by lesioning and viral tracing techniques in the rat. Endocrinology 141:3832–3841
Kalsbeek A, Ruiter M, La Fleur SE, Van Heijningen C, Buijs RM (2003) The diurnal modulation of hormonal responses in the rat varies with different stimuli. J Neuroendocrinol 15:1144–1155
Karatsoreos IN (2012) Effects of circadian disruption on mental and physical health. Curr Neurol Neurosci Rep 12:218–225
Karatsoreos IN (2014) Links between circadian rhythms and psychiatric disease. Front Behav Neurosci 8:162
Karatsoreos IN, McEwen BS (2011) Psychobiological allostasis: resistance, resilience and vulnerability. Trends Cogn Sci 15:576–584
Karatsoreos IN, Silver R (2007) Minireview: the neuroendocrinology of the suprachiasmatic nucleus as a conductor of body time in mammals. Endocrinology 148:5640–5647
Karatsoreos IN, Wang A, Sasanian J, Silver R (2007) A role for androgens in regulating circadian behavior and the suprachiasmatic nucleus. Endocrinology 148:5487–5495
Khan AR, Kauffman AS (2012) The role of kisspeptin and RFamide-related peptide-3 neurones in the circadian-timed preovulatory luteinising hormone surge. J Neuroendocrinol 24:131–143
Klein DC, Moore RY (1979) Pineal N-acetyltransferase and hydroxyindole-O-methyltransferase: control by the retinohypothalamic tract and the suprachiasmatic nucleus. Brain Res 174:245–262
Koike N, Yoo SH, Huang HC, Kumar V, Lee C, Kim TK, Takahashi JS (2012) Transcriptional architecture and chromatin landscape of the core circadian clock in mammals. Science 338:349–354
Kovanen L, Saarikoski ST, Aromaa A, Lonnqvist J, Partonen T (2010) ARNTL (BMAL1) and NPAS2 gene variants contribute to fertility and seasonality. PLoS One 5:e10007
Kriegsfeld LJ (2006) Driving reproduction: RFamide peptides behind the wheel. Horm Behav 50:655–666

Kriegsfeld LJ, Silver R, Gore AC, Crews D (2002) Vasoactive intestinal polypeptide contacts on gonadotropin-releasing hormone neurones increase following puberty in female rats. J Neuroendocrinol 14:685–690
Kriegsfeld LJ, Leak RK, Yackulic CB, LeSauter J, Silver R (2004) Organization of suprachiasmatic nucleus projections in Syrian hamsters (Mesocricetus auratus): an anterograde and retrograde analysis. J Comp Neurol 468:361–379
Lamont EW, Robinson B, Stewart J, Amir S (2005) The central and basolateral nuclei of the amygdala exhibit opposite diurnal rhythms of expression of the clock protein Period2. Proc Natl Acad Sci U S A 102:4180–4184
Leak RK, Moore RY (2001) Topographic organization of suprachiasmatic nucleus projection neurons. J Comp Neurol 433:312–334
Lehman MN, Silver R, Gladstone WR, Kahn RM, Gibson M, Bittman EL (1987) Circadian rhythmicity restored by neural transplant. Immunocytochemical characterization of the graft and its integration with the host brain. J Neurosci 7:1626–1638
Lehman MN, Coolen LM, Goodman RL (2010a) Minireview: kisspeptin/neurokinin B/dynorphin (KNDy) cells of the arcuate nucleus: a central node in the control of gonadotropin-releasing hormone secretion. Endocrinology 151:3479–3489
Lehman MN, Merkley CM, Coolen LM, Goodman RL (2010b) Anatomy of the kisspeptin neural network in mammals. Brain Res 1364:90–102
Lehman MN, Hileman SM, Goodman RL (2013) Neuroanatomy of the kisspeptin signaling system in mammals: comparative and developmental aspects. Adv Exp Med Biol 784:27–62
Levi F (2006) Chronotherapeutics: the relevance of timing in cancer therapy. Cancer Causes Control 17:611–621
Lewy AJ (2007) Melatonin and human chronobiology. Cold Spring Harb Symp Quant Biol 72:623–636
Lewy AJ, Sack RL (1989) The dim light melatonin onset as a marker for circadian phase position. Chronobiol Int 6:93–102
Liston C, Gan WB (2011) Glucocorticoids are critical regulators of dendritic spine development and plasticity in vivo. Proc Natl Acad Sci U S A 108:16074–16079
Liu RY, Unmehopa UA, Zhou JN, Swaab DF (2006) Glucocorticoids suppress vasopressin gene expression in human suprachiasmatic nucleus. J Steroid Biochem Mol Biol 98:248–253
Liu X, Lee K, Herbison AE (2008) Kisspeptin excites gonadotropin-releasing hormone neurons through a phospholipase C/calcium-dependent pathway regulating multiple ion channels. Endocrinology 149:4605–4614
Lok R, van Koningsveld MJ, Gordijn MCM, Beersma DGM, Hut RA (2019) Daytime melatonin and light independently affect human alertness and body temperature. J Pineal Res 67:e12583
Long MA, Jutras MJ, Connors BW, Burwell RD (2005) Electrical synapses coordinate activity in the suprachiasmatic nucleus. Nat Neurosci 8:61–66
Lucas RJ, Hattar S, Takao M, Berson DM, Foster RG, Yau KW (2003) Diminished pupillary light reflex at high irradiances in melanopsin-knockout mice. Science 299:245–247
Maurel D, Sage D, Mekaouche M, Bosler O (2000) Glucocorticoids up-regulate the expression of glial fibrillary acidic protein in the rat suprachiasmatic nucleus. Glia 29:212–221
Maywood ES, Chesham JE, O'Brien JA, Hastings MH (2011) A diversity of paracrine signals sustains molecular circadian cycling in suprachiasmatic nucleus circuits. Proc Natl Acad Sci U S A 108:14306–14311
Messager S, Chatzidaki EE, Ma D, Hendrick AG, Zahn D, Dixon J, Thresher RR, Malinge I, Lomet D, Carlton MB, Colledge WH, Caraty A, Aparicio SA (2005) Kisspeptin directly stimulates gonadotropin-releasing hormone release via G protein-coupled receptor 54. Proc Natl Acad Sci U S A 102:1761–1766
Meyer-Bernstein EL, Jetton AE, Matsumoto SI, Markuns JF, Lehman MN, Bittman EL (1999) Effects of suprachiasmatic transplants on circadian rhythms of neuroendocrine function in golden hamsters. Endocrinology 140:207–218

Miller BH, Olson SL, Turek FW, Levine JE, Horton TH, Takahashi JS (2004) Circadian clock mutation disrupts estrous cyclicity and maintenance of pregnancy. Curr Biol 14:1367–1373
Miller BH, Olson SL, Levine JE, Turek FW, Horton TH, Takahashi JS (2006) Vasopressin regulation of the proestrous luteinizing hormone surge in wild-type and clock mutant mice. Biol Reprod 75:778–784
Model Z, Butler MP, LeSauter J, Silver R (2015) Suprachiasmatic nucleus as the site of androgen action on circadian rhythms. Horm Behav 73:1–7
Moore RY, Eichler VB (1972) Loss of a circadian adrenal corticosterone rhythm following suprachiasmatic lesions in the rat. Brain Res 42:201–206
Moore RY, Klein DC (1974) Visual pathways and the central neural control of a circadian rhythm in pineal serotonin N-acetyltransferase activity. Brain Res 71:17–33
Moore RY, Lenn NJ (1972) A retinohypothalamic projection in the rat. J Comp Neurol 146:1–14
Morin LP (2013) Neuroanatomy of the extended circadian rhythm system. Exp Neurol 243:4–20
Morin LP, Fitzgerald KM, Zucker I (1977) Estradiol shortens the period of hamster circadian rhythms. Science 196:305–307
Nishiwaki-Ohkawa T, Yoshimura T (2016) Molecular basis for regulating seasonal reproduction in vertebrates. J Endocrinol 229:R117–R127
Nunez AA, Stephan FK (1977) The effects of hypothalamic knife cuts on drinking rhythms and the estrus cycle of the rat. Behav Biol 20:224–234
Oishi K, Shirai H, Ishida N (2005) CLOCK is involved in the circadian transactivation of peroxisome-proliferator-activated receptor alpha (PPARalpha) in mice. Biochem J 386:575–581
Oliver PL, Sobczyk MV, Maywood ES, Edwards B, Lee S, Livieratos A, Oster H, Butler R, Godinho SI, Wulff K, Peirson SN, Fisher SP, Chesham JE, Smith JW, Hastings MH, Davies KE, Foster RG (2012) Disrupted circadian rhythms in a mouse model of schizophrenia. Curr Biol 22:314–319
Ono D, Honma K, Honma S (2015) Circadian and ultradian rhythms of clock gene expression in the suprachiasmatic nucleus of freely moving mice. Sci Rep 5:12310
Oster H, Damerow S, Kiessling S, Jakubcakova V, Abraham D, Tian J, Hoffmann MW, Eichele G (2006) The circadian rhythm of glucocorticoids is regulated by a gating mechanism residing in the adrenal cortical clock. Cell Metab 4:163–173
Ottenweller JE, Tapp WN, Pitman DL, Natelson BH (1987) Adrenal, thyroid, and testicular hormone rhythms in male golden hamsters on long and short days. Am J Phys 253:R321–R328
Padilla SL, Qiu J, Nestor CC, Zhang C, Smith AW, Whiddon BB, Ronnekleiv OK, Kelly MJ, Palmiter RD (2017) AgRP to Kiss1 neuron signaling links nutritional state and fertility. Proc Natl Acad Sci U S A 114:2413–2418
Padilla SL, Perez JG, Ben-Hamo M, Johnson CW, Sanchez REA, Bussi IL, Palmiter RD, de la Iglesia HO (2019) Kisspeptin neurons in the Arcuate nucleus of the hypothalamus orchestrate circadian rhythms and metabolism. Curr Biol 29:592–604 e4
Panda S, Antoch MP, Miller BH, Su AI, Schook AB, Straume M, Schultz PG, Kay SA, Takahashi JS, Hogenesch JB (2002a) Coordinated transcription of key pathways in the mouse by the circadian clock. Cell 109:307–320
Panda S, Sato TK, Castrucci AM, Rollag MD, DeGrip WJ, Hogenesch JB, Provencio I, Kay SA (2002b) Melanopsin (Opn4) requirement for normal light-induced circadian phase shifting. Science 298:2213–2216
Panda S, Provencio I, Tu DC, Pires SS, Rollag MD, Castrucci AM, Pletcher MT, Sato TK, Wiltshire T, Andahazy M, Kay SA, Van Gelder RN, Hogenesch JB (2003) Melanopsin is required for non-image-forming photic responses in blind mice. Science 301:525–527
Pauls S, Foley NC, Foley DK, LeSauter J, Hastings MH, Maywood ES, Silver R (2014) Differential contributions of intra-cellular and inter-cellular mechanisms to the spatial and temporal architecture of the suprachiasmatic nucleus circadian circuitry in wild-type, cryptochrome-null and vasoactive intestinal peptide receptor 2-null mutant mice. Eur J Neurosci 40:2528–2540
Piet R, Dunckley H, Lee K, Herbison AE (2016) Vasoactive intestinal peptide excites GnRH neurons in male and female mice. Endocrinology 157:3621–3630
Prosser RA, Edgar DM, Heller HC, Miller JD (1994) A possible glial role in the mammalian circadian clock. Brain Res 643:296–301

Ralph MR, Foster RG, Davis FC, Menaker M (1990) Transplanted suprachiasmatic nucleus determines circadian period. Science 247:975–978

Rash JE, Olson CO, Pouliot WA, Davidson KG, Yasumura T, Furman CS, Royer S, Kamasawa N, Nagy JI, Dudek FE (2007) Connexin36 vs. connexin32, "miniature" neuronal gap junctions, and limited electrotonic coupling in rodent suprachiasmatic nucleus. Neuroscience 149:350–371

Reddy AB, Maywood ES, Karp NA, King VM, Inoue Y, Gonzalez FJ, Lilley KS, Kyriacou CP, Hastings MH (2007) Glucocorticoid signaling synchronizes the liver circadian transcriptome. Hepatology 45:1478–1488

Reddy TE, Gertz J, Crawford GE, Garabedian MJ, Myers RM (2012) The hypersensitive glucocorticoid response specifically regulates period 1 and expression of circadian genes. Mol Cell Biol 32:3756–3767

Reiter RJ, Tan DX (2002) Role of CSF in the transport of melatonin. J Pineal Res 33:61

Revel FG, Ansel L, Klosen P, Saboureau M, Pevet P, Mikkelsen JD, Simonneaux V (2007) Kisspeptin: a key link to seasonal breeding. Rev Endocr Metab Disord 8:57–65

Robertson JL, Clifton DK, de la Iglesia HO, Steiner RA, Kauffman AS (2009) Circadian regulation of Kiss1 neurons: implications for timing the preovulatory gonadotropin-releasing hormone/luteinizing hormone surge. Endocrinology 150:3664–3671

Roland AV, Moenter SM (2011a) Glucosensing by GnRH neurons: inhibition by androgens and involvement of AMP-activated protein kinase. Mol Endocrinol 25:847–858

Roland AV, Moenter SM (2011b) Regulation of gonadotropin-releasing hormone neurons by glucose. Trends Endocrinol Metab 22:443–449

Roseboom PH, Namboodiri MA, Zimonjic DB, Popescu NC, Rodriguez IR, Gastel JA, Klein DC (1998) Natural melatonin 'knockdown' in C57BL/6J mice: rare mechanism truncates serotonin N-acetyltransferase. Brain Res Mol Brain Res 63:189–197

Rosenfeld P, Van Eekelen JA, Levine S, De Kloet ER (1988) Ontogeny of the type 2 glucocorticoid receptor in discrete rat brain regions: an immunocytochemical study. Brain Res 470:119–127

Ruby NF, Brennan TJ, Xie X, Cao V, Franken P, Heller HC, O'Hara BF (2002) Role of melanopsin in circadian responses to light. Science 298:2211–2213

Russell GM, Kalafatakis K, Lightman SL (2015) The importance of biological oscillators for hypothalamic-pituitary-adrenal activity and tissue glucocorticoid response: coordinating stress and neurobehavioural adaptation. J Neuroendocrinol 27:378–388

Russo KA, La JL, Stephens SB, Poling MC, Padgaonkar NA, Jennings KJ, Piekarski DJ, Kauffman AS, Kriegsfeld LJ (2015) Circadian control of the female reproductive axis through gated responsiveness of the RFRP-3 system to VIP signaling. Endocrinology 156:2608–2618

Sachar EJ, Hellman L, Roffwarg HP, Halpern FS, Fukushima DK, Gallagher TF (1973) Disrupted 24-hour patterns of cortisol secretion in psychotic depression. Arch Gen Psychiatry 28:19–24

Schafer D, Kane G, Colledge WH, Piet R, Herbison AE (2018) Sex- and sub region-dependent modulation of arcuate kisspeptin neurones by vasopressin and vasoactive intestinal peptide. J Neuroendocrinol 30:e12660

Schibler U, Gotic I, Saini C, Gos P, Curie T, Emmenegger Y, Sinturel F, Gosselin P, Gerber A, Fleury-Olela F, Rando G, Demarque M, Franken P (2015) Clock-talk: interactions between central and peripheral circadian oscillators in mammals. Cold Spring Harb Symp Quant Biol 80:223–232

Schirman-Hildesheim TD, Ben-Aroya N, Koch Y (2006) Daily GnRH and GnRH-receptor mRNA expression in the ovariectomized and intact rat. Mol Cell Endocrinol 252:120–125

Schneider JE, Benton NA, Russo KA, Klingerman CM, Williams WP 3rd, Simberlund J, Abdulhay A, Brozek JM, Kriegsfeld LJ (2017) RFamide-related peptide-3 and the trade-off between reproductive and ingestive behavior. Integr Comp Biol 57:1225–1239

Schwartz WJ, Coleman RJ, Reppert SM (1983) A daily vasopressin rhythm in rat cerebrospinal fluid. Brain Res 263:105–112

Segall LA, Perrin JS, Walker CD, Stewart J, Amir S (2006) Glucocorticoid rhythms control the rhythm of expression of the clock protein, Period2, in oval nucleus of the bed nucleus of the stria terminalis and central nucleus of the amygdala in rats. Neuroscience 140:753–757

Seminara SB, Messager S, Chatzidaki EE, Thresher RR, Acierno JS Jr, Shagoury JK, Bo-Abbas Y, Kuohung W, Schwinof KM, Hendrick AG, Zahn D, Dixon J, Kaiser UB, Slaugenhaupt SA, Gusella JF, O'Rahilly S, Carlton MB, Crowley WF Jr, Aparicio SA, Colledge WH (2003) The GPR54 gene as a regulator of puberty. N Engl J Med 349:1614–1627

Shibata S, Oomura Y, Kita H, Hattori K (1982) Circadian rhythmic changes of neuronal activity in the suprachiasmatic nucleus of the rat hypothalamic slice. Brain Res 247:154–158

Shinohara K, Funabashi T, Mitushima D, Kimura F (2000) Effects of gap junction blocker on vasopressin and vasoactive intestinal polypeptide rhythms in the rat suprachiasmatic nucleus in vitro. Neurosci Res 38:43–47

Shinohara K, Funabashi T, Nakamura TJ, Kimura F (2001) Effects of estrogen and progesterone on the expression of connexin-36 mRNA in the suprachiasmatic nucleus of female rats. Neurosci Lett 309:37–40

Shughrue PJ, Lane MV, Merchenthaler I (1997) Comparative distribution of estrogen receptor-alpha and -beta mRNA in the rat central nervous system. J Comp Neurol 388:507–525

Silver R, Lehman MN, Gibson M, Gladstone WR, Bittman EL (1990) Dispersed cell suspensions of fetal SCN restore circadian rhythmicity in SCN-lesioned adult hamsters. Brain Res 525:45–58

Silver R, LeSauter J, Tresco PA, Lehman MN (1996) A diffusible coupling signal from the transplanted suprachiasmatic nucleus controlling circadian locomotor rhythms. Nature 382:810–813

Simonneaux V, Ribelayga C (2003) Generation of the melatonin endocrine message in mammals: a review of the complex regulation of melatonin synthesis by norepinephrine, peptides, and other pineal transmitters. Pharmacol Rev 55:325–395

Simonneaux V, Bur I, Ancel C, Ansel L, Klosen P (2012) A kiss for daily and seasonal reproduction. Prog Brain Res 199:423–437

Simonneaux V, Bahougne T, Angelopoulou E (2017) Daily rhythms count for female fertility. Best Pract Res Clin Endocrinol Metab 31:505–519

Skinner DC, Caraty A (2002) Measurement and possible function of GnRH in cerebrospinal fluid in ewes. Reprod Suppl 59:25–39

Skinner DC, Malpaux B (1999) High melatonin concentrations in third ventricular cerebrospinal fluid are not due to Galen vein blood recirculating through the choroid plexus. Endocrinology 140:4399–4405

Smith JT (2008) Kisspeptin signalling in the brain: steroid regulation in the rodent and ewe. Brain Res Rev 57:288–298

Smith JT, Clay CM, Caraty A, Clarke IJ (2007) KiSS-1 messenger ribonucleic acid expression in the hypothalamus of the ewe is regulated by sex steroids and season. Endocrinology 148:1150–1157

Spiga F, Waite EJ, Liu Y, Kershaw YM, Aguilera G, Lightman SL (2011) ACTH-dependent ultradian rhythm of corticosterone secretion. Endocrinology 152:1448–1457

Spiga F, Walker JJ, Terry JR, Lightman SL (2014) HPA axis-rhythms. Compr Physiol 4:1273–1298

Stephan FK, Zucker I (1972) Rat drinking rhythms: central visual pathways and endocrine factors mediating responsiveness to environmental illumination. Physiol Behav 8:315–326

Swann JM, Turek FW (1985) Multiple circadian oscillators regulate the timing of behavioral and endocrine rhythms in female golden hamsters. Science 228:898–900

Takahashi JS (2017) Transcriptional architecture of the mammalian circadian clock. Nat Rev Genet 18:164–179

Takayanagi Y, Onaka T (2010) Roles of prolactin-releasing peptide and RFamide related peptides in the control of stress and food intake. FEBS J 277:4998–5005

Talbi R, Laran-Chich MP, Magoul R, El Ouezzani S, Simonneaux V (2016) Kisspeptin and RFRP-3 differentially regulate food intake and metabolic neuropeptides in the female desert jerboa. Sci Rep 6:36057

Tricoire H, Moller M, Chemineau P, Malpaux B (2003) Origin of cerebrospinal fluid melatonin and possible function in the integration of photoperiod. Reprod Suppl 61:311–321

Tsutsui K, Osugi T (2009) Evolutionary origin and divergence of GnIH and its homologous peptides. Gen Comp Endocrinol 161:30–33

Tsutsui K, Saigoh E, Ukena K, Teranishi H, Fujisawa Y, Kikuchi M, Ishii S, Sharp PJ (2000) A novel avian hypothalamic peptide inhibiting gonadotropin release. Biochem Biophys Res Commun 275:661–667

Tsutsui K, Saigoh E, Yin H, Ubuka T, Chowdhury VS, Osugi T, Ukena K, Sharp PJ, Wingfield JC, Bentley GE (2009) A new key neurohormone controlling reproduction, gonadotrophin-inhibitory hormone in birds: discovery, progress and prospects. J Neuroendocrinol 21:271–275

Tsutsui K, Osugi T, Son YL, Ubuka T (2018) Review: structure, function and evolution of GnIH. Gen Comp Endocrinol 264:48–57

Ungar F, Halberg F (1962) Circadian rhythm in the in vitro response of mouse adrenal to adrenocorticotropic hormone. Science 137:1058–1060

Van Cauter E, Refetoff S (1985) Multifactorial control of the 24-hour secretory profiles of pituitary hormones. J Endocrinol Investig 8:381–391

van der Beek EM, Wiegant VM, van der Donk HA, van den Hurk R, Buijs RM (1993) Lesions of the suprachiasmatic nucleus indicate the presence of a direct vasoactive intestinal polypeptide-containing projection to gonadotrophin-releasing hormone neurons in the female rat. J Neuroendocrinol 5:137–144

Van der Beek EM, Horvath TL, Wiegant VM, Van den Hurk R, Buijs RM (1997) Evidence for a direct neuronal pathway from the suprachiasmatic nucleus to the gonadotropin-releasing hormone system: combined tracing and light and electron microscopic immunocytochemical studies. J Comp Neurol 384:569–579

van der Horst GT, Muijtjens M, Kobayashi K, Takano R, Kanno S, Takao M, de Wit J, Verkerk A, Eker AP, van Leenen D, Buijs R, Bootsma D, Hoeijmakers JH, Yasui A (1999) Mammalian Cry1 and Cry2 are essential for maintenance of circadian rhythms. Nature 398:627–630

Vida B, Hrabovszky E, Kalamatianos T, Coen CW, Liposits Z, Kallo I (2008) Oestrogen receptor alpha and beta immunoreactive cells in the suprachiasmatic nucleus of mice: distribution, sex differences and regulation by gonadal hormones. J Neuroendocrinol 20:1270–1277

von Gall C, Lewy A, Schomerus C, Vivien-Roels B, Pevet P, Korf HW, Stehle JH (2000) Transcription factor dynamics and neuroendocrine signalling in the mouse pineal gland: a comparative analysis of melatonin-deficient C57BL mice and melatonin-proficient C3H mice. Eur J Neurosci 12:964–972

Vrang N, Larsen PJ, Mikkelsen JD (1995) Direct projection from the suprachiasmatic nucleus to hypophysiotrophic corticotropin-releasing factor immunoreactive cells in the paraventricular nucleus of the hypothalamus demonstrated by means of Phaseolus vulgaris-leucoagglutinin tract tracing. Brain Res 684:61–69

Walker JJ, Terry JR, Lightman SL (2010) Origin of ultradian pulsatility in the hypothalamic-pituitary-adrenal axis. Proc Biol Sci 277:1627–1633

Walker JJ, Spiga F, Waite E, Zhao Z, Kershaw Y, Terry JR, Lightman SL (2012) The origin of glucocorticoid hormone oscillations. PLoS Biol 10:e1001341

Weitzman ED, Fukushima D, Nogeire C, Roffwarg H, Gallagher TF, Hellman L (1971) Twenty-four hour pattern of the episodic secretion of cortisol in normal subjects. J Clin Endocrinol Metab 33:14–22

Welsh DK, Logothetis DE, Meister M, Reppert SM (1995) Individual neurons dissociated from rat suprachiasmatic nucleus express independently phased circadian firing rhythms. Neuron 14:697–706

Williams WP 3rd, Jarjisian SG, Mikkelsen JD, Kriegsfeld LJ (2011) Circadian control of kisspeptin and a gated GnRH response mediate the preovulatory luteinizing hormone surge. Endocrinology 152:595–606

Wong CC, Dohler KD, Geerlings H, von zur Muhlen A (1983) Influence of age, strain and season on circadian periodicity of pituitary, gonadal and adrenal hormones in the serum of male laboratory rats. Horm Res 17:202–215

Wotus C, Lilley TR, Neal AS, Suleiman NL, Schmuck SC, Smarr BL, Fischer BJ, de la Iglesia HO (2013) Forced desynchrony reveals independent contributions of suprachiasmatic oscillators to the daily plasma corticosterone rhythm in male rats. PLoS One 8:e68793

Yamazaki S, Numano R, Abe M, Hida A, Takahashi R, Ueda M, Block GD, Sakaki Y, Menaker M, Tei H (2000) Resetting central and peripheral circadian oscillators in transgenic rats. Science 288:682–685

Yan L, Karatsoreos I, Lesauter J, Welsh DK, Kay S, Foley D, Silver R (2007) Exploring spatiotemporal organization of SCN circuits. Cold Spring Harb Symp Quant Biol 72:527–541

Zhao D, Yu Y, Shen Y, Liu Q, Zhao Z, Sharma R, Reiter RJ (2019) Melatonin synthesis and function: evolutionary history in animals and plants. Front Endocrinol (Lausanne) 10:249

Neuroendocrine Control of Sleep

Philip C. Smith and Jessica A. Mong

Contents

1 Sleep Behavior Consists of Multiple Distinguishable States 354
2 The Circadian and Homeostatic Systems Drive Sleep Pressure 355
3 Diverse Chemical Mechanisms and Circuitry Govern Sleep-Pressure Systems 357
4 Several Hypotheses Exist as to the Functions of Sleep 359
5 Sleep Disorders Have a Broad Clinical Impact 361
6 Sleep Behavior and Circadian Timing Impact Multiple Hormonal Functions 363
7 Non-gonadal Endocrine Factors Impact Sleep 365
8 Biological Sex Differences and Ovarian Steroids Impact Sleep Behavior 366
9 Diverse Models Exist for the Investigation of Endocrine Regulation of Sleep Behavior 367
10 Animal Studies Illuminate Mechanisms Connecting Female Gonadal Hormones and Sleep 369
11 Conclusion 370
References 371

Abstract Sleep is a phenomenon in animal behavior as enigmatic as it is ubiquitous, and one deeply tied to endocrine function. Though there are still many unanswered questions about the neurochemical basis of sleep and its functions, extensive interactions have been identified between sleep and the endocrine system, in both the endocrine system's effect on sleep and sleep's effect on the endocrine system. Unfortunately, until recent years, much research on sleep behavior largely disregarded its connections with the endocrine system. Use of both clinical studies and rodent models to investigate interactions between neuroendocrine function, including biological sex, and sleep therefore presents a promising area of further exploration. Further investigation of the neurobiological and neuroendocrine basis of sleep could have wide impact on a number of clinical and basic science fields. In this review, we summarize the state of basic sleep biology and its connections to the field of neuroendocrine biology, as well as suggest key future directions for the

P. C. Smith (✉) and J. A. Mong
Department of Pharmacology, University of Maryland School of Medicine, Baltimore, MD, USA
e-mail: philip.smith@som.umaryland.edu

Curr Topics Behav Neurosci (2019) 43: 353–378
DOI 10.1007/7854_2019_107
Published Online: 9 August 2019

neuroendocrine regulation of sleep that may significantly impact new therapies for sleep disorders in women and men.

Keywords Endocrine · Sex differences · Sleep

Despite being among life's most common behavioral states, sleep remains a phenomenon that resists easy explanation. Sleep is generally defined as being characterized by an increased threshold for response to sensory input, a decrease in motor function, and a lack of consciousness. Though humans spend on average a third of their lifespan in the sleep state (Aminoff et al. 2011), there are still many unanswered questions about the neurochemical basis of sleep and its functions. Extensive interactions have been identified between sleep and the endocrine system, in both the endocrine system's effect on sleep and sleep's effect on the endocrine system. Numerous endocrine factors can affect sleep quantity and quality, while studies have shown a profound effect of sleep behavior on overall endocrine function and stability (Morgan and Tsai 2015; Spiegel et al. 1999). Further investigation of the neurobiological and neuroendocrine basis of sleep could have wide impact on a number of clinical and basic science fields, from treatment of insomnia to exploration of sex-distinct sleep differences to investigation of the pathogenesis of neurodegenerative diseases. In this review, we summarize the state of basic sleep biology and its connections to the field of neuroendocrine biology, as well as suggest key future directions for the neuroendocrine regulation of sleep that may significantly impact new therapies for sleep disorders in women and men.

1 Sleep Behavior Consists of Multiple Distinguishable States

Sleep consists of several distinct states, which can be distinguished by their patterns of brain activity (Saper et al. 2010). The most important distinction between sleep states is between Rapid Eye Movement (REM) and non-Rapid Eye Movement (non-REM) sleep. Non-REM sleep predominates at the outset of a particular sleep bout, and is distinguished by an ordering and synchronizing of brain activity (Mong and Cusmano 2016). This synchronization leads to a decrease in the frequency and increase in the amplitude of brain waves, causing waves in the delta (0–4 Hz) range to predominate; those waves are considered synonymous with slow wave activity (SWA) (Lanquart et al. 2018). During non-REM sleep, muscle activity is decreased relative to the wake state, but paralysis of skeletal muscles is not present (Mong and Cusmano 2016). In humans, further refinement of non-REM sleep can be achieved by separating it into distinct stages, numbered 1–3 in order of increasing depth of sleep. Stage 3 (redefined in 2007 from the prior stages 3 and 4) (Moser et al. 2009) is referred to as slow-wave sleep and represents the deepest sleep states.

In contrast, REM sleep, also known as paradoxical sleep, consists of highly disordered brain activity that somewhat mimics brain activity in the wake state. In this state, waves in the theta (4–8 Hz) range dominate (Hutchison and Rathore 2015), and skeletal muscles are paralyzed. REM sleep does not occur at the onset of a sleep bout in healthy animals, instead appearing later in the sleep bout after a period of slow-wave sleep has been completed (Saper et al. 2010). The differing functions of REM and non-REM sleep are poorly understood, and to date many sleep studies have focused on the aggregate time spent in sleep versus wake as their main metric. However, methods do exist for isolating REM or non-REM sleep. For example, the flowerpot method, in which an experimental animal is allowed to sleep on a small shelf, such as an upside down flowerpot, above a pool of water, selectively deprives the experimental subject of REM sleep only by prohibiting sleep during periods of muscle paralysis (Aalto and Kiianmaa 1984). These methods may become more prominent as further differences between the two states are elucidated.

2 The Circadian and Homeostatic Systems Drive Sleep Pressure

The biological circuitry of sleep is an area of intense inquiry, with many questions remaining on both the neuroanatomy and neurochemistry of the relevant pathways. This question is complicated by the existence of two distinct systems governing aspects of sleep regulation. These systems' net output is generally described as sleep pressure. Sleep pressure has been defined as the intrinsic need for sleep of a given animal at a given time (Eban-Rothschild et al. 2017). Beyond the familiar intrinsic feeling of sleepiness as a manifestation of sleep pressure, quantitative markers derived from EEG outputs exist that can approximate sleep pressure in a reproducible fashion (Mong and Cusmano 2016). The two sleep-pressure systems, known as the circadian wake system and the homeostatic sleep-pressure system, operate in parallel and in concert to generate an overall sleep pressure that is responsive to both the animal's intrinsic homeostatic needs as well as external factors such as the light–dark cycle.

The better-understood of the two systems which combine to govern sleep pressure is the circadian wake system. The circadian system orients sleep to the light and dark cycle, as well as consolidates sleep and wake into larger blocks. Circadian timing has two key properties. First, it has an endogenous rhythm with a period of approximately (though not exactly) 24 h (Abbott et al. 2015). Second, that rhythm can be shifted in response to external cues (Abbott et al. 2015). Light–dark cycles are both the most prevalent and potent of these cues, but other factors such as exercise, feeding, temperature, and certain pharmacological agents have been shown to entrain the system as well, in some cases maladaptively (Abbott et al. 2015). The key neurobiological regulator of the circadian sleep system is the suprachiasmatic nucleus (SCN) of the hypothalamus. Animals with lesions of the SCN have been shown to have as much total sleep time as controls, but sleep in unconsolidated random bouts unrelated to the light–dark cycle (Mouret et al. 1978). Transplantation of a donor SCN has been shown to rescue a normal phenotype in that regard (Sawaki et al. 1984). Additionally, studies of humans isolated from the natural light–dark cycle and left to sleep ad libitum show that those humans settle into a diurnal sleep pattern that approximates, but does not exactly mimic, the 24-h day, showing the intrinsic rhythmicity of the SCN. The SCN receives its principal entraining inputs from environmental light cues through specialized photosensitive ganglion cells in the retina. Importantly, the photoreceptors and cortical areas responsible for conscious vision are not involved, meaning that the circadian rhythm is reasonably well entrained in most blind animals (Squarcini et al. 2013). Importantly for the endocrine system, the circadian system serves to create "biological day" and "biological night". Hormones and other biological properties have been shown to fluctuate on a 24-h cycle according to stereotypical patterns; lesions of the circadian system have been shown to disrupt the daily fluctuation of hormone levels such as growth hormone (Steyn and Ngo 2017), cortisol (Challet 2015), and leptin (Challet 2015), among others (see Sect. 6).

The second sleep system, quite distinct from the circadian system, is the homeostatic pressure system. As the name suggests, the homeostatic sleep system governs the amount of sleep needed after a given period of wake to maintain homeostasis (Allada et al. 2017). The total amount of sleep needed for an animal in a given period of time tends to be quite consistent, and independent of both the circadian system and the light–dark cycle (Donlea 2017). This phenomenon is further exemplified by the need for recovery sleep, which is nearly always necessary after periods of sleep deprivation (Donlea 2017). Similarly, to other homeostatic systems such as temperature, extreme loss of homeostasis (such as through prolonged sleep deprivation) is fatal (Greene and Siegel 2004). Homeostatic sleep pressure increases roughly linearly with increasing wake time, reaching a maximum at the onset of the sleep state, and then decreases roughly linearly with time spent asleep (Donlea 2017). There is debate over whether homeostatic sleep pressure has a measurable biological correlate, though studies have shown correlations with both molecular markers such as adenosine (Reichert et al. 2016; Zeitzer et al. 2006) and behavioral markers such as delta power during recovery non-REM sleep (Alam et al. 2014).

The two pathways, circadian and homeostatic, work in concert to generate an overall sleep pressure (Fig. 1). Homeostatic sleep pressure increases monotonically throughout the day; however, during the daylight hours in humans, circadian wake drive is also high and increasing. By the onset of the dark phase of the cycle, both homeostatic sleep drive and circadian wake drive are high, canceling the effect of either. However, after the onset of the dark phase, circadian wake drive begins to decrease, while homeostatic sleep drive remains high, stimulating the onset of the sleep state. By the end of the sleep state in the early morning hours, circadian wake drive is still low, but homeostatic sleep drive is low enough to compensate, causing the switch to flip again and the onset of the wake state to commence. This dual cycle has important impacts on situations where sleep occurs outside of the normal rhythm; for example, in shift workers, falling asleep in the early morning is not a

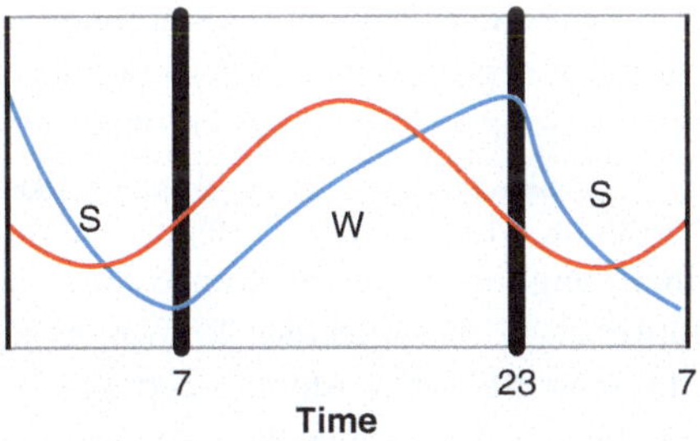

Fig. 1 Schematic of the circadian and homeostatic systems. The homeostatic sleep process (blue) creates sleep pressure roughly linearly in response to time awake, and decreases with time asleep. The circadian wake process (red) is most active during the daylight hours and provides a method of maintaining wake. After sunset, melatonin stimulates a breakdown in the circadian process, which allows the pro-sleep homeostatic process to become dominant, stimulating the sleep state. After a period of sleep, homeostatic pressure has lowered enough that even a low circadian wake drive becomes dominant, stimulating the wake state

problem as circadian wake drive is low and sleep pressure is high (Wagner 1999). However, remaining asleep through the day can be difficult as circadian wake drive increases while sleep pressure falls.

3 Diverse Chemical Mechanisms and Circuitry Govern Sleep-Pressure Systems

The circadian system is highly centralized, with the SCN being the key center of its action. The SCN exerts circadian control in myriad ways, both neurologically and hormonally. Neurons in the SCN are home to a set of clock transcription factors that stimulate their own repressors (Aryal et al. 2017), forming a daily oscillating cycle of gene expression. By a mechanism not fully understood, this mechanism governs the differing release of neurotransmitters depending on the time of the cycle (Golombek and Rosenstein 2010). Thus, the SCN releases different waves of neurotransmitters, such as glutamate, GABA, and vasopressin (Shinohara et al. 1998). The main neurological projections of the SCN radiate to the medial hypothalamus, but for hormonal control, the epithalamus is a key site of SCN effects. The pineal gland of the epithalamus releases melatonin, perhaps the best known hormonal circadian modulator. Release of melatonin is dependent on environmental inputs, most notably the light cycle; melatonin increases several hours after the onset of the dark phase of the light cycle, and remains high until the restoration of the light phase the next day (Bedrosian et al. 2013). While melatonin is not required for the initiation or maintenance of circadian rhythms overall (Gandhi et al. 2015), it serves as a key link to entrain biological circadian processes to light cues (Sack et al. 2000), and may stimulate sleep in humans and some animal species, possibly through adenosine signaling (Gandhi et al. 2015; Zhdanova et al. 2001). In humans, melatonin's effects are the principal mediator for normal sleep onset several hours into the dark phase of the diurnal cycle. Melatonin receptors have been shown to exist in myriad tissues (Morgan et al. 1994), potentially providing a system-wide mechanism for circadian synchronicity that drives an acute need for sleep at a stereotypical time each day.

Unlike the circadian system, which is largely dependent on environmental inputs via the specialized retinal ganglion cells, the homeostatic system appears to have multiple inputs. An entire class of molecules known as somnogens have been identified that appear to increase homeostatic sleep pressure. Among the most important of these molecules is the nucleoside adenosine (Lazarus et al. 2017; Huang et al. 2014), produced in the brain both purposefully as a neurotransmitter and as a waste product of ATP metabolism. Indeed, many laypersons are familiar with the hypnogenic effects of adenosine through the widespread use of the non-specific adenosine receptor antagonist caffeine (Yanik et al. 1987). Adenosine has been shown to accumulate in the brain, particularly in the basal forebrain (Blanco-Centurion et al. 2006), with increasing wake time and decrease in the sleep state (Blanco-Centurion et al. 2006). Additionally, low (nanomolar) concentrations of adenosine have been shown to enhance wake neurotransmission due to activation of the inhibitory A1 receptor in the preoptic hypothalamus, while high (micromolar) adenosine concentrations have been shown to inhibit wake neurotransmission through activation of stimulatory A2A receptors in the same nuclei (Methippara et al. 2005; Kumar et al. 2013). Several other molecules, such as prostaglandin D2 (Zhang et al. 2017), IL-1 (Obal et al. 1990), and TNF-alpha (Kapás et al. 1992), have been hypothesized to act in similar fashions to stimulate the sleep-pressure homeostat. Additionally, an emerging area of research in mechanisms of the homeostatic sleep-pressure system may be through the process of protein phosphorylation. A family of proteins, sleep need index phosphorylation proteins (SNIPPs), have been found to become steadily phosphorylated during wake time and dephosphorylated during sleep (Wang et al. 2018). The kinase Sik3 has been shown to aid in this phosphorylation; a constitutively active mutant of this kinase has been shown to induce sleep pressure artificially, resulting in mice with higher sleep times and delta power during their non-REM sleep (Honda et al. 2018).

Downstream of these somnogen initiators, the homeostatic sleep-pressure system contains multiple neurotransmitters, including orexin, acetylcholine, monoamines, and glutamate (Stenberg 2007). The sleep circuitry is a complex and multi-faceted system from a neuroanatomical perspective, with separate wake-promoting and sleep-promoting networks. Studies of wake-promoting systems historically focused on monoaminergic and cholinergic neurons of the upper brainstem, including noradrenaline from the locus coeruleus, 5HT from the raphe, and acetyl choline from the tegmentum, among others (Saper and Fuller 2017). These systems project broadly to the cortex by way of the thalamus, ventral portions of hypothalamus, and basal forebrain. This system is also augmented by peptidergic orexin inputs joining at the hypothalamus. Paradoxically though, lesions of these pathways had little effect on total sleep and wake time (Fuller et al. 2011). Thus, in recent years, the importance of glutamatergic and GABAergic networks on the wake-promoting circuitry has grown. GABAergic inputs from the basal forebrain, lateral hypothalamus, and supramammillary nucleus have been shown to promote wake, as have glutamatergic inputs from the supramammillary, parabrachial, and pedunculopontine nuclei (Saper and Fuller 2017).

Several nuclei are thought to stimulate sleep in the homeostatic sleep-pressure system, with two nuclei of the preoptic hypothalamus of key importance. The ventral lateral preoptic nucleus (VLPO) (Wagner 1999) and median preoptic nucleus (MnPN) (Mong and Cusmano 2016) are thought to be key originators of this pathway. These nuclei send GABAergic projections to key mediators of the wake state, particularly nuclei in the lateral hypothalamus governing the orexinergic wake system (Mong and Cusmano 2016). A feed-forward loop has been identified in which the MnPN both inhibits the orexinergic wake system and stimulates the VLPO, which itself serves as an inhibitor of the orexinergic wake nuclei in the lateral hypothalamus (Szymusiak and McGinty 2008). Orexinergic wake nuclei have widespread projections to the cortex and brainstem and are primarily active during the wake phase (Fulcher et al. 2014); lesions of these nuclei have been shown to induce a narcoleptic-like phenotype (Ocampo-Garcés et al. 2011). The orexin system is important from an endocrine perspective in its dual importance to both sleep–wake circuitry and feeding behavior. Orexin projections from the lateral hypothalamus project broadly across the brain to centers important for feeding, such as the paraventricular nucleus, and centers important for maintenance of wake, such as the locus coeruleus (Grafe and Bhatnagar 2018). However, the exact structure and function relationship of these pathways, and what neuronal pathways may exist connecting between feeding and sleep behavior, is incompletely known. Beyond the orexinergic system, the MnPN and VLPO send broad GABAergic inhibitory projections to many of the same nuclei involved in the wake system, including the supramammillary, tuberomammillary, and parabrachial nuclei, as well as monoaminergic nuclei such as the raphe and locus coeruleus. Additionally, projections from the MnPN and VLPO stimulate sleep by way of other brainstem nuclei such as the ventral periaqueductal gray (Saper and Fuller 2017). The MnPN and VLPO have been shown to receive circadian inputs from the SCN by indirect projections via the dorsal medial hypothalamus and/or supraventricular zone, suggesting a potential pathway for the integration of the circadian and homeostatic systems (Deurveilher and Semba 2003; Sun et al. 2001).

4 Several Hypotheses Exist as to the Functions of Sleep

Despite its ubiquitous nature, very little is known about why sleep is necessary; from an evolutionary standpoint, the necessity for an animal to spend such a large portion of its lifespan in a position both vulnerable and seemingly of little use to the animal would seem a poor adaptation. The question of sleep's function in animal physiology is a hotly debated one, with several working hypotheses. In particular, three well-developed hypotheses have formed of key sleep functions: as a method of brain microenvironmental homeostasis, a mechanism for memory consolidation and cognition, and as a regulator of metabolism and energy balance.

The evidence for sleep as a homeostatic process is well-established. Like other homeostatic systems, sleep pressure responds in an analog fashion to the relative distance from its homeostatic mean; in essence, sleep pressure directly increases with wake time. The recently identified glymphatic system may provide a mechanism for control of brain microenvironmental homeostasis that is sleep-dependent. The glymphatic system is a fluid-dynamic model of cerebrospinal (CSF) and interstitial (ISF) fluid flow around the brain and through the brain parenchyma. This flow has been shown to be important for clearance of metabolites and other waste products, including amyloid beta (Xie et al. 2013) from the brain. Additionally, glymphatic flow has been shown to be upregulated by as much as a factor of ten in the sleep state (Xie et al. 2013). Thus, a model has emerged in which waste products of metabolism and brain activity build up in the wake state due to inadequate glymphatic flow, but are cleared from the parenchyma in the sleep state when flow is increased (Plog and Nedergaard 2018). These findings could suggest that brain clearance is a key function of the sleep state and a key purpose of the homeostatic sleep function.

Another hypothesis for the function of sleep involves the process of memory consolidation (Cellini 2017). Multiple studies have shown that memories are enhanced during sleep; in particular, declarative memories have been shown to be enhanced after non-REM sleep (Krause et al. 2017; Ackermann and Rasch 2014), while non-declarative or emotional memories have shown enhancement after REM sleep (Sun et al. 2001). Sleep has also been shown to be important for synaptic downscaling, in which synapses are uniformly lessened in strength during sleep (Raven et al. 2018). This uniform downscaling prevents or relieves the saturation of synapse receptor patches. Relief of saturation allows for further long-term potentiation and depression at the same synapses, in order for more differentiation of synaptic strength (and thus memory formation) to proceed.

Perhaps most importantly from an endocrine perspective, sleep has been hypothesized as an important mechanism for the regulation of metabolism and systemic energy balance. Overall metabolism declines only modestly in sleep (Fraser and Nordin 1955), suggesting that energy conservation per se is not a key function of the sleep state. One working hypothesis for the purpose of slower metabolism is that it may allow free radical scavengers more freedom to reduce reactive molecules that can damage the brain (Villafuerte et al. 2015). More broadly though, sleep has been shown to induce the fluctuation of hormones important for the regulation of normal metabolism. Sleep disruption has been shown to enhance ghrelin (Copinschi et al. 2014) and decrease leptin (Allada et al. 2017) levels, stimulating appetite; it is also associated with a state similar to insulin resistance, possibly due to dysregulation of growth hormone (GH) levels (Rasmussen et al. 2008). As a result, disruptions in sleep have been shown to be a risk factor and exacerbating factor for metabolic syndrome and related disorders (Rasmussen et al. 2008).

5 Sleep Disorders Have a Broad Clinical Impact

Sleep disorders are both quite prevalent and generally thought to be underdiagnosed. While we will not attempt to give an exhaustive overview of all sleep disorders here, this section will profile some common conditions to illustrate the myriad interactions with the endocrine system. Sleep disorders can take two forms: disorders of sleep quantity and quality per se, and related comorbidities which can be introduced or exacerbated by sleep disruption.

Insomnia is the most common sleep disorder; it is defined as the persistent inability to sleep despite the opportunity to do so. It has been estimated that insomnia is clinically present in 6% of the population, and as many as a third of individuals may show some symptoms (Ohayon 2002). Insomnia may be a primary condition, or secondary to a multitude of other neurological, psychiatric, and physical disorders. A multitude of pharmaceutical and non-pharmaceutical interventions exists to combat insomnia, of varying effectiveness. Importantly for the endocrine field, insomnia shows a sex difference, as it is significantly more common in women than men (Mong and Cusmano 2016). Additionally, insomnia pharmacology is an arena where sex differences have become a prominent issue. One of the most common anti-insomnia medications, the benzodiazepine-like drug zolpidem (popularly known by the brand name Ambien), was also the first major drug to show a strong interaction with the patient's biological sex, as in 2013 the FDA reduced the recommended dose for women to half that of men. Though the case of zolpidem sexual interaction was due to differing rates of liver metabolism (Greenblatt et al. 2014), the brain has shown an interaction between pharmacology and biological sex in the context of insomnia as well. A study administering olanzapine (a second-generation anti-psychotic) showed sex differences in its effect on sleep, as slow-wave sleep increased in women and decreased in men (Giménez et al. 2011).

Another prominent sleep disorder is restless legs syndrome (RLS). RLS is the uncontrollable urge to move one's legs when at rest, which leads to an inhibition of deep sleep states (Gamaldo and Earley 2006). RLS is quite common, with estimates of between 2 and 15% of the population displaying symptoms. The causes are unclear, but several underlying diseases and medication side-effects, most notably iron deficiency (Trotti and Becker 2019), have been speculated as potential causes. Similar to insomnia, RLS is far more prevalent in women than men, for reasons that remain unclear (Berger et al. 2004).

A prominent sleep disorder about which more is known surrounding the etiology is obstructive sleep apnea (OSA). OSA is marked by closure of the airway during the night, leading to periods of cessation or attenuation of breathing and hypoxia (White 2017). OSA is often underdiagnosed, as it presents with very non-specific symptoms, such as daytime sleepiness, fatigue, and impaired cognition, to the sufferer. As such, it is often only noticed by bed partners due to nighttime snoring (Punjabi 2008). OSA is often a comorbidity of obesity, due to the presence of additional fatty tissue in structures surrounding the airway (Schwab et al. 2015). Unlike many sleep disorders, OSA is more commonly diagnosed in males (Punjabi 2008), though there is speculation that it may simply be underdiagnosed in women. OSA is usually managed mechanically through the use of continuous positive airway pressure (CPAP) machines or mandibular splint devices, which both physically open the airway (Bratton et al. 2015).

Disorders of circadian regulation are also widespread. Delayed sleep phase disorder (DSPD) is a chronic circadian dysregulation that pushes the onset of sleep and the onset of wake much later relative to societal norms (Pavlova 2017), often due to genetics (Matheson and Hainer 2017). It is a form of "social jet lag", a broader term that also encompasses a delayed circadian phase due to behavioral and environmental factors (Kayaba et al. 2018). DSPD and social jet lag are relatively rare in adults, with a prevalence of under 2 in 1000, but are common in adolescents, with studies showing a prevalence of 5% (Danielsson et al. 2016) and some estimates being even higher. Treatments of DSPD with melatonin (Auld et al. 2017) and analogues have shown some success in combating the symptoms, though relapse can be a concern (Micic et al. 2016). The converse of this condition, advanced sleep phase disorder, is significantly rarer, though more common in the elderly. Both DSPD and ASPD have shown a strong genetic component in familial studies (Matheson and Hainer 2017).

Mood disruptions have shown a sleep and circadian component. Disruption of the circadian rhythms of melatonin has been linked to seasonal affective disorder (SAD), a common form of depression. SAD is prevalent in the winter months, when daylight cycles are shorter and cause disruption of melatonin secretion (Wirz-Justice 2018); some SAD patients experience melatonin secretion well into the morning hours, when levels should be low. Morning bright-light therapy to resynchronize melatonin levels has been shown to mitigate some effects of SAD (Wirz-Justice 2018). Major depression has also been linked to melatonin release, and some melatonin receptor agonists have been approved to treat depression (Hickie and Rogers 2011).

Sleep disruption has been shown to be tied to many serious pathologies, both cognitive and physical. Sleep changes have been shown to be correlated with Alzheimer's pathology as well as a potential leading sign of the disease (Peter-Derex et al. 2015). Alzheimer's patients have shown decreased sleep at night and increased sleep during the day, as well as an overall loss of REM sleep (Pase et al. 2017). Additionally, self-reported sleep problems, most notably sleep fragmentation, have been associated with a higher risk of Alzheimer's years later (Macedo et al. 2017). The loss of sleep-dependent microenvironmental homeostasis may be a contributing factor to the accumulation of brain metabolites in such dementia. Amyloid Beta, the key protein which aggregates in Alzheimer's, has been shown to display a diurnal rhythm that increases during wake time and decreases during sleep (Lucey and Bateman 2014). The impact of sleep on memory formation and consolidation may also explain portions of this connection.

The strong endocrine connections between sleep and metabolism also present a possible explanation for population-level correlations between sleep disruption and the key public health issue of metabolic syndrome. Obesity is a key public health concern, and sleep has been shown by multiple studies to both impact and have an impact on metabolic and feeding behaviors. Most notably, obesity is a major risk factor for OSA (Schwab et al. 2015) as described above. Conversely, sleep loss has been shown to have a stimulating effect on appetite (Schmid et al. 2015) and has been correlated with increased obesity. However, the molecular mechanisms of these interactions, particularly the connections between clinical phenotypes and neuroendocrine mechanisms, are still ill defined and multi-faceted. Sleep disruption has been shown to increase oxidative stress (Villafuerte et al. 2015), enhance pro-inflammatory mediators such as IL-1 and TNF-alpha (Obal et al. 1990; Kapás et al. 1992), activate the sympathetic nervous system (Schlaich et al. 2015), and stimulate cortisol secretion (Wright et al. 2015). Activation of these pathways has been shown to be risk factors for obesity, metabolic syndrome, and sequelae such as type 2 diabetes.

6 Sleep Behavior and Circadian Timing Impact Multiple Hormonal Functions

Multiple different hormones have been shown to be impacted by the sleep–wake cycle, though it remains something of an open question what roles the intrinsic circadian timekeeper and sleep behavioral cycle play in governing the differential release. Indeed, there is evidence to suggest that different hormones may display different mechanisms of diurnal entrainment vis-à-vis homeostatic sleep pressure, circadian timing, and sleep behavior (Pietrowsky et al. 1994).

Among the most important hormonal pathways regulated by sleep is the hypothalamic–pituitary–adrenal (HPA) axis (Fig. 2). The SCN sends projections to the pituitary, which result in an oscillating hormone secretion rhythm in line with the diurnal cycle. The SCN also sends direct neuronal projections to the pituitary, further

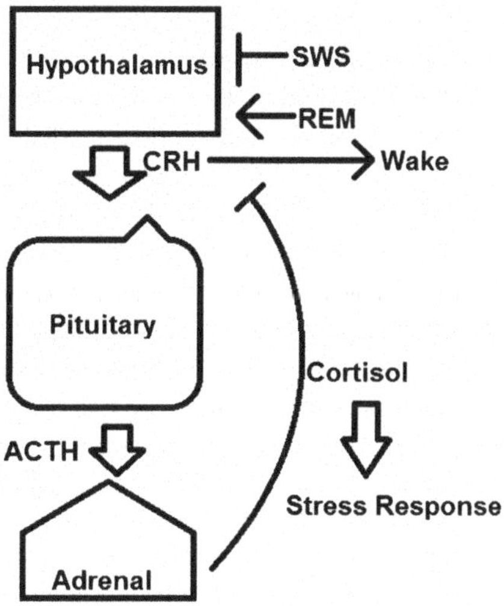

Fig. 2 Schematic of sleep impacts on the HPA axis. Slow-wave sleep is an inhibitor of the HPA axis stress pathway, while REM sleep stimulates cortisol production. Corticotropin releasing hormone (CRH) has been shown to stimulate wake, while exogenous glucocorticoids have been shown to counteract this effect and stimulate sleep by feedback inhibition

entraining the release of glucocorticoids in a stereotypical diurnal rhythm. In humans, cortisol levels peak in the early morning just after the onset of the wake phase (Allada et al. 2017), a phenomenon thought adapted to prepare the body for wake-time activity. This cortisol rhythm is light-entrained, meaning that significant disruptions have been shown in shift workers and others with circadian rhythm disruption (Allada et al. 2017; James et al. 2004).

The HPA axis and cortisol play a major role in regulation of the stress response, and sleep has long been shown to have a potent inhibitory influence on this pathway. Sympathetic nervous activity and its downstream effects, including cardiovascular function, display a dependence on sleep state (Miki and Yoshimoto 2013). Slow-wave sleep in particular is an inhibitor of HPA axis activity; consequently, cortisol is elevated in the later portions of a sleep bout and during REM sleep (Born and Wagner 2004). Furthermore, sleep displays a modulation of the adrenal response of cortisol production to the action of adrenocorticotropic hormone; adrenal ACTH sensitivity has also been shown to vary with the diurnal cycle (Späth-Schwalbe et al. 1991). Further downstream, cortisol rhythms have been shown to affect the immune system; as cortisol is a potent immune and inflammatory suppressor, circadian disruption has been shown to increase inflammatory cytokines and inflammatory pathologies, including cancers (Schlaich et al. 2015; Born and Wagner 2004).

Insulin is another hormone entrained rhythmically. Insulin, a pancreatic hormone, has a principal function of promoting the absorption of glucose from the blood. However, in experiments of clamping glucose concentration, insulin still rises in the late phase of the sleep cycle (Copinschi et al. 2014), most likely to restrain hepatic glucose production in order to prevent a glycemic peak. This rhythm appears to be SCN-based by studies of dysregulated feeding in SCN-lesioned rats (Marchant and Mistlberger 1997); however, there is also evidence that the pancreatic beta cells responsible for insulin secretion have their own set of clock transcription factors that may operate independently (Perelis et al. 2016). Insulin response is also circadian-modulated, with insulin sensitivity in adipose tissue being significantly higher in daytime hours (Carrasco-Benso et al. 2016). Sleep deprivation has been shown to be sufficient to induce insulin resistance by multiple pathways (Donga et al. 2010).

Similarly, another key hormone for metabolism, growth hormone (GH), which promotes lipolysis and muscle growth, displays cycles entrained to sleep activity (Steyn and Ngo 2017). GH is elevated in the earlier portions of a sleep bout, particularly in slow-wave sleep, and decreases in later sleep phases (van Cauter and Plat 1996). These hormonal changes, as well as similar sleep-dependent fluctuation in the appetite stimulating and repressing hormones ghrelin (Rasmussen et al. 2008) and leptin (Challet 2015), may in part explain the correlation between sleep disruption and obesity and metabolic syndrome.

Circadian factors have also been shown to have an effect on reproductive hormones. However, studies of the circadian cycle and reproductive-related hormones have been complicated by the fact that the connection appears much stronger in rodents than in humans (Mong and Cusmano 2016). For example, in rodents, the surge of luteinizing hormone occurs just before the onset of the dark phase (and thus mating activity), a surge that is mediated by the SCN (Ramírez et al. 2017). Thus, circadian rhythms are very stereotypical in rodent mating behaviors; however, humans do not display any equivalent circadian rhythm in mating activity. Despite that phenomenon, there is clinical evidence to suggest that sleep–wake and circadian disruption may play a role in certain human reproductive disorders. For example, shiftwork in pregnant women may be associated with a greater risk of preterm birth (Nurminen 1998), though other analyses dispute this conclusion (van Melick et al. 2014).

Sleep has also been shown to have an effect on sex hormones, particularly testosterone. Testosterone secretion is linked to sleep cycles, with peak levels occurring in the middle of the sleep cycle, often near the time of REM sleep onset (Wittert 2014). Insufficient or fragmented sleep, which reduces the amount of REM, blocks the nocturnal increase in testosterone. Therefore, sleep disruption could be a risk factor for low testosterone levels (Wittert 2014).

7 Non-gonadal Endocrine Factors Impact Sleep

A multitude of different hormones have been shown to impact the sleep state, as well as the quantity and quality of sleep achieved. The key circadian mediator melatonin is one of the best-known and most directly sleep-impacting hormones; it is classically low during the daytime and increases after the onset of darkness, stimulating sleep in humans (Saper and Fuller 2017). Melatonin administration, which is available as an over-the-counter pharmaceutical, has been shown to increase total sleep time and sleep maintenance (White 2017). Though melatonin may be sedative in some species such as zebrafish (Zhdanova et al. 2001), it appears to not have direct sedative effects in humans (Azeez et al. 2018). Instead, melatonin in humans appears to be a link between environmental cues, most notably light cues, and the circadian synchronization of biological processes (Gandhi et al. 2015). This effect can be recapitulated pharmacologically, as exogenous melatonin has also been shown to replicate EEG changes from the circadian pacemaker (Dijk and Cajochen 1997).

Beyond melatonin, the HPA axis has been shown to have an impact on sleep. Administration of corticotropin releasing hormone (CRH), the first hormone in the HPA axis, increases wake time (Held et al. 2005), and conversely, administration of a specific CRH receptor blockade decreases wake time (Chang and Opp 1999). Similarly, insomnia has been associated with an all-day increase in HPA axis activity and cortisol secretion (D'Aurea et al. 2015). However, paradoxically, exogenous glucocorticoids have been shown to be stimulators of slow-wave sleep and inhibitors of REM sleep, particularly in the context of the very high cortisol levels of late pregnancy (Santiago et al. 2001). It has been hypothesized that this effect is due to feedback inhibition by cortisol of CRH release (Steiger 2002).

8 Biological Sex Differences and Ovarian Steroids Impact Sleep Behavior

As sleep is highly evolutionarily conserved, the suggestion of biological sex differences may be counterintuitive, but reproductive and sex hormones have also been shown to interact with the sleep–wake system. Women and men have long been clinically shown to have differing sleep patterns (Mong and Cusmano 2016). In particular, women paradoxically sleep longer than men, but generally self-report a lower sleep quality (Mong and Cusmano 2016). Objective data suggest that women should have higher sleep quality than men; women have longer total sleep time and less total wake time, a shorter latency to sleep onset, and higher sleep efficiency than men (Bixler et al. 2009; Goel et al. 2005). EEG studies have also shown a higher proportion of deep slow-wave sleep (stage 3) and less light sleep (stage 1 and 2) in women than men (Redline et al. 2004). However, clinical evidence is not in agreement with those findings, showing that women have been consistently diagnosed with insomnia and other sleep disorders, including RLS, at a markedly higher rate than men (Mong and Cusmano 2016; Berger et al. 2004). It is unclear if male sex steroids, mainly testosterone, affect sleep in men, as paradoxically both low testosterone levels and testosterone replacement have been shown to be risk factors for sleep deprivation (Wittert 2014). In animal studies, castration does not significantly change sleep time in male rodents, suggesting little impact of testosterone levels on male sleep (Cusmano et al. 2014). However, there is a larger complement of scientific literature suggesting an impact of female hormones on sleep behavior.

Overall sex differences in sleep patterns may be due to sleep-independent factors, including psychosocial ones, which may be tied to a higher presence of anxiety in females (Mong and Cusmano 2016); inversely, sleep loss may also be more potently anxiogenic in females as well (Goldstein-Piekarski et al. 2018). However, there is strong evidence that hormonal complement plays a role; most strikingly, the sex difference in sleep quality emerges in females with puberty (Johnson et al. 2006) and disappears at menopause. There are specific distinctions in the circadian timing of sleep with biological sex and female hormonal state as well. The endogenous circadian rhythm may have an interaction with biological sex, though the mechanism has not been fully explored (Eastman et al. 2017). However, there is clinical evidence for a sex difference in circadian timing. Premenopausal women go to bed earlier than men and have much earlier melatonin peaks (Cain et al. 2010) until menopause, when the sex difference in sleep onset disappears but melatonin peaks become even earlier (Mong and Cusmano 2016). The mechanisms of these circadian differences are incompletely understood.

There is little evidence to suggest major primary changes in sleep patterns in women between the different phases of the menstrual cycle (Baker and Lee 2018). However, there are some minor cyclic differences in sleep attributable to differing hormone levels; for example, a specific type of non-REM sleep waveform, sleep spindles, is elevated in the post-ovulatory luteal phase of the menstrual cycle (Baker and Lee 2018). High progesterone and estrogen levels also correlate with a lower amount of REM sleep (Lancel et al. 1996). Clinically, the luteal phase has also been shown to be a time of particularly pronounced sleep disruption during menopause (De Zambotti et al. 2015). Finally, another sexually differentiated hormone, prolactin, has also been shown to increase slow-wave sleep based on limited studies of patients with hyperprolactinemia (Frieboes et al. 1998). The broader significance of these changes in humans has not fully been established.

Sleep disorders have also been shown to be particularly prevalent in women at times of hormonal flux, including puberty, pregnancy, and menopause. Exogenous hormones, most notably oral contraception, have been shown to affect sleep in young women; oral contraceptive use increases REM sleep and light sleep, while reducing deep slow-wave sleep (Baker et al. 2001; Burdick et al. 2002); the mechanism for this change is unclear. Pregnancy has been shown to be a time of high levels of sleep change and disruption, though it is difficult to differentiate direct effects of hormonal change from physiological changes due to growth and development of the fetus. While there appear to be few changes in melatonin levels or circadian rhythms during pregnancy (Santiago et al. 2001), changes in the levels of estrogen, progesterone, cortisol, and oxytocin may contribute to disruption in sleep, particularly the consistent finding of lower REM sleep times in the third trimester (Santiago et al. 2001). Many women report sleep disruption in the perimenopausal period. Studies on using hormone replacement therapies to combat sleep disruption in menopause have shown improvement in self-reporting of subjective sleep quality. There is evidence that hormones play a role in consolidating sleep at night (Mong and Cusmano 2016), possibly leading to increased sleep quality with their replacement. However, more objective metrics have shown differing effects, with inconsistent findings in objective sleep quality measures with hormonal therapy (Cintron et al. 2017). These situations may be due to differing hormonal formulations between studies; alternatively, hormone therapy may exert its main impact in relieving the non-sleep symptoms of menopause, particularly vasomotor symptoms (Cintron et al. 2017), making women more comfortable and sleep easier to obtain.

Apart from the clinical finding that sex steroids may affect sleep behavior and architecture, the mechanisms underlying how sex steroids influence the sleep circuitry remain a significant gap in our knowledge. The use of animal models is critical for advancing our understanding of the potential endocrine–sleep nexus.

9 Diverse Models Exist for the Investigation of Endocrine Regulation of Sleep Behavior

As with many investigations of neurobiological behavior phenomena, animal models are a key tool in the sleep field. Rats are the most commonly used model of sleep behavior, with mice as a secondary rodent organism. Rats are widespread models for sleep studies as the circuitry and neurochemistry of sleep share similarities with humans, and pharmacologic manipulation and EEG measurement of sleep states are both feasible. Use of mice in the sleep field is generally limited to exploration of sleep in the context of specific genetic backgrounds (Mong and Cusmano 2016), which are more readily available in mice than rats. A key consideration in the use of models for sleep is the different patterns of sleep between animals. Sleep is generally entrained to the light–dark cycle through the circadian sleep system, but the specific pattern of sleep differs dramatically between species. Under normal conditions, humans are monophasic sleepers, with a single consolidated period of 7–8 h per day, concentrated in the dark portion of the light/dark cycle. Conversely, however, rodents are polyphasic sleepers, with many periods of sleep and wake throughout the day (Acerbi et al. 2008). While rodents do have some periods of consolidated wake or sleep of an hour or more, they can also experience bouts as short as a few seconds in duration (Simasko and Mukherjee 2009). Additionally, while rodents do preferentially sleep during the light period of the light–dark cycle, they exhibit significant periods of both wake and sleep throughout the entire light–dark cycle. Unlike rodents, the other major animal model system for sleep, drosophila, are largely monophasic sleepers, with a single consolidated period of sleep entrained to the dark portion of the cycle (Dubowy and Sehgal 2017). As a result, drosophila may be a more useful model of some aspects of the circadian system where light–dark dependence and sleep consolidation are key points of experimental investigation.

Historically, the majority of sleep studies have been performed in men or male animals (Mong and Cusmano 2016), a deeply unfortunate occurrence that has led the impact female animal models can have on illuminating ties between ovarian steroids and sleep. The paucity of basic studies investigating sex differences in sleep has resulted in an unclear picture on the nature of those differences. Gonadally intact female rodents generally spend less time in sleep states compared with males (Paul et al. 2006), but females, despite accumulating less total sleep, have more consolidated sleep bouts, consisting of longer bout durations with less state transitions and fewer arousals (Paul et al. 2006). Moreover, delta power, a quantitative measure of sleep intensity, is higher in females during baseline sleep as well as in recovery sleep following deprivation, a finding in agreement with human clinical data (Paul et al. 2006).

Perhaps more striking is that in the absence of circulating sex steroids, these sex differences in sleep behavior and architecture are eliminated, suggesting that sex differences in sleep are in part dependent on sex steroids. Sleep patterns in female rats are exquisitely sensitive to the natural fluctuations of ovarian steroids (Paul et al. 2006; Koehl et al. 2006). Multiple studies in rats show that during proestrus, when estrogen and progesterone are elevated, sleep time is significantly reduced compared with other phases of the estrous cycle (Schwartz and Mong 2013; Schwierin et al. 1998). Exogenous hormone replacement is observed to recapitulate this phenotype (Cusmano et al. 2014) in both rats and mice. In these studies, estrogen predominately suppresses dark phase sleep and has little or no effect in the light phase. Thus, a key paradigm for studies of hormonal modulation of sleep has been the use of hormone replacement in oophorectomized rodents (Mong and Cusmano 2016) which can provide hormonal stability that bypasses the rapid hormonal changes inherent to the 4-day menstrual cycle in rats.

10 Animal Studies Illuminate Mechanisms Connecting Female Gonadal Hormones and Sleep

In rodents, estrogens have been broadly shown to increase wake and decrease spontaneous sleep, particularly in the active phase of the light–dark cycle; exogenous replacement of estrogen in females decreases dark phase sleep by 55% (Mong and Cusmano 2016). Furthermore, estrogens have been shown to consolidate wake and fragment sleep. However, estrogen-treated rats also have more consolidated slow-wave sleep following sleep deprivation by gentle handling (Schwartz and Mong 2013), and thus estrogen may be acting to facilitate recovery from sleep deprivation. Estradiol, the most potent estrogen, may interact with the circadian system, as it is shown to have a time-of-day-dependent effect; estradiol has been shown to decrease sleep in the active phase and increase it following deprivation in the stereotypical sleep phase (Mong and Cusmano 2016; Cusmano et al. 2014). Thus, hormonal impact may be to improve circadian fealty. Supporting that contention, aromatase knockout mice, which are deficient in the formation of estrogen from testosterone, have similar duration of sleep but sleep that is more fragmented and less entrained to the stereotypical (light) phase of the light–dark cycle (Vyazovskiy et al. 2006).

The molecular mechanisms of hormone impact on sleep are poorly understood, and studies investigating where and how female steroids act on the brain are only an emerging area of investigation. Sexual differentiation of the rodent brain occurs during a brief window of early development. Exposure to sex steroids around the day of birth results in the masculinization and defeminization of the rodent brain, while absence of sex steroids leads to a feminization process (Nugent et al. 2015; Sato et al. 2004). Production of sex steroids in adults cements appropriate behaviors specific to the sex of the animals. Studies in rats have suggested that estradiol effects on sleep are established by the first phase of this process, the early programming effects of sex steroids (Cusmano et al. 2014). Female rats exposed to a masculinizing dose of testosterone during the sensitive window for brain sexual differentiation exhibit male-like responsivity to estradiol and testosterone in adulthood and exhibited male signatures in the sleep-active VLPO nucleus (Saper et al. 2010).

Steroid receptors, particularly for estrogen and progesterone, are present throughout the brain and prevalent on multiple sleep-regulating nuclei such as the hypothalamus (Rønnekleiv and Kelly 2005) and basal forebrain (Donahue et al. 2000). Previous work in rodents implicates the VLPO in particular as a key site of mediating estradiol actions over sleep. In adult oophorectomized females, estradiol decreases activation of sleep-active VLPO neurons (Hadjimarkou et al. 2008) and downregulates levels of lipocalin-type prostaglandin D synthase (L-PGDS), the enzyme responsible for the production of prostaglandin D2 that potently promotes sleep (Devidze et al. 2010; Mong et al. 2003). Estrogens also decrease expression of wake-inhibiting adenosine 2A receptors (Ribeiro et al. 2009), suggesting a potential alternate mechanism for an inhibitory impact on homeostatic sleep pressure. However, these findings are complicated by studies showing that the VLPO is not a major site of estrogen sensitivity (Bailey and Silver 2014). Instead, it is an upstream nucleus, the median preoptic (MnPN), which may be most responsible for mediating estradiol action over homeostatic sleep pressure. Blocking estradiol action directly in the median preoptic nucleus of female rats attenuates estradiol suppression of sleep (Hadjimarkou et al. 2008). Downstream, the orexinergic wake-promoting system of the lateral hypothalamus receives inputs from the MnPN and VLPO and is highly sensitive to fluctuations in endogenous and exogenous ovarian steroids (Mong and Cusmano 2016), suggesting that this section of the homeostatic sleep/wake circuitry may be a key site for estrogen action.

While estrogen receptors are present in the key circadian nucleus of the SCN (Vida et al. 2008), there is not a great deal of evidence suggesting an important function for estrogen in circadian rhythms. By contrast, androgens appear to be important to the activity of the SCN, increasing the fealty of certain behaviors to the circadian clock by mediating its response to light (Karatsoreos et al. 2011; Model et al. 2015). Finally, it is important to note that certain sex differences may be more impacted by chromosomal complement than hormonal status. Female mice had a higher level of slow-wave activity in their active phase than male mice when both were ovariectomized or gonadectomized, respectively (Ehlen et al. 2013). Additionally, anatomically female mice engineered to have an XY chromosomal compliment in the "four core genotypes" model acquire more sleep during their active phase and have higher NREM delta power than XX females, suggesting processes mediating recovery from sleep loss are partially dependent on sex chromosomes (Arnold 2004).

11 Conclusion

Sleep behavior demonstrates myriad neuroendocrine interactions and has broad implications for human health. While there is much that is unknown about the reasons for sleep, evidence exists that it is important for homeostasis of a diverse array of biological functions, including memory process, brain microenvironment homeostasis, and systemic metabolic function. Disorders of sleep regulation are extremely prevalent and are both a major cause of primary morbidity and an exacerbating factor to many health conditions. Sleep and the endocrine system exhibit a bi-directional interaction, with sleep behavior having a strong influence on endocrine factors and endocrine factors reciprocally influencing sleep behavior. In particular, biological sex and sex hormones have been shown to have a significant impact on sleep function. Unfortunately, until recent years much research on sleep behavior largely disregarded its connections with the endocrine system. Use of both clinical studies and rodent models to investigate interactions between neuroendocrine function, including biological sex, and sleep therefore presents a promising area of further exploration.

References

Aalto J, Kiianmaa K (1984) Increased voluntary alcohol drinking concurrent with REM-sleep deprivation. Alcohol 1(1):77–79. PMID: 6543613

Abbott SM, Reid KJ, Zee PC (2015) Circadian rhythm sleep-wake disorders. Psychiatr Clin North Am 38(4):805–823. https://doi.org/10.1016/j.psc.2015.07.012. PMID: 26600110

Acerbi A, McNamara P, Nunn CL (2008) To sleep or not to sleep: the ecology of sleep in artificial organisms. BMC Ecol 8:10. https://doi.org/10.1186/1472-6785-8-10. PMID: 18479523

Ackermann S, Rasch B (2014) Differential effects of non-REM and REM sleep on memory consolidation? Curr Neurol Neurosci Rep 14(2):430. https://doi.org/10.1007/s11910-013-0430-8. PMID: 24395522

Alam MA, Kumar S, McGinty D, Alam MN, Szymusiak R (2014) Neuronal activity in the preoptic hypothalamus during sleep deprivation and recovery sleep. J Neurophysiol 111(2):287–299. https://doi.org/10.1152/jn.00504.2013. PMID: 24174649

Allada R, Cirelli C, Sehgal A (2017) Molecular mechanisms of sleep homeostasis in flies and mammals. Cold Spring Harb Perspect Biol 9(8):a027730. https://doi.org/10.1101/cshperspect.a027730. PMID: 28432135

Aminoff MJ, Boller F, Swaab DF (2011) Handb Clin Neurol 98:vii. https://doi.org/10.1016/B978-0-444-52006-7.00047-2. PMID: 21056174

Arnold AP (2004) Sex chromosomes and brain gender. Nat Rev Neurosci 5(9):701–708. https://doi.org/10.1038/nrn1494. PMID: 15322528

Aryal RP, Kwak PB, Tamayo AG, Gebert M, Chiu PL, Walz T et al (2017) Macromolecular assemblies of the mammalian circadian clock. Mol Cell 67(5):770–782.e6. https://doi.org/10.1016/j.molcel.2017.07.017. PMID: 28886335

Auld F, Maschauer EL, Morrison I, Skene DJ, Riha RL (2017) Evidence for the efficacy of melatonin in the treatment of primary adult sleep disorders. Sleep Med Rev 34:10–22. https://doi.org/10.1016/j.smrv.2016.06.005. PMID: 28648359

Azeez IA, Del Gallo F, Cristino L, Bentivoglio M (2018) Daily fluctuation of orexin neuron activity and wiring: the challenge of "chronoconnectivity". Front Pharmacol 9:1061

Bailey M, Silver R (2014) Sex differences in circadian timing systems: implications for disease. Front Neuroendocrinol 35(1):111–139. https://doi.org/10.1016/j.yfrne.2013.11.003. PMID: 24287074

Baker FC, Lee KA (2018) Menstrual cycle effects on sleep. Sleep Med Clin 13(3):283–294. https://doi.org/10.1016/j.jsmc.2018.04.002. PMID: 30098748

Baker FC, Mitchell D, Driver HS (2001) Oral contraceptives alter sleep and raise body temperature in young women. Pflugers Arch 442(5):729–737. PMID: 11512029

Bedrosian TA, Herring KL, Walton JC, Fonken LK, Weil ZM, Nelson RJ (2013) Evidence for feedback control of pineal melatonin secretion. Neurosci Lett 542:123–125. https://doi.org/10.1016/j.neulet.2013.03.021. PMID: 23528860

Berger K, Luedemann J, Trenkwalder C, John U, Kessler C (2004) Sex and the risk of restless legs syndrome in the general population. Arch Intern Med 164(2):196–202. https://doi.org/10.1001/archinte.164.2.196. PMID: 14744844

Bixler EO, Papaliaga MN, Vgontzas AN, Lin HM, Pejovic S, Karataraki M et al (2009) Women sleep objectively better than men and the sleep of young women is more resilient to external stressors: effects of age and menopause. J Sleep Res 18(2):221–228. https://doi.org/10.1111/j.1365-2869.2008.00713.x. PMID: 19302341

Blanco-Centurion C, Xu M, Murillo-Rodriguez E, Gerashchenko D, Shiromani AM, Salin-Pascual RJ et al (2006) Adenosine and sleep homeostasis in the basal forebrain. J Neurosci 26(31):8092–8100. https://doi.org/10.1523/JNEUROSCI.2181-06.2006. PMID: 16885223

Born J, Wagner U (2004) Memory consolidation during sleep: role of cortisol feedback. Ann N Y Acad Sci 1032:198–201. https://doi.org/10.1196/annals.1314.020. PMID: 15677410

Bratton DJ, Gaisl T, Wons AM, Kohler M (2015) CPAP vs mandibular advancement devices and blood pressure in patients with obstructive sleep apnea: a systematic review and meta-analysis. JAMA 314(21):2280–2293. https://doi.org/10.1001/jama.2015.16303. PMID: 26624827
Burdick RS, Hoffmann R, Armitage R (2002) Short note: oral contraceptives and sleep in depressed and healthy women. Sleep 25(3):347–349. PMID: 12003166
Cain SW, Dennison CF, Zeitzer JM, Guzik AM, Khalsa SB, Santhi N et al (2010) Sex differences in phase angle of entrainment and melatonin amplitude in humans. J Biol Rhythm 25(4):288–296. https://doi.org/10.1177/0748730410374943. PMID: 20679498
Carrasco-Benso MP, Rivero-Gutierrez B, Lopez-Minguez J, Anzola A, Diez-Noguera A, Madrid JA et al (2016) Human adipose tissue expresses intrinsic circadian rhythm in insulin sensitivity. FASEB J 30(9):3117–3123. https://doi.org/10.1096/fj.201600269RR. PMID: 27256623
Cellini N (2017) Memory consolidation in sleep disorders. Sleep Med Rev 35:101–112. https://doi.org/10.1016/j.smrv.2016.09.003. PMID: 27765468
Challet E (2015) Keeping circadian time with hormones. Diabetes Obes Metab 17:176–183. https://doi.org/10.1111/dom.12516. PMID: 26332971
Chang FC, Opp MR (1999) Pituitary CRH receptor blockade reduces waking in the rat. Physiol Behav 67(5):691–696. PMID: 10604839
Cintron D, Lipford M, Larrea-Mantilla L, Spencer-Bonilla G, Lloyd R, Gionfriddo MR et al (2017) Efficacy of menopausal hormone therapy on sleep quality: systematic review and meta-analysis. Endocrine 55(3):702–711. https://doi.org/10.1007/s12020-016-1072-9. PMID: 27515805
Copinschi G, Leproult R, Spiegel K (2014) The important role of sleep in metabolism. Front Horm Res 42:59–72. https://doi.org/10.1159/000358858. PMID: 24732925
Cusmano DM, Hadjimarkou MM, Mong JA (2014) Gonadal steroid modulation of sleep and wakefulness in male and female rats is sexually differentiated and neonatally organized by steroid exposure. Endocrinology 155(1):204–214. https://doi.org/10.1210/en.2013-1624. PMID: 24189140
D'Aurea C, Poyares D, Piovezan RD, Passos G, Tufik S, Mello MT (2015) Objective short sleep duration is associated with the activity of the hypothalamic-pituitary-adrenal axis in insomnia. Arq Neuropsiquiatr 73(6):516–519. https://doi.org/10.1590/0004-282X20150053. PMID: 26083888
Danielsson K, Markström A, Broman JE, von Knorring L, Jansson-Fröjmark M (2016) Delayed sleep phase disorder in a Swedish cohort of adolescents and young adults: prevalence and associated factors. Chronobiol Int 33(10):1331–1339. https://doi.org/10.1080/07420528.2016.1217002. PMID: 27537980
De Zambotti M, Willoughby AR, Sassoon SA, Colrain IM, Baker FC (2015) Menstrual cycle-related variation in physiological sleep in women in the early menopausal transition. J Clin Endocrinol Metab 100(8):2918–2926. https://doi.org/10.1210/jc.2015-1844. PMID: 26079775
Deurveilher S, Semba K (2003) Indirect projections from the suprachiasmatic nucleus to the median preoptic nucleus in rat. Brain Res 987(1):100–106. PMID: 14499951
Devidze N, Fujimori K, Urade Y, Pfaff DW, Mong JA (2010) Estradiol regulation of lipocalin-type prostaglandin D synthase promoter activity: evidence for direct and indirect mechanisms. Neurosci Lett 474(1):17–21. https://doi.org/10.1016/j.neulet.2010.02.064. PMID: 20193744
Dijk DJ, Cajochen C (1997) Melatonin and the circadian regulation of sleep initiation, consolidation, structure, and the sleep EEG. J Biol Rhythm 12(6):627–635. https://doi.org/10.1177/074873049701200618. PMID: 9406038
Donahue JE, Stopa EG, Chorsky RL, King JC, Schipper HM, Tobet SA et al (2000) Cells containing immunoreactive estrogen receptor-alpha in the human basal forebrain. Brain Res 856(1–2):142–151. PMID: 10677621
Donga E, van Dijk M, van Dijk JG, Biermasz NR, Lammers GJ, van Kralingen KW et al (2010) A single night of partial sleep deprivation induces insulin resistance in multiple metabolic pathways in healthy subjects. J Clin Endocrinol Metab 95(6):2963–2968. https://doi.org/10.1210/jc.2009-2430. PMID: 20371664

Donlea JM (2017) Neuronal and molecular mechanisms of sleep homeostasis. Curr Opin Insect Sci 24:51–57. https://doi.org/10.1016/j.cois.2017.09.008. PMID: 29208223

Dubowy C, Sehgal A (2017) Circadian rhythms and sleep in Drosophila melanogaster. Genetics 205(4):1373–1397. https://doi.org/10.1534/genetics.115.185157. PMID: 28360128

Eastman CI, Tomaka VA, Crowley SJ (2017) Sex and ancestry determine the free-running circadian period. J Sleep Res 26(5):547–550. https://doi.org/10.1111/jsr.12521. PMID: 28332253

Eban-Rothschild A, Giardino WJ, de Lecea L (2017) To sleep or not to sleep: neuronal and ecological insights. Curr Opin Neurobiol 44:132–138. https://doi.org/10.1016/j.conb.2017.04.010. PMID: 28500869

Ehlen JC, Hesse S, Pinckney L, Paul KN (2013) Sex chromosomes regulate nighttime sleep propensity during recovery from sleep loss in mice. PLoS One 8(5):e62205. https://doi.org/10.1371/journal.pone.0062205. PMID: 23658713

Fraser R, Nordin BE (1955) The basal metabolic rate during sleep. Lancet 268(6863):532–533. PMID: 14354919

Frieboes RM, Murck H, Stalla GK, Antonijevic IA, Steiger A (1998) Enhanced slow wave sleep in patients with prolactinoma. J Clin Endocrinol Metab 83(8):2706–2710. https://doi.org/10.1210/jcem.83.8.5016. PMID: 9709935

Fulcher BD, Phillips AJ, Postnova S, Robinson PA (2014) A physiologically based model of orexinergic stabilization of sleep and wake. PLoS One 9(3):e91982. https://doi.org/10.1371/journal.pone.0091982. PMID: 24651580

Fuller PM, Fuller P, Sherman D, Pedersen NP, Saper CB, Lu J (2011) Reassessment of the structural basis of the ascending arousal system. J Comp Neurol 519(5):933–956

Gamaldo CE, Earley CJ (2006) Restless legs syndrome: a clinical update. Chest 130(5):1596–1604. https://doi.org/10.1378/chest.130.5.1596. PMID: 17099042

Gandhi AV, Mosser EA, Oikonomou G, Prober DA (2015) Melatonin is required for the circadian regulation of sleep. Neuron 85(6):1193–1199

Giménez S, Romero S, Gich I, Clos S, Grasa E, Antonijoan RM et al (2011) Sex differences in sleep after a single oral morning dose of olanzapine in healthy volunteers. Hum Psychopharmacol 26 (7):498–507. https://doi.org/10.1002/hup.1232. PMID: 21953682

Goel N, Kim H, Lao RP (2005) Gender differences in polysomnographic sleep in young healthy sleepers. Chronobiol Int 22(5):905–915. https://doi.org/10.1080/07420520500263235. PMID: 16298775

Goldstein-Piekarski AN, Greer SM, Saletin JM, Harvey AG, Williams LM, Walker MP (2018) Sex, sleep deprivation, and the anxious brain. J Cogn Neurosci 30(4):565–578. https://doi.org/10.1162/jocn_a_01225. PMID: 29244642

Golombek DA, Rosenstein RE (2010) Physiology of circadian entrainment. Physiol Rev 90 (3):1063–1102. https://doi.org/10.1152/physrev.00009.2009. PMID: 20664079

Grafe LA, Bhatnagar S (2018) Orexins and stress. Front Neuroendocrinol 51:132–145

Greenblatt DJ, Harmatz JS, Singh NN, Steinberg F, Roth T, Moline ML et al (2014) Gender differences in pharmacokinetics and pharmacodynamics of zolpidem following sublingual administration. J Clin Pharmacol 54(3):282–290. https://doi.org/10.1002/jcph.220. PMID: 24203450

Greene R, Siegel J (2004) Sleep: a functional enigma. NeuroMolecular Med 5(1):59–68. https://doi.org/10.1385/NMM:5:1:059. PMID: 15001813

Hadjimarkou MM, Benham R, Schwarz JM, Holder MK, Mong JA (2008) Estradiol suppresses rapid eye movement sleep and activation of sleep-active neurons in the ventrolateral preoptic area. Eur J Neurosci 27(7):1780–1792. https://doi.org/10.1111/j.1460-9568.2008.06142.x. PMID: 18371078

Held K, Antonijevic I, Murck H, Künzel H, Steiger A (2005) Alpha-helical CRH exerts CRH agonistic effects on sleep-endocrine activity in humans. Neuropsychobiology 52(2):62–67. https://doi.org/10.1159/000086606. PMID: 15990457

Hickie IB, Rogers NL (2011) Novel melatonin-based therapies: potential advances in the treatment of major depression. Lancet 378(9791):621–631. https://doi.org/10.1016/S0140-6736(11)60095-0. PMID: 21596429

Honda T, Fujiyama T, Miyoshi C, Ikkyu A, Hotta-Hirashima N, Kanno S et al (2018) A single phosphorylation site of SIK3 regulates daily sleep amounts and sleep need in mice. Proc Natl Acad Sci U S A 115(41):10458–10463. https://doi.org/10.1073/pnas.1810823115. PMID: 30254177

Huang ZL, Zhang Z, Qu WM (2014) Roles of adenosine and its receptors in sleep-wake regulation. Int Rev Neurobiol 119:349–371. https://doi.org/10.1016/B978-0-12-801022-8.00014-3. PMID: 25175972

Hutchison IC, Rathore S (2015) The role of REM sleep theta activity in emotional memory. Front Psychol 6:1439. https://doi.org/10.3389/fpsyg.2015.01439. PMID: 26483709

James FO, Walker CD, Boivin DB (2004) Controlled exposure to light and darkness realigns the salivary cortisol rhythm in night shift workers. Chronobiol Int 21(6):961–972. PMID: 15646242

Johnson EO, Roth T, Schultz L, Breslau N (2006) Epidemiology of DSM-IV insomnia in adolescence: lifetime prevalence, chronicity, and an emergent gender difference. Pediatrics 117(2): e247–e256. https://doi.org/10.1542/peds.2004-2629. PMID: 16452333

Kapás L, Hong L, Cady AB, Opp MR, Postlethwaite AE, Seyer JM et al (1992) Somnogenic, pyrogenic, and anorectic activities of tumor necrosis factor-alpha and TNF-alpha fragments. Am J Phys 263(3 Pt 2):R708–R715. https://doi.org/10.1152/ajpregu.1992.263.3.R708. PMID: 1357984

Karatsoreos IN, Butler MP, Lesauter J, Silver R (2011) Androgens modulate structure and function of the suprachiasmatic nucleus brain clock. Endocrinology 152(5):1970–1978. https://doi.org/10.1210/en.2010-1398. PMID: 21363939

Kayaba M, Sasai-sakuma T, Inoue Y (2018) Clinical significance of social jetlag in patients with excessive daytime sleepiness. Chronobiol Int 35(12):1637–1646

Koehl M, Battle S, Meerlo P (2006) Sex differences in sleep: the response to sleep deprivation and restraint stress in mice. Sleep 29(9):1224–1231. PMID: 17040010

Krause AJ, Simon EB, Mander BA, Greer SM, Saletin JM, Goldstein-Piekarski AN et al (2017) The sleep-deprived human brain. Nat Rev Neurosci 18(7):404–418. https://doi.org/10.1038/nrn.2017.55. PMID: 28515433

Kumar S, Rai S, Hsieh KC, McGinty D, Alam MN, Szymusiak R (2013) Adenosine A (2A) receptors regulate the activity of sleep regulatory GABAergic neurons in the preoptic hypothalamus. Am J Physiol Regul Integr Comp Physiol 305(1):R31–R41. https://doi.org/10.1152/ajpregu.00402.2012. PMID: 23637137

Lancel M, Faulhaber J, Holsboer F, Rupprecht R (1996) Progesterone induces changes in sleep comparable to those of agonistic GABAA receptor modulators. Am J Phys 271(4 Pt 1):E763–E772. https://doi.org/10.1152/ajpendo.1996.271.4.E763. PMID: 8897866

Lanquart JP, Nardone P, Hubain P, Loas G, Linkowski P (2018) The dichotomy between low frequency and delta waves in human sleep: a reappraisal. J Neurosci Methods 293:234–246. https://doi.org/10.1016/j.jneumeth.2017.09.019. PMID: 28982607

Lazarus M, Chen JF, Huang ZL, Urade Y, Fredholm BB (2017) Adenosine and sleep. Handb Exp Pharmacol 7:238–245. https://doi.org/10.1007/164_2017_36. PMID: 28646346

Lucey BP, Bateman RJ (2014) Amyloid-β diurnal pattern: possible role of sleep in Alzheimer's disease pathogenesis. Neurobiol Aging 35(Suppl 2):S29–S34. https://doi.org/10.1016/j.neurobiolaging.2014.03.035. PMID: 24910393

Macedo AC, Balouch S, Tabet N (2017) Is sleep disruption a risk factor for Alzheimer's disease? J Alzheimers Dis 58(4):993–1002. https://doi.org/10.3233/JAD-161287. PMID: 28550253

Marchant EG, Mistlberger RE (1997) Anticipation and entrainment to feeding time in intact and SCN-ablated C57BL/6j mice. Brain Res 765(2):273–282. PMID: 9313900

Matheson E, Hainer BL (2017) Insomnia: pharmacologic therapy. Am Fam Physician 96(1):29–35

Methippara MM, Kumar S, Alam MN, Szymusiak R, McGinty D (2005) Effects on sleep of microdialysis of adenosine A1 and A2a receptor analogs into the lateral preoptic area of rats. Am J Physiol Regul Integr Comp Physiol 289(6):R1715–R1723. https://doi.org/10.1152/ajpregu.00247.2005. PMID: 16109808

Micic G, Lovato N, Gradisar M, Ferguson SA, Burgess HJ, Lack LC (2016) The etiology of delayed sleep phase disorder. Sleep Med Rev 27:29–38. https://doi.org/10.1016/j.smrv.2015.06.004. PMID: 26434674

Miki K, Yoshimoto M (2013) Sympathetic nerve activity during sleep, exercise, and mental stress. Auton Neurosci 174(1–2):15–20. https://doi.org/10.1016/j.autneu.2012.12.007. PMID: 23332904

Model Z, Butler MP, LeSauter J, Silver R (2015) Suprachiasmatic nucleus as the site of androgen action on circadian rhythms. Horm Behav 73:1–7. https://doi.org/10.1016/j.yhbeh.2015.05.007. PMID: 26012711

Mong JA, Cusmano DM (2016) Sex differences in sleep: impact of biological sex and sex steroids. Philos Trans R Soc Lond Ser B Biol Sci 371(1688):20150110. https://doi.org/10.1098/rstb.2015.0110. PMID: 26833831

Mong JA, Devidze N, Frail DE, O'Connor LT, Samuel M, Choleris E et al (2003) Estradiol differentially regulates lipocalin-type prostaglandin D synthase transcript levels in the rodent brain: evidence from high-density oligonucleotide arrays and in situ hybridization. Proc Natl Acad Sci U S A 100(1):318–323. https://doi.org/10.1073/pnas.262663799. PMID: 12518068

Morgan D, Tsai SC (2015) Sleep and the endocrine system. Crit Care Clin 31(3):403–418. https://doi.org/10.1016/j.ccc.2015.03.004. PMID: 26118912

Morgan PJ, Barrett P, Howell HE, Helliwell R (1994) Melatonin receptors: localization, molecular pharmacology and physiological significance. Neurochem Int 24(2):101–146. PMID: 8161940

Moser D, Anderer P, Gruber G, Parapatics S, Loretz E, Boeck M et al (2009) Sleep classification according to AASM and Rechtschaffen & Kales: effects on sleep scoring parameters. Sleep 32 (2):139–149. PMID: 19238800

Mouret J, Coindet J, Debilly G, Chouvet G (1978) Suprachiasmatic nuclei lesions in the rat: alterations in sleep circadian rhythms. Electroencephalogr Clin Neurophysiol 45(3):402–408. PMID: 79478

Nugent BM, Wright CL, Shetty AC, Hodes GE, Lenz KM, Mahurkar A et al (2015) Brain feminization requires active repression of masculinization via DNA methylation. Nat Neurosci 18(5):690–697. https://doi.org/10.1038/nn.3988. PMID: 25821913

Nurminen T (1998) Shift work and reproductive health. Scand J Work Environ Health 24(Suppl 3):28–34. PMID: 9916814

Obal F, Opp M, Cady AB, Johannsen L, Postlethwaite AE, Poppleton HM et al (1990) Interleukin 1 alpha and an interleukin 1 beta fragment are somnogenic. Am J Phys 259(3 Pt 2):R439–R446. https://doi.org/10.1152/ajpregu.1990.259.3.R439. PMID: 2396703

Ocampo-Garcés A, Ibáñez F, Perdomo G, Torrealba F (2011) Orexin-B-saporin lesions in the lateral hypothalamus enhance photic masking of rapid eye movement sleep in the albino rat. J Sleep Res 20(1 Pt 1):3–11. https://doi.org/10.1111/j.1365-2869.2010.00864.x. PMID: 20626614

Ohayon MM (2002) Epidemiology of insomnia: what we know and what we still need to learn. Sleep Med Rev 6(2):97–111. PMID: 12531146

Pase MP, Himali JJ, Grima NA, Beiser AS, Satizabal CL, Aparicio HJ et al (2017) Sleep architecture and the risk of incident dementia in the community. Neurology 89 (12):1244–1250. https://doi.org/10.1212/WNL.0000000000004373. PMID: 28835407

Paul KN, Dugovic C, Turek FW, Laposky AD (2006) Diurnal sex differences in the sleep-wake cycle of mice are dependent on gonadal function. Sleep 29(9):1211–1223. PMID: 17040009

Pavlova M (2017) Circadian rhythm sleep-wake disorders. Continuum 23(4):1051–1063. https://doi.org/10.1212/CON.0000000000000499. PMID: 28777176

Perelis M, Ramsey KM, Marcheva B, Bass J (2016) Circadian transcription from beta cell function to diabetes pathophysiology. J Biol Rhythm 31(4):323–336. https://doi.org/10.1177/0748730416656949. PMID: 27440914

Peter-Derex L, Yammine P, Bastuji H, Croisile B (2015) Sleep and Alzheimer's disease. Sleep Med Rev 19:29–38. https://doi.org/10.1016/j.smrv.2014.03.007. PMID: 24846773

Pietrowsky R, Meyrer R, Kern W, Born J, Fehm HL (1994) Effects of diurnal sleep on secretion of cortisol, luteinizing hormone, and growth hormone in man. J Clin Endocrinol Metab 78 (3):683–687. https://doi.org/10.1210/jcem.78.3.8126142. PMID: 8126142

Plog BA, Nedergaard M (2018) The glymphatic system in central nervous system health and disease: past, present, and future. Annu Rev Pathol 13:379–394. https://doi.org/10.1146/annurev-pathol-051217-111018. PMID: 29195051

Punjabi NM (2008) The epidemiology of adult obstructive sleep apnea. Proc Am Thorac Soc 5 (2):136–143. https://doi.org/10.1513/pats.200709-155MG. PMID: 18250205

Ramírez DA, Vieyra E, González AI, Morán C, Domínguez R, Morales-Ledesma L (2017) Both the suprachiasmatic nucleus and the superior ovarian nerve contribute to the processes of ovulation and steroid hormone secretion on proestrus. Reprod Sci 24(6):844–855. https://doi.org/10.1177/1933719116670307. PMID: 27688242

Rasmussen MH, Wildschiødtz G, Juul A, Hilsted J (2008) Polysomnographic sleep, growth hormone insulin-like growth factor-I axis, leptin, and weight loss. Obesity (Silver Spring) 16 (7):1516–1521. https://doi.org/10.1038/oby.2008.249. PMID: 18464752

Raven F, Van der Zee EA, Meerlo P, Havekes R (2018) The role of sleep in regulating structural plasticity and synaptic strength: implications for memory and cognitive function. Sleep Med Rev 39:3–11. https://doi.org/10.1016/j.smrv.2017.05.002. PMID: 28641933

Redline S, Kirchner HL, Quan SF, Gottlieb DJ, Kapur V, Newman A (2004) The effects of age, sex, ethnicity, and sleep-disordered breathing on sleep architecture. Arch Intern Med 164 (4):406–418. https://doi.org/10.1001/archinte.164.4.406. PMID: 14980992

Reichert CF, Maire M, Schmidt C, Cajochen C (2016) Sleep-wake regulation and its impact on working memory performance: the role of adenosine. Biology 5(1):E11. https://doi.org/10.3390/biology5010011. PMID: 26861410

Ribeiro AC, Pfaff DW, Devidze N (2009) Estradiol modulates behavioral arousal and induces changes in gene expression profiles in brain regions involved in the control of vigilance. Eur J Neurosci 29(4):795–801. https://doi.org/10.1111/j.1460-9568.2009.06620.x. PMID: 19200066

Rønnekleiv OK, Kelly MJ (2005) Diversity of ovarian steroid signaling in the hypothalamus. Front Neuroendocrinol 26(2):65–84. https://doi.org/10.1016/j.yfrne.2005.05.001. PMID: 16009409

Sack RL, Brandes RW, Kendall AR, Lewy AJ (2000) Entrainment of free-running circadian rhythms by melatonin in blind people. N Engl J Med 343(15):1070–1077

Santiago JR, Nolledo MS, Kinzler W, Santiago TV (2001) Sleep and sleep disorders in pregnancy. Ann Intern Med 134(5):396–408

Saper CB, Fuller PM (2017) Wake-sleep circuitry: an overview. Curr Opin Neurobiol 44:186–192

Saper CB, Fuller PM, Pedersen NP, Lu J, Scammell TE (2010) Sleep state switching. Neuron 68 (6):1023–1042. https://doi.org/10.1016/j.neuron.2010.11.032. PMID: 21172606

Sato T, Matsumoto T, Kawano H, Watanabe T, Uematsu Y, Sekine K et al (2004) Brain masculinization requires androgen receptor function. Proc Natl Acad Sci U S A 101(6):1673–1678. https://doi.org/10.1073/pnas.0305303101. PMID: 14747651

Sawaki Y, Nihonmatsu I, Kawamura H (1984) Transplantation of the neonatal suprachiasmatic nuclei into rats with complete bilateral suprachiasmatic lesions. Neurosci Res 1(1):67–72. PMID: 6536887

Schlaich M, Straznicky N, Lambert E, Lambert G (2015) Metabolic syndrome: a sympathetic disease? Lancet Diabetes Endocrinol 3(2):148–157. https://doi.org/10.1016/S2213-8587(14)70033-6. PMID: 24731670

Schmid SM, Hallschmid M, Schultes B (2015) The metabolic burden of sleep loss. Lancet Diabetes Endocrinol 3(1):52–62. https://doi.org/10.1016/S2213-8587(14)70012-9. PMID: 24731536

Schwab RJ, Kim C, Bagchi S, Keenan BT, Comyn FL, Wang S et al (2015) Understanding the anatomic basis for obstructive sleep apnea syndrome in adolescents. Am J Respir Crit Care Med 191(11):1295–1309. https://doi.org/10.1164/rccm.201501-0169OC. PMID: 25835282

Schwartz MD, Mong JA (2013) Estradiol modulates recovery of REM sleep in a time-of-day-dependent manner. Am J Physiol Regul Integr Comp Physiol 305(3):R271–R280. https://doi.org/10.1152/ajpregu.00474.2012. PMID: 23678032

Schwierin B, Borbély AA, Tobler I (1998) Sleep homeostasis in the female rat during the estrous cycle. Brain Res 811(1–2):96–104. PMID: 9804908

Shinohara K, Honma S, Katsuno Y, Abe H, Honma K (1998) Circadian release of amino acids in the suprachiasmatic nucleus in vitro. Neuroreport 9(1):137–140. PMID: 9592063

Simasko SM, Mukherjee S (2009) Novel analysis of sleep patterns in rats separates periods of vigilance cycling from long-duration wake events. Behav Brain Res 196(2):228–236. https://doi.org/10.1016/j.bbr.2008.09.003. PMID: 18835301

Späth-Schwalbe E, Gofferje M, Kern W, Born J, Fehm HL (1991) Sleep disruption alters nocturnal ACTH and cortisol secretory patterns. Biol Psychiatry 29(6):575–584. PMID: 1647222

Spiegel K, Leproult R, Van Cauter E (1999) Impact of sleep debt on metabolic and endocrine function. Lancet 354(9188):1435–1439. https://doi.org/10.1016/S0140-6736(99)01376-8. PMID: 10543671

Squarcini CF, Pires ML, Lopes C, Benedito-Silva AA, Esteves AM, Cornelissen-Guillaume G et al (2013) Free-running circadian rhythms of muscle strength, reaction time, and body temperature in totally blind people. Eur J Appl Physiol 113(1):157–165. https://doi.org/10.1007/s00421-012-2415-8. PMID: 22618303

Steiger A (2002) Sleep and the hypothalamo-pituitary-adrenocortical system. Sleep Med Rev 6 (2):125–138. PMID: 12531148

Stenberg D (2007) Neuroanatomy and neurochemistry of sleep. Cell Mol Life Sci 64 (10):1187–1204. https://doi.org/10.1007/s00018-007-6530-3. PMID: 17364141

Steyn FJ, Ngo ST (2017) Endocrine rhythms of growth hormone release: insights from animal studies. Best Pract Res Clin Endocrinol Metab 31(6):521–533. https://doi.org/10.1016/j.beem.2017.10.009. PMID: 29224665

Sun X, Whitefield S, Rusak B, Semba K (2001) Electrophysiological analysis of suprachiasmatic nucleus projections to the ventrolateral preoptic area in the rat. Eur J Neurosci 14(8):1257–1274. PMID: 11703455

Szymusiak R, McGinty D (2008) Hypothalamic regulation of sleep and arousal. Ann N Y Acad Sci 1129:275–286. https://doi.org/10.1196/annals.1417.027. PMID: 18591488

Trotti LM, Becker LA (2019) Iron for the treatment of restless legs syndrome. Cochrane Database Syst Rev 1:CD007834. https://doi.org/10.1002/14651858.CD007834.pub3. PMID: 30609006

van Cauter E, Plat L (1996) Physiology of growth hormone secretion during sleep. J Pediatr 128 (5 Pt 2):S32–S37. PMID: 8627466

Van Melick MJ, Van Beukering MD, Mol BW, Frings-Dresen MH, Hulshof CT (2014) Shift work, long working hours and preterm birth: a systematic review and meta-analysis. Int Arch Occup Environ Health 87(8):835–849. https://doi.org/10.1007/s00420-014-0934-9. PMID: 24584887

Vida B, Hrabovszky E, Kalamatianos T, Coen CW, Liposits Z, Kalló I (2008) Oestrogen receptor alpha and beta immunoreactive cells in the suprachiasmatic nucleus of mice: distribution, sex differences and regulation by gonadal hormones. J Neuroendocrinol 20(11):1270–1277. https://doi.org/10.1111/j.1365-2826.2008.01787.x. PMID: 18752649

Villafuerte G, Miguel-Puga A, Rodríguez EM, Machado S, Manjarrez E, Arias-Carrión O (2015) Sleep deprivation and oxidative stress in animal models: a systematic review. Oxid Med Cell Longev 2015:234952. https://doi.org/10.1155/2015/234952. PMID: 25945148

Vyazovskiy VV, Kopp C, Wigger E, Jones ME, Simpson ER, Tobler I (2006) Sleep and rest regulation in young and old oestrogen-deficient female mice. J Neuroendocrinol 18(8):567–576. https://doi.org/10.1111/j.1365-2826.2006.01452.x. PMID: 16867177

Wagner DR (1999) Circadian rhythm sleep disorders. Curr Treat Options Neurol 1(4):299–308. PMID: 11096717

Wang Z, Ma J, Miyoshi C, Li Y, Sato M, Ogawa Y et al (2018) Quantitative phosphoproteomic analysis of the molecular substrates of sleep need. Nature 558(7710):435–439. https://doi.org/10.1038/s41586-018-0218-8. PMID: 29899451

White DP (2017) Advanced concepts in the pathophysiology of obstructive sleep apnea. Adv Otorhinolaryngol 80:7–16. https://doi.org/10.1159/000470522. PMID: 28738382

Wirz-Justice A (2018) Seasonality in affective disorders. Gen Comp Endocrinol 258:244–249. https://doi.org/10.1016/j.ygcen.2017.07.010. PMID: 28711512

Wittert G (2014) The relationship between sleep disorders and testosterone in men. Asian J Androl 16(2):262–265. https://doi.org/10.4103/1008-682X.122586. PMID: 24435056

Wright KP, Drake AL, Frey DJ, Fleshner M, Desouza CA, Gronfier C et al (2015) Influence of sleep deprivation and circadian misalignment on cortisol, inflammatory markers, and cytokine balance. Brain Behav Immun 47:24–34. https://doi.org/10.1016/j.bbi.2015.01.004. PMID: 25640603

Xie L, Kang H, Xu Q, Chen MJ, Liao Y, Thiyagarajan M et al (2013) Sleep drives metabolite clearance from the adult brain. Science 342(6156):373–377. https://doi.org/10.1126/science.1241224. PMID: 24136970

Yanik G, Glaum S, Radulovacki M (1987) The dose-response effects of caffeine on sleep in rats. Brain Res 403(1):177–180. PMID: 3828812

Zeitzer JM, Morales-Villagran A, Maidment NT, Behnke EJ, Ackerson LC, Lopez-Rodriguez F et al (2006) Extracellular adenosine in the human brain during sleep and sleep deprivation: an in vivo microdialysis study. Sleep 29(4):455–461. PMID: 16676778

Zhang BJ, Huang ZL, Chen JF, Urade Y, Qu WM (2017) Adenosine A2A receptor deficiency attenuates the somnogenic effect of prostaglandin D2 in mice. Acta Pharmacol Sin 38 (4):469–476. https://doi.org/10.1038/aps.2016.140. PMID: 28112177

Zhdanova IV, Wang SY, Leclair OU, Danilova NP (2001) Melatonin promotes sleep-like state in zebrafish. Brain Res 903(1–2):263–268

Hormonal Regulation of Hippocampal Neurogenesis: Implications for Depression and Exercise

Ana Gheorghe, Wansu Qiu, and Liisa A. M. Galea

Contents

1 Neurogenesis in the Hippocampus 380
2 Steroid Hormones and Neurogenesis 383
 2.1 Estrogens and Neurogenesis 384
 2.2 Testosterone and Neurogenesis 386
 2.3 Corticosterone and Neurogenesis 387
 2.4 Why Do Steroid Hormones Modulate Neurogenesis in the Hippocampus? 389
 2.5 Neurogenesis in the Subventricular Zone 390
3 Depression and Hippocampal Neurogenesis 391
4 Antidepressants and Neurogenesis 392
 4.1 Antidepressants Effects of Endogenous Gonadal Hormones 394
 4.2 Postpartum Depression and Neurogenesis 395
5 The Involvement of Brain-Derived Neurotrophic Factor (BDNF) in Neurogenesis and Depression 396
6 Exercise and Neurogenesis in the Hippocampus 399
7 Type of Exercise 399
 7.1 Voluntary vs Forced Running 399
 7.2 Voluntary Exercise 401
 7.3 Aerobic Exercise: Does Intensity Matter? 402
 7.4 Resistance Training 403
8 Sex Differences in Exercise Studies 403
 8.1 Sex Differences in Distance Run 404

Ana Gheorghe and Wansu Qiu contributed equally.

A. Gheorghe and W. Qiu
Djavad Mowafaghian Centre for Brain Health, University of British Columbia, Vancouver, BC, Canada

Graduate Program in Neuroscience, University of British Columbia, Vancouver, BC, Canada

L. A. M. Galea (✉)
Djavad Mowafaghian Centre for Brain Health, University of British Columbia, Vancouver, BC, Canada

Graduate Program in Neuroscience, University of British Columbia, Vancouver, BC, Canada

Department of Psychology, University of British Columbia, Vancouver, BC, Canada
e-mail: liisa.galea@ubc.ca

Curr Topics Behav Neurosci (2019) 43: 379–422
DOI 10.1007/7854_2018_62
Published Online: 10 November 2018

9 Ovarian Hormones, Physical Activity, and Neurogenesis 405
9.1 Sex Differences in the Hypothalamic-Pituitary-Adrenal (HPA) Axis 406
10 Androgens and Exercise in the Modulation of Neurogenesis 407
11 Physical Activity in Animal Models of Depression 407
12 Conclusions and Future Research 408
References 409

Abstract Adult hippocampal neurogenesis exists in all mammalian species, including humans, and although there has been considerable research investigating the function and regulation of neurogenesis, there remain many open questions surrounding the complexity of this phenomenon. This stems partially from the fact that neurogenesis is a multistage process that involves proliferation, differentiation, migration, survival, and eventual integration of new cells into the existing hippocampal circuitry, each of which can be independently influenced. The function of adult neurogenesis in the hippocampus is related to stress regulation, behavioral efficacy of antidepressants, long-term spatial memory, forgetting, and pattern separation. Steroid hormones influence the regulation of hippocampal neurogenesis, stress regulation, and cognition and differently in males and females. In this chapter, we will briefly tap into the complex network of steroid hormone modulation of neurogenesis in the hippocampus with specific emphasis on stress, testosterone, and estrogen. We examine the possible role of neurogenesis in the etiology of depression and influencing treatment by examining the influence of both pharmacological (selective serotonin reuptake inhibitors, tricyclic antidepressants) treatments and non-pharmacological (exercise) remedies.

Keywords Depression · Exercise · Hippocampus · Neurogenesis · Steroid hormones · Stress

1 Neurogenesis in the Hippocampus

Prior to the 1960s, it was common to believe that mammals were born with a terminal number of neurons. However, Altman and Das (1965) provided compelling evidence for the existence of postnatal cell genesis in the hippocampus of adult mammals. Today, the phenomenon of adult hippocampal neurogenesis is well established, though the precise mechanisms and implications continue to be an area of exciting research. The hippocampus is divided into the dorsal and ventral regions, which may serve different functions. The dorsal region projects to retrosplenial and anterior cingulate cortices in rodents and may be involved in processing cognitive visuospatial information as well as spatial memory (reviewed by Fanselow and Dong 2010). The ventral region projects to the nucleus accumbens, prefrontal cortex, amygdala, and hypothalamus and may be more involved in regulating stress, emotions, and reward behavior (Sahay and Hen 2007; Fanselow and Dong 2010). Examining neurogenesis in the dorsal versus ventral regions can provide us with some cues as to how it is regulated and what functions adult neurogenesis in the hippocampus may serve. However, the separate contributions of dorsal versus ventral regions of the dentate gyrus have been less well studied (Lacar et al. 2014; Tanti and Belzung 2013; O'Leary and Cryan 2014).

Neurogenesis in the hippocampus during adulthood originates from neural stem/progenitor cells in the subgranular zone in the dentate gyrus that when stimulated to proliferate form two daughter cells, at least one of which will become a mature granule neuron. The dentate gyrus receives its major input from the entorhinal cortex via the perforant path, but it also receives some input from the locus coeruleus, the ventral tegmental area, septal nuclei in the basal forebrain, and supramammillary nuclei in the hypothalamus (reviewed in Amaral et al. 2007). Glial-like cells, putative neuronal stem cells, in the subgranular zone go through multiple stages to develop into new granule cells that, if they survive to maturity, will form synapses with pyramidal cells in the CA3 region of the hippocampus. These new neurons become integrated into the functional hippocampal circuitry (Kempermann et al. 2015; Fig. 1). However, whether new neurons form appropriate connections or contribute in a positive way to the function of the dentate gyrus depends on a

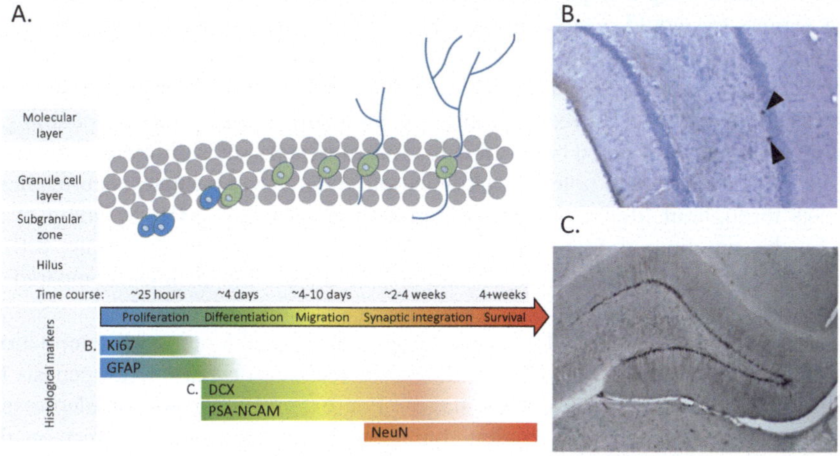

Fig. 1 (**a**) Diagram description of the stages of neurogenesis alongside a non-exhaustive list of endogenous markers such as Ki67, DCX, and NeuN. These markers are used to indicate at which stage of cell maturity is each marked cell at during the time of the staining. Two daughter cells (blue) are produced (cell proliferation), and at least one (green) will mature into a neuron. Markers such as Ki67 are specifically used to measure cell proliferation. Whereas DCX can be used to measure the number of immature cells within the granule cell layer, NeuN is a marker expressed in mature neurons only, often used in conjunction with an immature cell marker to indicate a the portion of proliferated cells that has progressed into neurons. *GFAP* glial fibrillary acidic protein, *DCX* doublecortin, *NeuN* neuronal nuclei. (**b**) Photomicrograph of Ki67-expressing cells. Arrows indicated Ki67-expressing newly proliferated cells. (**c**) DCX-expressing immature cells along the granule cell layer with branches of dendrites extending out from the cell body. Figure (**a**) is reprinted with permission from Mahmoud et al. (2016b)

number of factors including the environment into which the new neuron was produced and matured (Jakubs et al. 2006).

New cells can be visualized using endogenous or exogenous markers. Endogenous proteins such as Ki67, a nuclear protein that is expressed exclusively during cell proliferation (with the exception of G_0), or doublecortin (DCX), a microtubule-associated protein that is expressed in immature neurons, are often used depending on the question of interest. Exogenous markers such as thymidine analogs are used to label cells that have divided during a short time window. Bromodeoxyuridine (BrdU), a synthetic nucleoside, is incorporated into dividing cells during a 2-h-long period after injection and can be used to quantify cell proliferation and/or survival of new cells depending on the timeline between BrdU incorporation and sacrifice. However, it is important, when using BrdU or other synthetic nucleosides, to be aware of several caveats mainly around the dose, toxicity, and timing of BrdU exposure relative to any manipulations and sacrifice as these can affect outcome (see Taupin 2007). BrdU-immunoreactive staining itself is not able to identify cell phenotype. In order to identify cell fate, this exogenous marker staining must be coupled with endogenous markers such as glial fibrillary acidic protein (GFAP), NeuN (mature neuronal marker), or DCX. BrdU is advantageous when needing to explore the exact timeline of neurogenesis as it provides a "time stamp" of when a cell was proliferating (Taupin 2007). The endogenous marker Ki67 is limited by the time expression would occur (the past 24 h) and its expression is not restricted to neuronal lineages (Mahmoud et al. 2016b). DCX expression is however neuron-specific, but it does not provide an accurate representation of cell proliferation and is expressed for up to 2–3 weeks during the immature stage, prior to becoming a mature neuron, and as noted below, the timeline of expression is different in different species. It is important to note that while in both rodents and primates, neurogenesis persists throughout life (reviewed by Cameron and Mckay 2001; Spalding et al. 2013), the timeline of maturation of new neurons in the hippocampus differs between species including a longer timeline of maturation in mice compared to rats (Snyder et al. 2009).

Neurogenesis is comprised of at least four stages: proliferation, differentiation, migration, and survival (Fig. 1). Factors can increase or decrease neurogenesis by influencing any one of these four stages. For example, chronic antidepressant exposure increases cell proliferation, but there is no independent effect on the survival of new neurons (Malberg et al. 2000); that is, while the cells that proliferate in response to chronic antidepressants will survive, those cells produced *prior to* exposure to antidepressants do not show enhanced survival of new neurons. On the other hand, chronic testosterone can increase the survival of new neurons but not cell proliferation (Hamson et al. 2013). Furthermore, it is important to be aware that an increase in neurogenesis does not equate to better functioning of the hippocampus or dentate gyrus, as these new neurons need to make appropriate connections to be functionally relevant (Jakubs et al. 2006). Indeed, the work in Paul Frankland and Sheena Josselyn's labs has shown that neurogenesis is linked to forgetting and promoting proactive interference (Epp et al. 2016; Akers et al. 2014). For further information on this topic, the reader is directed to other reviews on the topic of neurogenesis and memory (Frankland et al. 2013; Epp et al. 2013b).

The function of adult hippocampal neurogenesis has been attributed to certain forms of learning and memory such as pattern separation (Gould et al. 1999; Clelland et al. 2009; Yau et al. 2015), anxiolytic effects (Revest et al. 2009; Hill et al. 2015), and stress regulation (Snyder et al. 2011). While there is little evidence that implies a direct role of adult neurogenesis on mood, new neurons in the hippocampus may regulate stress and antidepressant efficacy (Sahay and Hen 2007; Vollmayr et al. 2007). As is detailed below, neurogenesis in the hippocampus is upregulated by chronic exposure to antidepressants and may be involved in the efficacy of antidepressant treatment on anxiety (Malberg et al. 2000; David et al. 2009). Furthermore, physical exercise has antidepressant properties and can enhance neurogenesis in the hippocampus (reviewed by Ernst et al. 2006; Triviño-Paredes et al. 2016). Neurogenesis in the hippocampus is also linked to cognition. Intriguingly, cognitive deficits are seen in depression (Rock et al. 2014), and exercise can enhance cognition (Hötting and Röder 2013; Barha et al. 2017a, b), suggesting an association between cognition, depression, exercise, and neurogenesis, some of which will be discussed within this chapter. For the main purpose of this chapter, we will discuss how hormonal factors can influence neurogenesis in the hippocampus, depression, and exercise. First, we briefly review the influence of the steroid hormones corticosterone, estrogens, and testosterone in both male and female rodents on neurogenesis in the hippocampus and then outline the influence of depression and exercise on neurogenesis. This chapter will also explore the roles of pharmacological and non-pharmacological antidepressants in influencing neurogenesis with an emphasis on stress regulation, mood, and cognition.

2 Steroid Hormones and Neurogenesis

Steroid hormones are powerful regulators of neurogenesis in the hippocampus of adult rodents, and there are a number of sex differences seen in the hormonal regulation of neurogenesis. An extensive review of hormonal involvement on neurogenesis is beyond the scope of this chapter, but the reader is directed to a number of reviews on the subject (Mahmoud et al. 2016b; Schoenfeld and Gould 2012). Briefly, we discuss some of the major findings on estrogen, androgen, and corticosterone regulation of neurogenesis.

Both androgen receptor (AR) and estrogen receptor (ER) α mRNA expression are located throughout the rat brain with particularly high densities within the hypothalamus and areas projecting to the hypothalamus in both male and female rats (Simerly et al. 1990). In terms of the hippocampus, both ERα and AR expression are prominent within the CA1 region of the hippocampus. AR has sparse expression within the dentate gyrus of the hippocampus except in Wistar rats (Xiao and Jordan 2002; Hamson et al. 2013; Brännvall et al. 2005). As noted in the androgens section below, despite the low expression of AR in the dentate gyrus, AR activation is required for regulating neurogenesis within the adult male hippocampus, perhaps via a cell autonomous manner or via AR in the CA3 region (Hamson et al. 2013). The two subtypes of ERs (ERα and ERβ) have been identified within the dentate gyrus in both male and female rats (Weiland et al. 1997; Zhang et al. 2002). ERα is expressed within the dentate gyrus and the CA1 area with no sex differences reported (Weiland et al. 1997). ERβ is expressed within the dentate gyrus, CA3, and CA4 in both males and female rats, with higher overall expression in the female compared to male hippocampus (Zhang et al. 2002). In females, the G-protein-coupled estrogen receptor (GPER) expression is prominent within the CA3 and CA1 area of the hippocampus; it is also present in the dentate gyrus as well but to a lesser extent (Duarte-Workman et al. 2015). As outlined below, both ERα and ERβ activation increases cell proliferation in the dentate gyrus of adult female rats (Mazzucco et al. 2006), suggesting that ERs are directly affecting neurogenesis in the adult female hippocampus.

Within the male rat brain, glucocorticoid receptor (GR) mRNA is localized throughout the brain, but the hippocampus has prominent expression of GR mRNA (Morimoto et al. 1996; Pryce 2008). Mineralocorticoid receptor (MR) mRNA expression within the central nervous system is predominantly within the hippocampus (Pryce 2008). Within the dentate gyrus, CA1, and CA2 regions specifically, GR and MR are expressed at higher levels than the CA3 region in both mice and rats. While there are no sex differences reported for MR expression in rats, adult female (P60) rats show higher GR expression than males (Bohn et al. 1994). It is important to note that expression levels for both GR and MR within the subregions of the hippocampus do differ depending on the age, sex, species, and strain of animal (Pryce 2008). However, in general, the dentate gyrus expresses high levels of both GR and MR, suggesting GR and MR may be directly involved in modulating neurogenesis as well as indirectly regulating adult hippocampal neurogenesis as discussed below (Anacker et al. 2013; Wong and Herbert 2005).

2.1 Estrogens and Neurogenesis

Estrogens are comprised of estrone, estradiol, estriol, and estetrol. Estradiol is the most abundant of the estrogens in premenopausal women and has the highest affinity to ERs. Estrone is the more abundant of the estrogens postmenopausal and has a weaker affinity to ERs and can be secreted by adipose tissue. Estriol and estetrol are secreted mainly during pregnancy. Estriol and estetrol have been the subject of study for patients with multiple sclerosis, who find their symptoms subside during pregnancy (Voskuhl and Momtazee 2017). To date, studies have examined only estradiol and estrone on neurogenesis in the hippocampus.

Estrogens have different effects on cell proliferation versus the survival of new neurons in adult female rodents, with limited effects in adult male rodents. Acute and repeated (15 days or less) estradiol can increase cell proliferation in the dentate gyrus of adult female rats (Tanapat et al. 1999; Barker and Galea 2008). The ability of acute estradiol to increase cell proliferation is due to estradiol's effects on serotonin (Banasr et al. 2001) but not via NMDAr activation (Ormerod et al. 2003). Activation of both ERα and ERβ via agonists, PPT and DPN, respectively, enhances cell proliferation in the dentate gyrus of adult female rats but not to the same extent as estradiol alone (Mazzucco et al. 2006), suggesting that there may be other indirect pathways by which estradiol increases cell proliferation. Intriguingly this is not due to estradiol's ability to activate GPER, as G1, a GPER agonist, decreases rather than increases cell proliferation in adult female rats (Duarte-Guterman et al. 2015). Our laboratory has shown that various doses of acute estradiol or estrone rapidly (within 30 min) increase cell proliferation (Barha et al. 2009). However, the timing of examination of the proliferation of progenitor cells relative to estradiol treatment is important; although acute estradiol can initially increase cell proliferation, 48 h after estradiol exposure, cell proliferation is suppressed (Ormerod and Galea 2003). The effect of prolonged exposure to estradiol to suppress cell proliferation is due to the fact that estradiol also promotes the production of adrenal steroids in females (Ormerod and Galea 2003). The same relationship between estradiol and cell proliferation is seen across the estrous cycle in rats and sometimes mice with the proestrous phase, and thus acute estradiol increases, associated with increased cell proliferation (Tanapat et al. 1999; Rummel et al. 2010; Tzeng et al. 2014, but see Lagace et al. 2007; Overall et al. 2013). Thus, acute estradiol and estrone can rapidly increase, but subsequently decrease, cell proliferation in the hippocampus of adult female rodents.

Chronic repeated injections of estradiol (15 days) also result in increased cell proliferation in the dentate gyrus of adult female rats, but curiously, after 21 days of estradiol, there is no longer a significant increase in cell proliferation (Barker and Galea 2008; Chan et al. 2014; Tanapat et al. 2005). Contrary to the effects on cell proliferation, chronic high estradiol, independent of its effects on cell proliferation, suppresses the survival of new neurons (Barker and Galea 2008; Chan et al. 2014). However, if estradiol is given *prior to* the BrdU injection, given that estradiol can increase cell proliferation (Barha et al. 2007), it should come as no surprise that more new neurons survive to maturity compared to vehicle controls (McClure et al. 2013). Furthermore, other hormone formulations such as Premarin, a commonly prescribed hormone therapy for postmenopausal symptoms, can increase neurogenesis via the survival of new neurons in young adult females (Barha and Galea 2013) but decrease neurogenesis in middle-aged females (Galea et al. 2018). To make the story more complex, the context in which the hormone is administered matters as the ability of Premarin to enhance the survival of new neurons in adult females is not seen in cage controls. Only in animals that have undergone some kind of cognitive stimulation (Radial Arm Maze) during the time of exposure to Premarin show increases in the survival of new neurons in young adult females (Barha et al. 2015; Barha and Galea 2013). These findings suggest that it is not merely the hormone that will promote or suppress neurogenesis, but whether or not the hippocampus is engaged during the hormone treatment that will determine the outcome on neurogenesis. Estrogens have a number of neuroprotective properties, and it may seem counterintuitive that repeated estrogens can suppress hippocampal neurogenesis in adult female rodents (Barker and Galea 2008; Chan et al. 2014). However, given that increased neurogenesis is related to increased forgetting (Akers et al. 2014) or that new neurons can interfere with the existing circuitry if they do not make appropriate connections (Jakubs et al. 2006; Jessberger et al. 2007), these two concepts become aligned, as it is possible that the reduced neurogenesis is accompanied by better acquisition of new memories.

So far, we have only referred to studies in female rodents; the findings of estrogen effects on neurogenesis in males are limited, but researchers suggest that androgens, not estrogens, affect neurogenesis in males (Spritzer and Galea 2007). However, at least one study found that estradiol given 6–10 days after BrdU injection increased neurogenesis in the hippocampus of adult male meadow voles, with or without spatial training (Ormerod et al. 2004). These findings collectively suggest that estrogens are a potent modulator of neurogenesis in adult female rodents, but do not modulate neurogenesis in adult male rodents to the same extent as androgens.

2.2 *Testosterone and Neurogenesis*

Several lines of evidence have indicated that in adult male rodents, chronic exposure to testosterone and other androgens, such as dihydrotestosterone (DHT), increases the survival of new granule neurons in the hippocampus, but does not increase cell proliferation. Ormerod and Galea (2003) found that reproductively active male meadow voles (higher testosterone) had higher cell survival than reproductively inactive voles but that there was no significant difference in cell proliferation between groups. Accordingly, Spritzer and Galea (2007) confirmed that both chronic (30 days) testosterone and DHT, but not estradiol, enhanced the survival of new neurons in adult male rats. Interestingly, a shorter time period of testosterone exposure did not produce the same neurogenic effect in castrated rats (Spritzer et al. 2011). These collective findings suggest that long-term androgen treatment influences the survival of new neurons but not the proliferation of new cells.

In another set of experiments, Hamson et al. (2013) used male rats that carried the testicular feminization mutation (TFM), which made the androgen receptors unresponsive to testosterone. In TFM rats chronic testosterone treatment (30 days) did not influence cell survival rate compared to wild-type animals. Further, castrated wild-type males were given 30 days of either DHT, flutamide (a competitive AR antagonist), or DHT + flutamide. In accordance with Spritzer and Galea (2007), DHT injections significantly increased survival of new neurons in castrated adult males compared to vehicle-injected controls. However, when animals received flutamide concurrently with DHT, the increased survival of new neurons was no longer seen. Thus, both the genetic and the pharmacologic techniques in this study indicate that the increased survival of new neurons mediated by androgens occurs through ARs. In a subsequent study, male mice that overexpressed neural ARs did not show DHT-induced enhancement in the survival of new neurons (Swift-Gallant et al. 2018), suggesting again that neural AR are important in DHT's influence to increase neurogenesis and may do so via a nonlinear fashion. Both Hamson et al. (2013) and Swift-Gallant et al. (2018) found no evidence for AR on immature neurons (DCX-expressing cells), and thus the precise localization of androgen effects in the dentate gyrus is still under investigation. Given that new granule cells of the dentate gyrus extend their axons to terminate onto pyramidal cells of the CA3 region, this suggests that androgens may indirectly influence new granule cell survival through its effects on the target pyramidal cells in the CA3 region. Collectively, these studies indicated that greater than 25 days of testosterone treatment can increase the survival of new neurons.

We presented research to indicate a role of androgens in male adult hippocampal neurogenesis; however the effects of androgens on hippocampal neurogenesis in the female are not explored. Preliminary unpublished data from our laboratory indicates that androgens may boost neurogenesis in middle-aged but not in young adult female rats and mice (Duarte-Guterman et al. unpublished observations; Chaiton et al. unpublished observations). Androgens can modulate spine density in the CA1 region of the hippocampus in both male and female rodents. Ovariectomized female rats given either testosterone propionate or DHT supplements showed an increase in hippocampal CA1 spine density compared to controls (Leranth et al. 2004). Pretreatment of letrozole, an aromatase inhibitor that prevents synthesis of estradiol from testosterone, prior to testosterone treatment prevented the changes in CA1 synaptic densities but not in the DHT-treated group (Leranth et al. 2004), indicating that both metabolites (estradiol and DHT) are important regulators of spine density structure in the female CA1 region of the hippocampus. More work is needed to determine whether androgens work to modulate other forms of hippocampal structural modification in females across the lifespan.

2.3 Corticosterone and Neurogenesis

Generally, the effects of stress or high levels of corticosterone (CORT), the main stress hormone in rodents, reduce neurogenesis. However, there are notable sex, type of stress, and stress duration differences. After acute exposure (15 min) to predator odor (main component of fox feces trimethylthiazoline – TMT), male rats showed a reduction in cell proliferation, while female rats did not show this same reduction (Falconer and Galea 2003). The TMT-induced suppression in cell proliferation in males was not influenced by castration in adult male rats (Kambo and Galea 2006), nor was the lack of suppression by TMT influenced by ovariectomy in adult female rats (Falconer and Galea 2003). If the stress is more intense, 30 min of electric shocks (30 0.5 mA shocks) followed by 30 min of restraint stress, both male and female mice show reduced cell proliferation that is also not affected by gonadectomy (Tzeng et al. 2014). Furthermore, 2 days of high CORT injections (40 mg/kg) suppressed cell proliferation in male rats (Cameron and Gould 1994). These findings suggest that acute CORT or exposure to stress can suppress cell proliferation in adult males, but not affect cell proliferation in adult females, and with more intense longer exposure to stress, cell proliferation is reduced in both males and females.

In adult male rats, chronic restraint stress for 3 weeks significantly reduced cell proliferation (Pham et al. 2003). In that study, acute restraint stress for either 2 or 6 h did not significantly alter levels of cell proliferation (Pham et al. 2003). When looking at survival of new neurons, chronic restraint stress for 3 weeks did not lower the rate of survival, but after 6 weeks of restraint stress, there were significantly lower numbers of BrdU+ cells in male rats (Pham et al. 2003). Following restraint stress throughout adolescence, female, but not male, rats showed a significant decrease in cell proliferation and survival of new neurons in the dentate gyrus in adulthood (Barha et al. 2011). Indeed, after adolescent restraint stress, there was a slight trend for increased new neuron survival in adult males (Barha et al. 2011). This may reflect the higher vulnerability to develop stress-related disorders in women following the onset of puberty.

Studies manipulating chronic exposure to stress or stress hormones during adulthood typically find a reduction in hippocampal neurogenesis in both male and females, although males are more often studied. Male rats undergoing chronic mild stress exhibited less survival of new neurons (Lee et al. 2006; Mineur et al. 2007) and impaired hippocampal-dependent contextual fear conditioning (Mineur et al. 2007). Chronic high CORT administration induces depressive-like behaviors (increased immobility during a forced swim test) after 21 days (Kalynchuk et al. 2004). This is associated with reduced levels of cell proliferation and survival of immature neurons in the dorsal and ventral hippocampus in male and female rats (Brummelte and Galea 2010b). These latter two findings suggest that depressive-like behavior is associated with reduced levels of neurogenesis in the hippocampus.

The length of exposure to CORT influences the reduction in the amount of neurogenesis in the dentate gyrus. When adult male rats received high CORT (40 mg/kg) for either 9 or 18 days, there was a reduction of survival of new neurons (Wong and Herbert 2006). Indeed, Murray et al. (2008) found that 7 days of high CORT (40 mg/kg) elicited depressive-like behaviors, but it was not until after 14 days of high CORT was there a reduction cell proliferation and in the volume of the hippocampus in adult male mice. Workman et al. (2015) found that 18 days of high CORT was able to reduce the number of immature neurons in male rats and increase depressive-like behaviors, but perhaps paradoxically facilitate learning and memory in the Morris water maze. However, they also found a reduction in the number of new neurons that were activated during memory retrieval in the CORT-treated male rats (Workman et al. 2015). This study indicates that even when new neurons survive under high CORT conditions, they may not be functionally integrated into the neural circuitry as well as those new neurons created without high CORT conditions. These studies collectively indicate that chronic CORT and stress exposure can reduce neurogenesis in the dentate gyrus via a reduction in survival of new neurons and cell proliferation, dependent on duration of stress, when during the lifespan the stress exposure occurred, and the sex of animal.

Does the reduction in neurogenesis following stress result in reductions in hippocampal volume? Recent work by Schoenfeld et al. (2017) sheds light on the relationship between hippocampal neurogenesis reductions and total hippocampal volume. Chronic restraint stress for 4 weeks in adult male rats was able to reduce total hippocampal volume and volume of the dentate gyrus, CA1, and CA3 sub-regions. However, when inhibiting neurogenesis, via a transgenic rat model to induce the expression of thymidine kinase in glial fibrillary acidic protein cells (see Snyder et al. 2016), restraint stress reduced dentate gyrus volume after 4 weeks and reduced CA1 and CA3 volumes after 8 weeks (Schoenfeld et al. 2017). This study indicates that a reduction in the volume of the hippocampus is seen after a reduction in neurogenesis in the dentate gyrus after 8 weeks of exposure to restraint stress.

The effects of high CORT exposure on reducing neurogenesis are exerted through its interactions with GRs. Mifepristone, a GR antagonist, was able to prevent the CORT-induced reduction in neurogenesis (Mayer et al. 2006). This suggests that the activation of GRs is required for the stress-induced reduction in neurogenesis. When adult male rats underwent adrenalectomy to abolish endogenous adrenal steroids, there were significant increases in the number of new cells born within the granule cell layer of the hippocampus (Cameron and Gould 1994). Similarly, when aged male rats underwent adrenalectomy to abolish endogenous adrenal hormones, there was a significant increase in neurogenesis when compared to sham-operated controls (Montaron et al. 2006). Subsequent CORT replacement therapy of 25 μg/mL in drinking water did not attenuate the adrenalectomy-induced increase in the number of newly born cells (Cameron and Gould 1994). This may be due to the changes in the number of GR induced by adrenalectomy after surgery, thus affecting the efficacy of agonists, or it may be due to the dose of CORT. Overall, the activation of GR is required to mediate the reduction in neurogenesis after stress.

2.4 *Why Do Steroid Hormones Modulate Neurogenesis in the Hippocampus?*

Here, we present many lines of evidence to support the notion that steroid hormones play important roles in modulating adult hippocampal neurogenesis. The reasons why this effect is present and seemingly preserved across different species are unknown. Perhaps, with gonadal hormone modulations allowing sex steroid to both promote and suppress neurogenesis, it may aid in pruning and refining synaptic connections in an effort to influence survival via influencing effects on cognition and/or reproduction. New neurons may help to better identify potential mates via increased spatial ability (Sherry et al. 1992), better recognize conspecifics (Monteiro et al. 2014), and/or better understand their relationship to their spatial environment to determine when and where safe locations are to mate or feed and to avoid predators. In the wild, as noted above there are fluctuations in neurogenesis levels, along with changes in space use and spatial ability in promiscuous voles (Galea and McEwen 1999), and we can speculate that fluctuations in neurogenesis with sex steroids evolved to aid in the survival of the host. During the postpartum, reductions seen in neurogenesis (Pawluski and Galea 2007) may be related to reduced spatial memory (Darnaudéry et al. 2007). Intriguingly, better reproductive success in pregnant and postpartum female meadow voles is seen when they are site tenacious and less likely to roam from their nest (Sheridan and Tamarin 1988). As it will be introduced in the latter half of this chapter, the hippocampus and hippocampal neurogenesis are influenced by reproductive experience.

The influence of stress and stress hormones to modulate neurogenesis in both sexes may be reflective of the living conditions animals have endured throughout evolution, such that in periods of acute stress (appearance of predators or lack of immediate food resources), the suppression of neurogenesis acutely may have once had positive effects in the survival of the animal. Acute stress may promote better cognitive abilities, via increased forgetting (Akers et al. 2014), leading to cognitive spatial advantages such as finding better hiding spaces or gaining the ability to venture further from home to scavenge for food. The hippocampus is involved in the negative feedback of the HPA response, and new neurons contribute to negative feedback (Surget et al. 2011), and thus the role that hippocampal neurogenesis plays could be due to promoting allostasis.

2.5 Neurogenesis in the Subventricular Zone

The scope of this chapter focusses on adult hippocampal neurogenesis, but another neurogenic niche in the mammalian brain is the subventricular zone (SVZ) within the forebrain (Alvarez-Buylla and Garcıa-Verdugo 2002). Neural stem cells that originate from the SVZ can proliferate, and these new cells migrate via the rostral migratory stream into the olfactory bulb. Briefly we outline how steroid hormones also influence SVZ neurogenesis. In adult female ovariectomized mice, one study reported that a single estradiol injection or an estradiol capsule lasting for 2 days decreased the levels of cell proliferation within the SVZ (Brock et al. 2010), much like what is seen after 48 h in adult female rats and voles in the hippocampus (Ormerod and Galea 2001; Tanapat et al. 2005). Both short-term (2 h after cease of treatment) and long-term (21 days after treatment) estradiol effects show decreased neurogenesis in the SVZ (Brock et al. 2010), consistent with the effects in the hippocampus (Barker and Galea 2008). Interestingly, pregnancy in mice increased SVZ neurogenesis, an effect mediated by prolactin (Shingo et al. 2003). However, studies in the hippocampus indicate no significant changes in neurogenesis during early pregnancy (Pawluski and Galea 2007) and reductions during the early postpartum in the dentate gyrus (Leuner et al. 2007; Pawluski and Galea 2007). These findings indicate both similarities and differences in neurogenesis in the SVZ versus dentate gyrus in response to estrogens and pregnancy in females.

In male but not female rats, both acute estradiol and testosterone increased cell proliferation in the SVZ 24 h after treatment (Farinetti et al. 2015), inconsistent with findings in the dentate gyrus. CORT injections for 14 days lowered cell proliferation levels within the SVZ compared to control male rats (Lau et al. 2007). This effect was attenuated by the antidepressant paroxetine giving in co-treatment with corticosterone (Lau et al. 2007). This effect of both CORT and antidepressants to lower and increase neurogenesis, respectively, is similar to the effects observed in the hippocampus (Wong and Herbert 2006; Workman et al. 2015; Malberg et al. 2000; see sections above and below). Thus, overall there are similarities in the stress modulation as stress can suppress neurogenesis in both niches. However, the effects on neurogenesis after sex hormone modulation show some dissimilarities between the hippocampus and the SVZ, in both males and females.

3 Depression and Hippocampal Neurogenesis

This section focuses on the relationship between neurogenesis within the hippocampus and its hypothesized role in modulating mood after exposure to stress. The lifetime prevalence of major depressive episodes is approximately 20% of the general population (Otte et al. 2016). The integrity of the hippocampus has been implicated in regulating mood and depression as meta-analyses indicate that depressed patients have smaller hippocampal volumes than healthy controls (MacQueen et al. 2003; Videbech and Ravnkilde 2004; Campbell et al. 2004; McKinnon et al. 2009). Specifically, Videbech and Ravnkilde (2004) reported in a meta-analysis that in patients diagnosed with depression, there was an average reduction of hippocampal volume by 8% within the left hemisphere and 10% within the right hemisphere. Furthermore, hippocampal volume decreases as a function of disease progression and duration, as patients who have been diagnosed for longer than 2 years, and have experienced more than one episode of depression, show significant decreases in hippocampal volume (McKinnon et al. 2009). Decreased hippocampal volume in depressed patients may result from decreased neurogenesis within the dentate gyrus and/or changes to neuropil as Schoenfeld et al. (2017) show correlation between neurogenesis and hippocampal volume in male rodents.

Women are more than twice as likely to develop depression than men, particularly during the reproductive years (Gutiérrez-Lobos et al. 2002). Even in community samples, women present with more depression symptoms than men (Angst et al. 2002). It is imperative that when working to understand the etiology and treatment of depression, we leverage the knowledge of the biological mechanisms behind the sex difference in incidence of depression. Indeed, many studies have capitalized on the influence of sex hormone modulation to understand the contributions of sex hormones to depression (Bloch et al. 2000; Frokjaer et al. 2015; Galea et al. 2001; Wainwright et al. 2011). In addition, stress hormones are implicated in depression, as most depressed individuals cite stress as a factor in the etiology of their depression (Angst et al. 2002) and a meta-analysis indicates that depressed patients have increased levels of cortisol (Stetler and Miller 2011).

Most animal models of depression rely on chronic stress exposure (Katz et al. 1981; Willner et al. 1987), stress hormone exposure (Kalynchuk et al. 2004; Gregus et al. 2005; Brummelte et al. 2006), or the modulation of sex hormones to induce depressive-like behavior (Bernardi et al. 1989a, b; Galea et al. 2001; Wainwright et al. 2011; Mahmoud et al. 2016a). Intriguingly, every animal model of depression tested to date, which includes learned helplessness (inescapable shock), olfactory bulbectomy, chronic unpredictable stress, chronic CORT treatment, ovarian hormone withdrawal after a hormone-simulated pregnancy, and castration coupled with chronic unpredictable stress, is associated with a reduction of neurogenesis within the hippocampus (Malberg and Duman 2003; Jaako-Movits et al. 2006; Green and Galea 2008; Bessa et al. 2009; David et al. 2009; Brummelte and Galea 2010a; Elizalde et al. 2010; Wainwright et al. 2011). Together these findings suggest that neurogenesis in the hippocampus is coincident with depressive-like endophenotypes in animal models. Indeed, in humans, postmortem studies show reduced neurogenesis in the dentate gyrus of untreated depressed patients (Boldrini et al. 2009).

One possible mechanism for stress to mediate the onset of depression or depressive-like behaviors is through a decrease in neurogenesis within the hippocampus. The neurogenic theory of depression and anxiety (reviewed by Miller and Hen 2015) suggests that because exposure to stressors reduces neurogenesis in the dentate gyrus, then this reduction of neurogenesis leads to a greater vulnerability to develop depression and/or anxiety (Schoenfeld et al. 2017).

4 Antidepressants and Neurogenesis

Another indication that hippocampal neurogenesis is involved in regulating mood is the ability of antidepressants to influence hippocampus neurogenesis and volume (reviewed in Warner-Schmidt and Duman 2006; Sahay and Hen 2007; Anacker 2014). In an early study conducted by Malberg et al. (2000), chronic, but not acute, administration of the several classes of pharmacological antidepressants, as well as electroconvulsive therapy, increased cell proliferation in the dentate gyrus of male rats. Perera et al. (2007) found similar effects of electroconvulsive therapy in nonhuman young adult (3–6 years) male bonnet monkeys. Malberg et al. (2000) also examined the effect of the selective serotonin reuptake inhibitor (SSRI), fluoxetine, on neurogenesis as a function of time and found the increase of cell proliferation was dependent on the duration of exposure to fluoxetine. They found that 14, but not 1 or 5, days of fluoxetine injections increased cell proliferation in adult male rats (Malberg et al. 2000). The ability of chronic fluoxetine to increase cell proliferation in the dentate gyrus has been repeated in other studies using male rats (Malberg and Duman 2003; Kodama et al. 2004), but more recently, studies have begun to investigate the effects of antidepressants to increase neurogenesis in models of depression, as an important proof of principle, and in female rodents.

Using models of depression with chronic exposure to CORT or in chronic unpredictable stress, pharmacological antidepressants can promote neurogenesis in males and sometimes females. For example, the reduction of neurogenesis induced by CORT can be attenuated with the administration of the tricyclic antidepressants such as imipramine (Diniz et al. 2013) and clomipramine in adult male rats (Liu et al. 2008). Agomelatine, a melatonergic antidepressant (Dagytė et al. 2011), increased the rate of neuronal maturation and survival within the hippocampus in male rats exposed to chronic mild stress when given at a dose of 40 mg/kg for 21 days, but not in nonstressed control animals. This latter finding is intriguing as it shows that the effects of agomelatine are dependent on context (the stress hormone state of the animal). Furthermore, in an animal model of postpartum depression, chronic exposure to imipramine reversed the suppression in cell proliferation in adult intact, but to a lesser extent in ovariectomized, female rats (Green and Galea 2008). Interestingly, the effect of chronic fluoxetine to increase neurogenesis is not as robust in female rats (Mahmoud et al. 2016a) as it is in male rodents and is absent in postpartum rats (see below; Workman et al. 2016; Gobinath et al. 2018). Thus, it is clear that when examining a number of pharmacological antidepressants in models of depression, chronic exposure can increase neurogenesis in the hippocampus in male rats and to a certain extent in female rats, although this may depend on ovarian hormone status.

Patients diagnosed with major depressive disorder, who were treated with SSRIs or tricyclic antidepressants, had higher levels of neural progenitor cells and cell proliferation (indicated by Nestin and Ki67, respectively) in postmortem tissue (Boldrini et al. 2009, 2012). The effects of antidepressant exposure differ by sex and age of the patient (Epp et al. 2013a; Lucassen et al. 2010). Prescribed antidepressants increased the ratio of immature to mature neurons in women compared to men, specifically in younger adults (<50) compared to older adults. Depression with or without antidepressant use in older patients increased rates of cell cycle inhibition (p21), and this effect was more pronounced in older women (Epp et al. 2013a). Furthermore, older patients (≥ 45, mean age $= 68$) do not show an antidepressant-induced increase in cell proliferation (Lucassen et al. 2010). Together, these studies suggest that pharmaceutical treatments for depression can increase neurogenesis within the hippocampus, dependent on sex and age of the patient.

Does inhibition of neurogenesis increase depressive-like behaviors? One study using methylazoxymethanol acetate (MAM) to reduce neurogenesis found that 2 weeks of exposure to MAM increased depressive-like behavior in adult male rats 4 weeks later (Mateus-Pinheiro et al. 2013). Curiously, using the same procedures but adding chronic unpredictable stress did not result in increased depressive-like behaviors, indicating a context-specific effect of MAM to induce depressive-like behaviors. In another study, using a transgenic Cre-Lox mouse model to increase neurogenesis (conditionally deletion of the pro-apoptotic gene Bax from neural stem cells) made male mice more resilient to chronic unpredictable stress (Hill et al. 2015), an effect that was independent of changes to the hypothalamic-pituitary-adrenal (HPA) axis. These findings suggest that levels of neurogenesis may be related to vulnerability to develop depressive-like endophenotypes in rodents.

Other studies have examined whether reducing neurogenesis can interfere with antidepressant efficacy. In at least three studies, using multiple methods of reducing neurogenesis (via x-radiation or MAM) resulted in the inability of antidepressants such as fluoxetine and imipramine to reduce anxiety in the novelty-suppressed feeding task (Santarelli et al. 2003; Mateus-Pinheiro et al. 2013; Bessa et al. 2009). However, inhibiting neurogenesis did not influence the ability of fluoxetine or imipramine to reduce depressive-like behaviors (forced swim test, sucrose preference; Bessa et al. 2009; David et al. 2009). Overall, there are neurogenesis-dependent and neurogenesis-independent influences of antidepressants to modulate behavior in male rats. More recently, another study found that the polysialylated neural cell adhesion molecule (PSA-NCAM) afforded antidepressant efficacy on sucrose anhedonia, passive coping in the forced swim test, and latency to feed in the novelty-suppressed feeding test as elimination of PSA-NCAM abolished the ability of fluoxetine to exert its effects on these behaviors (Wainwright et al. 2016a). These studies indicated that different forms of neuroplasticity, other than neurogenesis, may mediate different aspects of depressive-like endophenotypes.

4.1 Antidepressants Effects of Endogenous Gonadal Hormones

Estrogens and androgens may have antidepressant-like effects. Specifically, in both younger and older men, lower levels of testosterone, as a result of hypogonadism, can lead to increased depression or depressive symptoms (Shores et al. 2004, 2005; Veras and Nardi 2010). Clinical evidence shows that testosterone therapy reduces depressive symptoms in both young (Veras and Nardi 2010) and older patients (Shores et al. 2009). Both acute testosterone and estradiol treatments reduced depressive-like behavior in the forced swim test in intact male and female mice (Frye and Walf 2009). However, this begs the question, are the same effects of estrogens and androgens seen within an animal model of depression? Wainwright et al. (2011) found that castrated male rats that were also exposed to 3 weeks of chronic unpredictable stress showed depressive-like behaviors and reduced levels of neurogenesis in the hippocampus compared to intact male rats. This indicates that testicular hormones promote resilience in an animal model of depression. Further studies in castrated rats showed that social isolation induced depressive-like anhedonia (Carrier and Kabbaj 2012), which was abolished by treatment with testosterone and/or imipramine (Carrier and Kabbaj 2012). While imipramine alone was able to increase the number of BrdU+ cells, imipramine with testosterone heightened the effect on neurogenesis (Carrier and Kabbaj 2012; Wainwright et al. 2016b). Ovariectomized females, however, were not affected by any of the treatments, and the authors did not look at neurogenic effects in female rats. In a separate experiment using chronic unpredictable stress, the combination of both testosterone and imipramine given to castrated males elicited antidepressant effects in both the sucrose preference test and forced swim test (Wainwright et al. 2016b). Imipramine on its own or co-administered with testosterone was able to increase cell proliferation in stressed adult male rats (Wainwright et al. 2016b). In general, removal of endogenous gonadal hormones can increase susceptibility to depression and depressive-like symptoms in both humans and rodents (Wainwright et al. 2011; Galea et al. 2001; Green and Galea 2008; Frokjaer et al. 2015; Bloch et al. 2000). For males and females, testosterone or estradiol supplementation to hypogonadal animals is effective in alleviating depressive-like endophenotypes with/without conjunction of tricyclic antidepressants.

In perimenopausal depressed women, treatment with estradiol (100 μg) for 12 weeks showed greater remission after termination of estradiol treatment when compared to placebo controls (de Novaes Soares et al. 2001). Similarly, in middle-aged rats, ovariectomy increased susceptibility to depressive-like endophenotypes (sucrose anhedonia, passive coping in the forced swim test, HPA dysregulation, increased anxiety-like behavior) after chronic unpredictable stress compared to sham controls (Mahmoud et al. 2016a). In ovariectomized rats, surgery itself increased depressive-like behaviors 1 week after but not 2 or 3 weeks later (Estrada-Camarena et al. 2011). A single injection of high estradiol (10 μg/rat) 48 h prior to forced swim test decreased immobility 1 week and 3 weeks after ovariectomy indicating that estradiol can exert antidepressant-like effects, but these were not tested in a model of depression (Estrada-Camarena et al. 2011). In middle-aged ovariectomized rats undergoing chronic unpredictable stress, a single injection of estradiol valerate (2 mg/rat) increased sucrose preference when compared to vehicle (Romano-Torres and Fernandez-Guasti 2010). Here, the effect of estrogens in stressed animals was comparative to nonstressed animals that also received estradiol valerate injections (Romano-Torres and Fernandez-Guasti 2010). In another study, fluoxetine treatment (10 mg/kg) or estradiol (10 μg/rat) treatment were both able to increase swimming behavior in the forced swim test in ovariectomized rats (Vega-Rivera et al. 2015). Whereas suboptimal doses of either fluoxetine (1.25 mg/kg) or estradiol (2.3 μg/rat) did not increase swimming on their own, together the cocktail was able to increase swimming behavior (Vega-Rivera et al. 2015). Furthermore, the suboptimal estradiol and fluoxetine co-treatment was able to increase both cell proliferation and the number of immature cells within the dentate gyrus. This later study suggests that either stabilizing hormone levels or antidepressants are viable treatments, but a conjunction of both may be more effective for females, although these findings may depend on the age of animal and the presence of a model of depression, as this study was not conducted within an animal model of depression.

4.2 Postpartum Depression and Neurogenesis

The postpartum period is a time of great risk to develop depression. Work in the Galea laboratory has created two animal models of postpartum depression in rodents (Galea et al. 2001; Brummelte et al. 2006). One model is based on ovarian steroid withdrawal, and the other is based on high CORT injections during the postpartum period. Both models show a reduction in neurogenesis in the affected females (Green and Galea 2008; Brummelte and Galea 2010a). High CORT, and to a lesser extent low CORT, administration to dams either during pregnancy or after birth can lower cell proliferation in the dentate gyrus, and high corticosterone postpartum reduced the number of immature neurons in the dentate gyrus (Brummelte and Galea 2010a; Workman et al. 2016). These findings suggest that both cell proliferation and the survival of new neurons are reduced with high CORT during the postpartum, similar to those effects seen in nulliparous females as outlined in a section above.

As noted above, chronic antidepressant treatment can enhance neurogenesis in male rats and mice, but the same effects are not seen in the postpartum. In rat dams treated with CORT, chronic fluoxetine during the postpartum did not significantly influence neurogenesis (Workman et al. 2016; Gobinath et al. 2018). However, fluoxetine increased neurogenesis in the ventral hippocampus in both oil- and CORT-treated nulliparous rats (Workman et al. 2016). This finding indicates that fluoxetine does not have the same neurogenic effect in females during the postpartum period. Indeed, recent work indicated that exercise during pregnancy and postpartum boosts the efficacy of fluoxetine to increase neurogenesis in the hippocampus (Gobinath et al. 2018). This suggests that during the postpartum period, exposure to multiple antidepressants is needed to bolster the postpartum physiological milieu to allow for greater efficacy of SSRIs to increase neurogenesis in the hippocampus.

In the ovarian steroid withdrawal model of postpartum depression, Green and Galea (2008) found that following a hormone-simulated pregnancy, rats experienced a reduction in cell proliferation much as they do with a natural pregnancy (Pawluski and Galea 2007). The withdrawal from the hormone-simulated pregnancy induced a reduction in neurogenesis that may lead to an increased vulnerability to depressive-like endophenotypes. Furthermore, Green and Galea (2008) found that changes in gonadal hormones modulate the efficacy of antidepressants. In gonadally intact females, imipramine treatment induced an increase in cell proliferation, which was not seen in ovariectomized rats. However, chronic tricyclic antidepressants restored levels of neurogenesis in rats under the ovarian steroid withdrawal model (Green and Galea 2008). Intriguingly, chronic fluoxetine did not show the same effect on neurogenesis in intact versus ovariectomized middle-aged females (Mahmoud et al. 2016a). Overall, these studies suggest that ovarian hormonal milieu influences the effects of pharmacological antidepressants to increase neurogenesis and this may have implications for treating depression.

5 The Involvement of Brain-Derived Neurotrophic Factor (BDNF) in Neurogenesis and Depression

BDNF is one of the many neurotrophic growth factors that affect cell differentiation, development, survival, and regulation of neuronal plasticity (reviewed in Castrén et al. 2007). In particular, BDNF binds to tropomyosin receptor kinase B (TrkB) to elicit effects in facilitating cell survival. Specifically, Scharfman et al. (2005)

reported BDNF infusions increased neurogenesis within the dentate gyrus in the male rat. Both hippocampal neurogenesis and depression are regulated by BDNF. Intriguingly, the effect of BDNF on regulating neurogenesis may be sensitive to endogenous CORT in males (Pinnock and Herbert 2008). If the diurnal rhythm of CORT is flattened, where hormone levels remained consistent throughout the day (Gartside et al. 2003), the effect of BDNF to increase neurogenesis is abolished (Pinnock and Herbert 2008). In addition, L-NAME, a pharmacological inhibitor of nitric oxide synthase, can increase neurogenesis and levels of BDNF in control rats but is ineffective when animals have undergone adrenalectomy (Pinnock and Herbert 2008). Thus, the influence of BDNF to increase neurogenesis in the hippocampus depends on circulating levels of endogenous CORT to exert its actions.

Patients diagnosed with major depressive disorder have significantly lower levels of serum and plasma BDNF when compared to healthy controls (Karege et al. 2005). Shimizu et al. (2003) reported lower levels of BDNF within the hippocampus in depressed patients that were antidepressant naïve and that BDNF levels were inversely correlated with depression symptoms. This negative correlation between BDNF levels and severity suggests BDNF levels may be important in depression pathology (Shimizu et al. 2003). A meta-analysis reported that patients with depression show significantly lower levels of BDNF in serum when compared to healthy controls (Sen et al. 2008), whereas antidepressants were able to increase BDNF levels in treated patients. It is important to note that while studies have looked at the effects of depression on BDNF levels in both men and women, they have neglected to compare directly the sexes. Thus, it is not known whether these effects are seen equally in both men and women. Intriguingly, a meta-analysis looking at the effects of the Val66Met, a functional polymorphism, found that the polymorphism only increased the risk for major depression in men (Met or Met/Met), but not in women (Verhagen et al. 2010). Within the polymorphism, the Met allele is associated with lower brain BDNF levels, higher risks for depression (Youssef et al. 2018), and aberrant BDNF protein levels in neuronal cell cultures (Chen et al. 2004). During the postpartum period for a subset of Chinese women, higher depressive symptoms correlated with lower levels of serum BDNF 3 months after birth (Gao et al. 2016). Furthermore, only women with the BDNF Val66Met polymorphism showed higher depressive symptoms when birth occurred in the autumn/winter seasons compared to Val66Val homozygous genotype (Comasco et al. 2011). For a comprehensive meta-analysis of the role of BDNF in depression, see Brunoni et al. (2008).

Direct infusion of BDNF into the hippocampus was able to reduce depressive-like behavior, and co-administration of K252a, a nonspecific TrkB receptor inhibitor, was able to block the effects of BDNF alone in adult male rats (Shirayama et al. 2002). In young adult male rats, viral knockdown of BDNF within the dentate gyrus increased anhedonia (Taliaz et al. 2010) and lowered levels of cell proliferation and the number of immature cells. These studies show a direct link between BDNF, hippocampal neurogenesis, and depressive-like behavior at least in males.

Other groups have also found that antidepressants might contribute toward regulating BDNF signaling by acting on its receptor, TrkB. Saarelainen et al. (2003) showed that transgenic male mice (ones that overexpress a truncated form of the TrkB receptor (TrkB.T1) leading to reduced TrkB activation) failed to respond to both imipramine and fluoxetine treatments, such that they showed no reduction of depressive-like behaviors when compared to controls. Antidepressants increase TrkB activation with both acute and chronic treatment in the anterior cingulate but only after chronic treatment in the hippocampus in adult male mice in the TrkB transgenic model (Saarelainen et al. 2003). The inability of pharmacological antidepressants to exert behavioral effects but increase TrkB activation was also found in BDNF single-allele knockdown ($BDNF^{+/-}$) animals. Therefore, these findings suggest a BDNF-independent pathway for antidepressants to regulate BDNF signaling (Rantamäki et al. 2011). Overall, the data suggest that while BDNF itself is not required for antidepressants to be effective, signaling downstream of BDNF is required for antidepressant efficacy.

Estradiol may exert antidepressant effects in young adult female rats through BDNF (Su et al. 2016). Using a rodent model of post-stroke depression, estradiol treatment increased sucrose preference and decreased immobility in the forced swim test after chronic unpredictable stress. Indeed, estradiol treatment increased BDNF levels within the hippocampus relative to vehicle-treated animals (Su et al. 2016). There are a few pathways by which estrogens can interact with BDNF signaling. One method for estrogens to regulate BDNF is through direct induction of gene transcriptions as an estrogen response element (ERE) is present on the BDNF gene (Scharfman and MacLusky 2006). Estrogens can also bind to membrane receptors, and BDNF signaling may work together to cohesively influence the hippocampus (Scharfman and MacLusky 2006). The effect of estrogens to induce BDNF mRNA expression is outlined and reviewed in Sohrabji and Lewis (2006). Currently, there are no receptor labelling studies to indicate overlapping of localized ER and TrkB receptors within the hippocampus. ERs are present on cultured neural stem cells (Brännvall et al. 2002), and there are likely sex differences in ER on neural progenitor cells in the hippocampus (Isgor and Watson 2005; Mazzucco et al. 2006). Significant co-labelling of Ki67- and DCX-expressing cells is seen with both ERα and ERβ, indicating both proliferating cells and immature neurons express ER in male rats (Isgor and Watson 2005). However, lower levels of co-expression of Ki67 and ERα and ERβ mRNA were seen in female rats in the dentate gyrus (Mazzucco et al. 2006). TrkB receptors were co-labelled with DCX-expressing type 3 immature neurons in mice of which the sex was not specified (Donovan et al. 2008). Thus, both ERs and TrkB receptors are present on newly born cells, further implicating estrogens may influence adult hippocampal neurogenesis via its influence on BDNF. The mechanism of how estrogens influence BDNF remains unclear. Further research is needed to determine this relationship between BDNF, estradiol, and subsequently depression. It does seem appropriate to speculate that estrogens can drive the expression of BDNF, thus leading to the maturation and potential survival of newborn neurons within the hippocampus, to then alleviate depressive symptoms, but sex differences may be present. However, considering redundancy and the presence of ER subtypes, it is also plausible that signaling from estrogens can promote neurogenesis on its own accord, especially when there are many other genes that are also sensitive to estrogen-ERE activation. In general, when considering the effects of hormonal regulation on neurogenesis and depression, the BDNF signaling pathway is a key factor to consider, especially when investigating the efficacy of antidepressant treatments.

6 Exercise and Neurogenesis in the Hippocampus

Aerobic exercise can regulate hippocampal neurogenesis and acts as a mild antidepressant (Duman et al. 2008). Van Praag et al. (1999) first showed that running drastically increases neurogenesis in the hippocampus of female mice via increases in both cell proliferation and subsequent survival of new neurons. The effect of voluntary running to increase neurogenesis is also observed in both young and older male mice (Van Praag et al. 2005). Aerobic exercise improves several domains of cognitive function including spatial memory in males (Harburger et al. 2007; Gibbons et al. 2014; Van Praag et al. 2005) conditioned fear avoidance (Kim et al. 2010 (sex not mentioned)), and set-shifting (Brockett et al. 2015). However, as it is described in this chapter, the effects of exercise on neurogenesis and resultant cognitive outcomes depend on several factors including type of exercise, social housing, stress induced by the exercise paradigm, and, perhaps, sex. The first part of this section will describe the effects of exercise on neurogenesis and cognition, with a focus on the differences between voluntary exercise, forced exercise of different intensities, and resistance training. The second part of this chapter will describe how sex and sex hormones can modulate the effect of exercise on neurogenesis.

7 Type of Exercise

7.1 Voluntary vs Forced Running

Rodent exercise can be classified as either voluntary or forced. Voluntary running paradigms usually use a running wheel, while forced running paradigms usually use a motorized treadmill (Ang et al. 2006; Liu et al. 2009) or motorized running wheel (Leasure and Jones 2008). Voluntary running allows animals to run ad libitum, thus permitting them to control both the amount and intensity of exercise. However, this makes it challenging for researchers to control for total distance, time, and intensity of running across animals, which could lead to high variability in the exercise group. This variability can be circumvented by using a forced exercise paradigm. With the motorized treadmill or motorized running wheel, researchers can control the intensity and duration of exercise (Leasure and Jones 2008; Ang et al. 2006; Liu et al. 2009). However, in some forced exercise paradigms, mild electric shocks are administered if the animal steps off the treadmill prematurely to motivate continued running (Ang et al. 2006; Hayes et al. 2008). This added stress will undoubtedly influence the outcome measures and must be considered when comparing results between voluntary and forced exercise paradigms. In addition, time of day of exercise is an important consideration given that the circadian phase in which running occurs can influence neurogenesis (Holmes et al. 2004). Holmes et al. (2004) found that running significantly increased cell proliferation and survival of new neurons in adult male mice, but only if it was done in the middle of the active phase of their daily cycle. They also found that this effect was dose-dependent; 3 h of exercise increased neurogenesis, but 1 h of exercise did not (Fig. 2). Interestingly, 3 h of exercise did not increase neurogenesis if the exercise was offered during the non-active phase. These findings suggest that if animals are given limited access to aerobic exercise, the circadian timing and duration of the exercise can also affect the efficacy of exercise to promote neurogenesis in the dentate gyrus.

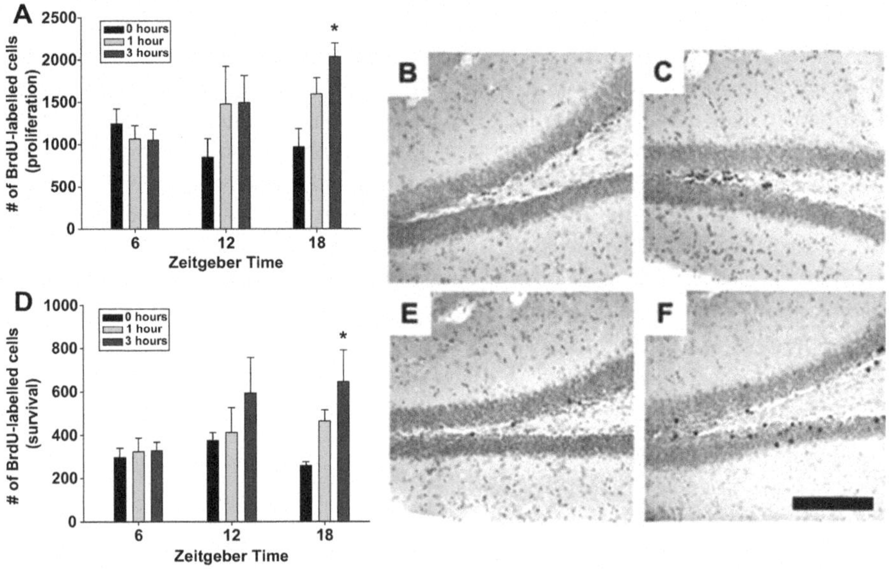

Fig. 2 Number of BrdU-positive cells in the dentate gyrus after voluntary running in male mice. (**a**, **d**) Total number of BrdU-positive cells 1 day (**a**) and 3 weeks (**d**) following the last BrdU injection. Zeitgeber time (ZT) *ZT6* mice receiving BrdU injections and wheel access at the middle of the light period (inactive phase), *ZT12* mice receiving BrdU injections and wheel access at the onset of the dark period (active period), *ZT18* mice receiving BrdU injections and wheel access at the middle of the dark period. 0 h, mice that received no access to a running wheel; 1 h, mice that received 1 h of daily wheel access; 3 h, mice that received 3 h of daily wheel access. *Significantly different ($P < 0.01$) from 0 h wheel access control group. (**b**, **c**) Photomicrographs of BrdU-positive cells 1 day following the last BrdU injection with 0 h wheel access (**b**) and 3 h wheel access (**c**) at ZT18. (**e**, **f**) Photomicrographs of BrdU-positive cells 3 weeks following the last BrdU injection with 0 h wheel access (**e**) and 3 h wheel access (**f**) at ZT18. Scale bar 150 μm. Reprinted with permission from Holmes et al. (2004)

7.2 Voluntary Exercise

Increased neurogenesis in the dentate gyrus after voluntary exercise is seen in juvenile male (Lou et al. 2008), adult male and female Sprague Dawley rats (Boehme et al. 2011; Redila and Christie 2006; Eadie et al. 2005), and C57BL/6 mice (Creer et al. 2010; Fischer et al. 2014; Van Praag et al. 1999). The effects of voluntary running to increase cell proliferation in the dentate gyrus begin within 5 days after the onset of voluntary running in female C57BL/6 mice (Fischer et al. 2014). Although not often studied, the effects of exercise increase neurogenesis in the dorsal dentate gyrus more so than the ventral dentate gyrus (Vivar et al. 2016). The beneficial effect of exercise on neurogenesis is not limited to young or adult animals, as it is also seen in older animals. van Praag et al. (2005) showed that male mice who began a 4-week voluntary exercise program at an age of 19 months in mice, equivalent to ~70 years in humans, exhibited increased neurogenesis compared to age-matched sedentary controls. Similarly, Speisman et al. (2013) found that aged male rats, which voluntarily exercised for 18 weeks, showed increased neurogenesis compared to sedentary controls. In both studies, the increase in neurogenesis was accompanied by improved water maze performance, suggesting that exercise can improve cognition even in old age. These findings are important given that the risk of memory loss, dementia, and Alzheimer's disease and the loss of antidepressant efficacy increase with age (Lindsay et al. 2002; Lucassen et al. 2010), while hippocampal neurogenesis decreases with age (Kuhn et al. 1996; Siwak-Tapp et al. 2007).

Although voluntary exercise is a potent inducer of neurogenesis, it is worth mentioning that Stranahan et al. (2006) found that social isolation delayed exercise-induced hippocampal proliferation, as increased neurogenesis became apparent after 48 days of voluntary running compared to group-housed runners who exhibited proliferation after only 12 days of running. In fact, after 12 days of running, individually housed runners showed a significant decrease in cell proliferation. Subsequent experiments showed that it was a combination of social isolation, the physiological stress of exercise, and the additional stress of daily handling that led to decreased proliferation in these male Sprague Dawley rats (Stranahan et al. 2006). Leasure and Decker (2009) also found that individually housed female Long-Evans rats did not show the expected increased neurogenesis with exercise that was nonetheless observed in the group-housed rats, even though both groups ran similar distances to the isolated animals. While Stranahan et al. (2006) found that social isolation both inhibited and reversed the neurogenic effect of exercise in males, Leasure and Decker (2009) found that the social isolation solely inhibited neurogenesis in females. A similar study by Kannangara et al. (2009) in male mice found that both group-housed and single-housed animals showed increased neurogenesis. However, when mice were given 15 min of restraint stress during the BrdU injection to measure cell proliferation, this prevented the exercise-induced neurogenesis in the socially housed, but not the isolated, animals. This suggests that stress and/or housing conditions are meditators of exercise-induced neurogenesis.

Indeed, social housing affects stress responses in part by modulating the HPA axis (DeVries et al. 2007), and as noted above high CORT inhibits cell proliferation. Further work had shown that in female mice, social housing and exercise buffer CORT levels with age to enhance cell proliferation (Kannangara et al. 2011). These studies indicate the relationship between stress, the HPA axis, and social housing can play a role in the neurogenic effect of exercise.

Social isolation stress notwithstanding, voluntary running paradigms are otherwise minimally stressful as the animals can choose the time, duration, and speed at which they exercise. Forced exercise, however, relies on the animal running at a specified time, for a predetermined duration or distance, and at a set speed. To motivate animals to remain on the treadmill, researchers often administer foot shocks if the animal steps off the treadmill, thus ensuring their compliance (Ang et al. 2006; Hayes et al. 2008). Although treadmill running paradigms are an excellent means of determining the effect of different exercise intensities on the brain, it is important to recognize the difficulty in dissociating the physiological stress associated with higher-intensity running from the stress associated with foot shocks. The next part of this chapter will describe research that has looked at the effect of forced treadmill running and of different exercise intensities on neurogenesis and brain function.

7.3 Aerobic Exercise: Does Intensity Matter?

Aerobic exercise is metabolically taxing. Running increases cellular metabolism and necessitates a higher turnover of electron accepting oxygen molecules for continued adenosine triphosphate (ATP) production. Lactic acid is produced during exercise; the point at which the rate of accumulation is greater than the rate of clearance out of the blood is termed the lactate threshold. Exercise at speeds above this threshold is fatiguing, cannot be sustained for long, and is generally considered "high-intensity" or anaerobic. Treadmill running at a low to moderate intensity increases neurogenesis in male rodents (Glasper et al. 2010; Lou et al. 2008; Inoue et al. 2015; Kim et al. 2016; So et al. 2017), while high-intensity exercise may not (Inoue et al. 2015).

Inoue et al. (2015) found that a long-term paradigm (6 weeks) of mild- but not high-intensity treadmill exercise increased cell proliferation and neuronal differentiation in adult male Wistar rats. In this study, the researchers measured CORT values in all groups and found that the animals that underwent 6 weeks (60 min/day) of high-intensity exercise (40 m/min) had significantly higher plasma CORT at euthanasia than animals that underwent a low-intensity (15 m/min) exercise paradigm. The CORT level of the low-intensity exercised animals was not significantly higher than that of control sedentary animals. However, while the low-intensity group only received electric shocks during the 1-week habituation phase, the higher-intensity group continued to receive electric shocks throughout the training phase. Thus, the effects of increased electric shock-induced stress may be a mediator of diminished neurogenesis in the high-intensity group (Inoue et al. 2015).

7.4 *Resistance Training*

There is a plethora of research looking at the effects of aerobic exercise on neurogenesis, but less is known about the impact of resistance training on the brain. Of the two studies, 1 month, but not 2 months, of a resistance training paradigm given to 8-week-old male rats showed increased cell proliferation compared to sedentary controls (Nokia et al. 2016; Gomes et al. 2014). Aerobic exercise increases neurogenesis by stimulating growth factor pathways, including BDNF signaling. Although Gomes et al. (2014) found increased Ki67-expressing cells with resistance training, BDNF levels remained unchanged compared to controls, suggesting that if resistance training does, in fact, increase neurogenesis, it may be via a different pathway than aerobic exercise. In 2012, Cassilhas et al. investigated the different pathways through which resistance training and aerobic exercise improve spatial memory in adult male Wistar rats. In accordance with Gomes et al. (2014), they found that resistance training did not alter hippocampal levels of BDNF, suggesting that resistance training affects the brain differently than aerobic exercise. One possibility is that resistance training exerts positive effects on the brain through the insulin growth factor (IGF)-1 signaling pathway. While both types of exercise increased hippocampal IGF-1, only resistance training concomitantly increased expression of the IGF-1 receptor and the downstream signaling proteins including phosphorylated AKT/protein kinase B (Cassilhas et al. 2012). On the other hand, only aerobic exercise increased hippocampal BDNF and its receptor TrkB (Cassilhas et al. 2012). Despite these differences, both aerobic and resistance training improved acquisition learning and memory in the water maze compared to controls. Unfortunately, Cassilhas et al. (2012) did not look at any neurogenesis markers. At present, there is insufficient research to fully understand the effects of resistance training on the brain and neurogenesis.

8 Sex Differences in Exercise Studies

As evident in many of the studies described above, sex is often not studied when investigating the effects of exercise on the brain. In meta-analyses of both rodent (Barha et al. 2017a) and human (Barha et al. 2017b) exercise interventions, sex differences in exercise efficacy on cognition were observed. In rodents, males tend to benefit more from voluntary aerobic exercise on certain domains of cognitive function that do not involve the integrity of the hippocampus, including conditioned avoidance memory and nonspatial memory (Barha et al. 2017a). The meta-analysis also showed that males and females respond differently to the type of exercise paradigm, with females showing greater effects on hippocampus-dependent memory tasks with the forced exercise paradigm compared to males (Barha et al. 2017a). In the meta-analysis of human exercise interventions (Barha et al. 2017b), larger effect sizes for executive function improvements were seen in studies comprised of a higher percentage of women, suggesting that women may experience more cognitive benefits than men with exercise.

Given the findings of the meta-analyses described above, it is evident that generalizing across sexes will lead to an incorrect, inconsistent, or at best, incomplete understanding of how exercise affects the brain and cognition. These sex differences can arise from various sources including differences in motivation to exercise, differences in HPA axis reactivity in response to stressors associated with exercise paradigms, and differences in the molecular pathways altered by exercise. This final section describes how these differences may be underlying the sex differences in response to exercise.

8.1 Sex Differences in Distance Run

Males and females differ in the amount of running they do voluntarily. Wang et al. (1925) described this sex difference, noting that female rats ran an average of 6,000–12,000 revolutions per day, while male rats only ran an average of 2,000–8,000 revolutions per day. Hitchcock (1925) also found that male rats ran on average only 56% of the distance traveled by female rats; a sex difference also was seen in several strains of mice (Lightfoot et al. 2004). Interestingly, Lightfoot et al. (2004) also showed that while females ran significantly farther than males, males spent an average of 15% more time running indicating that females ran at a significantly higher velocity than males. While there was a significant correlation between body mass and velocity for females, mass exhibited a poor predictive fit for velocity ($r^2 = 0.09$), suggesting that weight alone does not explain the sex differences in running velocity. These findings are especially important for interpreting voluntary aerobic exercise studies. Studies show that hippocampal BDNF production (Neeper et al. 1995; Oliff et al. 1998; Johnson and Mitchell 2003) and hippocampal neurogenesis (Allen et al. 2001; Rhodes et al. 2003 in control mice) were both positively correlated with distance run. If females run more than males, it would be expected that they would show more BDNF production and neurogenesis, simply due to greater distances run, and not necessarily because of any physiological differences in response to exercise. Preliminary data from our laboratory show that in mice, males and females exhibit elevated DCX-expressing cells, despite the greater distance run by females. However, females may have a greater increase of immature neurons in the dorsal dentate gyrus, while males have a greater increase of DCX-expressing cells in the ventral dentate gyrus in response to 28 days of ad libitum wheel running (see Fig. 3). The possible sex differences in regional increases in neurogenesis with running may be an important variable to consider in future research.

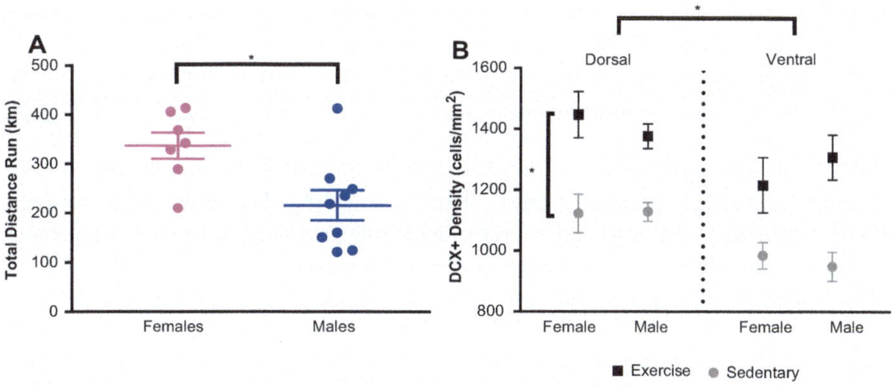

Fig. 3 (**a**) Total distance run voluntarily by C57BL/6J mice with ad libitum access to a running wheel in their home cage for 28 days. Females ran significantly greater distance than males. (**b**) Exercise increased the density of DCX-expressing cells in both the dorsal and the ventral region of the hippocampus for both males and females; no sex difference in the density of DCX-expressing cells was observed, despite females running a significantly greater distance over 28 days. The dorsal region had a higher density of DCX expression than the ventral region, irrespective of exercise intervention or sex. *$p < 0.05$

9 Ovarian Hormones, Physical Activity, and Neurogenesis

Ovarian hormones play a role in modulating the amount of distance a female will run voluntarily. OVX adult female mice and rats reduced their running amount to ~14% of intact animals (Forstner et al. 2006; Jones et al. 2016). Estradiol-treated OVX mice ran significantly more compared to vehicle-treated animals (Jones et al. 2016). However, Eddy et al. (2013) found OVX females did not run significantly less than intact females, though they suggested that this lack of effect could potentially be attributed to the food deprivation. Food deprivation is a stressor that paradoxically increases wheel running (Finger 1965; Price 1976; Epling and Pierce 1988) and thus could mask the OVX-associated decrease in activity.

In terms of cognitive effects and exercise with ovarian hormones, Kim et al. (2016) found that OVX adult rats exhibited reduced hippocampal neurogenesis, BDNF, and TrkB protein expression. Furthermore, these rats performed worse on memory tasks (Morris water maze and a step-down avoidance task) than sham females. However, ovariectomized animals that underwent a combination of moderate treadmill exercise and resistance training for 8 weeks showed increased neurogenesis in the dentate gyrus compared to the ovariectomized-sedentary controls and sham-operated controls (Kim et al. 2016). Additionally, only the ovariectomized but not sham animals showed an improvement in cognition with exercise, though this could likely be due to a floor effect on performance in the sham animals. Further, 1 week of running increased cell proliferation and DCX-expressing cells in the hippocampus of ovariectomized mice (Jin et al. 2008), indicating that ovariectomy does not preclude the effects of running on neurogenesis in female mice.

However, future studies should determine how different types and doses of ovarian hormones interact with different exercise paradigms to modulate neurogenesis.

Overall, these studies collectively indicate that ovarian hormones may not directly mediate the relationship between exercise-induced neurogenesis. In ovariectomized animals, exercise was able to increase BDNF and TrkB protein expression as well as hippocampal neurogenesis (as mentioned above) compared to both sedentary and sham-operated controls (Kim et al. 2016). The interactions between BDNF, estradiol signaling, and exercise are more prevalent in females given that they have higher levels of endogenous estrogens compared to males (Ho et al. 1987). Although there is less known on how resistance training affects the brain, as noted in a previous section, there is some indication that it exerts positive effects through insulin growth factor (IGF-1) signaling (Cassilhas et al. 2012) and there is evidence for interactions between IGF-1 and estrogen signaling pathways (Kahlert et al. 2000; Perez-Martin et al. 2003).

9.1 Sex Differences in the Hypothalamic-Pituitary-Adrenal (HPA) Axis

Exercise paradigms, whether they are voluntary or forced, can be stressful and activate the HPA axis. CORT inhibits neurogenesis, and thus levels of stress must be considered when investigating the effects of exercise on neurogenesis. There are major sex differences in the HPA axis (reviewed extensively in Goel et al. 2014). Female rodents have higher basal levels CORT than males (Weinstock et al. 1998). Females also show a faster rise in CORT in response to stressors and a longer delay in return to baseline CORT levels after the cessation of stress (Weinstock et al. 1998; Iwasaki-Sekino et al. 2009). Of relevance to interpreting rodent exercise studies are the sex differences in response to electric foot shocks and social isolation and stressors, which are often used in forced and voluntary exercise paradigms, respectively. Iwasaki-Sekino et al. (2009) found that females had a larger and more rapid increase in plasma adrenocorticotropic hormone (ACTH) and CORT and a higher expression of corticotrophin release hormone (CRH) mRNA in the paraventricular nuclei (PVN) of the hypothalamus than males in response to electric foot shocks. Interestingly, as mentioned earlier, a meta-analysis of rodent studies showed that females tend to perform better than males on hippocampus-dependent memory tasks if subjected to forced exercise, a finding that was not seen with voluntary exercise (Barha et al. 2017a). Given that stress hormones affect neurogenesis (see section above), it is important to account for sex differences in response to stress when designing and interpreting exercise studies in rodents.

10 Androgens and Exercise in the Modulation of Neurogenesis

The interaction between androgens and exercise in mediating hippocampal neurogenesis is still a new area of research. Eddy et al. (2013) did not find a significant difference in wheel running activity between castrated and intact males, but it is possible that the food restriction used in their study may have been a potential confounding factor, as discussed above. One study showed that running was able to increase cell proliferation, differentiation, and cell survival in both intact and orchiectomized male rats (Okamoto et al. 2012). However, in both intact and castrated animals, when androgen receptors are blocked via flutamide, running no longer increased survival of new neurons in the dentate gyrus compared to their sedentary counterparts (Okamoto et al. 2012). The increase in cell proliferation was still observed in flutamide-treated rats. Not only does this finding agree with previous research that androgens are involved in increasing survival, not proliferation, of new hippocampal neurons (see section above), but it also possibly indicates that brain-derived sources of androgens may be mediating the neurogenic response to exercise. In males, blocking ER activity using the antagonist tamoxifen did not abolish the effects of exercise-induced increase in cell proliferation, cell survival, and differentiation of neurons (Okamoto et al. 2012). Considering that hormones such as estrogens and androgens are implicated in exercise-induced effects on the brain, it is of utmost importance to consider males and females in studies of exercise on neurogenesis to influence mood and cognition.

11 Physical Activity in Animal Models of Depression

As stated above, exercise can influence various hormonal factors and trophic factors to regulate neurogenesis. This relationship is important to consider as exercise is a potent antidepressant (Ernst et al. 2006; Erickson et al. 2012). Looking more specifically in animal models of depression, several studies have shown that exercise can reduce depressive-like endophenotypes in both adult male rats (Marais et al. 2009; Liu and Zhou 2012; Liu et al. 2013) and in adult mice, sex not specified (Solberg et al. 1999). When comparing between a CORT injection-based model of depression and a chronic unpredictable stress based model of depression in male rats, exercise was only able to reduce depressive-like behavior in the chronic unpredictable stress model (Liu and Zhou 2012). However, in female rats, in a model of postpartum depression using chronic CORT injections to induce depressive-like endophenotypes, exercise decreased time spent immobile in the forced swim test (Gobinath et al. 2018). In that latter study, the CORT-induced decrements in maternal care behavior were not ameliorated by exercise alone (Gobinath et al. 2018). Exercise was also able to reduce plasma CORT levels in a single-prolonged stress model of posttraumatic stress disorder (Patki et al. 2014) and serum CORT level in a model of maternal separation-induced depression in adult male rats (Marais et al. 2009). Overall, exercise can reduce depressive-like endophenotypes in rodent models of depression as well as in human patients diagnosed with depression. In animal models, this effect may differ depending on sex, species, and the model of depression.

12 Conclusions and Future Research

Adult neurogenesis in the hippocampus is regulated by a number of factors including steroid hormones and sex. Not only do sex and sex hormones modulate neurogenesis, but they also interact with several other factors such as age, stress, and physical activity, to affect the production and survival of new neurons in the hippocampus. The function of adult neurogenesis is linked to cognition and stress regulation, both of which show sensitivity to sex hormones and sex. Several paradoxes must be explored to understand the role of hormones in adult hippocampal neurogenesis. For example, ovarian hormones promote greater physical activity in females (Forstner et al. 2006; Jones et al. 2016), and although exercise increases the production of new neurons, ovarian hormones suppress neurogenesis (Barker and Galea 2008; Chan et al. 2014). Similarly, physical exercise can increase corticosterone levels (Adlard and Cotman 2004; Stranahan et al. 2006), and although exercise increases neurogenesis, chronic (or acute) corticosterone suppresses it. These findings suggest that the effects of these hormones (estrogens and corticosterone) may be altered under different contextual environments. It is also possible that physical exercise may circumvent the influence of corticosterone and estrogens to suppress neurogenesis by influencing different mechanistic pathways, perhaps via growth factors such as BDNF and IGF. Furthermore, an underexplored area of research is whether changes in neurogenesis are region specific, and studies referred to in this chapter show a greater effect of pharmaceutical antidepressants to increase neurogenesis in the ventral hippocampus but a greater effect of physical exercise to increase neurogenesis in the dorsal hippocampus. These results further suggest that similar and different pathways may be at play with pharmaceutical versus non-pharmaceutical antidepressants to increase neurogenesis and potentially alleviate depression (see Fig. 4). Research elucidating how these different interactions modulate neurogenesis in different hippocampal regions and, ultimately, how these new neurons are incorporated into the existing brain circuitry will help form our understanding of the functional role of neurogenesis in the brain and on behavior and may aid to develop new therapeutic advances for psychiatric disorders such as depression.

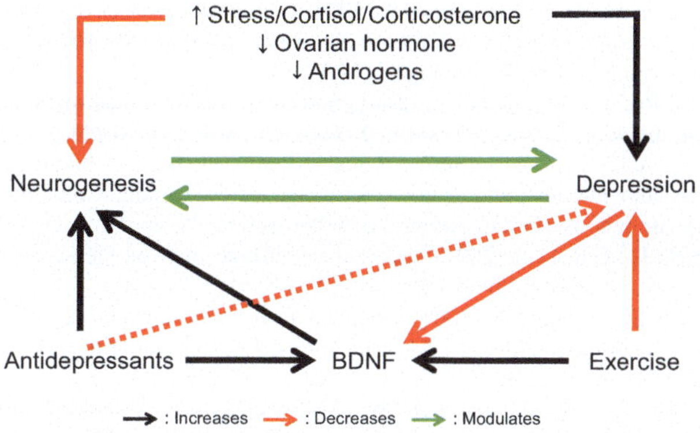

Fig. 4 A simplified model of the effects of neurogenesis on depression, along with endogenous influences such as BDNF, stress hormones, and gonadal hormones and exogenous influences such as exercise and antidepressants. The dotted red line depicts that the efficacy of antidepressants is reduced over time, such that the majority of patients do not achieve remission with continued exposure (de Sousa et al. 2015). There are a number of sex differences not depicted on the figure as women show stronger responses to selective serotonin reuptake inhibitors (SSRIs) (Khan et al. 2005), while men tend to show greater responses to tricyclic antidepressants (Kornstein et al. 2000; Baca et al. 2004)

References

Adlard PA, Cotman CW (2004) Voluntary exercise protects against stress-induced decreases in brain-derived neurotrophic factor protein expression. Neuroscience 124(4):985–992

Akers KG, Martinez-Canabal A, Restivo L, Yiu AP, De Cristofaro A, Hsiang HL, Wheeler AL, Guskjolen A, Niibori Y, Shoji H, Ohira K, Richards BA, Miyakawa T, Josselyn SA, Frankland PW (2014) Hippocampal neurogenesis regulates forgetting during adulthood and infancy. Science 344(6184):598–602

Allen DM, van Praag H, Ray J, Weaver Z, Winrow CJ, Carter TA, Braquet R, Harrington E, Ried T, Brown KD, Gage FH, Barlow C (2001) Ataxia telangiectasia mutated is essential during adult neurogenesis. Genes Dev 15(5):554–566

Altman J, Das GD (1965) Autoradiographic and histological evidence of postnatal hippocampal neurogenesis in rats. J Comp Neurol 124(3):319–335

Alvarez-Buylla A, Garcıa-Verdugo JM (2002) Neurogenesis in adult subventricular zone. J Neurosci 22(3):629–634

Amaral DG, Scharfman HE, Lavenex P (2007) The dentate gyrus: fundamental neuroanatomical organization (dentate gyrus for dummies). Prog Brain Res 163:3–790

Anacker C (2014) Adult hippocampal neurogenesis in depression: behavioral implications and regulation by the stress system. In: Behavioral neurobiology of stress-related disorders. Springer, Berlin, pp 25–43

Anacker C, Cattaneo A, Luoni A, Musaelyan K, Zunszain PA, Milanesi E, Rybka J, Berry A, Cirulli F, Thuret S, Price J, Riva MA, Gennarelli M, Pariante CM (2013) Glucocorticoid-related molecular signalling pathways regulating hippocampal neurogenesis. Neuropsychopharmacology 38:872–883

Ang ET, Dawe GS, Wong PT, Moochhala S, Ng YK (2006) Alterations in spatial learning and memory after forced exercise. Brain Res 1113(1):186–193

Angst J, Gamma A, Gastpar M, Lépine JP, Mendlewicz J, Tylee A (2002) Gender differences in depression. Eur Arch Psychiatry Clin Neurosci 252(5):201–209

Baca E, Garcia-Garcia M, Porras-Chavarino A (2004) Gender differences in treatment response to sertraline versus imipramine in patients with nonmelancholic depressive disorders. Prog Neuro-Psychopharmacol Biol Psychiatry 28(1):57–65

Banasr M, Hery M, Brezun JM, Daszuta A (2001) Serotonin mediates oestrogen stimulation of cell proliferation in the adult dentate gyrus. Eur J Neurosci 14(9):1417–1424

Barha CK, Galea LA (2013) The hormone therapy, Premarin, impairs hippocampus-dependent spatial learning and memory and reduces activation of new granule neurons in response to memory in female rats. Neurobiol Aging 34(3):986–1004

Barha CK, Pawluski JL, Galea LA (2007) Maternal care affects male and female offspring working memory and stress reactivity. Physiol Behav 92:939–950

Barha CK, Lieblich SE, Galea LA (2009) Different forms of oestrogen rapidly upregulate cell proliferation in the dentate gyrus of adult female rats. J Neuroendocrinol 21(3):155–166

Barha CK, Brummelte S, Lieblich SE, Galea LA (2011) Chronic restraint stress in adolescence differentially influences hypothalamic-pituitary-adrenal axis function and adult hippocampal neurogenesis in male and female rats. Hippocampus 21(11):1216–1227

Barha CK, Lieblich SE, Chow C, Galea LA (2015) Multiparity-induced enhancement of hippocampal neurogenesis and spatial memory depends on ovarian hormone status in middle age. Neurobiol Aging 36(8):2391–2405

Barha CK, Falck RS, Davis JC, Nagamatsu LS, Liu-Ambrose T (2017a) Sex differences in aerobic exercise efficacy to improve cognition: a systematic review and meta-analysis of studies in older rodents. Front Neuroendocrinol 46:86–105

Barha CK, Davis JC, Falck RS, Nagamatsu LS, Liu-Ambrose T (2017b) Sex differences in exercise efficacy to improve cognition: a systematic review and meta-analysis of randomized controlled trials in older humans. Front Neuroendocrinol 46:71–85

Barker JM, Galea LA (2008) Repeated estradiol administration alters different aspects of neurogenesis and cell death in the hippocampus of female, but not male, rats. Neuroscience 152(4):888–902

Bernardi M, Vergoni AV, Sandrini M, Tagliavini S, Bertolini A (1989a) Influence of ovariectomy, estradiol and progesterone on the behavior of mice in an experimental model of depression. Physiol Behav 45(5):1067–1068

Bernardi M, Genedani S, Tagliavini S, Bertolini A (1989b) Effect of castration and testosterone in experimental models of depression in mice. Behav Neurosci 103(5):1148

Bessa JM, Ferreira D, Melo I, Marques F, Cerqueira JJ, Palha JA, Almeida OF, Sousa N (2009) The mood-improving actions of antidepressants do not depend on neurogenesis but are associated with neuronal remodeling. Mol Psychiatry 14(8):764

Bloch M, Schmidt PJ, Danaceau M, Murphy J, Nieman L, Rubinow DR (2000) Effects of gonadal steroids in women with a history of postpartum depression. Am J Psychiatr 157(6):924–930

Boehme F, Gil-Mohapel J, Cox A, Patten A, Giles E, Brocardo PS, Christie BR (2011) Voluntary exercise induces adult hippocampal neurogenesis and BDNF expression in a rodent model of fetal alcohol spectrum disorders. Eur J Neurosci 33(10):1799–1811

Bohn MC, Dean D, Hussain S, Giuliano R (1994) Development of glucocorticoid and mineralocorticoid receptors in rat hippocampus. Dev Brain Res 77:157–162

Boldrini M, Underwood MD, Hen R, Rosoklija GB, Dwork AJ, Mann JJ, Arango V (2009) Antidepressants increase neural progenitor cells in the human hippocampus. Neuropsychopharmacology 34(11):2376

Boldrini M, Hen R, Underwood MD, Rosoklija GB, Dwork AJ, Mann JJ, Arango V (2012) Hippocampal angiogenesis and progenitor cell proliferation are increased with antidepressant use in major depression. Biol Psychiatry 72(7):562–571

Brännvall K, Korhonen L, Lindholm D (2002) Estrogen-receptor-dependent regulation of neural stem cell proliferation and differentiation. Mol Cell Neurosci 21(3):512–520

Brännvall K, Bogdanovic N, Korhonen L, Lindholm D (2005) 19-Nortestosterone influences neural stem cell proliferation and neurogenesis in the rat brain. Eur J Neurosci 21(4):871–878

Brock O, Keller M, Veyrac A, Douhard Q, Bakker J (2010) Short term treatment with estradiol decreases the rate of newly generated cells in the subventricular zone and main olfactory bulb of adult female mice. Neuroscience 166(2):368–376
Brockett AT, LaMarca EA, Gould E (2015) Physical exercise enhances cognitive flexibility as well as astrocytic and synaptic markers in the medial prefrontal cortex. PLoS One 10(5):e0124859
Brummelte S, Galea LA (2010a) Chronic corticosterone during pregnancy and postpartum affects maternal care, cell proliferation and depressive-like behavior in the dam. Horm Behav 58(5):769–779
Brummelte S, Galea LA (2010b) Chronic high corticosterone reduces neurogenesis in the dentate gyrus of adult male and female rats. Neuroscience 168(3):680–690
Brummelte S, Pawluski JL, Galea LA (2006) High post-partum levels of corticosterone given to dams influence postnatal hippocampal cell proliferation and behavior of offspring: a model of post-partum stress and possible depression. Horm Behav 50(3):370–382
Brunoni AR, Lopes M, Fregni F (2008) A systematic review and meta-analysis of clinical studies on major depression and BDNF levels: implications for the role of neuroplasticity in depression. Int J Neuropsychopharmacol 11(8):1169–1180
Cameron HA, Gould E (1994) Adult neurogenesis is regulated by adrenal steroids in the dentate gyrus. Neuroscience 61(2):203–209
Cameron HA, Mckay RD (2001) Adult neurogenesis produces a large pool of new granule cells in the dentate gyrus. J Comp Neurol 435(4):406–417
Campbell S, Marriott M, Nahmias C, MacQueen GM (2004) Lower hippocampal volume in patients suffering from depression: a meta-analysis. Am J Psychiatr 161(4):598–607
Carrier N, Kabbaj M (2012) Testosterone and imipramine have antidepressant effects in socially isolated male but not female rats. Horm Behav 61(5):678–685
Cassilhas RC, Lee KS, Fernandes J, Oliveira MG, Tufik S, Meeusen R, De Mello MT (2012) Spatial memory is improved by aerobic and resistance exercise through divergent molecular mechanisms. Neuroscience 202:309–317
Castrén E, Võikar V, Rantamäki T (2007) Role of neurotrophic factors in depression. Curr Opin Pharmacol 7(1):18–21
Chan M, Chow C, Hamson DK, Lieblich SE, Galea LA (2014) Effects of chronic oestradiol, progesterone and medroxyprogesterone acetate on hippocampal neurogenesis and adrenal mass in adult female rats. J Neuroendocrinol 26(6):386–399
Chen Z-Y, Patel PD, Sant G, Meng C-X, Teng KK, Hempstead BL, Lee FS (2004) Variant brain-derived neurotrophic factor (BDNF) (Met66) alters the intracellular trafficking and activity-dependent secretion of wild-type BDNF in neurosecretory cells and cortical neurons. J Neurosci 24(18):4401–4411
Clelland CD, Choi M, Romberg C, Clemenson GD, Fragniere A, Tyers P, Jessberger S, Saksida LM, Barker RA, Gage FH, Bussey TJ (2009) A functional role for adult hippocampal neurogenesis in spatial pattern separation. Science 325(5937):210–213
Comasco E, Sylvén SM, Papadopoulos FC, Oreland L, Sundström-Poromaa I, Skalkidou A (2011) Postpartum depressive symptoms and the BDNF Val66Met functional polymorphism: effect of season of delivery. Arch Womens Ment Health 14(6):453–463
Creer DJ, Romberg C, Saksida LM, van Praag H, Bussey TJ (2010) Running enhances spatial pattern separation in mice. Proc Natl Acad Sci 107(5):2367–2372
Dagytė G, Crescente I, Postema F, Seguin L, Gabriel C, Mocaër E, Den Boer JA, Koolhaas JM (2011) Agomelatine reverses the decrease in hippocampal cell survival induced by chronic mild stress. Behav Brain Res 218(1):121–128
Darnaudéry M, Perez-Martin M, Del Favero F, Gomez-Roldan C, Garcia-Segura LM, Maccari S (2007) Early motherhood in rats is associated with a modification of hippocampal function. Psychoneuroendocrinology 32:803–812
David DJ, Samuels BA, Rainer Q, Wang JW, Marsteller D, Mendez I, Drew M, Craig DA, Guiard BP, Guilloux JP, Artymyshyn RP, Gardier AM, Gerald C, Antonijevic IA, Leonardo ED, Hen R (2009) Neurogenesis-dependent and-independent effects of fluoxetine in an animal model of anxiety/depression. Neuron 62(4):479–493

de Novaes Soares C, Almeida OP, Joffe H, Cohen LS (2001) Efficacy of estradiol for the treatment of depressive disorders in perimenopausal women: a double-blind, randomized, placebo-controlled trial. Arch Gen Psychiatry 58(6):529–534

de Sousa RT, Zanetti MV, Brunoni AR, Machado-Vieira R (2015) Challenging treatment-resistant major depressive disorder: a roadmap for improved therapeutics. Curr Neuropharmacol 13(5):616–635

DeVries AC, Craft TK, Glasper ER, Neigh GN, Alexander JK (2007) 2006 Curt P. Richter award winner: social influences on stress responses and health. Psychoneuroendocrinology 32(6):587–603

Diniz L, dos Santos TB, Britto LR, Céspedes IC, Garcia MC, Spadari-Bratfisch RC, Medalha CC, de Castro GM, Montesano FT, Viana MB (2013) Effects of chronic treatment with corticosterone and imipramine on fos immunoreactivity and adult hippocampal neurogenesis. Behav Brain Res 238:170–177

Donovan MH, Yamaguchi M, Eisch AJ (2008) Dynamic expression of TrkB receptor protein on proliferating and maturing cells in the adult mouse dentate gyrus. Hippocampus 18(5):435–439

Duarte-Guterman P, Lieblich SE, Chow C, Galea LAM (2015) Estradiol and GPER activation differentially affect cell proliferation but not GPER expression in the hippocampus of adult female rats. PLoS One 10(6):e0129880

Duman CH, Schlesinger L, Russell DS, Duman RS (2008) Voluntary exercise produces antidepressant and anxiolytic behavioral effects in mice. Brain Res 1199:148–158

Eadie BD, Redila VA, Christie BR (2005) Voluntary exercise alters the cytoarchitecture of the adult dentate gyrus by increasing cellular proliferation, dendritic complexity, and spine density. J Comp Neurol 486(1):39–47

Eddy MC, Rifken KM, Toufexis DJ, Green JT (2013) Gonadal hormones and voluntary exercise interact to improve discrimination ability in a set-shift task. Behav Neurosci 127(5):744

Elizalde N, García-García AL, Totterdell S, Gendive N, Venzala E, Ramirez MJ, Del Rio J, Tordera RM (2010) Sustained stress-induced changes in mice as a model for chronic depression. Psychopharmacology 210(3):393–406

Epling WF, Pierce WD (1988) Activity-based anorexia: a biobehavioral perspective. Int J Eat Disord 7(4):475–485

Epp JR, Beasley CL, Galea LA (2013a) Increased hippocampal neurogenesis and p21 expression in depression: dependent on antidepressants, sex, age, and antipsychotic exposure. Neuropsychopharmacology 38(11):2297

Epp JR, Chow C, Galea LA (2013b) Hippocampus-dependent learning influences hippocampal neurogenesis. Front Neurosci 7:57

Epp JR, Mera RS, Köhler S, Josselyn SA, Frankland PW (2016) Neurogenesis-mediated forgetting minimizes proactive interference. Nat Commun 7:10838

Erickson KI, Miller DL, Roeckein KA (2012) The aging hippocampus: interactions between exercise, depression, and BDNF. Neuroscientist 18(1):82–97

Ernst C, Olson AK, Pinel JP, Lam RW, Christie BR (2006) Antidepressant effects of exercise: evidence for an adult-neurogenesis hypothesis? J Psychiatry Neurosci 31:84–92

Estrada-Camarena E, López-Rubalcava C, Hernández-Aragón A, Mejía-Mauries S, Picazo O (2011) Long-term ovariectomy modulates the antidepressant-like action of estrogens, but not of antidepressants. J Psychopharmacol 25(10):1365–1377

Falconer EM, Galea LA (2003) Sex differences in cell proliferation, cell death and defensive behavior following acute predator odor stress in adult rats. Brain Res 975(1–2):22–36

Fanselow MS, Dong HW (2010) Are the dorsal and ventral hippocampus functionally distinct structures? Neuron 65(1):7–19

Farinetti A, Tomasi S, Foglio B, Ferraris A, Ponti G, Gotti S, Peretto P, Panzica GC (2015) Testosterone and estradiol differentially affect cell proliferation in the subventricular zone of young adult gonadectomized male and female rats. Neuroscience 286:162–170

Finger FW (1965 Jun) Effect of food deprivation on running-wheel activity in naive rats. Psychol Rep 16(3):753–757

Fischer TJ, Walker TL, Overall RW, Brandt MD, Kempermann G (2014) Acute effects of wheel running on adult hippocampal precursor cells in mice are not caused by changes in cell cycle length or S phase length. Front Neurosci 8:314

Forstner JP, Gorzek JF, Hendrickson KC, Rixen JL, Moran AL, Lowe DA (2006) Effects of estrogen replacement on voluntary aerobic exercise and body weight in an animal model of menopause. J Women's Health Phys Ther 30(2):28–29

Frankland PW, Köhler S, Josselyn SA (2013) Hippocampal neurogenesis and forgetting. Trends Neurosci 36(9):497–503

Frokjaer VG, Pinborg A, Holst KK, Overgaard A, Henningsson S, Heede M, Larsen EC, Jensen PS, Agn M, Nielsen AP, Stenbæk DS, Cunha-Bang S, Lehel S, Siebner HR, Mikkelsen JD, Svarer C, Knudsen GM (2015) Role of serotonin transporter changes in depressive responses to sex-steroid hormone manipulation: a positron emission tomography study. Biol Psychiatry 78(8):534–543

Frye CA, Walf AA (2009) Depression-like behavior of aged male and female mice is ameliorated with administration of testosterone or its metabolites. Physiol Behav 97(2):266–269

Galea LA, McEwen BS (1999) Sex and seasonal differences in the rate of cell proliferation in the dentate gyrus of adult wild meadow voles. Neuroscience 89(3):955–964

Galea LA, Wide JK, Barr AM (2001) Estradiol alleviates depressive-like symptoms in a novel animal model of post-partum depression. Behav Brain Res 122(1):1–9

Galea LA, Roes MM, Dimesh C, Chow C, Mahmoud R, Lieblich SL, Duarte-Guterman P (2018) Premarin has opposing effects on spatial learning, neural activation and serum cytokine levels in middle aged female rats depending on reproductive history. Neurobiol Aging 70:291–301

Gao X, Wang J, Yao H, Cai Y, Cheng R (2016) Serum BDNF concentration after delivery is associated with development of postpartum depression: a 3-month follow up study. J Affect Disord 200:25–30

Gartside SE, Leitch MM, McQuade R, Swarbrick DJ (2003) Flattening the glucocorticoid rhythm causes changes in hippocampal expression of messenger RNAs coding structural and functional proteins: implications for aging and depression. Neuropsychopharmacology 28(5):821

Gibbons TE, Pence BD, Petr G, Ossyra JM, Mach HC, Bhattacharya TK, Perez S, Martin SA, McCusker RH, Kelley KW, Rhodes JS, Johnson RW, Woods JA (2014) Voluntary wheel running, but not a diet containing (−)-epigallocatechin-3-gallate and β-alanine, improves learning, memory and hippocampal neurogenesis in aged mice. Behav Brain Res 272:131–140

Glasper ER, Llorens-Martin MV, Leuner B, Gould E, Trejo JL (2010) Blockade of insulin-like growth factor-I has complex effects on structural plasticity in the hippocampus. Hippocampus 20(6):706–712

Gobinath AR, Richardson RJ, Chow C, Workman JL, Lieblich SE, Barr AM, Galea LA (2018) Voluntary running influences the efficacy of fluoxetine in a model of postpartum depression. Neuropharmacology 128:106–118

Goel N, Workman JL, Lee TT, Innala L, Viau V (2014) Sex differences in the HPA axis. Compr Physiol 4:1121–1155

Gomes FG, Fernandes J, Campos DV, Cassilhas RC, Viana GM, D'Almeida V, de Moraes Rêgo MK, Buainain PI, Cavalheiro EA, Arida RM (2014) The beneficial effects of strength exercise on hippocampal cell proliferation and apoptotic signaling is impaired by anabolic androgenic steroids. Psychoneuroendocrinology 50:106–117

Gould E, Beylin A, Tanapat P, Reeves A, Shors TJ (1999) Learning enhances adult neurogenesis in the hippocampal formation. Nat Neurosci 2(3):260

Green AD, Galea LA (2008) Adult hippocampal cell proliferation is suppressed with estrogen withdrawal after a hormone-simulated pregnancy. Horm Behav 54(1):203–211

Gregus A, Wintink AJ, Davis AC, Kalynchuk LE (2005) Effect of repeated corticosterone injections and restraint stress on anxiety and depression-like behavior in male rats. Behav Brain Res 156(1):105–114

Gutiérrez-Lobos K, Scherer M, Anderer P, Katschnig H (2002) The influence of age on the female/male ratio of treated incidence rates in depression. BMC Psychiatry 2(1):3

Hamson DK, Wainwright SR, Taylor JR, Jones BA, Watson NV, Galea LA (2013) Androgens increase survival of adult-born neurons in the dentate gyrus by an androgen receptor-dependent mechanism in male rats. Endocrinology 154(9):3294–3304

Harburger LL, Nzerem CK, Frick KM (2007) Single enrichment variables differentially reduce age-related memory decline in female mice. Behav Neurosci 121(4):679

Hayes K, Sprague S, Guo M, Davis W, Friedman A, Kumar A, Jimenez DF, Ding Y (2008) Forced, not voluntary, exercise effectively induces neuroprotection in stroke. Acta Neuropathol 115(3):289–296

Hill AS, Sahay A, Hen R (2015) Increasing adult hippocampal neurogenesis is sufficient to reduce anxiety and depression-like behaviors. Neuropsychopharmacology 40(10):2368

Hitchcock FA (1925) Studies in vigor: V. The comparative activity of male and female albino rats. Am J Physiol Legacy Content 75(1):205–210

Ho KY, Evans WS, Blizzard RM, Veldhuis JD, Merriam GR, Samojlik E, Furlanetto R, Rogol AD, Kaiser DL, Thorner MO (1987) Effects of sex and age on the 24-hour profile of growth hormone secretion in man: importance of endogenous estradiol concentrations. J Clin Endocrinol Metabol 64(1):51–58

Holmes MM, Galea LA, Mistlberger RE, Kempermann G (2004) Adult hippocampal neurogenesis and voluntary running activity: circadian and dose-dependent effects. J Neurosci Res 76(2):216–222

Hötting K, Röder B (2013) Beneficial effects of physical exercise on neuroplasticity and cognition. Neurosci Biobehav Rev 37(9):2243–2257

Inoue K, Okamoto M, Shibato J, Lee MC, Matsui T, Rakwal R, Soya H (2015) Long-term mild, rather than intense, exercise enhances adult hippocampal neurogenesis and greatly changes the transcriptomic profile of the hippocampus. PLoS One 10(6):e0128720

Isgor C, Watson SJ (2005) Estrogen receptor α and β mRNA expressions by proliferating and differentiating cells in the adult rat dentate gyrus and subventricular zone. Neuroscience 134(3):847–856

Iwasaki-Sekino A, Mano-Otagiri A, Ohata H, Yamauchi N, Shibasaki T (2009) Gender differences in corticotropin and corticosterone secretion and corticotropin-releasing factor mRNA expression in the paraventricular nucleus of the hypothalamus and the central nucleus of the amygdala in response to footshock stress or psychological stress in rats. Psychoneuroendocrinology 34(2):226–237

Jaako-Movits K, Zharkovsky T, Pedersen M, Zharkovsky A (2006) Decreased hippocampal neurogenesis following olfactory bulbectomy is reversed by repeated citalopram administration. Cell Mol Neurobiol 26(7–8):1557

Jakubs K, Nanobashvili A, Bonde S, Ekdahl CT, Kokaia Z, Kokaia M, Lindvall O (2006) Environment matters: synaptic properties of neurons born in the epileptic adult brain develop to reduce excitability. Neuron 52(6):1047–1059

Jessberger S, Nakashima K, Clemenson GD, Mejia E, Mathews E, Ure K, Ogawa S, Sinton CM, Gage FH, Hsieh J (2007) Epigenetic modulation of seizure-induced neurogenesis and cognitive decline. J Neurosci 27(22):5967–5975

Jin J, Jing H, Choi G, Oh MS, Ryu JH, Jeong JW, Huh Y, Park C (2008) Voluntary exercise increases the new cell formation in the hippocampus of ovariectomized mice. Neurosci Lett 439(3):260–263

Johnson RA, Mitchell GS (2003) Exercise-induced changes in hippocampal brain-derived neurotrophic factor and neurotrophin-3: effects of rat strain. Brain Res 983(1–2):108–114

Jones AB, Gupton R, Curtis KS (2016) Estrogen and voluntary exercise interact to attenuate stress-induced corticosterone release but not anxiety-like behaviors in female rats. Behav Brain Res 311:279–286

Kahlert S, Nuedling S, van Eickels M, Vetter H, Meyer R, Grohé C (2000) Estrogen receptor α rapidly activates the IGF-1 receptor pathway. J Biol Chem 275(24):18447–18453

Kalynchuk LE, Gregus A, Boudreau D, Perrot-Sinal TS (2004) Corticosterone increases depression-like behavior, with some effects on predator odor-induced defensive behavior, in male and female rats. Behav Neurosci 118(6):1365

Kambo JS, Galea LA (2006) Activational levels of androgens influence risk assessment behaviour but do not influence stress-induced suppression in hippocampal cell proliferation in adult male rats. Behav Brain Res 175(2):263–270

Kannangara TS, Webber A, Gil-Mohapel J, Christie BR (2009) Stress differentially regulates the effects of voluntary exercise on cell proliferation in the dentate gyrus of mice. Hippocampus 19(10):889–897

Kannangara TS, Lucero MJ, Gil-Mohapel J, Drapala RJ, Simpson JM, Christie BR, van Praag H (2011) Running reduces stress and enhances cell genesis in aged mice. Neurobiol Aging 32(12):2279–2286

Karege F, Bondolfi G, Gervasoni N, Schwald M, Aubry JM, Bertschy G (2005) Low brain-derived neurotrophic factor (BDNF) levels in serum of depressed patients probably results from lowered platelet BDNF release unrelated to platelet reactivity. Biol Psychiatry 57(9):1068–1072

Katz RJ, Roth KA, Carroll BJ (1981) Acute and chronic stress effects on open field activity in the rat: implications for a model of depression. Neurosci Biobehav Rev 5(2):247–251

Kempermann G, Song H, Gage FH (2015) Neurogenesis in the adult hippocampus. Cold Spring Harb Perspect Biol 7(9):a018812

Khan A, Brodhead AE, Schwartz KA, Kolts RL, Brown WA (2005) Sex differences in antidepressant response in recent antidepressant clinical trials. J Clin Psychopharmacol 25(4):318–324

Kim SE, Ko IG, Kim BK, Shin MS, Cho S, Kim CJ, Kim SH, Baek SS, Lee EK, Jee YS (2010) Treadmill exercise prevents aging-induced failure of memory through an increase in neurogenesis and suppression of apoptosis in rat hippocampus. Exp Gerontol 45(5):357–365

Kim TW, Kim CS, Kim JY, Kim CJ, Seo JH (2016) Combined exercise ameliorates ovariectomy-induced cognitive impairment by enhancing cell proliferation and suppressing apoptosis. Menopause 23(1):18–26

Kodama M, Fujioka T, Duman RS (2004) Chronic olanzapine or fluoxetine administration increases cell proliferation in hippocampus and prefrontal cortex of adult rat. Biol Psychiatry 56(8):570–580

Kornstein SG, Schatzberg AF, Thase ME, Yonkers KA, McCullough JP, Keitner GI, Gelenberg AJ, Davis SM, Harrison WM, Keller MB (2000) Gender differences in treatment response to sertraline versus imipramine in chronic depression. Am J Psychiatr 157(9):1445–1452

Kuhn HG, Dickinson-Anson H, Gage FH (1996) Neurogenesis in the dentate gyrus of the adult rat: age-related decrease of neuronal progenitor proliferation. J Neurosci 16(6):2027–2033

Lacar B, Parylak SL, Vadodaria KC, Sarkar A, Gage FH (2014) Increasing the resolution of the adult neurogenesis picture. F1000Prime Rep 6:8

Lagace DC, Fischer SJ, Eisch AJ (2007) Gender and endogenous levels of estradiol do not influence adult hippocampal neurogenesis in mice. Hippocampus 17(3):175–180

Lau WM, Qiu G, Helmeste DM, Lee T, Tang SW, So KF (2007) Corticosteroid decreases subventricular zone cell proliferation, which could be reversed by paroxetine. Restor Neurol Neurosci 25(1):17–23

Leasure JL, Decker L (2009) Social isolation prevents exercise-induced proliferation of hippocampal progenitor cells in female rats. Hippocampus 19(10):907–912

Leasure JL, Jones M (2008) Forced and voluntary exercise differentially affect brain and behavior. Neuroscience 156(3):456–465

Lee KJ, Kim SJ, Kim SW, Choi SH, Shin YC, Park SH, Moon BH, Cho E, Lee MS, Choi SH, Chun BG, Shin KH (2006) Chronic mild stress decreases survival, but not proliferation, of new-born cells in adult rat hippocampus. Exp Mol Med 38(1):44

Leranth C, Hajszan T, MacLusky NJ (2004) Androgens increase spine synapse density in the CA1 hippocampal subfield of ovariectomized female rats. J Neurosci 24(2):495–499

Leuner B, Mirescu C, Noiman L, Gould E (2007) Maternal experience inhibits the production of immature neurons in the hippocampus during the postpartum period through elevations of adrenal steroids. Hippocampus 17:434–442

Lightfoot JT, Turner MJ, Daves M, Vordermark A, Kleeberger SR (2004) Genetic influence on daily wheel running activity level. Physiol Genomics 19(3):270–276

Lindsay J, Laurin D, Verreault R, Hébert R, Helliwell B, Hill GB, McDowell I (2002) Risk factors for Alzheimer's disease: a prospective analysis from the Canadian Study of Health and Aging. Am J Epidemiol 156(5):445–453

Liu W, Zhou C (2012) Corticosterone reduces brain mitochondrial function and expression mitofusin, BDNF in depression-like rodents regardless of exercise preconditioning. Psychoneuroendocrinology 37:1057–1070

Liu Q, Yu J, Mao-Ying QL, Mi WL, Li B, Wang YQ, Wang J, Wu GC (2008) Repeated clomipramine treatment reversed the inhibition of cell proliferation in adult hippocampus induced by chronic unpredictable stress. Pharmacogenomics J 8(6):375

Liu YF, Chen HI, Wu CL, Kuo YM, Yu L, Huang AM, Wu FS, Chuang JI, Jen CJ (2009) Differential effects of treadmill running and wheel running on spatial or aversive learning and memory: roles of amygdalar brain-derived neurotrophic factor and synaptotagmin I. J Physiol 587(13):3221–3231

Liu W, Sheng H, Xu Y, Liu Y, Lu J, Ni X (2013) Swimming exercise ameliorates depression-like behavior in chronically stressed rats: relevant to proinflammatory cytokines and IDO activation. Behav Brain Res 242:110–116

Lou SJ, Liu JY, Chang H, Chen PJ (2008) Hippocampal neurogenesis and gene expression depend on exercise intensity in juvenile rats. Brain Res 1210:48–55

Lucassen PJ, Meerlo P, Naylor AS, Van Dam AM, Dayer AG, Fuchs E, Oomen CA, Czeh B (2010) Regulation of adult neurogenesis by stress, sleep disruption, exercise and inflammation: implications for depression and antidepressant action. Eur Neuropsychopharmacol 20(1):1–7

MacQueen GM, Campbell S, McEwen BS, Macdonald K, Amano S, Joffe RT, Nahmias C, Young LT (2003) Course of illness, hippocampal function, and hippocampal volume in major depression. Proc Natl Acad Sci 100(3):1387–1392

Mahmoud R, Wainwright SR, Chaiton JA, Lieblich SE, Galea LA (2016a) Ovarian hormones, but not fluoxetine, impart resilience within a chronic unpredictable stress model in middle-aged female rats. Neuropharmacology 107:278–293

Mahmoud R, Wainwright SR, Galea LA (2016b) Sex hormones and adult hippocampal neurogenesis: regulation, implications, and potential mechanisms. Front Neuroendocrinol 41:129–152

Malberg JE, Duman RS (2003) Cell proliferation in adult hippocampus is decreased by inescapable stress: reversal by fluoxetine treatment. Neuropsychopharmacology 28(9):1562

Malberg JE, Eisch AJ, Nestler EJ, Duman RS (2000) Chronic antidepressant treatment increases neurogenesis in adult rat hippocampus. J Neurosci 20(24):9104–9110

Marais L, Stein DJ, Daniels WMU (2009) Exercise increases BDNF levels in the striatum and decreases depressive-like behavior in chronically stressed rats. Metab Brain Dis 24:587–597

Mateus-Pinheiro A, Pinto L, Bessa JM, Morais M, Alves ND, Monteiro S, Patricio P, Almeida OF, Sousa N (2013) Sustained remission from depressive-like behavior depends on hippocampal neurogenesis. Transl Psychiatry 3(1):e210

Mayer JL, Klumpers L, Maslam S, De Kloet ER, Joels M, Lucassen PJ (2006) Brief treatment with the glucocorticoid receptor antagonist mifepristone normalises the corticosterone-induced reduction of adult hippocampal neurogenesis. J Neuroendocrinol 18(8):629–631

Mazzucco CA, Lieblich SE, Bingham BI, Williamson MA, Viau V, Galea LA (2006) Both estrogen receptor alpha and estrogen receptor beta agonist enhance cell proliferation in the dentate gyrus of adult female rats. Neuroscience 141(4):1793–1800

McClure RE, Barha CK, Galea LA (2013) 17β-estradiol, but not estrone, increases the survival and activation of new neurons in the hippocampus in response to spatial memory in adult female rats. Horm Behav 63(1):144–157

McKinnon MC, Yucel K, Nazarov A, MacQueen GM (2009) A meta-analysis examining clinical predictors of hippocampal volume in patients with major depressive disorder. J Psychiatry Neurosci 34(1):41
Miller BR, Hen R (2015) The current state of the neurogenic theory of depression and anxiety. Curr Opin Neurobiol 30:51–58
Mineur YS, Belzung C, Crusio WE (2007) Functional implications of decreases in neurogenesis following chronic mild stress in mice. Neuroscience 150(2):251–259
Montaron MF, Drapeau E, Dupret D, Kitchener P, Aurousseau C, Le Moal M, Piazza PV, Abrous DN (2006) Lifelong corticosterone level determines age-related decline in neurogenesis and memory. Neurobiol Aging 27(4):645–654
Monteiro BMM, Moreira FA, Massensini AR, Moraes MFD, Pereira GS (2014) Enriched environment increases neurogenesis and improves social memory persistence in socially isolated adult mice. Hippocampus 24(2):239–248
Morimoto M, Morita N, Ozawa H, Yokoyama K, Kawata M (1996) Distribution of glucocorticoid receptor immunoreactivity and mRNA in the rat brain: an immunohistochemical and in situ hybridization study. Neurosci Res 26(3):235–269
Murray F, Smith DW, Hutson PH (2008) Chronic low dose corticosterone exposure decreased hippocampal cell proliferation, volume and induced anxiety and depression like behaviours in mice. Eur J Pharmacol 583(1):115–127
Neeper SA, Góauctemez-Pinilla F, Choi J, Cotman C (1995) Exercise and brain neurotrophins. Nature 373(6510):109
Nokia MS, Lensu S, Ahtiainen JP, Johansson PP, Koch LG, Britton SL, Kainulainen H (2016) Physical exercise increases adult hippocampal neurogenesis in male rats provided it is aerobic and sustained. J Physiol 594(7):1855–1873
O'Leary OF, Cryan JF (2014) A ventral view on antidepressant action: roles for adult hippocampal neurogenesis along the dorsoventral axis. Trends Pharmacol Sci 35(12):675–687
Okamoto M, Hojo Y, Inoue K, Matsui T, Kawato S, McEwen BS, Soya H (2012) Mild exercise increases dihydrotestosterone in hippocampus providing evidence for androgenic mediation of neurogenesis. Proc Natl Acad Sci 109(32):13100–13105
Oliff HS, Berchtold NC, Isackson P, Cotman CW (1998) Exercise-induced regulation of brain-derived neurotrophic factor (BDNF) transcripts in the rat hippocampus. Mol Brain Res 61(1–2):147–153
Ormerod BK, Galea LA (2001) Reproductive status influences cell proliferation and cell survival in the dentate gyrus of adult female meadow voles: a possible regulatory role for estradiol. Neuroscience 102(2):369–379
Ormerod BK, Galea LA (2003) Reproductive status influences the survival of new cells in the dentate gyrus of adult male meadow voles. Neurosci Lett 346(1–2):25–28
Ormerod BK, Falconer EM, Galea LA (2003) N-methyl-D-aspartate receptor activity and estradiol: separate regulation of cell proliferation in the dentate gyrus of adult female meadow vole. J Endocrinol 179(2):155–163
Ormerod BK, Lee TY, Galea LA (2004) Estradiol enhances neurogenesis in the dentate gyri of adult male meadow voles by increasing the survival of young granule neurons. Neuroscience 128(3):645–654
Otte C, Gold SM, Penninx BW, Pariante CM, Etkin A, Fava M, Mohr DC, Schatzberg AF (2016) Major depressive disorder. Nat Rev Disease Primers 2:16065
Overall RW, Walker TL, Leiter O, Lenke S, Ruhwald S, Kempermann G (2013) Delayed and transient increase of adult hippocampal neurogenesis by physical exercise in DBA/2 mice. PLoS One 8(12):e83797
Patki G, Li L, Allam F, Solanki N, Dao AT, Alkadhi K, Salim S (2014) Moderate treadmill exercise rescues anxiety and depression-like behavior as well as memory impairment in a rat model of posttraumatic stress disorder. Physiol Behav 130:47–53
Pawluski JL, Galea LA (2007) Reproductive experience alters hippocampal neurogenesis during the postpartum period in the dam. Neuroscience 149(1):53–67

Perera TD, Coplan JD, Lisanby SH, Lipira CM, Arif M, Carpio C, Spitzer G, Santarelli L, Scharf B, Hen R, Rosoklija G, Sackeim HA, Dwork AJ (2007) Antidepressant-induced neurogenesis in the hippocampus of adult nonhuman primates. J Neurosci 27(18):4894–4901

Perez-Martin M, Azcoitia I, Trejo JL, Sierra A, Garcia-Segura LM (2003) An antagonist of estrogen receptors blocks the induction of adult neurogenesis by insulin-like growth factor-I in the dentate gyrus of adult female rats. Eur J Neurosci 18:923–930

Pham K, Nacher J, Hof PR, McEwen BS (2003) Repeated restraint stress suppresses neurogenesis and induces biphasic PSA-NCAM expression in the adult rat dentate gyrus. Eur J Neurosci 17(4):879–886

Pinnock SB, Herbert J (2008) Brain-derived neurotropic factor and neurogenesis in the adult rat dentate gyrus: interactions with corticosterone. Eur J Neurosci 27(10):2493–2500

Price EO (1976) Food deprivation effects on the running-wheel activity of wild and domestic Norway rats. Behav Process 1(2):191–196

Pryce CR (2008) Postnatal ontogeny of expression of the corticosteroid receptor genes in mammalian brains: inter-species and intra-species differences. Brain Res Rev 57(2):596–605

Rantamäki T, Vesa L, Antila H, Di Lieto A, Tammela P, Schmitt A, Lesch KP, Rios M, Castrén E (2011) Antidepressant drugs transactivate TrkB neurotrophin receptors in the adult rodent brain independently of BDNF and monoamine transporter blockade. PLoS One 6(6):e20567

Redila VA, Christie BR (2006) Exercise-induced changes in dendritic structure and complexity in the adult hippocampal dentate gyrus. Neuroscience 137(4):1299–1307

Revest JM, Dupret D, Koehl M, Funk-Reiter C, Grosjean N, Piazza PV, Abrous DN (2009) Adult hippocampal neurogenesis is involved in anxiety-related behaviors. Mol Psychiatry 14(10):959

Rhodes JS, Van Praag H, Jeffrey S, Girard I, Mitchell GS, Garland T Jr, Gage FH (2003) Exercise increases hippocampal neurogenesis to high levels but does not improve spatial learning in mice bred for increased voluntary wheel running. Behav Neurosci 117(5):1006

Rock PL, Roiser JP, Riedel WJ, Blackwell AD (2014) Cognitive impairment in depression: a systematic review and meta-analysis. Psychol Med 44(10):2029–2040

Romano-Torres M, Fernandez-Guasti A (2010) Estradiol valerate elicits antidepressant-like effects in middle-aged female rats under chronic mild stress. Behav Pharmacol 21(2):104–111

Rummel J, Epp JR, Galea LA (2010) Estradiol does not influence strategy choice but place strategy choice is associated with increased cell proliferation in the hippocampus of female rats. Horm Behav 58(4):582–590

Saarelainen T, Hendolin P, Lucas G, Koponen E, Sairanen M, MacDonald E, Agerman K, Haapasalo A, Nawa H, Aloyz R, Ernfors P, Castren E (2003) Activation of the TrkB neurotrophin receptor is induced by antidepressant drugs and is required for antidepressant-induced behavioral effects. J Neurosci 23(1):349–357

Sahay A, Hen R (2007) Adult hippocampal neurogenesis in depression. Nat Neurosci 10(9):1110

Santarelli L, Saxe M, Gross C, Surget A, Battaglia F, Dulawa S, Weisstaub N, Lee J, Duman R, Arancio O, Belzung C, Hen R (2003) Requirement of hippocampal neurogenesis for the behavioral effects of antidepressants. Science 301(5634):805–809

Scharfman HE, MacLusky NJ (2006) Estrogen and brain-derived neurotrophic factor (BDNF) in hippocampus: complexity of steroid hormone-growth factor interactions in the adult CNS. Front Neuroendocrinol 27(4):415–435

Scharfman H, Goodman J, Macleod A, Phani S, Antonelli C, Croll S (2005) Increased neurogenesis and the ectopic granule cells after intrahippocampal BDNF infusion in adult rats. Exp Neurol 192(2):348–356

Schoenfeld TJ, Gould E (2012) Stress, stress hormones, and adult neurogenesis. Exp Neurol 233(1):12–21

Schoenfeld TJ, McCausland HC, Morris HD, Padmanaban V, Cameron HA (2017) Stress and loss of adult neurogenesis differentially reduce hippocampal volume. Biol Psychiatry 82(12):914–923

Sen S, Duman R, Sanacora G (2008) Serum brain-derived neurotrophic factor, depression, and antidepressant medications: meta-analyses and implications. Biol Psychiatry 64(6):527–532

Sheridan M, Tamarin RH (1988) Space use, longevity, and reproductive success in meadow voles. Behav Ecol Sociobiol 22(2):85–90
Sherry DF, Jacobs LF, Gaulin SJC (1992) Spatial memory and adaptive specialization of the hippocampus. Trends Neurosci 15(8):298–303
Shimizu E, Hashimoto K, Okamura N, Koike K, Komatsu N, Kumakiri C, Nakazato M, Watanabe H, Shinoda N, Okada SI, Iyo M (2003) Alterations of serum levels of brain-derived neurotrophic factor (BDNF) in depressed patients with or without antidepressants. Biol Psychiatry 54(1):70–75
Shingo T, Gregg C, Enwere E, Fujikawa H, Hassam R, Geary C, Cross JC, Weiss S (2003) Pregnancy-stimulated neurogenesis in the adult female forebrain mediated by prolactin. Science 299(5,603):117–120
Shirayama Y, Chen AC, Nakagawa S, Russell DS, Duman RS (2002) Brain-derived neurotrophic factor produces antidepressant effects in behavioral models of depression. J Neurosci 22(8):3251–3261
Shores MM, Sloan KL, Matsumoto AM, Moceri VM, Felker B, Kivlahan DR (2004) Increased incidence of diagnosed depressive illness in hypogonadal older men. Arch Gen Psychiatry 61(2):162–167
Shores MM, Moceri VM, Sloan KL, Matsumoto AM, Kivlahan DR (2005) Low testosterone levels predict incident depressive illness in older men: effects of age and medical morbidity. J Clin Psychiatry 66:7–14
Shores MM, Kivlahan DR, Sadak TI, Li EJ, Matsumoto AM (2009) A randomized, double-blind, placebo-controlled study of testosterone treatment in hypogonadal older men with subthreshold depression (dysthymia or minor depression). J Clin Psychiatry 70(7):1009–1016
Simerly RB, Swanson LW, Chang C, Muramatsu M (1990) Distribution of androgen and estrogen receptor mRNA-containing cells in the rat brain: an in situ hybridization study. J Comp Neurol 294(1):76–95
Siwak-Tapp CT, Head E, Muggenburg BA, Milgram NW, Cotman CW (2007) Neurogenesis decreases with age in the canine hippocampus and correlates with cognitive function. Neurobiol Learn Mem 88(2):249–259
Snyder JS, Choe JS, Clifford MA, Jeurling SI, Hurley P, Brown A, Kamhi JF, Cameron HA (2009) Adult-born hippocampal neurons are more numerous, faster maturing, and more involved in behavior in rats than in mice. J Neurosci 29(46):14484–14495
Snyder JS, Soumier A, Brewer M, Pickel J, Cameron HA (2011) Adult hippocampal neurogenesis buffers stress responses and depressive behaviour. Nature 476(7,361):458
Snyder JS, Grigereit L, Russo A, Seib DR, Brewer M, Pickel J, Cameron HA (2016) A transgenic rat for specifically inhibiting adult neurogenesis. eNeuro 3(3):ENEURO-0064. https://doi.org/10.1523/ENEURO.0064-16.2016
So JH, Huang C, Ge M, Cai G, Zhang L, Lu Y, Mu Y (2017) Intense exercise promotes adult hippocampal neurogenesis but not spatial discrimination. Front Cell Neurosci 11:13
Sohrabji F, Lewis DK (2006) Estrogen–BDNF interactions: implications for neurodegenerative diseases. Front Neuroendocrinol 27(4):404–414
Solberg LC, Horton TH, Turek FW (1999) Circadian rhythms and depression: effects of exercise in an animal model. Am J Physiol Regul Integr Comp Physiol 276:152–161
Spalding KL, Bergmann O, Alkass K, Bernard S, Salehpour M, Huttner HB, Boström E, Westerlund I, Vial C, Buchholz BA, Possnert G, Mash DC, Druid H, Frisen J (2013) Dynamics of hippocampal neurogenesis in adult humans. Cell 153(6):1219–1227
Speisman RB, Kumar A, Rani A, Foster TC, Ormerod BK (2013) Daily exercise improves memory, stimulates hippocampal neurogenesis and modulates immune and neuroimmune cytokines in aging rats. Brain Behav Immun 28:25–43
Spritzer MD, Galea LA (2007) Testosterone and dihydrotestosterone, but not estradiol, enhance survival of new hippocampal neurons in adult male rats. Dev Neurobiol 67(10):1321–1333
Spritzer MD, Ibler E, Inglis W, Curtis MG (2011) Testosterone and social isolation influence adult neurogenesis in the dentate gyrus of male rats. Neuroscience 195:180–190
Stetler C, Miller GE (2011) Depression and hypothalamic-pituitary-adrenal activation: a quantitative summary of four decades of research. Psychosom Med 73(2):114–126

Stranahan AM, Khalil D, Gould E (2006) Social isolation delays the positive effects of running on adult neurogenesis. Nat Neurosci 9(4):526

Su Q, Cheng Y, Jin K, Cheng J, Lin Y, Lin Z, Wang L, Shao B (2016) Estrogen therapy increases BDNF expression and improves post-stroke depression in ovariectomy-treated rats. Exp Ther Med 12(3):1843–1848

Surget A, Tanti A, Leonardo ED, Laugeray A, Rainer Q, Touma C, Plame R, Griebel G, Ibarguen-Vargas Y, Hen R, Belzung C (2011) Antidepressants recruit new neurons to improve stress response regulation. Mol Psychiatry 16(12):1177–1188

Swift-Gallant A, Duarte-Guterman P, Hamson DK, Ibrahim M, Monks DA, Galea LAM (2018) Neural androgen receptors affect the number of surviving new neurones in the adult dentate gyrus of male mice. J Neuroendocrinol 30(4):e12578

Taliaz D, Stall N, Dar DE, Zangen A (2010) Knockdown of brain-derived neurotrophic factor in specific brain sites precipitates behaviors associated with depression and reduces neurogenesis. Mol Psychiatry 15(1):80

Tanapat P, Hastings NB, Reeves AJ, Gould E (1999) Estrogen stimulates a transient increase in the number of new neurons in the dentate gyrus of the adult female rat. J Neurosci 19(14):5792–5801

Tanapat P, Hastings NB, Gould E (2005) Ovarian steroids influence cell proliferation in the dentate gyrus of the adult female rat in a dose-and time-dependent manner. J Comp Neurol 481(3):252–265

Tanti A, Belzung C (2013) Neurogenesis along the septo-temporal axis of the hippocampus: are depression and the action of antidepressants region-specific? Neuroscience 252:234–252

Taupin P (2007) BrdU immunohistochemistry for studying adult neurogenesis: paradigms, pitfalls, limitations, and validation. Brain Res Rev 53(1):198–214

Triviño-Paredes J, Patten AR, Gil-Mohapel J, Christie BR (2016) The effects of hormones and physical exercise on hippocampal structural plasticity. Front Neuroendocrinol 41:23–43

Tzeng WY, Chen LH, Cherng CG, Tsai YN, Yu L (2014) Sex differences and the modulating effects of gonadal hormones on basal and the stressor-decreased newly proliferative cells and neuroblasts in dentate gyrus. Psychoneuroendocrinology 42:24–37

Van Praag H, Kempermann G, Gage FH (1999) Running increases cell proliferation and neurogenesis in the adult mouse dentate gyrus. Nat Neurosci 2(3):266

Van Praag H, Shubert T, Zhao C, Gage FH (2005) Exercise enhances learning and hippocampal neurogenesis in aged mice. J Neurosci 25(38):8680–8685

Vega-Rivera NM, Fernández-Guasti A, Ramírez-Rodríguez G, Estrada-Camarena E (2015) Effect of sub-optimal doses of fluoxetine plus estradiol on antidepressant-like behavior and hippocampal neurogenesis in ovariectomized rats. Psychoneuroendocrinology 57:113–124

Veras AB, Nardi AE (2010) The complex relationship between hypogonadism and major depression in a young male. Prog Neuro-Psychopharmacol Biol Psychiatry 34(2):421–422

Verhagen M, Van Der Meij A, Van Deurzen PA, Janzing JG, Arias-Vasquez A, Buitelaar JK, Franke B (2010) Meta-analysis of the BDNF Val66Met polymorphism in major depressive disorder: effects of gender and ethnicity. Mol Psychiatry 15(3):260

Videbech P, Ravnkilde B (2004) Hippocampal volume and depression: a meta-analysis of MRI studies. Am J Psychiatr 161(11):1957–1966

Vivar C, Peterson BD, van Praag H (2016) Running rewires the neuronal network of adult-born dentate granule cells. NeuroImage 131:29–41

Vollmayr B, Mahlstedt MM, Henn FA (2007) Neurogenesis and depression: what animal models tell us about the link. Eur Arch Psychiatry Clin Neurosci 257(5):300–303

Voskuhl R, Momtazee C (2017) Pregnancy: effect on multiple sclerosis, treatment considerations, and breastfeeding. Neurotherapeutics 14:974–984

Wainwright SR, Lieblich SE, Galea LA (2011) Hypogonadism predisposes males to the development of behavioural and neuroplastic depressive phenotypes. Psychoneuroendocrinology 36(9):1327–1341

Wainwright SR, Barha CK, Hamson DK, Epp JR, Chow C, Lieblich SE, Rutishauser U, Galea LA (2016a) Enzymatic depletion of the polysialic acid moiety associated with the neural cell adhesion molecule inhibits antidepressant efficacy. Neuropsychopharmacology 41(6):1670

Wainwright SR, Workman JL, Tehrani A, Hamson DK, Chow C, Lieblich SE, Galea LA (2016b) Testosterone has antidepressant-like efficacy and facilitates imipramine-induced neuroplasticity in male rats exposed to chronic unpredictable stress. Horm Behav 79:58–69

Wang GH, Richter CP, Guttmacher AF (1925) Activity studies on male castrated rats with ovarian transplants, and correlation of the activity with the histology of the grafts. Am J Physiol Legacy Content 73(3):581–599

Warner-Schmidt JL, Duman RS (2006) Hippocampal neurogenesis: opposing effects of stress and antidepressant treatment. Hippocampus 16(3):239–249

Weiland NG, Orikasa C, Hayashi S, McEwen BS (1997) Distribution and hormone regulation of estrogen receptor immunoreactive cells in the hippocampus of male and female rats. J Comp Neurol 388(4):603–612

Weinstock M, Razin M, Schorer-Apelbaum D, Men D, Mccarty R (1998) Gender differences in sympathoadrenal activity in rats at rest and in response to footshock stress. Int J Dev Neurosci 16(3–4):289–295

Willner P, Towell A, Sampson D, Sophokleous S, Muscat R (1987) Reduction of sucrose preference by chronic unpredictable mild stress, and its restoration by a tricyclic antidepressant. Psychopharmacology 93(3):358–364

Wong EYH, Herbert J (2005) Roles of mineralocorticoid and glucocorticoid receptors in the regulation of progenitor proliferation in the adult hippocampus. Eur J Neurosci 22(4):785–792

Wong EY, Herbert J (2006) The corticoid environment: a determining factor for neural progenitors' survival in the adult hippocampus. Eur J Neurosci 20(10):2491–2498

Workman JL, Chan MY, Galea LA (2015) Prior high corticosterone exposure reduces activation of immature neurons in the ventral hippocampus in response to spatial and nonspatial memory. Hippocampus 25(3):329–344

Workman JL, Gobinath AR, Kitay NF, Chow C, Brummelte S, Galea LA (2016) Parity modifies the effects of fluoxetine and corticosterone on behavior, stress reactivity, and hippocampal neurogenesis. Neuropharmacology 105:443–453

Xiao L, Jordan CL (2002) Sex differences, laterality, and hormonal regulation of androgen receptor immunoreactivity in rat hippocampus. Horm Behav 42(3):327–336

Yau SY, Li A, So KF (2015) Involvement of adult hippocampal neurogenesis in learning and forgetting. Neural Plast 2015:717958

Youssef MM, Underwood MD, Huang Y-Y, Hsung S-C, Liu Y, Simpson NR, Bakalian MJ, Rosoklija GB, Dwork AJ, Arango V, Mann JJ (2018) Association of BDNF Val66Met polymorphism and brain DBNF levels with major depression and suicide. Int J Neuropsychopharmacol 21(6):528–538

Zhang JQ, Cai WQ, Zhou DS, Su BY (2002) Distribution and differences of estrogen receptor beta immunoreactivity in the brain of adult male and female rats. Brain Res 935(1–2):73–80

Neuroimmune Impacts of Early-Life Stress on Development and Psychopathology

Heather C. Brenhouse, Andrea Danese, and Rodrigo Grassi-Oliveira

Contents

1 Introduction 424
2 Toward an Operational Definition of ELS 426
3 A Brief Overview of the Immune System 426
4 Neuroimmune Development 428
5 Immunity, Childhood Trauma, and Mental Illness in Humans 430
6 Evidence from Animal Models: Immune Alterations After ELS 433
7 The Importance of Individual Differences 435
8 How Do ELS-Induced Immune Changes Ultimately Affect Brain and Behavior? 436
9 Future Directions 437
10 Conclusion 438
References 439

Abstract Maltreatment and trauma in childhood, termed early-life stress (ELS), has long-term effects on the immune system. ELS impacts immune signaling at the time of exposure but also disrupts the developmental trajectory of certain immunological processes, both in the periphery and in the brain. One consequence of these early alterations is a heightened immune response to stressors later in life. However, chronic and sustained inflammatory response can also lead to excitotoxicity and prevent typical brain development. In this chapter, we discuss current progress toward understanding the contribution of neuroimmune signaling to ELS-attributable dysfunction or maladaptation with a focus on postnatal

H. C. Brenhouse (✉)
Psychology Department, Northeastern University, Boston, MA, USA
e-mail: h.brenhouse@neu.edu

A. Danese
Social, Genetic, and Developmental Psychiatry Centre, Department of Child and Adolescent Psychiatry, Institute of Psychiatry, Psychology, and Neuroscience, King's College London, London, UK

R. Grassi-Oliveira
Developmental Cognitive Neuroscience Lab (DCNL), Graduate Program in Psychology, Pontifical Catholic University of Rio Grande do Sul (PUCRS), Porto Alegre, RS, Brazil

Curr Topics Behav Neurosci (2019) 43: 423–448
DOI 10.1007/7854_2018_53
Published Online: 13 July 2018

experiences. To do so we first present an operational definition of ELS. Then, we offer a brief overview of the immune system and neuroimmune development, followed by a section discussing the interaction between immunity, childhood trauma, and mental disorders in humans. We present evidence from animal models about immune alterations after ELS and discuss the ways in which ELS-induced immune changes ultimately affect brain and behavior, as well as the importance of individual differences and future directions in this field. Taken together, we submit that when encountered with ELS, some core brain circuits could develop differently via various mechanisms involving dysfunctional immune reprograming. However, given the remarkable plasticity of both the brain and the immune system, many of the deleterious effects of ELS may be mitigated with interventions that account for sex and target neuroimmune interactions over the lifespan.

Keywords Brain · Childhood adversity · Inflammation · Mental illness

1 Introduction

Evolutionary biology describes how individual genotypes can produce different phenotypes when exposed to varying environmental conditions, termed "phenotypic plasticity" (Pigliucci et al. 2006). This includes the possibility to modify trajectories of biological development in response to specific environmental cues, which allows the organism to cope with environmental unpredictability and/or heterogeneity. As such, early-life experiences – both positive and negative – can have profound effects on brain development in mammals. Rearing environments that are enriched with good parental care, suitable protection and safety, and engaging sensory stimulation can offer resilience to insults later in life such as psychological stressors (Francis et al. 2002) or even pathological infection (Johnson et al. 2014). Not all plasticity is equally adaptive, however; as a developing organism is impacted by environmental stimuli, misdirected trajectories that are aimed for one type of environment sometimes yield physiology and behavior that is maladapted for later life stages. For example, early-life adversity (ELS) such as parental deprivation, neglect, abuse, or exposure to threats has been repeatedly shown to yield a myriad of deleterious deviations in brain circuitry, stress responsivity, cognitive function, and general health (Brown et al. 2010; Dube et al. 2009; Anda et al. 2008). While some physiological changes have been shown to be markers of resilience to illness (e.g., Silvers et al. 2016), others predict dysfunction and vulnerability to illness (e.g., Grassi-Oliveira et al. 2016; Danese et al. 2008).

Although the phenomena of developmental plasticity have been best studied with regard to the basic brain functions, such as vision (Hensch 2005), they are increasingly appreciated as key determinants of complex brain and immune functions (Danese and McEwen 2012; Rook et al. 2015). Importantly, there is a growing appreciation of a dynamic interaction between the nervous and immune systems that is critical for healthy behavioral responses to our environment. In this chapter, we will discuss current progress toward understanding the contribution of neuroimmune signaling to ELS-attributable dysfunction or maladaptation. We focus on postnatal experiences; however maternal (prenatal) stress exposure has also been repeatedly shown to affect neuroimmune development with pathological consequences (see Miller et al. 2017; Goldstein et al. 2016; Entringer et al. 2015). Notably, we present some evidence from rodent models, where the early (first 7–10 days) postnatal period coincides with late gestation in humans, with regard to immune development and the peak brain growth spurt (Semple et al. 2013). Therefore, while the line between effects of prenatal and postnatal stress exposure is slightly blurred when comparing human and animal models, we will discuss experiences of the offspring, not of the mother during pregnancy.

Excess psychological stress has been shown to cause unhealthy sensitization of the immune response (Hennessy et al. 2011). ELS impacts the immune system at the time of exposure (Hennessy et al. 2010, 2011) and can also alter the normal developmental trajectory of certain immunological processes (Coe et al. 1989; Coe 1996). One consequence of these early alterations is a heightened immune response to stressors later in life; the adaptive advantage of heightened immune function in response to stress can be seen from an evolutionary perspective, since a psychological stressor would typically occur alongside a threat to an animal's physical well-being (e.g., injury, predator). Therefore, a sensitized immune response to stressors could better prepare an animal for future threatening environments. However, in one well-characterized example, a behavioral consequence of heightened inflammation is the phenomenon referred to as "sickness behavior." This sickness behavior is characterized by lethargy, social avoidance, and anhedonia, which are also symptoms associated with exposure to an immunostimulant (e.g., a pathogen). These responses can be viewed as a part of the organism's effort to recruit all of its resources for fighting against the invading pathogen and overcoming the disease (Hartung et al. 1988). Additionally, excess inflammatory (Munhoz et al. 2010) or subsequent oxidative stress damage (Spiers et al. 2013; Manikandan et al. 2006; Madrigal et al. 2001) can lead to excitotoxicity (Moghaddam 1993) or prevent typical developmental patterns of neural innervation and receptor activity, which could be interpreted as adaptations to the environment in preparation for future challenges (Tottenham and Sheridan 2009). Indeed, ELS represents stressors that impact the brain during a time of rapid development and, importantly, during a time preceding the tumultuous period of adolescence. Since sickness behavior purportedly shares phenomenology and immunological physiology with major depressive disorder (Maes et al. 2012), we begin to see a role of immunity in ELS-attributable depression, which often first manifests during adolescence. This chapter explores how such perturbations to the adaptive maturation of neuroimmune signaling can play important roles in shaping brain and behavior development.

2 Toward an Operational Definition of ELS

Considering the vast number of children worldwide growing up under some form of chronic stress, there is certainly a need for translational research on mechanistic targets for prevention and intervention. While ELS has consistently been shown to alter developmental trajectories and lead to deleterious health-related consequences, different types of such experiences during different times of development can produce distinct outcomes (Lupien et al. 2009; Teicher and Samson 2013). That said, it is generally recognized that maltreatment by caretakers is the most robust and reproducible source of ELS-attributable pathologies, rather than other types of trauma like natural or man-made disasters (Teicher and Samson 2013). This is because developing mammals rely on the adults around them for survival and can typically endure adversity as long as the safety of a caretaker is present. However, the type, timing, and amount of maltreatment are sources of great variability for later consequences. With regard to the impact of differing amounts of maltreatment, a "dose-response" effect has been shown, with a graded relationship between the number of incidences of adverse childhood events (such as abuse, neglect, or household dysfunction) and the number of adult psychiatric and inflammatory disorders (Edwards et al. 2003; Felitti et al. 1998). With regard to other sources of variability, researchers are tasked with interpreting results in the context of each individual study and its representation of ELS, as well as the interaction with individual differences between participant groups, such as sex and genetic makeup.

In addition to the challenges just described, because many of the symptoms resulting from ELS often first emerge later in life (Kessler et al. 2001; Davey et al. 2008; Andersen and Teicher 2008), intervening variables found in clinical studies make the role that ELS plays in these diseases difficult to interpret. Animal models have therefore been helpful to clarify the causality of ELS; however they also bring new issues including the translatability to human experience, strain and species differences, and paradigm differences between laboratories. While faced with some ongoing challenges to reproducibility and cohesiveness, animal models of ELS have indeed confirmed the importance of parental care for healthy maturation of both brain and immunity. Taken together, an operational definition of ELS could be described as events occurring during early childhood that trigger the stress response in a way that diverts energy expenditure toward managing safety and survival, producing maladaptive allostatic load. These events can vary in both humans and animal models; therefore it will continue to be important for research to carefully parse variables in order to identify cause and effects on developing systems.

3 A Brief Overview of the Immune System

Our immune system has evolved to recognize and defend our bodies against invasion from viruses, bacteria, and other antigens. The immune system can eliminate the presence of a pathogen via a specialized, robust molecular and cellular response. As part of this response, immune cells secrete elevated levels of immune molecules, resulting in inflammation that coordinates a cellular attack against the pathogen.

Peripheral immune responses can be divided into two types, either adaptive or innate, both of which have the ability to impact neural function. Adaptive immune responses are acquired responses that result from exposure to specific components of bacteria or virus and require days to develop but confer an immunological memory for a lifetime. As a result of an adaptive immune response, a second exposure to the same bacteria or virus (or antigen) results in a very specific and immediate immune response (Berczi 1998). Adaptive immune responses are largely orchestrated by white blood cells called lymphocytes, which mature in the thymus or bone marrow. After they mature, T (thymus-derived) and B (bone-derived) cells circulate through the blood and lymph where they respond to the foreign antigen via antibody production (B cells). T-cell-specific functions include assisting other lymphocytes, eliminating infected host cells, and expressing unique subsets of cytokines and chemokines. After infection, these cells survey for re-exposure to the previously detected pathogens. As their name and function implies, adaptive immune responses develop throughout the lifespan and are dependent on their exposure to discrete signals. Thus, our adaptive immune system continues to develop immunity to, and memory for, environmental factors that can subsequently impact the function of the brain and associated behaviors.

In contrast, innate immune responses refer to nonspecific resistance to pathogens that we all possess even prior to birth. This type of immunity occurs via the recognition of pathogens (and other environmental factors) via highly conserved receptors and adapter proteins expressed on the surface of immune cells (e.g., Toll-like receptors, TLRs). These innate immune receptors respond to a specific molecular pattern expressed by a pathogen and allow an individual to mount a rapid, robust immune response to the invasion without dependence on prior exposure during the individual's lifetime (Berczi 1998; Brodsky and Medzhitov 2009). In contrast with the later developing adaptive immunity, we are all born with the capacity to mount an innate immune response. However, there is evidence suggesting that the innate immune response matures and changes throughout the lifespan, potentially even into adolescence, thereby differentially impacting the brain and behavior (Ellis et al. 2005; Levy 2007; Ortega et al. 2010). Innate immune responses are largely carried out by a number of cells including macrophages, monocytes, neutrophils, and other phagocytes, as well as microglia in the brain. Exposure to any insult that activates innate immune cells will result in the induction of innate immune molecules, such as cytokines and chemokines, a process known as the "inflammatory response."

Cytokines, chemokines, and related immune molecules are typically classified as either pro-inflammatory, meaning that they help to stimulate an immune response (e.g., interleukin (IL)-1, IL-6, or TNFα), or anti-inflammatory, meaning that they can control or attenuate an immune response (e.g., IL-10). Both the pro- and anti-inflammatory responses are necessary for proper immune function. If left unchecked, inflammation can cause significant tissue damage and cell death, particularly within the brain. Cytokines are often redundant in their function and yet rarely work alone, as they orchestrate a set of physiological changes throughout the entire body, including the brain and central nervous system (CNS) (Dantzer et al. 1998a, b), which is the focus of this chapter.

The CNS also has its own immune cells called microglia that are capable of responding not only to insult or injury via the expression of innate immune receptors but also to peripheral immune activation via their communication with circulating immune molecules. Additionally, astrocytes in the brain also produce cytokines and chemokines (McKimmie and Graham 2010) that – in conjunction with the immune molecules produced by microglia – are important for immune function but also can influence neuronal function and behavior [see Rivest (2009) for review].

4 Neuroimmune Development

Neuronal and immune mechanisms coordinate to regulate cognitive and behavioral function over the lifespan. Moreover, the central nervous and immune systems develop in concert throughout early life and into adolescence, allowing an organism to shape its responsiveness profile to its own unique environment and experiences. Therefore, the early-life environment critically shapes the development of the immune system and of neuroimmune interaction. The immune system is immature at birth and during childhood (see Simon et al. 2015), which yields a period when infection (Bilbo and Schwarz 2009), stress, or other sources of allostatic load can permanently alter the course of development. For example, although newborn humans can recognize common pathogens via a functioning innate immunity, their ability to process such antigens and respond to them is immature at birth and develops throughout childhood and adolescent years (Dowling and Levy 2014). This is partially due to the fact that newborns have fewer and less functional antigen-presenting cells, and their ability to produce pro-inflammatory cytokines in response to infectious stimuli only normalizes to adult levels in preschool children (or in teenage years for some cytokines). Neutrophils are one kind of antigen-presenting cell that particularly shows impaired functioning in the neonatal period, with reduced ability to adhere to blood vessel walls, to extravasate from the bloodstream, to follow chemotactic signals to the site of infection, and to produce cytotoxic substances. Furthermore, natural killer cells are lymphocytes that display reduced ability to destroy virus-infected cells or produce regulatory cytokines until school age (Ygberg and Nilsson 2012). Finally, levels of complement factors, which contribute to chemotaxis and facilitate phagocytosis, are very low at birth and only normalize in postnatal life.

In the brain, microglia also undergo crucial developmental processes during early postnatal life (see Schwarz and Bilbo 2011). Microglia colonize the human and rodent brain early in embryonic development, but during early postnatal life, they begin to establish a distinct morphology with short, stubby processes in infancy transitioning to longer, thinner processes in adulthood. Importantly, this maturation occurs in a region-dependent manner, with microglia in certain brain regions still showing immature morphology in juvenility and adolescence. Microglial morphology reflects differential activity of these cells; amoeboid-like morphology with short, stubby processes reflects a more immunogenic activation profile, accompanied by expression of pro-inflammatory molecules and their receptors. For example, in early brain development, pruning of spurious synapses occurs via microglial recognition of complement protein C1q that is localized to targeted synapses (Stevens et al. 2007; Schafer et al. 2012). In contrast, microglia with longer, thinner processes are referred to as "ramified," which reflects less pro-inflammatory function but rather constant surveillance of surrounding neurons and ability to maintain synaptic plasticity in the mature brain. Importantly, microglia originate early in life but are very long-lived, such that stressors such as trauma or maltreatment can program these cells to differentially affect brain function and plasticity throughout the lifespan. Microglia can become "sensitized" or "primed" by different stimuli including stress (Fig. 1), which results in the elicitation of an exaggerated immune response to an otherwise weak stimulus (Frank et al. 2007). In this way, a "second hit" – be it a stressor or immune challenge that otherwise would have been benign – can lead to increased immunogenic activation of microglia with consequential pro-inflammatory cytokine release and resulting neuronal changes and psychopathology. Here it is important to note that microglial maturation also follows sexually dimorphic trajectories, such that phagocytic activity (Nelson et al. 2017, p. 8729), transcriptomic changes (Hanamsagar et al. 2017, p. 8730), and responses to

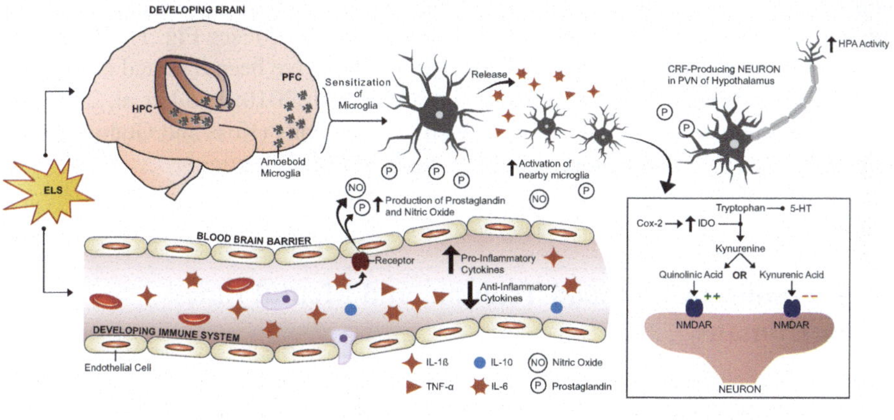

Fig. 1 Neuroimmune effects of early-life stress. Both developing brain (with existing evidence for effects on hippocampal and prefrontal cortex microglia) and developing peripheral immune cells are impacted by stress during early postnatal life. In the brain, microglia become sensitized and are more activated during later life, which could lead to altered microglia-mediated neurodevelopment and synapse formation. Microglial sensitization leads to release of pro-inflammatory cytokines that potentially affect HPA activity and the kynurenine pathway. In the periphery, pro-inflammatory cytokine expression is increased, and anti-inflammatory cytokine expression is decreased, potentially leading to endothelial cell secretion of prostaglandins and nitric oxide into the brain. Not shown: many of these effects occur on different developmental timelines and to a different extent in males and females

sensitizing stimuli (Ganguly et al. 2018, p. 8728) appear different between males and females at specific developmental time points. Therefore, disrupted homeostasis via ELS could potentially impact males and females differently depending on when (i.e., infancy, juvenility, adolescence) exposure takes place.

While microglia are essential regulators of neuroimmune signaling, circulating immune factors also impact brain activity and development. Passive leakage of immune cells or immune molecules into the brain is relatively rare and typically occurs only in severe pathogenic states (Muller and Ackenheil 1998). However, several pathways of active communication allow controlled signaling between the periphery and the brain, including vagal afferents and endothelial cell activity at the blood-brain barrier (BBB) or at circumventricular organs (Blatteis 1992; Stamatovic et al. 2008). The BBB particularly is a dynamic system that controls the access of a number of factors, including blood-borne immune mediators, into the brain. In this way, the BBB adapts to the changing needs of the brain that occur throughout development and aging. For example, the extent to which the BBB is permeable to various factors can change through activity of surrounding astrocytes and pericytes or through altered expression of tight junction protein expression between endothelial cells (Banks 2015). Additionally, cytokine binding on BBB endothelial cells yields release of vasoactive and immune molecules such as cytokines, prostaglandins, and nitric oxide into the parenchyma (Fig. 1) (Pan et al. 2011).

Circulating cytokines can also affect brain function via their influence endothelial cells. Binding of pro-inflammatory cytokines to their receptors on endothelial cells can cause production of prostaglandins, which can activate corticotropin-releasing hormone-producing neurons in the hypothalamus and thereby initiate hypothalamic pituitary adrenal axis (HPA) activity (Silverman et al. 2005) (see Fig. 1). Therefore, short-term and enduring changes to HPA activity that have been reported in animals exposed to postnatal maternal separation (Litvin et al. 2010; Ladd et al. 2004) or limited bedding (Walker et al. 2017) as well as in humans (Koss and Gunnar 2018) may be related to inflammatory actions at the blood-brain barrier.

5 Immunity, Childhood Trauma, and Mental Illness in Humans

Immunological disturbances have been implicated as being part of the pathophysiology of mental disorders including major depressive disorder (MDD) (Raison and Miller 2011; Miller et al. 2009), bipolar disorder (Hamdani et al. 2012; Grande et al. 2012; Berk et al. 2011; Goldstein et al. 2009; Barbosa et al. 2012), and schizophrenia (Mansur et al. 2012; Brietzke et al. 2012; Meyer et al. 2011), among others. Importantly, several groups have reported that childhood adversity promotes the clustering of psychopathologies and inflammation. For example, among females at high risk for depression, a transition to depression was accompanied by increases in the pro-inflammatory markers C-reactive protein and IL-6 *only* in those who were exposed to high childhood adversity, and not in those without such histories (Miller and Cole 2012). Childhood trauma also predicted increased levels of the pro-inflammatory cytokine TNFα in patients with schizophrenia (Dennison et al. 2012). These findings suggest that ELS may generate a population with increased vulnerability to mental illness through a pipeline of amplified signaling between the brain and immune system.

Growing evidence points to an association between the immunological profile and outcomes related to exposure to ELS. For example, there is a longitudinal association between childhood trauma and the incidence of chronic medical illnesses, such as diabetes and metabolic diseases in adulthood (Goodwin and Stein 2004; Karavanaki et al. 2008; McIntyre et al. 2008; Midei et al. 2010; Von Korff et al. 2009). One cohort study involving 9,310 individuals found a 20–50% risk of obesity among individuals with a history of childhood abuse and/or neglect (Thomas et al. 2008). The strongest association was found in participants reporting a history of physical abuse, which was directly associated with elevated levels of glycosylated hemoglobin. Considering that metabolic and cardiovascular diseases have been constantly linked to pro-inflammatory states (Giordano et al. 2011; Alikasifoglu et al. 2009; Mauras et al. 2010; Soczynska et al. 2011) and the high rate of medical problems in individuals with mental disorders (Zhao et al. 2012), one could hypothesize that the activation of inflammatory processes could be the mediator between childhood trauma and negative clinical outcomes.

The direct association between ELS and inflammatory biomarkers was investigated in humans as part of the Dunedin Multidisciplinary Health and Development Study (Poulton et al. 2015), in which 1,037 members were followed since their birth (between 1972 and 1973). This longitudinally designed study circumvented the limitations of many observational human studies that rely on retrospective subject reporting, which may be inaccurate (Maughan and Rutter 1997). Cumulative exposure to ELS in the form of childhood maltreatment was associated with elevated levels of inflammatory markers C-reactive protein and fibrinogen, as well as elevated white blood cell count in adulthood, approximately 20 years following ELS exposure (Danese et al. 2007). Importantly, the association was not explained by potential confounders or mediators such as low birth weight, socioeconomic conditions, IQ, or poor adult health. Since that report, the association between ELS and an inflammatory profile has been supported by several independent studies (Baumeister et al. 2016; Coelho et al. 2014). While an inflammatory profile endures into adulthood following ELS, markers of inflammation have been detectable during childhood as well – particularly in children who later developed depression (Danese et al. 2011; Slopen et al. 2013; Cicchetti et al. 2015). Similarly, preliminary evidence suggests that the functioning of the immune system in adolescents with MDD is dysregulated by an imbalance between T-helper 1 and 2 (Th1/Th2) lymphocytes expression, with a predominance of Th1 profile (Gabbay et al. 2009). These findings are consistent with previous results in adults with MDD (Simon et al. 2008).

In addition to its effects on circulating baseline immune molecules, ELS also predicts greater inflammatory reactivity to subsequent psychosocial challenges. Physiological response to acute psychosocial stressors can be tested through experimental procedures, such as in the Trier Social Stress Test (TSST). In this test, public speaking and mental arithmetical calculation tasks of subjects are socially evaluated. This and other types of social stress tests have been shown to induce a more pronounced inflammatory response in patients with major depression and history of childhood maltreatment, compared to healthy controls (Pace et al. 2006; Carpenter et al. 2010). Higher reactivity to later life stressors is an important substrate for vulnerability to several mental disorders with developmental and/or immune components.

Post-traumatic stress disorder (PTSD) was the first stress-related disorder that was studied from an immunological perspective. Effects of PTSD include elevation of different circulating lymphocyte subtypes (CD4+, CD8+, and NK cells) and increased plasma levels of pro-inflammatory cytokines (IL-6, IL-1 and TNFα) (Gill et al. 2009). These findings were supported by another study with participants suffering from PTSD, reporting that patients had hypocortisolemia, higher levels of the endogenous steroid dehydroepiandrosterone and increased production of IL-6 and TNFα compared to healthy controls. This effect on cytokines became more pronounced among those with PTSD and MDD, with higher levels in comparison to those with PTSD, but without MDD (Gill et al. 2008).

Higher levels of IL-6 after ELS have also been found to be *predictive* of PTSD development (Pervanidou et al. 2007; Sutherland et al. 2003). Increased serum IL-6 concentrations within the first 24 h after a traffic accident predicted development of PTSD in children, 6 months after the event (Pervanidou et al. 2007). These results were significantly correlated with elevated cortisol levels (Pervanidou et al. 2007). IL-6 plays an important role in the vulnerability of PTSD, being associated with non-recovery from traumatic episodes (Guo et al. 2012; Bob et al. 2010; Cohen et al. 2011; Gill et al. 2010). This cytokine acts to promote the survival of catecholaminergic neurons that are involved in elevating dopaminergic release in the hippocampus (Zalcman et al. 1994), which increases the risk of neuronal death in the hippocampus in animal models (Stam 2007), serving as a possible precursor for the development of PTSD in humans.

An important limitation of studies in the field of inflammation is the widespread lack of knowledge about the relationship between central and peripheral phenomena in mental disorders. A vast majority of studies report peripheral rather than central levels of inflammatory markers. Both need to be studied in conjunction, especially since the relationship between peripheral and central levels is still not well established. However, some evidence indicates that inflammatory mediators, including cytokines, are able to cross the blood-brain barrier and access the CNS (Brietzke et al. 2011). In addition, actions of cytokines on endothelial cells of the brain vasculature can incite pro-inflammatory signaling in the brain. Vagus afferents can also signal pro-inflammatory events to the brain. Taken together, while human studies provide evidence of peripheral immune changes only, animal studies can provide additional information about these potential mechanistic links between circulating inflammatory markers after ELS and psychopathological outcomes.

It is also unclear from human studies whether high inflammation levels in victimized children reflect causal effects of victimization or the result of genetic liability. Inflammation is moderately heritable (Pankow et al. 2001), and inflammatory genes might influence children's risk of experiencing victimization through gene-environment correlation (Jaffee and Price 2007; Danese and Baldwin 2017). For example, genetic pathways related to the immune system predict risk for several psychiatric diagnoses (Network and Pathway Analysis Subgroup of the Psychiatric Genomics Consortium 2015). In turn, early expressions of risk-based conditions, such as emotional dysregulation or oppositional behavior, might increase the probability of victimization in children (Danese and McCrory 2015). Similarly, psychopathology and impaired emotion regulation in parents have been linked to high risk of victimization in children (Danese and McCrory 2015). These examples of evocative and passive gene-environment correlations illustrate alternative and yet untested mechanisms underlying the association between ELS and inflammation. Genetic influences do not bias the association between ELS and inflammation in experimental animal models because the assignment to ELS is manipulated irrespective of the individual animal's characteristics, but this factor cannot be excluded in observational human studies.

6 Evidence from Animal Models: Immune Alterations After ELS

Pioneering work describing how ELS can alter immunological development was conducted in the 1980s in primates (Coe et al. 1987, 1988, 1989; Laudenslager et al. 1982, 1990). Since then, the growing appreciation for an immunological contribution to mental illness has led to a resurgence of interest in neuroimmune interactions. ELS in rodents and nonhuman primates is typically performed during 1–3 weeks between birth and weaning, via separation from the mother for several hours per day [maternal separation paradigm (Tractenberg et al. 2016)]. Variations of the paradigm include raising infants in a motherless environment (e.g., Massart et al. 2016), depleting resources such as cage bedding or nesting material (Bath et al. 2016), or utilizing natural variation of maternal care (Liu et al. 1997). While not all paradigms have yet been tested for the same outcome measures, to date, a pattern emerges suggesting that ELS leads to increased pro-inflammatory responsivity, despite stress paradigms, species/strain differences, and age differences. For example, guinea pigs that are separated from their mother exhibit a characteristic behavior that resembles sickness behavior and is reportedly blocked with anti-inflammatory treatment (Hennessy et al. 2007, 2011; Perkeybile et al. 2009). This early immune programming through ELS has been proposed to sensitize later pro-inflammatory processes and lead to greater vulnerability to depression and anxiety in adulthood (Hennessy et al. 2010). Moreover, nonhuman primate models have allowed a longitudinal and causational analysis of epigenetic alterations to peripheral immune cells following ELS (Massart et al. 2016). These studies have uncovered sex-specific effects of maternal separation on the developmental trajectory of DNA methylation profiles in

T lymphocytes. Specifically, although both sexes exhibited dynamic alterations to DNA methylation of immune cells in response to maternal separation early at birth, males exhibited larger differences than females in the pattern of methylation during adolescence (Massart et al. 2016). Additionally, there is a general increased expression of genes involved in inflammation and T-lymphocyte activation, in both males and females, which was also seen in infancy (Cole et al. 2012).

Rather than an overall increased pro-inflammatory profile, some studies suggest that aberrant immune development due to ELS may involve an early suppression of some immune mechanisms (reviewed by Ganguly and Brenhouse 2015), which may alter neuronal connectivity and function later in life. Behavioral consequences have also been noted early on, as suppressed neuroimmune signaling co-occurs with anhedonic and withdrawal behaviors in rodents (Hennessy et al. 2010) and altered fear learning (Callaghan and Richardson 2011). These behavioral changes are likely mediated by interactive changes in neuroimmune and neuroendocrine activity. Neuroendocrine responses to stress via the HPA axis have important bidirectional influences on developing immune activity and brain development. Glucocorticoids secreted via the HPA axis can have suppressive effects on circulating innate immune cells via actions at the glucocorticoid receptor (GR) (Scheinman et al. 1995). In turn, GR activation on microglia in the brain has a priming effect that increase pro-inflammatory neuroimmune signals (Frank et al. 2012). Comprising a feedback loop, pro-inflammatory cytokines themselves can stimulate corticotropin-releasing hormone release in the hypothalamus to activate the HPA axis (Silverman and Sternberg 2012), affecting stress responsivity (Fig. 1).

One recent study (Wei et al. 2012) conducted in the highly stress-sensitive and anxiety-prone BALB/cByj mouse strain investigated the effects of repeated brief (15 min) separations of male pups from their mothers between postnatal day (P)1 and 21 and revealed that these repeated brief daily separations prevented a typical transient rise of lipopolysaccharide (LPS)-binding protein in the hippocampus during the second and third weeks of life. In other words, BALB/cByj mice exposed to brief daily separations showed a blunted level of lipopolysaccharide-binding protein on P14 and P21, while control BALB/cByj mice displayed higher expression of this gene on P14 and 21 compared to younger and older ages. One acute 3 h separation from the mother on P14 also caused an immediate decrease of LPS-binding protein in this strain. LPS-binding protein is an acute-phase protein important for elicitation of immune responses by presenting the LPS to important cell surface pattern recognition receptors. Importantly, the developmental expression of LPS-binding protein, along with observations in LPS-binding protein knockout mice, suggests that this protein is also crucial for recruiting microglia for developmental processes in the hippocampus such as spine formation and synaptic pruning (Wei et al. 2012). Therefore, in this model, brief daily separations were reported to perturb normal neuroimmune processes necessary for hippocampal maturation. These findings may also highlight an important gene x environment interaction, since the same brief daily separation protocol has often been used in other rodent strains as a model of increased maternal care and enrichment and has been shown to *enhance* hippocampal function in Sprague Dawley rats (Fenoglio et al. 2005) and increased hippocampal plasticity in Wistar rats (Katsouli et al. 2014) and C57Bl/6NCrl mice (Gross et al. 2012).

7 The Importance of Individual Differences

Individual differences including genetic makeup, sex, and age all appear to influence the observed neuroimmune effects of ELS. For example, maternal separation in rats was recently found to decrease circulating levels of the anti-inflammatory cytokine IL-10 at P35 (peri-puberty), but not P25 (juvenility) nor P55 (late adolescence). This transient decrease occurred in peri-pubertal males, but not females (Grassi-Oliveira et al. 2016). In support of the idea that adolescent changes in circulating cytokines can predict altered behavioral outcomes, lower levels of circulating peripheral IL-10 at P35 predicted poor performance in the win-shift cognitive behavioral task 2 weeks later, only in maternally separated male rats (Grassi-Oliveira et al. 2016). Administration of IL-10 intracerebroventricularly during early adolescence to ELS-exposed males prevented prefrontal cortex (PFC) interneuron loss (Wieck et al. 2013), suggesting that peripheral immune changes may be reflected in the brain. The finding that female rats did not exhibit the same profile of circulating IL-10 in adolescence following ELS may suggest that either the female immune system is not as sensitive to this particular stressor or that, only in females, adolescent peripheral biomarkers may predict the risk of negative cognitive outcomes that have yet to be identified. Interestingly, decrease in peripheral IL-10 observed in adolescent male rats also co-occurred with significantly lower testosterone levels compared to controls (Grassi-Oliveira et al. 2016). Therefore, the early-life events linked to achieving important pubertal milestones may also involve and modulate the orchestrated maturation of the immune system during adolescence (in this example, the induction of IL-10 in the periphery that occurs specifically at this time). Conversely, ELS can also produce long-term changes in responses to circulating sex hormones via their impact on the immune system (Blaustein and Ismail 2013). Therefore, it will be important to determine exactly how these pubertal hormones influence the full maturation of the peripheral and central immune systems and how this may impact both physiological and behavioral outcomes later in life.

In addition to these peripheral effects, maternal separation in rats yields neuroinflammatory changes in the PFC that correlate with interneuron deficits and manifest in adolescence, but not before (Brenhouse and Andersen 2011). Specifically, the neuroinflammatory mediator cyclooxygenase-2 (COX-2) is upregulated in the PFC during juvenility in females but is first upregulated in adolescence in males (Holland et al. 2014). It is not clear whether development of the neuroinflammatory response underlies delayed effects of ELS; however the adolescent alteration in PFC interneurons is prevented with preadolescent inhibition of COX-2 (Brenhouse and Andersen 2011). These adolescent effects are reminiscent of clinical evidence which points to a delayed effect of ELS on depressive illnesses (Teicher et al. 2009), with an average of 11.2 years between sexual abuse exposure and depressive episodes.

Taken together, longitudinal and trajectory-based studies in male and female animals have begun to reveal that ELS differentially impact males and females. It appears likely that the timing of ELS determines its impact on the male or female brain, due to sex-specific trajectory of developmental events. Conversely, males and females display differential timing of neuroimmune effects, with females sometimes being affected at an earlier time point than males (Holland et al. 2014). Future research will shed light on whether these phenotypic responses to ELS are underpinned by sex-specific epigenetic responses such as those recently uncovered in lymphocytes (Massart et al. 2016).

8 How Do ELS-Induced Immune Changes Ultimately Affect Brain and Behavior?

As we have discussed in this chapter, ELS can enhance or sensitize pro-inflammatory activity, which may be due to the memory function of the adaptive immune system, derailed development of the innate immune system, endocrine regulation, or priming of microglia. These immune changes can affect neural function in several ways (see Fig. 1). For instance, inflammatory molecules that are secreted in the blood-brain barrier, expressed by extravasated phagocytes and lymphocytes, or by microglia, can initiate the COX-2 catalyzed indoleamine pathway. Indoleamine metabolizes tryptophan to kynurenine, thereby decreasing the availability of tryptophan for serotonin synthesis. Kynurenine is then converted to either the NMDA agonist quinolinic acid or the NMDA antagonist kynurenic acid, depending on the type of cytokines present. These glutamate receptor mediators have been repeatedly shown to play a role in schizophrenia (with a heightened kynurenic acid/quinolinic acid ratio) and in major depressive disorder (with a heightened quinolinic acid/kynurenic acid ratio) (Muller and Schwarz 2008). While the direct neural effects of ELS-induced changes are not yet well understood, some later pro-inflammatory effects of ELS have been found to co-occur with altered glutamatergic NMDA receptor expression (Wieck et al. 2013). Indeed, ELS has been shown in several species to disturb the balance between inhibitory and excitatory signaling in the prefrontal cortex (Bock et al. 2014).

These immune-regulated glutamatergic changes may involve microglial sensitization. One recent study investigated microglial effects in a model of ELS that involved postweaning isolation lasting through late adolescence, which notably a later time point in development than the pre-weaning maternal separation paradigms thus far described in this chapter. In this model, rats displayed depressive-like behaviors, along with increased hippocampal expression of pro-inflammatory cytokines and of the microglial marker Iba1 (Wang et al. 2017). These changes were mitigated by treatment with minocycline, an antibiotic that has been shown to suppress microglial activity. The brief daily separations model in stress-sensitive mice described above also yields microglial changes in the hippocampus, with increased microglial density during brief daily separation exposure at P14, and an atypically immunogenic activation profile of hippocampal microglia during later juvenility at P28 (Delpech et al. 2016). These findings corroborate other recent reports of increased microglial immunogenicity in the hippocampus during the juvenile period, following postnatal maternal separation in rats (Roque et al. 2016). While evidence of microglial activation within the hippocampus during and immediately following ELS is growing, less is known about other brain regions; this will be important to investigate especially given the regional specificity of microglial development and its related effects on brain development.

It is important to reiterate that some immune-mediated effects on brain function following ELS are due to a derailment of brain development during a time when immune signaling is crucial for processes such as synaptic pruning, myelination, and regulation of critical periods of plasticity. Interestingly, microglial colonization of the brain occurs much earlier in males than in females in several regions including the parietal cortex, hippocampus, and amygdala (Schwarz et al. 2012). It has been proposed that sex differences in the colonization and function of glia within the normal developing brain may contribute to distinct windows of vulnerability between males and females (Schwarz and Bilbo 2012). One might hypothesize that early stress exposure during a period when microglia have colonized these regions in males but not females could lead to preferential sensitization of the neuroimmune response in males; however this has not been determined.

9 Future Directions

Given the important role of the immune system in normal brain development, it is not surprising that an altered trajectory of inflammatory responses will lead to atypical brain development in response to ELS. However, the immune-related effects of ELS are not yet fully understood. The existing evidence provides only limited and mostly correlational support for the mediating role of the immune system in the association between ELS and mental disorders. It will be important to determine whether different types or time courses of ELS can differentially affect immunity depending on the presence of a secondary insult ("second hit"), the developmental stage of assessment, or the sex of the individual. This will facilitate targeted research into interventions that are timed appropriately and that manipulate affected pathways. In this regard, more work is needed to provide a causational link between ELS and later activation of the immune system.

Research has begun to uncover an ELS-attributable inflammatory profile in patients with MDD (Miller and Cole 2012; Danese et al. 2008, 2011), suggesting that this subpopulation with mental illness may require anti-inflammatory intervention and may therefore be more resistant to traditional therapies. Supporting this notion is that elevated inflammation levels have been linked to relative resistance to traditional therapies in mood disorders (Strawbridge et al. 2015). Therefore, systemic or target-oriented treatment with immunomodulators may be a valuable investigative path, including anti-inflammatory drugs, immune-related siRNAs, anti-inflammatory interleukins, and mood stabilizers and antidepressants with immunomodulatory effects. Importantly, precision medicine will need to account for experiential histories, since intervention or treatment strategies are likely to succeed based on discrete mechanistic origins. Current challenges to analyzing ELS history in both clinical practice and research include the common reliance on recollection, rather than identification of childhood trauma as it occurs, with longitudinal assessment. Another challenge is the vast array of traumatic or stressful experiences that fall under the umbrella of ELS, which might yield equally varied consequences on immune and brain development. Animal research has the potential for helping to parse these intervening variables.

Animal research will continue to guide intervention strategies by uncovering mechanisms by which ELS directly alters developmental trajectories. For instance, studies with the stress-sensitive Wistar Kyoto rat revealed that resilience to anxiety- and depressive-like behaviors following maternal separation ELS was mediated by reduced methylation of key nodes of insulin signaling pathways within the hippocampus (McCoy et al. 2016). Additionally, DNA methylation patterns within circulating T lymphocytes of rhesus monkeys were differentially affected by imposed weaning, depending on whether animals were reared with or without their mother (Massart et al. 2016), further suggesting that ELS and responsivity to future challenges can be mediated by overlapping immune genes. Indeed, it has been proposed that resilience to stress might be achieved by boosting the adaptive immune system, since mice who show greater lymphocyte trafficking in response to predator odor developed lower anxiety levels and showed faster recovery to prestress hippocampal neurotrophin levels (Lewitus and Schwartz 2009). Investigation of brain-derived immune signaling via microglia and astrocytes will likely also provide critical insight into the mechanistic links between ELS and psychopathology. New techniques that allow chemogenetic or optogenetic activation or inhibition (Cho et al. 2016) or targeted genetic manipulation will allow precise inquiries into the impact of ELS on microglia-driven development. These types of studies will be necessary to distinguish the bidirectional effects of ELS on glia versus circulating immune cells, since it is otherwise often difficult to determine whether immune signatures of ELS that are found in peripheral tissues reflect alterations in programming within the brain.

10 Conclusion

The developing brain aims to meet the changing demands of each stage-specific environment, and the immune system is a partner in that goal. Neuroimmune systems that are vulnerable to stress are impacted by ELS at a time when they are essential for circuit formation. Therefore, when encountered with day-to-day challenges (i.e., a new need for decision-making in adolescence), the circuits in an ELS-exposed brain are, at best, wired for a very different type of challenge or, at worst, mis-wired and dysfunctional. Given the remarkable plasticity of both the brain and the immune system, many of the deleterious effects of ELS can be treated with interventions that account for sex and target neuroimmune interactions over the lifespan. Notably, targeted prevention – not just treatment – should be a high-priority goal of research. ELS exposure may yield a vulnerable population with neurodevelopmental deficits that can particularly benefit from immune interventions, possibly during a sex-dependent critical period.

References

Alikasifoglu A, Gonc N, Ozon ZA, Sen Y, Kandemir N (2009) The relationship between serum adiponectin, tumor necrosis factor-alpha, leptin levels and insulin sensitivity in childhood and adolescent obesity: adiponectin is a marker of metabolic syndrome. J Clin Res Pediatr Endocrinol 1(5):233–239. https://doi.org/10.4274/jcrpe.v1i5.233

Anda RF, Brown DW, Dube SR, Bremner JD, Felitti VJ, Giles WH (2008) Adverse childhood experiences and chronic obstructive pulmonary disease in adults. Am J Prev Med 34 (5):396–403. https://doi.org/10.1016/j.amepre.2008.02.002

Andersen SL, Teicher MH (2008) Stress, sensitive periods and maturational events in adolescent depression. Trends Neurosci 31(4):183–191

Banks WA (2015) The blood-brain barrier in neuroimmunology: tales of separation and assimilation. Brain Behav Immun 44:1–8. https://doi.org/10.1016/j.bbi.2014.08.007

Barbosa IG, Rocha NP, Bauer ME, de Miranda AS, Huguet RB, Reis HJ, Zunszain PA, Horowitz MA, Pariante CM, Teixeira AL (2012) Chemokines in bipolar disorder: trait or state? Eur Arch Psychiatry Clin Neurosci. https://doi.org/10.1007/s00406-012-0327-6

Bath KG, Manzano-Nieves G, Goodwill H (2016) Early life stress accelerates behavioral and neural maturation of the hippocampus in male mice. Horm Behav 82:64–71. https://doi.org/10.1016/j.yhbeh.2016.04.010

Baumeister D, Akhtar R, Ciufolini S, Pariante CM, Mondelli V (2016) Childhood trauma and adulthood inflammation: a meta-analysis of peripheral C-reactive protein, interleukin-6 and tumour necrosis factor-alpha. Mol Psychiatry 21(5):642–649. https://doi.org/10.1038/mp.2015.67

Berczi I (1998) Neurohormonal host defense in endotoxin shock. Ann N Y Acad Sci 840:787–802

Berk M, Kapczinski F, Andreazza AC, Dean OM, Giorlando F, Maes M, Yucel M, Gama CS, Dodd S, Dean B, Magalhaes PV, Amminger P, McGorry P, Malhi GS (2011) Pathways underlying neuroprogression in bipolar disorder: focus on inflammation, oxidative stress and neurotrophic factors. Neurosci Biobehav Rev 35(3):804–817. https://doi.org/10.1016/j.neubiorev.2010.10.001

Bilbo SD, Schwarz JM (2009) Early-life programming of later-life brain and behavior: a critical role for the immune system. Front Behav Neurosci 3:14

Blatteis CM (1992) Role of the OVLT in the febrile response to circulating pyrogens. Prog Brain Res 91:409–412

Blaustein JD, Ismail N (2013) Enduring influence of pubertal stressors on behavioral response to hormones in female mice. Horm Behav 64(2):390–398. https://doi.org/10.1016/j.yhbeh.2013.01.015

Bob P, Raboch J, Maes M, Susta M, Pavlat J, Jasova D, Vevera J, Uhrova J, Benakova H, Zima T (2010) Depression, traumatic stress and interleukin-6. J Affect Disord 120(1–3):231–234. https://doi.org/10.1016/j.jad.2009.03.017

Bock J, Rether K, Groger N, Xie L, Braun K (2014) Perinatal programming of emotional brain circuits: an integrative view from systems to molecules. Front Neurosci 8:11. https://doi.org/10.3389/fnins.2014.00011

Brenhouse HC, Andersen SL (2011) Nonsteroidal anti-inflammatory treatment prevents delayed effects of early life stress in rats. Biol Psychiatry 70(5):434–440. https://doi.org/10.1016/j.biopsych.2011.05.006

Brietzke E, Scheinberg M, Lafer B (2011) Therapeutic potential of interleukin-6 antagonism in bipolar disorder. Med Hypotheses 76(1):21–23. https://doi.org/10.1016/j.mehy.2010.08.021

Brietzke E, Mansur RB, Grassi-Oliveira R, Soczynska JK, McIntyre RS (2012) Inflammatory cytokines as an underlying mechanism of the comorbidity between bipolar disorder and migraine. Med Hypotheses 78(5):601–605. https://doi.org/10.1016/j.mehy.2012.01.036

Brodsky IE, Medzhitov R (2009) Targeting of immune signalling networks by bacterial pathogens. Nat Cell Biol 11(5):521–526

Brown DW, Anda RF, Felitti VJ, Edwards VJ, Malarcher AM, Croft JB, Giles WH (2010) Adverse childhood experiences are associated with the risk of lung cancer: a prospective cohort study. BMC Public Health 10:20. https://doi.org/10.1186/1471-2458-10-20

Callaghan BL, Richardson R (2011) Maternal separation results in early emergence of adult-like fear and extinction learning in infant rats. Behav Neurosci 125(1):20–28. https://doi.org/10.1037/a0022008

Carpenter LL, Gawuga CE, Tyrka AR, Lee JK, Anderson GM, Price LH (2010) Association between plasma IL-6 response to acute stress and early-life adversity in healthy adults. Neuropsychopharmacology 35(13):2617–2623. https://doi.org/10.1038/npp.2010.159

Cho WH, Barcelon E, Lee SJ (2016) Optogenetic glia manipulation: possibilities and future prospects. Exp Neurobiol 25(5):197–204. https://doi.org/10.5607/en.2016.25.5.197

Cicchetti D, Handley ED, Rogosch FA (2015) Child maltreatment, inflammation, and internalizing symptoms: investigating the roles of C-reactive protein, gene variation, and neuroendocrine regulation. Dev Psychopathol 27(2):553–566. https://doi.org/10.1017/S0954579415000152

Coe CL (1996) Developmental psychoneuroimmunology revisited. Brain Behav Immun 10 (3):185–187

Coe CL, Rosenberg LT, Fischer M, Levine S (1987) Psychological factors capable of preventing the inhibition of antibody responses in separated infant monkeys. Child Dev 58(6):1420–1430

Coe CL, Rosenberg LT, Levine S (1988) Prolonged effect of psychological disturbance on macrophage chemiluminescence in the squirrel monkey. Brain Behav Immun 2(2):151–160

Coe CL, Lubach G, Ershler WB (1989) Immunological consequences of maternal separation in infant primates. New Dir Child Dev 45:65–91

Coelho R, Viola TW, Walss-Bass C, Brietzke E, Grassi-Oliveira R (2014) Childhood maltreatment and inflammatory markers: a systematic review. Acta Psychiatr Scand 129(3):180–192. https://doi.org/10.1111/acps.12217

Cohen M, Meir T, Klein E, Volpin G, Assaf M, Pollack S (2011) Cytokine levels as potential biomarkers for predicting the development of posttraumatic stress symptoms in casualties of accidents. Int J Psychiatry Med 42(2):117–131

Cole SW, Conti G, Arevalo JM, Ruggiero AM, Heckman JJ, Suomi SJ (2012) Transcriptional modulation of the developing immune system by early life social adversity. Proc Natl Acad Sci U S A 109(50):20578–20583. https://doi.org/10.1073/pnas.1218253109

Danese A, Baldwin JR (2017) Hidden wounds? Inflammatory links between childhood trauma and psychopathology. Annu Rev Psychol 68:517–544. https://doi.org/10.1146/annurev-psych-010416-044208

Danese A, McCrory E (2015) Child maltreatment. In: Thapar A, Pine DS, Leckman JF, Scott S, Snoling MJ, Taylor E (eds) Rutter's child and adolescent psychiatry. Wiley-Blackwell, London

Danese A, McEwen BS (2012) Adverse childhood experiences, allostasis, allostatic load, and age-related disease. Physiol Behav 106(1):29–39. https://doi.org/10.1016/j.physbeh.2011.08.019

Danese A, Pariante CM, Caspi A, Taylor A, Poulton R (2007) Childhood maltreatment predicts adult inflammation in a life-course study. Proc Natl Acad Sci U S A 104(4):1319–1324
Danese A, Moffitt TE, Pariante CM, Ambler A, Poulton R, Caspi A (2008) Elevated inflammation levels in depressed adults with a history of childhood maltreatment. Arch Gen Psychiatry 65 (4):409–415. https://doi.org/10.1001/archpsyc.65.4.409
Danese A, Caspi A, Williams B, Ambler A, Sugden K, Mika J, Werts H, Freeman J, Pariante CM, Moffitt TE, Arseneault L (2011) Biological embedding of stress through inflammation processes in childhood. Mol Psychiatry 16(3):244–246. https://doi.org/10.1038/mp.2010.5
Dantzer R, Bluthe RM, Gheusi G, Cremona S, Laye S, Parnet P, Kelley KW (1998a) Molecular basis of sickness behavior. Ann N Y Acad Sci 856:132–138
Dantzer R, Bluthe RM, Laye S, Bret-Dibat JL, Parnet P, Kelley KW (1998b) Cytokines and sickness behavior. Ann N Y Acad Sci 840:586–590
Davey C, Yucel M, Allen N (2008) The emergence of depression in adolescence: development of the prefrontal cortex and the representation of reward. Neurosci Biobehav Rev 32:1–19
Delpech JC, Wei L, Hao J, Yu X, Madore C, Butovsky O, Kaffman A (2016) Early life stress perturbs the maturation of microglia in the developing hippocampus. Brain Behav Immun 57:79–93. https://doi.org/10.1016/j.bbi.2016.06.006
Dennison U, McKernan D, Cryan J, Dinan T (2012) Schizophrenia patients with a history of childhood trauma have a pro-inflammatory phenotype. Psychol Med 42(9):1865–1871. https://doi.org/10.1017/s0033291712000074
Dowling DJ, Levy O (2014) Ontogeny of early life immunity. Trends Immunol 35(7):299–310. https://doi.org/10.1016/j.it.2014.04.007
Dube SR, Fairweather D, Pearson WS, Felitti VJ, Anda RF, Croft JB (2009) Cumulative childhood stress and autoimmune diseases in adults. Psychosom Med 71(2):243–250. https://doi.org/10.1097/PSY.0b013e3181907888
Edwards VJ, Holden GW, Felitti VJ, Anda RF (2003) Relationship between multiple forms of childhood maltreatment and adult mental health in community respondents: results from the adverse childhood experiences study. Am J Psychiatry 160(8):1453–1460
Ellis S, Mouihate A, Pittman QJ (2005) Early life immune challenge alters innate immune responses to lipopolysaccharide: implications for host defense as adults. FASEB J 19(11):1519–1521. https://doi.org/10.1096/fj.04-3569fje
Entringer S, Buss C, Wadhwa PD (2015) Prenatal stress, development, health and disease risk: a psychobiological perspective-2015 Curt Richter Award Paper. Psychoneuroendocrinology 62:366–375. https://doi.org/10.1016/j.psyneuen.2015.08.019
Felitti VJ, Anda RF, Nordenberg D, Williamson DF, Spitz AM, Edwards V, Koss MP, Marks JS (1998) Relationship of childhood abuse and household dysfunction to many of the leading causes of death in adults. The Adverse Childhood Experiences (ACE) study. Am J Prev Med 14 (4):245–258
Fenoglio KA, Brunson KL, Avishai-Eliner S, Stone BA, Kapadia BJ, Baram TZ (2005) Enduring, handling-evoked enhancement of hippocampal memory function and glucocorticoid receptor expression involves activation of the corticotropin-releasing factor type 1 receptor. Endocrinology 146(9):4090–4096. https://doi.org/10.1210/en.2004-1285
Francis DD, Diorio J, Plotsky PM, Meaney MJ (2002) Environmental enrichment reverses the effects of maternal separation on stress reactivity. J Neurosci 22(18):7840–7843
Frank MG, Baratta MV, Sprunger DB, Watkins LR, Maier SF (2007) Microglia serve as a neuroimmune substrate for stress-induced potentiation of CNS pro-inflammatory cytokine responses. Brain Behav Immun 21(1):47–59. https://doi.org/10.1016/j.bbi.2006.03.005
Frank MG, Thompson BM, Watkins LR, Maier SF (2012) Glucocorticoids mediate stress-induced priming of microglial pro-inflammatory responses. Brain Behav Immun 26(2):337–345. https://doi.org/10.1016/j.bbi.2011.10.005
Gabbay V, Klein RG, Alonso CM, Babb JS, Nishawala M, De Jesus G, Hirsch GS, Hottinger-Blanc PM, Gonzalez CJ (2009) Immune system dysregulation in adolescent major depressive disorder. J Affect Disord 115(1–2):177–182. https://doi.org/10.1016/j.jad.2008.07.022

Ganguly P, Brenhouse HC (2015) Broken or maladaptive? Altered trajectories in neuroinflammation and behavior after early life adversity. Dev Cogn Neurosci 11:18–30. https://doi.org/10.1016/j.dcn.2014.07.001

Ganguly P, Thompson V, Gildawie K, Brenhouse HC (2018) Adolescent food restriction in rats alters prefrontal cortex microglia in an experience-dependent manner. Stress 21(2):162–168

Gill J, Vythilingam M, Page GG (2008) Low cortisol, high DHEA, and high levels of stimulated TNF-alpha, and IL-6 in women with PTSD. J Trauma Stress 21(6):530–539. https://doi.org/10.1002/jts.20372

Gill JM, Saligan L, Woods S, Page G (2009) PTSD is associated with an excess of inflammatory immune activities. Perspect Psychiatr Care 45(4):262–277. https://doi.org/10.1111/j.1744-6163.2009.00229.x

Gill J, Luckenbaugh D, Charney D, Vythilingam M (2010) Sustained elevation of serum interleukin-6 and relative insensitivity to hydrocortisone differentiates posttraumatic stress disorder with and without depression. Biol Psychiatry 68(11):999–1006. https://doi.org/10.1016/j.biopsych.2010.07.033

Giordano P, Del Vecchio GC, Cecinati V, Delvecchio M, Altomare M, De Palma F, De Mattia D, Cavallo L, Faienza MF (2011) Metabolic, inflammatory, endothelial and haemostatic markers in a group of Italian obese children and adolescents. Eur J Pediatr 170(7):845–850. https://doi.org/10.1007/s00431-010-1356-7

Goldstein BI, Kemp DE, Soczynska JK, McIntyre RS (2009) Inflammation and the phenomenology, pathophysiology, comorbidity, and treatment of bipolar disorder: a systematic review of the literature. J Clin Psychiatry 70(8):1078–1090. https://doi.org/10.4088/JCP.08r04505

Goldstein JM, Holsen L, Huang G, Hammond BD, James-Todd T, Cherkerzian S, Hale TM, Handa RJ (2016) Prenatal stress-immune programming of sex differences in comorbidity of depression and obesity/metabolic syndrome. Dialogues Clin Neurosci 18(4):425–436

Goodwin RD, Stein MB (2004) Association between childhood trauma and physical disorders among adults in the United States. Psychol Med 34(3):509–520

Grande I, Magalhaes PV, Kunz M, Vieta E, Kapczinski F (2012) Mediators of allostasis and systemic toxicity in bipolar disorder. Physiol Behav 106(1):46–50. https://doi.org/10.1016/j.physbeh.2011.10.029

Grassi-Oliveira R, Holland FH, Honeycutt JA, Ganguly P, Brenhouse HC (2016) Cognitive impairment effects of early life stress in adolescents can be predicted with early biomarkers: impacts of sex, experience, and cytokines. Psychoneuroendocrinology 71:19–30

Gross CM, Flubacher A, Tinnes S, Heyer A, Scheller M, Herpfer I, Berger M, Frotscher M, Lieb K, Haas CA (2012) Early life stress stimulates hippocampal reelin gene expression in a sex-specific manner: evidence for corticosterone-mediated action. Hippocampus 22(3):409–420. https://doi.org/10.1002/hipo.20907

Guo M, Liu T, Guo JC, Jiang XL, Chen F, Gao YS (2012) Study on serum cytokine levels in posttraumatic stress disorder patients. Asian Pac J Trop Med 5(4):323–325. https://doi.org/10.1016/S1995-7645(12)60048-0

Hamdani N, Tamouza R, Leboyer M (2012) Immuno-inflammatory markers of bipolar disorder: a review of evidence. Front Biosci (Elite Ed) 4:2170–2182

Hanamsagar R, Alter MD, Block CS, Sullivan H, Bolton JL, Bilbo SD (2017) Generation of a microglial developmental index in mice and in humansreveals a sex difference in maturation and immune reactivity. Glia 65(9):1504–1520

Hartung HP, Heininger K, Schafer B, Fierz W, Toyka KV (1988) Immune mechanisms in inflammatory polyneuropathy. Ann N Y Acad Sci 540:122–161

Hennessy MB, Schiml-Webb PA, Miller EE, Maken DS, Bullinger KL, Deak T (2007) Anti-inflammatory agents attenuate the passive responses of guinea pig pups: evidence for stress-induced sickness behavior during maternal separation. Psychoneuroendocrinology 32 (5):508–515. https://doi.org/10.1016/j.psyneuen.2007.03.004

Hennessy MB, Deak T, Schiml-Webb PA (2010) Early attachment-figure separation and increased risk for later depression: potential mediation by proinflammatory processes. Neurosci Biobehav Rev 34(6):782–790

Hennessy MB, Paik KD, Caraway JD, Schiml PA, Deak T (2011) Proinflammatory activity and the sensitization of depressive-like behavior during maternal separation. Behav Neurosci 125 (3):426–433. https://doi.org/10.1037/a0023559

Hensch TK (2005) Critical period plasticity in local cortical circuits. Nat Rev Neurosci 6 (11):877–888. https://doi.org/10.1038/nrn1787

Holland FH, Ganguly P, Potter DN, Chartoff EH, Brenhouse HC (2014) Early life stress disrupts social behavior and prefrontal cortex parvalbumin interneurons at an earlier time-point in females than in males. Neurosci Lett 566:131–136. https://doi.org/10.1016/j.neulet.2014.02.023

Jaffee SR, Price TS (2007) Gene-environment correlations: a review of the evidence and implications for prevention of mental illness. Mol Psychiatry 12(5):432–442. https://doi.org/10.1038/sj.mp.4001950

Johnson RR, Maldonado Bouchard S, Prentice TW, Bridegam P, Rassu F, Young CR, Steelman AJ, Welsh TH Jr, Welsh CJ, Meagher MW (2014) Neonatal experience interacts with adult social stress to alter acute and chronic Theiler's virus infection. Brain Behav Immun. https://doi.org/10.1016/j.bbi.2014.03.002

Karavanaki K, Tsoka E, Liacopoulou M, Karayianni C, Petrou V, Pippidou E, Brisimitzi M, Mavrikiou M, Kakleas K, Dacou-Voutetakis C (2008) Psychological stress as a factor potentially contributing to the pathogenesis of type 1 diabetes mellitus. J Endocrinol Investig 31 (5):406–415

Katsouli S, Stamatakis A, Giompres P, Kouvelas ED, Stylianopoulou F, Mitsacos A (2014) Sexually dimorphic long-term effects of an early life experience on AMPA receptor subunit expression in rat brain. Neuroscience 257:49–64. https://doi.org/10.1016/j.neuroscience.2013.10.073

Kessler R, Avenevoli S, Ries Merikangas K (2001) Mood disorders in children and adolescents: an epidemiologic perspective. Biol Psychiatry 49(12):1002–1014

Koss KJ, Gunnar MR (2018) Annual research review: Early adversity, the hypothalamic-pituitary-adrenocortical axis, and child psychopathology. J Child Psychol Psychiatry 59(4):327–346. https://doi.org/10.1111/jcpp.12784

Ladd CO, Huot RL, Thrivikraman KV, Nemeroff CB, Plotsky PM (2004) Long-term adaptations in glucocorticoid receptor and mineralocorticoid receptor mRNA and negative feedback on the hypothalamo-pituitary-adrenal axis following neonatal maternal separation. Biol Psychiatry 55 (4):367–375. https://doi.org/10.1016/j.biopsych.2003.10.007

Laudenslager ML, Reite M, Harbeck RJ (1982) Suppressed immune response in infant monkeys associated with maternal separation. Behav Neural Biol 36(1):40–48

Laudenslager ML, Held PE, Boccia ML, Reite ML, Cohen JJ (1990) Behavioral and immunological consequences of brief mother-infant separation: a species comparison. Dev Psychobiol 23 (3):247–264. https://doi.org/10.1002/dev.420230305

Levy O (2007) Innate immunity of the newborn: basic mechanisms and clinical correlates. Nat Rev Immunol 7(5):379–390

Lewitus GM, Schwartz M (2009) Behavioral immunization: immunity to self-antigens contributes to psychological stress resilience. Mol Psychiatry 14(5):532–536. https://doi.org/10.1038/mp.2008.103

Litvin Y, Tovote P, Pentkowski NS, Zeyda T, King LB, Vasconcellos AJ, Dunlap C, Spiess J, Blanchard DC, Blanchard RJ (2010) Maternal separation modulates short-term behavioral and physiological indices of the stress response. Horm Behav 58(2):241–249. https://doi.org/10.1016/j.yhbeh.2010.03.010

Liu D, Diorio J, Tannenbaum B, Caldji C, Francis D, Freedman A, Sharma S, Pearson D, Plotsky PM, Meaney MJ (1997) Maternal care, hippocampal glucocorticoid receptors, and hypothalamic-pituitary-adrenal responses to stress. Science (New York, NY) 277 (5332):1659–1662

Lupien SJ, McEwen BS, Gunnar MR, Heim C (2009) Effects of stress throughout the lifespan on the brain, behaviour and cognition. Nat Rev Neurosci 10(6):434–445. https://doi.org/10.1038/nrn2639

Madrigal JL, Olivenza R, Moro MA, Lizasoain I, Lorenzo P, Rodrigo J, Leza JC (2001) Glutathione depletion, lipid peroxidation and mitochondrial dysfunction are induced by chronic stress in rat brain. Neuropsychopharmacology 24(4):420–429. https://doi.org/10.1016/s0893-133x(00)00208-6

Maes M, Berk M, Goehler L, Song C, Anderson G, Galecki P, Leonard B (2012) Depression and sickness behavior are Janus-faced responses to shared inflammatory pathways. BMC Med 10:66. https://doi.org/10.1186/1741-7015-10-66

Manikandan S, Padma MK, Srikumar R, Jeya Parthasarathy N, Muthuvel A, Sheela Devi R (2006) Effects of chronic noise stress on spatial memory of rats in relation to neuronal dendritic alteration and free radical-imbalance in hippocampus and medial prefrontal cortex. Neurosci Lett 399(1–2):17–22. https://doi.org/10.1016/j.neulet.2006.01.037

Mansur RB, Zugman A, de Miranda Asevedo Ede M, da Cunha GR, Bressan RA, Brietzke E (2012) Cytokines in schizophrenia: possible role of anti-inflammatory medications in clinical and preclinical stages. Psychiatry Clin Neurosci 66(4):247–260. https://doi.org/10.1111/j.1440-1819.2012.02354.x

Massart R, Nemoda Z, Suderman MJ, Sutti S, Ruggiero AM, Dettmer AM, Suomi SJ, Szyf M (2016) Early life adversity alters normal sex-dependent developmental dynamics of DNA methylation. Dev Psychopathol 28(4pt2):1259–1272. https://doi.org/10.1017/S0954579416000833

Maughan B, Rutter M (1997) Retrospective reporting of childhood adversity: issues in assessing long-term recall. J Personal Disord 11(1):19–33

Mauras N, Delgiorno C, Kollman C, Bird K, Morgan M, Sweeten S, Balagopal P, Damaso L (2010) Obesity without established comorbidities of the metabolic syndrome is associated with a proinflammatory and prothrombotic state, even before the onset of puberty in children. J Clin Endocrinol Metab 95(3):1060–1068. https://doi.org/10.1210/jc.2009-1887

McCoy CR, Rana S, Stringfellow SA, Day JJ, Wyss JM, Clinton SM, Kerman IA (2016) Neonatal maternal separation stress elicits lasting DNA methylation changes in the hippocampus of stress-reactive Wistar Kyoto rats. Eur J Neurosci 44(10):2829–2845. https://doi.org/10.1111/ejn.13404

McIntyre RS, Soczynska JK, Mancini D, Lam C, Woldeyohannes HO, Moon S, Konarski JZ, Kennedy SH (2008) The relationship between childhood abuse and suicidality in adult bipolar disorder. Violence Vict 23(3):361–372

McKimmie CS, Graham GJ (2010) Astrocytes modulate the chemokine network in a pathogen-specific manner. Biochem Biophys Res Commun 394(4):1006–1011. https://doi.org/10.1016/j.bbrc.2010.03.111

Meyer U, Schwarz MJ, Muller N (2011) Inflammatory processes in schizophrenia: a promising neuroimmunological target for the treatment of negative/cognitive symptoms and beyond. Pharmacol Ther 132(1):96–110. https://doi.org/10.1016/j.pharmthera.2011.06.003

Midei AJ, Matthews KA, Bromberger JT (2010) Childhood abuse is associated with adiposity in midlife women: possible pathways through trait anger and reproductive hormones. Psychosom Med 72(2):215–223. https://doi.org/10.1097/PSY.0b013e3181cb5c24

Miller GE, Cole SW (2012) Clustering of depression and inflammation in adolescents previously exposed to childhood adversity. Biol Psychiatry 72(1):34–40. https://doi.org/10.1016/j.biopsych.2012.02.034

Miller AH, Maletic V, Raison CL (2009) Inflammation and its discontents: the role of cytokines in the pathophysiology of major depression. Biol Psychiatry 65(9):732–741. https://doi.org/10.1016/j.biopsych.2008.11.029

Miller GE, Borders AE, Crockett AH, Ross KM, Qadir S, Keenan-Devlin L, Leigh AK, Ham P, Ma J, Arevalo JMG, Ernst LM, Cole SW (2017) Maternal socioeconomic disadvantage is associated with transcriptional indications of greater immune activation and slower tissue maturation in placental biopsies and newborn cord blood. Brain Behav Immun. https://doi.org/10.1016/j.bbi.2017.04.014

Moghaddam B (1993) Stress preferentially increases extraneuronal levels of excitatory amino acids in the prefrontal cortex: comparison to hippocampus and basal ganglia. J Neurochem 60 (5):1650–1657

Muller N, Ackenheil M (1998) Psychoneuroimmunology and the cytokine action in the CNS: implications for psychiatric disorders. Prog Neuro-Psychopharmacol Biol Psychiatry 22 (1):1–33

Muller N, Schwarz MJ (2008) A psychoneuroimmunological perspective to Emil Kraepelins dichotomy: schizophrenia and major depression as inflammatory CNS disorders. Eur Arch Psychiatry Clin Neurosci 258(Suppl 2):97–106. https://doi.org/10.1007/s00406-008-2012-3

Munhoz CD, Sorrells SF, Caso JR, Scavone C, Sapolsky RM (2010) Glucocorticoids exacerbate lipopolysaccharide-induced signaling in the frontal cortex and hippocampus in a dose-dependent manner. J Neurosci 30(41):13690–13698. https://doi.org/10.1523/jneurosci.0303-09.2010

Nelson LH, Warden S, Lenz KM (2017) Sex differences in microglial phagocytosis in the neonatal hippocampus. Brain Behav Immun 64:11–22

Network and Pathway Analysis Subgroup of Psychiatric Genomics Consortium (2015) Psychiatric genome-wide association study analysesimplicate neuronal, immune and histone pathways. Nat Neurosci 18(2):199–209

Ortega A, Jadeja V, Zhou H (2010) Postnatal development of lipopolysaccharide-induced inflammatory response in the brain. Inflamm Res. https://doi.org/10.1007/s00011-010-0252-y

Pace TW, Mletzko TC, Alagbe O, Musselman DL, Nemeroff CB, Miller AH, Heim CM (2006) Increased stress-induced inflammatory responses in male patients with major depression and increased early life stress. Am J Psychiatry 163(9):1630–1633. https://doi.org/10.1176/ajp.2006.163.9.1630

Pan W, Stone KP, Hsuchou H, Manda VK, Zhang Y, Kastin AJ (2011) Cytokine signaling modulates blood-brain barrier function. Curr Pharm Des 17(33):3729–3740

Pankow JS, Folsom AR, Cushman M, Borecki IB, Hopkins PN, Eckfeldt JH, Tracy RP (2001) Familial and genetic determinants of systemic markers of inflammation: the NHLBI family heart study. Atherosclerosis 154(3):681–689

Perkeybile AM, Schiml-Webb PA, O'Brien E, Deak T, Hennessy MB (2009) Anti-inflammatory influences on behavioral, but not cortisol, responses during maternal separation. Psychoneuroendocrinology 34(7):1101–1108. https://doi.org/10.1016/j.psyneuen.2009.02.014

Pervanidou P, Kolaitis G, Charitaki S, Margeli A, Ferentinos S, Bakoula C, Lazaropoulou C, Papassotiriou I, Tsiantis J, Chrousos GP (2007) Elevated morning serum interleukin (IL)-6 or evening salivary cortisol concentrations predict posttraumatic stress disorder in children and adolescents six months after a motor vehicle accident. Psychoneuroendocrinology 32 (8–10):991–999. https://doi.org/10.1016/j.psyneuen.2007.07.001

Pigliucci M, Murren CJ, Schlichting CD (2006) Phenotypic plasticity and evolution by genetic assimilation. J Exp Biol 209(Pt 12):2362–2367. https://doi.org/10.1242/jeb.02070

Poulton R, Moffitt TE, Silva PA (2015) The Dunedin Multidisciplinary Health and Development study: overview of the first 40 years, with an eye to the future. Soc Psychiatry Psychiatr Epidemiol 50(5):679–693. https://doi.org/10.1007/s00127-015-1048-8

Raison CL, Miller AH (2011) Is depression an inflammatory disorder? Curr Psychiatry Rep 13 (6):467–475. https://doi.org/10.1007/s11920-011-0232-0

Rivest S (2009) Regulation of innate immune responses in the brain. Nat Rev Immunol 9 (6):429–439. https://doi.org/10.1038/nri2565

Rook GA, Lowry CA, Raison CL (2015) Hygiene and other early childhood influences on the subsequent function of the immune system. Brain Res 1617:47–62. https://doi.org/10.1016/j.brainres.2014.04.004

Roque A, Ochoa-Zarzosa A, Torner L (2016) Maternal separation activates microglial cells and induces an inflammatory response in the hippocampus of male rat pups, independently of hypothalamic and peripheral cytokine levels. Brain Behav Immun 55:39–48. https://doi.org/10.1016/j.bbi.2015.09.017

Schafer DP, Lehrman EK, Kautzman AG, Koyama R, Mardinly AR, Yamasaki R, Ransohoff RM, Greenberg ME, Barres BA, Stevens B (2012) Microglia sculpt postnatal neural circuits in an activity and complement-dependent manner. Neuron 74(4):691–705. https://doi.org/10.1016/j.neuron.2012.03.026

Scheinman RI, Cogswell PC, Lofquist AK, Baldwin AS Jr (1995) Role of transcriptional activation of I kappa B alpha in mediation of immunosuppression by glucocorticoids. Science (New York, NY) 270(5234):283–286

Schwarz JM, Bilbo SD (2011) The immune system and the developing brain. In: McCarthy MM (ed) Colloquium series on the developing brain. Morgan & Claypool Life Sciences, San Rafael

Schwarz JM, Bilbo SD (2012) Sex, glia, and development: interactions in health and disease. Horm Behav 62(3):243–253. https://doi.org/10.1016/j.yhbeh.2012.02.018

Schwarz JM, Sholar PW, Bilbo SD (2012) Sex differences in microglial colonization of the developing rat brain. J Neurochem 120(6):948–963. https://doi.org/10.1111/j.1471-4159.2011.07630.x

Semple BD, Blomgren K, Gimlin K, Ferriero DM, Noble-Haeusslein LJ (2013) Brain development in rodents and humans: identifying benchmarks of maturation and vulnerability to injury across species. Prog Neurobiol 106–107:1–16. https://doi.org/10.1016/j.pneurobio.2013.04.001

Silverman MN, Sternberg EM (2012) Glucocorticoid regulation of inflammation and its functional correlates: from HPA axis to glucocorticoid receptor dysfunction. Ann N Y Acad Sci 1261:55–63. https://doi.org/10.1111/j.1749-6632.2012.06633.x

Silverman MN, Pearce BD, Biron CA, Miller AH (2005) Immune modulation of the hypothalamic-pituitary-adrenal (HPA) axis during viral infection. Viral Immunol 18(1):41–78. https://doi.org/10.1089/vim.2005.18.41

Silvers JA, Lumian DS, Gabard-Durnam L, Gee DG, Goff B, Fareri DS, Caldera C, Flannery J, Telzer EH, Humphreys KL, Tottenham N (2016) Previous institutionalization is followed by broader amygdala-hippocampal-PFC network connectivity during aversive learning in human development. J Neurosci 36(24):6420–6430. https://doi.org/10.1523/jneurosci.0038-16.2016

Simon NM, McNamara K, Chow CW, Maser RS, Papakostas GI, Pollack MH, Nierenberg AA, Fava M, Wong KK (2008) A detailed examination of cytokine abnormalities in Major Depressive Disorder. Eur Neuropsychopharmacol 18(3):230–233. https://doi.org/10.1016/j.euroneuro.2007.06.004

Simon AK, Hollander GA, McMichael A (2015) Evolution of the immune system in humans from infancy to old age. Proc Biol Sci 282(1821):20143085. https://doi.org/10.1098/rspb.2014.3085

Slopen N, Kubzansky LD, McLaughlin KA, Koenen KC (2013) Childhood adversity and inflammatory processes in youth: a prospective study. Psychoneuroendocrinology 38(2):188–200. https://doi.org/10.1016/j.psyneuen.2012.05.013

Soczynska JK, Kennedy SH, Woldeyohannes HO, Liauw SS, Alsuwaidan M, Yim CY, McIntyre RS (2011) Mood disorders and obesity: understanding inflammation as a pathophysiological nexus. NeuroMolecular Med 13(2):93–116. https://doi.org/10.1007/s12017-010-8140-8

Spiers JG, Chen HJ, Bradley AJ, Anderson ST, Sernia C, Lavidis NA (2013) Acute restraint stress induces rapid and prolonged changes in erythrocyte and hippocampal redox status. Psychoneuroendocrinology 38(11):2511–2519. https://doi.org/10.1016/j.psyneuen.2013.05.011

Stam R (2007) PTSD and stress sensitisation: a tale of brain and body Part 2: animal models. Neurosci Biobehav Rev 31(4):558–584. https://doi.org/10.1016/j.neubiorev.2007.01.001

Stamatovic SM, Keep RF, Andjelkovic AV (2008) Brain endothelial cell-cell junctions: how to "open" the blood brain barrier. Curr Neuropharmacol 6(3):179–192. https://doi.org/10.2174/157015908785777210

Stevens B, Allen NJ, Vazquez LE, Howell GR, Christopherson KS, Nouri N, Micheva KD, Mehalow AK, Huberman AD, Stafford B, Sher A, Litke AM, Lambris JD, Smith SJ, John

SW, Barres BA (2007) The classical complement cascade mediates CNS synapse elimination. Cell 131(6):1164–1178. https://doi.org/10.1016/j.cell.2007.10.036

Strawbridge R, Arnone D, Danese A, Papadopoulos A, Herane Vives A, Cleare AJ (2015) Inflammation and clinical response to treatment in depression: a meta-analysis. Eur Neuropsychopharmacol 25(10):1532–1543. https://doi.org/10.1016/j.euroneuro.2015.06.007

Sutherland AG, Alexander DA, Hutchison JD (2003) Disturbance of pro-inflammatory cytokines in post-traumatic psychopathology. Cytokine 24(5):219–225

Teicher MH, Samson JA (2013) Childhood maltreatment and psychopathology: a case for ecophenotypic variants as clinically and neurobiologically distinct subtypes. Am J Psychiatry 170(10):1114–1133. https://doi.org/10.1176/appi.ajp.2013.12070957

Teicher MH, Samson JA, Polcari A, Andersen SL (2009) Length of time between onset of childhood sexual abuse and emergence of depression in a young adult sample: a retrospective clinical report. J Clin Psychiatry 70(5):684–691

Thomas C, Hypponen E, Power C (2008) Obesity and type 2 diabetes risk in midadult life: the role of childhood adversity. Pediatrics 121(5):e1240–e1249. https://doi.org/10.1542/peds.2007-2403

Tottenham N, Sheridan MA (2009) A review of adversity, the amygdala and the hippocampus: a consideration of developmental timing. Front Hum Neurosci 3:68. https://doi.org/10.3389/neuro.09.068.2009

Tractenberg SG, Levandowski ML, de Azeredo LA, Orso R, Roithmann LG, Hoffmann ES, Brenhouse H, Grassi-Oliveira R (2016) An overview of maternal separation effects on behavioural outcomes in mice: evidence from a four-stage methodological systematic review. Neurosci Biobehav Rev 68:489–503. https://doi.org/10.1016/j.neubiorev.2016.06.021

Von Korff M, Alonso J, Ormel J, Angermeyer M, Bruffaerts R, Fleiz C, de Girolamo G, Kessler RC, Kovess-Masfety V, Posada-Villa J, Scott KM, Uda H (2009) Childhood psychosocial stressors and adult onset arthritis: broad spectrum risk factors and allostatic load. Pain 143 (1–2):76–83. https://doi.org/10.1016/j.pain.2009.01.034

Walker CD, Bath KG, Joels M, Korosi A, Larauche M, Lucassen PJ, Morris MJ, Raineki C, Roth TL, Sullivan RM, Tache Y, Baram TZ (2017) Chronic early life stress induced by limited bedding and nesting (LBN) material in rodents: critical considerations of methodology, outcomes and translational potential. Stress 20(5):421–448. https://doi.org/10.1080/10253890.2017.1343296

Wang HT, Huang FL, Hu ZL, Zhang WJ, Qiao XQ, Huang YQ, Dai RP, Li F, Li CQ (2017) Early-life social isolation-induced depressive-like behavior in rats results in microglial activation and neuronal histone methylation that are mitigated by minocycline. Neurotox Res 31(4):505–520. https://doi.org/10.1007/s12640-016-9696-3

Wei L, Simen A, Mane S, Kaffman A (2012) Early life stress inhibits expression of a novel innate immune pathway in the developing hippocampus. Neuropsychopharmacology 37(2):567–580. https://doi.org/10.1038/npp.2011.239

Wieck A, Andersen SL, Brenhouse HC (2013) Evidence for a neuroinflammatory mechanism in delayed effects of early life adversity in rats: relationship to cortical NMDA receptor expression. Brain Behav Immun 28:218–226. https://doi.org/10.1016/j.bbi.2012.11.012

Ygberg S, Nilsson A (2012) The developing immune system – from foetus to toddler. Acta Paediatr (Oslo, Norway: 1992) 101(2):120–127. https://doi.org/10.1111/j.1651-2227.2011.02494.x

Zalcman S, Green-Johnson JM, Murray L, Nance DM, Dyck D, Anisman H, Greenberg AH (1994) Cytokine-specific central monoamine alterations induced by interleukin-1, -2 and -6. Brain Res 643(1–2):40–49

Zhao SJ, Guo CN, Wang MQ, Chen WJ, Zhao YB (2012) Serum levels of inflammation factors and cognitive performance in amnestic mild cognitive impairment: a Chinese clinical study. Cytokine 57(2):221–225. https://doi.org/10.1016/j.cyto.2011.11.006

The manufacturer's authorised representative in the EU is Springer Nature Customer Service Centre GmbH, Europaplatz 3, 69115 Heidelberg, Germany. If you have any concerns regarding our products, please contact ProductSafety@springernature.com

Printed and bound by CPI Group (UK) Ltd, Croydon, CR0 4YY
30/09/2026
02225458-0016